RESEARCH IN PHOTOBIOLOGY

RESEARCH IN PHOTOBIOLOGY

Edited by

Amleto Castellani

Division of Radioprotection
CNEN, Centro Studi Nucleari Casaccia
Rome, Italy

PLENUM PRESS · NEW YORK AND LONDON

Library of Congress Cataloging in Publication Data

International Congress on Photobiology, 7th, Rome, 1976.
  Research in photobiology.

  Includes index.
  1. Photobiology—Congresses. I. Castellani, A. II. Title.
  QH515.I48 1976   574.1'9153   77-2189
  ISBN 0-306-31034-1

Proceedings of the Seventh International Congress on Photobiology held in Rome,
Italy, August 29—September 3, 1976

© 1977 Plenum Press, New York
A Division of Plenum Publishing Corporation
227 West 17th Street, New York, N.Y. 10011

Printed in the United States of America

*The Seventh International Congress on Photobiology was
organized by an Italian Committee under the auspices of the
Comité International de Photobiologie*

Patron
The President of the Italian Republic
Giovanni Leone

President
F. Pocchiari

Vice-Presidents
A. Caputo, B. Nicoletti, G. Rodighiero

Secretary-General
A. Castellani

Treasurer
P. Misiti

Comité International de Photobiologie

President
R.B. Setlow (USA)

Vice-Presidents
A. Castellani (Italy), **I. Honje** (Japan)
G.O. Schenck (Germany), **D. Vince-Prue** (U.K.)

Secretary-General
D.O. Hall (U.K.)

Treasurer
A. Wiskemann (Germany)

Italian National Committee in the CIP

President
L. Musajo

Vice-President
G. Rodighiero

Secretary-General
A. Castellani

FOREWORD

Every four years the photobiologists of the world get together
in an International Congress. They discuss and learn not only research details and findings in their own, often narrow, fields but
educate one another broadly in the many biological systems that
interact with light. It is this latter purpose that is exemplified
by these proceedings - the Symposium papers and Workshop summaries
of the VIIth International Congress on Photobiology held in Rome,
August 29 - September 3, 1976.

Photobiology is one of the few true interdisciplinary fields.
It has an air of excitement about it. A glance at the table of
contents indicates clearly that photobiology and its practitioners
(individuals whose primary interests are in medicine, plant sciences, animal sciences, molecular properties, and energy conversion)
interact with the entire and diverse world of living creatures. We
supply not only the basic research background to help evaluate many
present-day environmental problems but are also evaluating and
pointing the way toward solutions to a number of these problems.
Photobiological research plays a direct role in such diverse fields
as photosynthesis, solar energy conversion, and skin cancer. It
has supplied the basic and applied information that ultimately permits us to understand the variables necessary to evaluate effects
of stratospheric ozone depletion on the biosphere. Moreover,
photochemical and photobiological investigations have led to the
discovery of cellular mechanisms that can repair ultraviolet induced damage to DNA and as a result have opened up completely new
ways of looking at and investigating hazardous environmental
chemicals.

It is obvious from these Symposium papers that fundamental
research in photobiology has had a tremendous impact on applied
problems. Hence, it is clear that a strong, long range, basic
research program is necessary if we are to solve short range, applied, problems in biology.

It is a pleasure, on behalf of the Comité International de
Photobiologie (now Association Internationale de Photobiologie) to
thank the Italian Organizing Committee and, in particular, the
President of the Congress, F. Pocchiari and the Secretary General,
A. Castellani for their successful efforts in organizing a com-
prehensive and stimulating Congress.

R. B. Setlow

Biology Department
Brookhaven National Laboratory
Upton, New York

PREFACE

"Research in Photobiology" is the complete collection of the 75
papers presented by the invited lecturers to the VII International
Congress on Photobiology, held in Rome (August 29 - September 3,
1976).
The topics presented in the 15 symposia show how comprehensively
the Conference covered the major scientific issues that are the
object of concern today in the field of photobiology. Clearly,
photobiology can truly be seen as an interdisciplinary field of
research involving any biological and biophysical events affected
by exciting radiations. Many natural phenomena are correlated with
the effects of sunlight - life itself is dependent upon the sun.
In this light photobiology includes photophysiology which studies
the natural conditions required to make some of the physiological
processes possible.
When the natural condition changes, as in the case of the stratospheric ozone reduction, the effect on life can be disastrous, and
photobiology then becomes photopathology, which opens up the whole
field of prevention and protection, amply covered by several
speakers.
The repair of radiation damage at cellular level, as a natural
defence, was given due emphasis as a specific topic on the field
of photobiology with many correlations with radiation cancerogen-
esis and mutagenesis.
We devoted a symposium to the "Comparative Effect of Exciting
and Ionizing Radiations" and many speakers made comparative references to the effects of ionizing and exciting radiations, because
of the complementary nature of photobiology and radiobiology,which
together cover the whole field of radiation biology.
This volume is dedicated to Professor Luigi Musajo, former President of the Congress, and of the Italian Group of Photobiology,
who passed away while we were in the preparatory phase of this

Congress. His place was taken by the Vice-President of the Congress, Professor Francesco Pocchiari, General Director of the National Institute of Health, whose chairmanship ensured the success of the Congress.

The Congress is also indebted to the sponsors, the Comité International de Photobiologie and to the organizers, CNEN, the National Institute of Health and the Italian Group of Photobiology.

We finally express our appreciation to the Ministry of Health, CNEN, CNR, UNEP, IUBS and IARR for having sponsored some of the symposia and for having provided grants to enable some of the invited speakers to attend.

AMLETO CASTELLANI

Division of Radiation Protection
CNEN, Centro Studi Nucleari, Casaccia
Rome, Italy

CONTENTS

SYMPOSIA

PHOTOPHYSICAL AND PHOTOCHEMICAL PROPERTIES OF EXCITED STATES

PHOTOREACTIONS IN BIOLOGICAL MACROMOLECULAR COMPLEXES

PHOTOMOVEMENT IN MICROORGANISMS

PHOTOSYNTHESIS

COMPARATIVE EFFECTS OF EXCITING AND IONIZING RADIATIONS

PHOTOSENSITIZED REACTIONS OF NUCLEIC ACIDS AND PROTEINS

REPAIR OF RADIATION DAMAGE

SOLAR ENERGY CONVERSION SYSTEMS

PHOTOBIOLOGY IN MEDICINE

CANCEROGENIC EFFECTS OF RADIATION

LIGHT AND DEVELOPMENT

LIGHT INDUCED DEGENERATION OF SKIN: CHRONIC ACTINIC DERMATOSIS

ENVIRONMENT-SPACE INTERACTIONS: PHOTOBIOLOGICAL IMPLICATIONS

VISION

MUTAGENIC EFFECTS OF RADIATION

ROUND TABLES

SOLAR ENERGY

FARRINGTON DANIELS MEMORIAL LECTURE

SIR GEORGE PORTER

THE ROYAL INSTITUTION

21 ALBEMARLE STREET, LONDON W1X 4BS

Farrington Daniels was a pioneer as well as an evangelist in the
field of solar energy, the importance of whose work will be
increasingly recognised in the future.   It seems appropriate
therefore, tonight, to look at solar energy in its broadest
context; in the past as an essential for the creation of life,
in the present as our source of food and energy and in the
future as a potential source of energy income when our capital of
fossil fuels has been exhausted.

    It is with a sense of wonder and admiration that we
contemplate the remarkable series of chemical reactions which
have given us life.   The reactants were unpromising, stormy seas
and fiery earth, but the seas gave forth life, simple at first
and gradually becoming more complex.   The land produced the herb
with its seed and the tree with its fruit, which were able to
provide later for still higher, more complex, forms of life, the
fowl of the air, the beasts of the field and eventually man
himself, who has recently taken more than his fair share of the
harvest.

    The most obvious characteristic of this sequence of changes
is that, contrary to general experience of spontaneous processes
in the physical world, it proceeds from the simple to the complex,
from chaos to elaborate organisation, from disorder to order.
Since we would not wish to forsake the established laws of
thermodynamics, even in the biological world, we can only conclude
that we are not here dealing with a series of spontaneous changes
in a closed system but that some outside influence is at work
which makes possible a localised decrease of entropy within the
evolving biosphere.   It was natural that the first men who sought

1

to explain these things should identify this outside source as a person.   In the first chapter of Genesis, God is seen as One who brings order from chaos; He is the source of negative entropy. But the authors of the Book of Genesis take us further and tell us, in the third verse, how God provided this source. 'And God said, Let there be light, and there was light'. Without that light on the first day the rest could not have happened, and the source of that light was the Sun.

The sun is a nuclear fusion reactor which derives its energy principally from the fusion of hydrogen atoms into helium. This reaction has been going on for about 5 billion years and will continue for about as long again before the fuel begins to run out. The inside temperature is several million degrees but the surface temperature corresponds very approximately to that of a black body at $6,000^{\circ}$ C. The energy maximum in the radiation which reaches the earth's surface occurs near the middle of the visible region in the green, as one would expect of a well adapted eye; 40% of the total radiation is in the visible, 51% in the infra-red and 9% in the ultra-violet region below 400 nm.

It is these radiations which have created life, which maintain it today and which have provided all our fossil fuels, except nuclear fuels.

The origins of life, which seemed to present an insuperable 'chicken and egg' problem to men over the centuries, are now broadly understood.   The problem had been that living things can only arise from simpler organic matter and yet all organic matter on earth arises from living things. Furthermore, such organic matter tends to be rapidly destroyed by biological decomposition or by oxidation. Finally if, as seems likely, the sun is the source of energy from which organic matter is to be built from inorganic molecules, there is the difficulty that such molecules as carbon dioxide and water absorb only in the far ultra-violet region of the spectrum where there is no significant amount of radiation reaching the earth's surface, because of the strong absorption of these wavelengths by the oxygen of the atmosphere. Oxygen is therefore an obstacle in two respects and the whole thinking about the origins of life was transformed when the suggestion was made, nearly 50 years ago, independently by A.I. Oparin and J.B.S. Haldane, that our atmosphere was originally a reducing one, consisting of such components as methane, ammonia and water. Since oxygen was absent, the short-wave ultra-violet light could penetrate the atmosphere and photochemical reactions would take place, leading to larger molecules.   These would dissolve in the seas where it would then be possible to build up more complex molecules. In 1953 these suggestions were made far more plausible by an experiment performed by Stanley Miller, working in Harold Urey's laboratory in Chicago. Miller passed an

electrical discharge through the gases methane, ammonia, water
and hydrogen and, after a week, the organic compounds which had
been formed and were dissolved in the water were analysed. Miller
found among them four of the common amino-acids of protein;
glycine, alanine, aspartic acid and glutamic acid. Later
experiments by Cycil Ponnamperuma and others have shown that
ultra-violet irradiation of these gases also produces amino-acids
as well as a host of other substances; indeed most of the basic
'building bricks' of living matter have now been produced in this
way. The origins of the first organic molecules and how they
evolved into proteins, nucleic acids and cell membranes is now a
very active and increasingly sophisticated field of research, and
the first experiment of Miller is regarded as a classic. It is
interesting to contemplate that Miller's experiment could have
been carried out by almost any chemist at that time with quite
modest resources and that the idea on which it was based required
only modest theoretical understanding not beyond the reach of an
intelligent schoolboy. Even today there are still great
discoveries to be made by individuals who do not have immense team
and financial resources behind them ... provided of course that
they have a good idea which no man has had before them and the
confidence to pursue it.

These processes of organic chemical evolution were taking
place over 3½ billion years ago, less than a million years after
the earth was formed. Gradually, as the lighter molecules of
hydrogen were lost from the earth's gravitational field, the
atmosphere changed to become the oxidising one we know today.
The biosphere could no longer use the short ultra-violet wavelengths since they were now intercepted by oxygen and ozone and a
new scheme of things had to be developed if life was to continue
with the sun as its source of energy. This new scheme of things
was photosynthesis, which appeared about 2.7 billion years ago.
It utilises the chlorophyll molecule, which absorbs very strongly
in the visible and near infra-red regions of the spectrum, to
combine water and carbon dioxide from the atmosphere into organic
molecules, such as sugars, which provide the store of chemical
energy needed to sustain life. When sugars are metabolised with
oxygen, water and carbon dioxide are returned to the atmosphere
so providing an elegant cycle of energy, driven by the sun.

Today life in all its forms is wholly dependent on
photosynthesis and the details of its mechanism are being
unravelled in laboratories throughout the world. The later stages
of photosynthesis occur in the dark and are rather well understood
as a result, particularly, of the work of Melvin Calvin and his
school. The earlier, photochemical, stages are less well
understood and provide the greatest of all challenges to the

photochemist. For example, it is generally accepted that the light is first absorbed by an array of chlorophyll molecules, a light harvesting unit, and that the energy is rapidly transferred between these molecules until it is trapped at a centre where chemical reaction can begin. But all attempts to construct a unit of this kind in vitro run into the problem that in highly concentrated solutions of chlorophyll, such as those which must exist in vivo, excitation is rapidly lost by a process known as concentration quenching, before it can reach the photochemical trap. Another process about which almost nothing is known is the one by which water molecules yield oxygen; this must involve the transfer of four electrons before one oxygen molecule can be liberated.

One of the characteristics of the primary processes which take place in photosynthesis is that they are extremely rapid. It has been known for some time, from fluorescence studies in the nanosecond region, that the excited states of chlorophyll first formed by the action of light disappear in one or two nanoseconds at the most. We have recently been able to look at these states in the picosecond region of time, using the mode locked laser and a streak camera which is capable of resolving times down to about ten picoseconds. We find that, in the early stages after light absorption, the excited states of chlorophyll have a lifetime of about one hundred picoseconds and in this time they have probably transferred their energy many times from one molecule to another, reached the trap and carried out the first step of the chemical reaction. Most of the action which is interesting to a photochemist, and which is most characteristic of the photosynthetic process as opposed to the biochemical electron transport reactions in general, is complete in less than a billionth of a second so here is a field where fast pulse techniques become quite essential.

As well as providing the energy needed for our bodies in the form of food, photosynthesis has always provided the fuels which we need increasingly to maintain our civilisation. Until 1850 man lived on his income of solar energy; 90% of his fuel was wood and his other sources of power were wind, water and animal labour, all derived ultimately from the sun. By 1900 he was living on capital, as stored solar energy in the form of coal became the dominant (70%) fuel. Today oil and gas have taken over and represent 70% of our fuel energy consumption.

Since the second world war there has been little interest in using our solar energy income; oil and gas have been cheap until recently and, although we know they must run out in another hundred years or so, coal will keep us going a little longer and then we have the promise of abundant nuclear energy. Over the last two or three years things have changed again. Those who own our supplies of oil are valuing it more realistically and the

price of coal is rising almost as rapidly because people are
becoming less willing to dig it out. Furthermore, the bright
promise of nuclear power is now seen to be clouded with problems.
Supplies of uranium, if it is to be burned in non-breeder reactors,
are not much greater than those of coal; breeder reactors are
beset with technical difficulties and no power has yet been
produced from fusion, except in a bomb. Even more serious, some
are beginning to question whether mankind is yet socially and
politically adult enough to be trusted with nuclear technology.
Recent events have adequately shown that there is little difference
between the nuclear reactor and the nuclear bomb in the hands of
those who have the former and wish to use the latter. If we are
making a mistake, it will be the worst we have ever made because
it is virtually irreversible. The half life of plutonium is
24,000 years ... a time much longer than that of civilised man.

There is, therefore, some case for seeking an alternative
and the possibilities of using our traditional nuclear fusion
reactor, situated at a safe distance of 93 million miles, deserve
serious consideration. Its output may be impressively stated in
several ways:

(1) The amount of solar energy falling on the earth's surface
    in ten days is equivalent to all the fossil fuel reserves
    on earth.

(2) The present average energy demand per person can be met,
    with 10% recovery, by an area 3 yards square between
    latitudes 40° N and 40° S where 80% of the world's
    population lives and where the greatest needs are found.

(3) An area 330 miles square is sufficient to supply the
    present-day energy needs of mankind.

A large part of our energy is used as low grade heat and this
is the most promising way of utilising solar energy in the near
future. After all, most of our heating is solar anyway and we
only use fuels to top up this supply when it is not enough or to
pump it away when it becomes too much. The difficulty is, of
course, that the sun is an intermittent source and we need some
way of storing its energy if ever it is to be used for more
general purposes. The only practical way of storing large amounts
of energy, preferably in transportable form, is as chemical
potential and the chemical fuel which burns in air has unique
advantages as is shown by its predominant place in our energy
structure. The prospect of driving an uphill, endothermic,
photochemical reaction by the sun and so providing a daily income
of chemical fuels for mankind is, therefore, extremely attractive.
Although no chemical reaction is known at present which approaches
an efficiency or economy sufficient to merit commercial development,
we have the continued stimulation and encouragement of nature's

photosynthetic process. The only photochemical route to fuels
which is available to us now is through this natural photosynthesis.
Sugar cane, for example, can be grown with an efficiency of solar
energy conversion of about 1% and the sugar can be fermented to
alcohol. Two tons of alcohol/year/acre is possible and alcohol
is a perfectly good substitute for petrol. But even this yield
is low and looks even worse when the complete energy balance,
including the manufacture of fertilisers, is taken into account.
Work on the production of hydrogen by algae, which is in progress,
may provide a more efficient energy conversion.

Hydrogen from water or organic substances such as methanol
from carbon dioxide are the most attractive processes for a
reversible fuel cycle utilising solar energy. It is perfectly
possible today to produce hydrogen from water by electrolysis with
reasonable efficiency by using the solar cells which were developed
to supply energy continuously for space vehicles. The silicon
solar cell is a barrier layer photovoltaic device with an
efficiency of conversion of solar energy to electrical energy
better than 15% and silicon is the second most abundant element
on the earth. Unfortunately, the high purity silicon crystals
which are needed cost about one hundred times more than the price
which would bring them into a competitive position with
conventional power stations. This gap will be reduced but it is
doubtful whether it can ever be low enough to make silicon, or
even cadmium sulphide and other solid state materials, economic
except for special purposes.

Against the impressive figures of the amount of solar energy
available which I gave earlier, it is salutary to note that the
energy is so dispersed that the cost per area of the collecting
device cannot exceed £10 per square metre if the capital outlay
per unit of power is to be comparable with that of a conventional
power station. The materials which can be made to cover one metre
for ten pounds are strictly limited .... one is driven to think
in terms of old newspapers! Or leaves. Alternatively, we might
dispense with solid support materials altogether and carry out
the photochemical reaction in solution or suspension in water.

A promising system for the photolysis of water has recently
been reported by Fujishima and Honda in Japan. Two electrodes,
one composed of a pure crystal of titanium dioxide and the other
of an inert material such as platinum are connected together and
immersed in water. When the titanium dioxide is irradiated with
near ultra-violet light, oxygen bubbles off from the irradiated
electrode and hydrogen from the inert anode. It seems now to
have been established that the hydrogen and oxygen arise from
the decomposition of water alone and that the titanium dioxide
is unchanged. When the TiO_2 is irradiated, an electron passes
into the conduction band leaving a positive hole in the valence

band, which reacts with water to form oxygen and protons. The
protons pick up an electron from the platinum electrode and are
liberated as hydrogen. Unfortunately, ultra-violet light is
necessary, though sensitisation by dyes is being studied;
titanium dioxide crystals are expensive but there seems to be no
reason why polycrystalline materials should not be effective if
the problem of their internal resistance can be overcome.

Another approach is the photogalvanic cell using, for example,
the well-known reaction between a dye, thionine, and ferrous ions.
If a mixture of these substances is irradiated, the excited dye
reacts to form the colourless leuco thionine and ferric ions.
If two electrodes connected via a galvanometer, are placed in the
cell and the solution near one electrode only is irradiated, a
current flows. The illuminated electrode is the anode since the
electrode reaction of the leuco dye, $D^{--} = D + 2e$, is faster than
the reaction $Fe^{3+} + e = Fe^{2+}$.

Power can be drawn during illumination and for a very short
time afterwards but is rapidly lost owing to the reversible dark
reaction between ferric ions and leuco dye. This reversibility
of reactions of the electron-transfer type in solution is the
main problem in all systems of this kind and none of them has yet
given an efficiency greater than 1%. In principle the attraction
of such photogalvanic cells is that they can provide an electrical
storage battery which is rechargeable in sunlight and, for many
purposes, the use of a photochemically produced storage fuel to
produce electricity directly, rather than a combustible substance,
has many advantages.

The simplest of all systems would be one which took place
entirely in aqueous solution, with no electrodes or other
complexities, and which produced a gaseous, solid or easily
separable liquid fuel. As its raw materials,there seem to be no
alternatives to water and carbon dioxide and the absorption would
have to be in the visible region of the spectrum, probably by a
dyestuff. There seems to be no example of the photochemical
reduction of carbon dioxide but there are a number of examples of
the photochemical oxidation of water. The earliest experiments
of Heidt and others used inorganic ions such as ceric and cerous,
but yields were low and ultra-violet light is essential. We have
experimented with organic molecules related to dyes such as ketones
and quinones and, although water is decomposed, the quantum yield
is low, no hydrogen is produced and the organic material is
eventually removed by reaction. It will probably be necessary to
use more sophisticated systems, based on what we have learned of
the photosynthetic unit of plants and, perhaps, using micelles or
vesicles to separate the oxidised and reduced products. This is a
very active field of research at the present time.

Altogether, man's efforts to produce stable chemical fuels by means of visible light absorption have been strikingly unsuccessful when one considers the vast scale and success of natural photosynthesis. But the research effort has been very small because the incentive has been small. It has been said that, if sunbeams were weapons of war, solar energy would have been developed long ago. But, apart from some apocryphal reports about the military efforts of Archimedes at Syracuse, the sun's energy has always been used benevolently. It is true that the efficiency of photosynthesis is not high on an energy basis but the green plant has much more to do than merely provide energy. It has to look to its survival under extreme climatic conditions, its reproduction and continual repair. It will be surprising if we cannot design a system, based on the photosynthetic process, which is more efficient for the single purpose of solar energy collection and storage. If we can, there will be less need to fear that the lights may go out in the twenty-first century.

SYMPOSIUM I

PHOTOPHYSICAL AND PHOTOCHEMICAL PROPERTIES
OF EXCITED STATES

PHOTOPHYSICAL AND PHOTOCHEMICAL PROPERTIES OF EXCITED STATES

INTRODUCTION

Sir George Porter

The Royal Institution

21 Albemarle Street, London W1X 4BS

When there is an interval of four years between international congresses, it is long enough to expect some significant advances. Among the main developments since the VIth Congress in Bochum in 1972 I would mention the following:

1. A new appreciation of the potential practical importance of the field. The fuel crisis led to a new appraisal of alternative sources of energy and, as the only long term alternative to nuclear energy, solar energy research and the fundamental photophysics and chemistry which lie behind it, are being encouraged in many countries.

2. Kinetic absorption and fluorescence spectroscopy have been extended into the picosecond region which is near to the ultimate requirement for the study of primary photophysical processes.

3. The theory of radiationless conversion between two electronic energy levels in a molecule has been developed to the stage where, although quantitative predictions still cannot be made with much confidence, the essential factors governing the rate of these processes are now broadly understood.

4. The properties of molecular associates, particularly those of excited states such as exciplexes and excimers but also those of the ground state, e.g. the dimers and oligomers of chlorophyll, have been intensively studied.

5. The photophysics of model systems relevant to photobiology,

11

particularly monolayers and multilayers on solid supports,
bilayer lipid membranes, vesicles and liposomes, has become
a very active field of research. At the same time there has
been a steady advance in our knowledge of all areas of excited
states behaviour and energy transfer.

In the midst of this gain, one great loss has been sustained
... that of Theodor Forster, a pioneer in so many of these fields
and recipient of the Finsen Medal at the VIth International Congress.
His influence on our subject will be very clear in each of the
papers of this morning's session, which deal respectively with the
photophysical and photochemical aspects of the subject.

I should like to devote the rest of my remarks to the advances
which have taken place recently in extending flash photolysis and
time resolved spectroscopy into the picosecond region of time. The
source of picosecond flashes is the mode-locked laser which provides
a train of pulses, separated by twice the transit time of the laser
cavity (typically 6ns), each of duration about 5 ps. It is usually
necessary to extract a single pulse from the train by means of a
Pockells cell between crossed polarizers.

Until very recently, real time recording methods were not
possible in the picosecond region since photocells and oscilloscopes
with adequate time resolutions were not available as they are for
nanoseconds work. A variety of two pulse methods has been used, all
of which have one factor in common; the single pulse is divided into
two parts by a beam splitter: one part is used for excitation and
the other is used for monitoring after an optical delay. If a
broad band spectrum is required for monitoring, it can be produced
by focusing the monochromatic pulse into glass or a liquid such as
water – this produces a spectrum covering the whole of the visible
region through self focusing and self phase-modulation effects.

For picosecond fluorescence studies the second part of the beam
can be used to operate an optical shutter the principle of which
is the optical Kerr effect. A cell containing a liquid such as
carbon disulphite between two crossed polarizers forms the shutter.
The shutter is opened when a laser pulse passes through the carbon
disulphite since its electric field produces, through the Kerr
effect, a rotation of the plane of polarization of the fluorescent
light. By repeating the experiment with a series of delays between
the excitation of fluorescence and the opening of the shutter by
the second part of the pulse, the complete decay curve can be
derived. All these techniques require a number of separate
operations of the laser and suffer from irreproducibility from
pulse to pulse.

It is now possible to make real time studies, either in absorption or emission, using a single picosecond pulse, by use of the streak camera. This instrument operates by rapidly accelerating the electrons liberated at a photocathode and deflecting them by an electric field onto a phosphorescent screen. The intensity on the screen is proportional to the intensity of the image focused on the photocathode and time is displayed as a linear displacement.

The image, after amplification by an image converter tube, may be photographed or, preferably, detected by a vidicon camera and stored in the five hundred channels of an optical multichannel analyzer. In this way digital information is obtained of much higher precision than is available from any other picosecond techniques at the present time.

Recent work has shown, however, that the high intensity of picosecond pulses may result in second order effects and interpretation of kinetic processes must be made with caution and awareness of this possibility. Capillo and Shapiro, Breton and Giacinto and, in our laboratory, Tredwell and Synowiec have shown that the initial decay of phosphorescence is faster and the fluorescence yield is reduced at high intensities, this is probably to be attributed to singlet annihilation within the photosynthetic unit but the decay does not follow the expected second order decay law. Furthermore, Tredwell and Synowiec have shown that the yield decreases not only with the intensity of a single pulse but also with that of earlier pulses. We tentatively attribute this effect to triplet formation in the singlet annihilation process followed by triplet quenching of excited singlet by the processes

$$S_1 + S_1 \qquad\qquad T_1 + T_1$$

$$S_1 + T_1 \qquad\qquad T_1 + T_1$$

Since quenchers are not destroyed in these processes the decay law will not follow second order kinetics.

If the photon density is reduced to less than 10^{13} photons/cm^2/ pulse the probability of absorption of more than a photon in a photosynthetic unit becomes small and second order excited state processes are eliminated. With care, therefore, the powerful techniques of picosecond spectroscopy should provide important information about the very earliest stages of the photosynthetic process.

THE RULES OF ORGANIC MOLECULAR FLUORESCENCE

J.B. Birks

The Schuster Laboratory

University of Manchester, Manchester, U.K.

Summary

The four principal rules of organic molecular fluorescence are Stokes' law, Vavilov's law, Kasha's fluorescence rule, and the radiative lifetime relation. Recent experiments have shown that none of these rules are universally valid, although they are reasonable approximations in most cases. Each deviation from the rules has revealed new photophysical phenomena.

Introduction

Nearly 500 years ago Leonardo da Vinci, the great Italian scientist, artist and scholar, wrote the following words:

"Experiments never deceive, it is our judgement that sometimes deceives itself because it expects results that experiment refuses. We must consult experiment, changing the circumstances, until we have deduced reliable rules".

These words express the basic philosophy of experimental science and they are as valid today as when they were written.

The first symposium of this Congress is devoted to photophysics and photochemistry, the two sister sciences to which photobiology is so closely related. This

15

is therefore an appropriate occasion to consider, in
the light of Leonardo's words, the current status of
certain rules of photophysics, namely those relating
to the fluorescence of organic molecules. Are the rules
reliable and universally valid, or do they require re-
vision in the light of recent experimental consultation?

Fluorescence rules

The four principal rules of organic molecular fluorescence that have been proposed are (a) Stokes' law, (b)
Vavilov's law, (c) Kasha's fluorescence rule, and (d)
the radiative lifetime relation.

Stokes' law[1], formulated in 1852, states that the
fluorescence wavelength λ_F is equal to or greater than
the excitation wavelength λ_{ex}. For over 50 years this
empirical law lacked any theoretical explanation, but
this was provided in 1905 by Einstein's quantum theory
of radiation[2]. In terms of quantum theory Stokes' law
becomes

$$h \, \nu_F \leq h \, \nu_{ex}$$

where ν_F is the fluorescence frequency, ν_{ex} is the excitation frequency, and h is Planck's constant, a relation consistent with the energy conservation principle.

Vavilov's law, formulated in about 1930, states that
the fluorescence quantum yield $\emptyset_F$ is independent of the
excitation wavelength λ_{ex} for excitation with non-ionizing radiation.

Kasha's rules[3], formulated in 1950, state that in a
complex molecule luminescence occurs only from the lowest excited state of a given multiplicity, i.e. fluorescence corresponds to the $S_1 \rightarrow S_0$ radiative transition and phosphorescence to the $T_1 \rightarrow S_0$ radiative
transition, where S_1, T_1 and S_0 are the first excited
singlet, first excited triplet, and ground singlet states of the molecule, respectively. Kasha extended the
Jablonski energy level diagram to include spin-allowed
(internal conversion) and spin-forbidden (intersystem
crossing) radiationless transitions between electronic
states, followed by vibrational relaxation in which the
excess vibrational energy is dissipated to the solvent.
From the available experimental and theoretical eviden-

ce he concluded that the rate of internal conversion
between adjacent excited states, e.g. $S_2 \rightarrow S_1$, was much
more rapid than radiative transitions from higher excited states, e.g. $S_2 \rightarrow S_0$, so that the luminescence
yield of the latter was negligible. Rapid internal conversion and vibrational relaxation within the singlet
manifold would also account for Stokes' and Vavilov's
laws.

The Einstein radiation law[4] considers transitions
between a pair of energy levels of energy difference
$h\nu$. The emission probability A is related to the absorption probability $B\rho(\nu)$, where $\rho(\nu)$ is the radiation density, by the relation

$$A = \frac{8\pi h\nu^3}{c^3} \, B \, n^3$$

where c is the speed of light in vacuo, and n is the
refractive index of the medium. Strickler and Berg[5] and
Birks and Dyson[6] independently extended the Einstein
relation to organic molecules, and they obtained the
relation

$$k_F^{th} = 2.88 \times 10^{-9} n^2 f \int \frac{\epsilon(\bar{\nu})d\bar{\nu}}{\bar{\nu}}$$

where k_F^{th} is the theoretical radiative (fluorescence)
transition probability, $\bar{\nu}$ is the wavenumber, $\epsilon(\bar{\nu})$ is
the decadic molar extinction coefficient with the inte-
gral taken over the $S_1 \leftarrow S_0$ absorption spectrum, and

$$f = \left\langle \bar{\nu}_F^{-3} \right\rangle_{Av}^{-1}$$

is a mean fluorescence wavenumber. The theoretical value of k_F^{th} is to be compared with the experimental
value

$$k_F = \phi_F / \tau$$

where τ is the S_1 (fluorescence) lifetime.

Deviations from Vavilov's law

The first major deviation from Vavilov's law was observed in 1963 in benzene, when Braun, Kato and Lipsky[7]
reported that the relative fluorescence yield

$$\beta = \phi_F / \phi_{F0}$$

where $\emptyset_{FO}$ is the fluorescence quantum yield for excitation into S_1, of liquid benzene decreases to 0.6 and
0.4 for excitation into S_2 and S_3, respectively. For
benzene in solution the values of β are reduced even
further depending on the solvent and the concentration.
For benzene vapour the behaviour is even more dramatic,
in that $\emptyset_F$ and the triplet quantum yield $\emptyset_T$ each fall
to zero at an excitation energy of about 3000 cm^{-1} above the zero point level of S_1. The behaviour is attri-
buted to the onset of a fast (about 10^{12} s^{-1}) $S_1 \rightarrow S_0$
internal conversion process, known as Channel 3 since
its mechanism is not fully understood. In the condensed
phase Channel 3 manifests itself as a thermally activated radiationless transition (frequency factor about
10^{12} s^{-1}, activation energy about 3000 cm^{-1}) competing
with the $S_1 \rightarrow S_0$ fluorescence and $S_1 \rightarrow T_1$ intersystem
crossing.

The Channel 3 rate is slower than the vibrational
relaxation rate, but comparable with the rates of
$S_2 \rightarrow S_1$ and $S_3 \rightarrow S_2$ internal conversion. Hence in the
condensed phase the fluorescence excitation spectrum
remains constant ($\beta = 1$) over the $S_1 \leftarrow S_0$ absorption
band system, but decreases in steps at the $S_2 \leftarrow S_0$ and
$S_3 \leftarrow S_0$ absorption edges.

Similar deviations from Vavilov's law are observed
in toluene and other alkyl benzenes. Tatischeff and
Klein[8] have recently reported similar behaviour (dependence of β on λ_{ex}) in the three aromatic amino-acids,
tryptophan, tyrosine and phenylalanine and their results
have been confirmed by other workers. Thus deviations
from Vavilov's law are of direct photobiological interest.

Deviations from Kasha's fluorescence rule

During recent years, fluorescence has been observed
from higher excited singlet states of several aromatic
molecules, in violation of Kasha's fluorescence rule.
Apart from azulene and its derivatives which primarily
emit $S_2 \rightarrow S_0$ fluorescence, fluorescence has been observed from higher excited singlet states of the following molecules in addition to their normal $S_1 \rightarrow S_0$
fluorescence[9],

Benzene: S_2 (liquid, vapour)

Toluene: S_2 (liquid, vapour)

S_3 (vapour)

p-Xylene: S_2 (liquid, vapour)

S_3 (liquid, vapour)

Mesitylene: S_2 (vapour)

S_3 (solution, vapour)

Naphthalene: S_2 (vapour)

S_3 (solution, vapour)

1:2-Benzanthracene: S_2 (solution)

S_3 (solution, delayed
        fluorescence)

Pyrene: S_2 (vapour)

S_3, S_4, S_5 (solution)

3:4-Benzopyrene: S_2 (solution, vapour)
1:12-Benzoperylene: S_2 (solution, vapour)

Fluoranthene: S_2, S_3, S_4 (solution, delayed
                                  fluorescence).

An ingenious technique used for the observation of higher fluorescences was that of Nickel[10] who employed triplet-triplet interaction,

$$T_1 + T_1 \longrightarrow S_p + S_o$$

to generate molecules in a higher excited state S_p, the delayed fluorescence of which was then observed in the complete absence of the exciting light.

The quantum yields of the fluorescence from higher excited states are about $10^{-4} - 10^{-5}$ (except for azulene, where $\emptyset_F = 0.03$). These values are similar in magnitude to those predicted by Kasha[3] in 1950, and which he approximated to zero in formulating his rules. Similar behaviour is to be expected in other organic molecules, including those of photobiological interest. Higher excited state fluorescence has been reported from various porphyrins and phthalocyanins, but the evidence is less definitive than for the aromatic hydrocarbons listed above.

Deviations from Stokes' law

An interesting example of anti-Stokes fluorescence has been observed in 1:12-benzoperylene in 10^{-6} M n-heptane solution from -90 to 90°C by Birks, Easterly and Christophorou (in press). The notation S_E^V is used to indicate an electronic state E in a vibrational level V, of energy $\bar{\nu}_V$.

The normal (Stokes) fluorescence spectrum consists of five bands 0, 1, 2, 3, and 4, and symmetric vibrational $\bar{\nu}_5$ (= 1350 cm^{-1}) progressions originating therefrom. Bands 0, 1, 2, 3, and 4 are assigned to $S_1^0 \to S_0^0$, S_0^1, S_0^2, S_0^3 and S_0^4 transitions, where $\bar{\nu}_1$, $\bar{\nu}_2$, $\bar{\nu}_3$ and $\bar{\nu}_4$ are different antisymmetric vibrations.

There is also an anti-Stokes fluorescence spectrum, beyond the 0-0 transition, consisting of four bands 1', 2', 3' and S_2^0. Comparison with the corresponding absorption spectrum, measured over the same temperature range, shows that these correspond to S_1^1, S_1^2, S_1^3, $S_2^0 \to S_0^0$ transitions, respectively. These four bands are also the origins of symmetric vibrational $\bar{\nu}_5$ progressions which underlie the normal fluorescence spectrum.

Of the observed fluorescence origins only S_2^0 is symmetry allowed. The remainder require their oscillator strength by vibronic coupling to allowed transitions as follows:

$$0 \quad (S_3^0 \text{ or } S_2^1 \longrightarrow S_0^0)$$
$$1 \quad (S_2^1 \longrightarrow S_0^1)$$
$$2 \quad (S_2^2 \longrightarrow S_0^2)$$
$$3 \quad (S_2^3 \longrightarrow S_0^3)$$
$$4 \quad (S_2^4 \longrightarrow S_0^4)$$
$$1', 2', 3' \quad (S_2^0 \longrightarrow S_0^0)$$

The magnitude of the vibronic coupling is proportional to $(\Delta E)^{-2}$, where ΔE is the energy gap between the initial states of the two coupled transitions (their final states are common).

The intensity i_n of the three anti-Stokes (hot) fluorescence bands 1', 2' and 3' is given by

$$i_n \propto e^{-W/kT} (\Delta E)^{-2}$$

where $W = S_1^n - S_1^o$ is the Boltzmann activation energy
which determines the population of the vibrationally
excited state S_1^n, and $\Delta E = S_2^o - S_1^n$ is the energy gap
which determines the magnitude of the vibronic coupling
factor. This relation has been verified experimentally
from -90 to 90°C. The strong vibronic coupling resulting
from the low values of ΔE (ΔE decreases with T) partially offsets the Boltzmann factor and this increases
the intensity of the anti-Stokes fluorescence. Other molecules in which S_2 $(= {}^1L_a)$ and S_1 $(= {}^1L_b)$ are adjacent
exhibit similar behaviour.

Deviations from the radiative
lifetime relation

The radiative lifetime relation has been verified
experimentally by Strickler and Berg[5], Birks and Dyson[6],
Ware and Baldwin[11], Berlman[12] and others. Writing

$$R = k_F^{th}/k_F$$

values of R = 1 have been observed for anthracene, 9,10-
-diphenylanthracene, perylene, fluorescein and many o-
ther molecules.

One group of molecules which do not conform to the
radiative lifetime relations are the all-trans diphenyl
polyenes $\emptyset-(CH=CH)_n-\emptyset$ and their biological relations,
the retinol polyenes. For DPH (n=3) R = 6 to 40, and
for DPO (n=4) R = 20 to 90, depending on the solvent
and temperature[13]. For four retinol polyenes Dalle and
Rosenberg[14] observed values of R between 75 and 470.

A partial explanation for this anomalous behaviour
in DPH and the higher polyenes has been provided by experimental and theoretical evidence for a fluorescent
${}^1A_g^*$ state lying below the ${}^1B_u^*$ state observed in absorption. The ${}^1B_u^* \leftarrow {}^1A_g$ absorption transition, which
is allowed, is used to evaluate k_F^{th}. The ${}^1A_g^* \rightarrow {}^1A_g$ fluorescence transition, for which k_F is determined, is for-
bidden and it is considered to be induced by vibronic
coupling to the ${}^1B_u^* \rightarrow {}^1A_g$ transition. Hence

$$k_F = \frac{K^2}{\Delta E^2} \, k_F^{th}$$

where K is the vibronic coupling matrix element, and ΔE is the $^1B_u^* - {}^1A_g^*$ energy gap. Hence

$$R = \frac{\Delta E^2}{K^2}$$

This model, proposed by Birks and Birch[13], provides a qualitative explanation of the dependence of R on the temperature and solvent, in terms of the influence of the latter on ΔE.

The model is not, however, applicable to trans-stilbene (n=1), which also exhibits radiative lifetime anomalies. In trans-stilbene the fluorescence corresponds to the allowed $^1B_u^* -> {}^1A_g$ transition at all temperatures. At -100°C R=1, but R increases with temperature to R=13 at 25°C. Birch and Birks[15] have explained the behaviour using the theoretical stilbene model of Orlandi and Siebrand[16]. According to this model, the S_1 potential of stilbene has minima in the trans ($\emptyset$=0°), perp ($\emptyset$=90°) and cis ($\emptyset$=180°) configurations, where $\emptyset$ is the angle of rotation about the ethylenic bond. The symmetries in the three configurations are t ($^1B_u^*$), p ($^1A_g^*$) and c ($^1B_u^*$), due to a crossing of the $^1B_u^*$ and $^1A_g^*$ potentials. The photophysical behaviour of trans-stilbene is explained in terms of thermally-activated internal conversion (rotation) between t($^1B_u^*$) which is fluorescent (rate k_{FB}) and p($^1A_g^*$) which is non-fluorescent and decays by internal conversion (rate k_{GA}) to p(^1A_g), where cis isomerization occurs.

The reaction scheme is as follows

$$t(^1B_u^*) \quad \underset{k_{BA}}{\overset{k_{AB}}{\rightleftharpoons}} \quad p(^1A_g^*)$$

$$\downarrow k_{FB} \qquad\qquad\qquad\qquad \downarrow k_{GA}$$

$$t(^1A_g) \qquad\qquad\qquad\qquad p(^1A_g)$$

At normal temperatures, where k_{AB}, $k_{BA} >> k_{F'B}$, k_{GA}, there is dynamic equilibrium between the two excited states, and the observed fluorescence and internal conversion rate parameters are

$$k_F = f_B k_{FB} = f_B k_F^{th}$$

$$k_G = f_A k_{GA}$$

where f_B (=1-f_A) and f_A are the fractions of excited molecules in $t({}^1B_u{}^*)$ and $p({}^1A_g{}^*)$. Above 0°C $f_B < 0.1$, so that R > 10.

Internal rotation in the excited state between a fluorescent and a non-fluorescent molecular configuration is a new effect which merits further study. It probably occurs in the higher diphenyl and retinol polyenes, so that the vibronic coupling model described previously requires extension to take account of internal rotation.

Conclusion

None of the rules of organic molecular fluorescence are universally valid, although they are reasonable approximations in most cases. Their most useful function is to serve not as a dogma but as a criterion against which new experimental data can be assessed. Each deviation from the rules has revealed new phenomena: unsuspected mechanisms of internal conversion, fluorescence from higher excited states, hot fluorescence transitions induced by vibronic coupling, and internal rotation in the excited state between fluorescent and non-fluorescent configurations. There is no inherent reason why biological molecules should be exempt from any of these effects. "We must consult experiment, changing the circumstances, until we have deduced reliable rules".

References

1. G. Q. Stokes, Phil. Trans. _142_, II, 463 (1852); _143_, III, 383 (1853).
2. A. Einstein, Ann. Physik _17_, 132 (1905).
3. M. Kasha, Disc. Faraday Soc. _9_, 14 (1950).
4. A. Einstein, Physik. Z. _18_, 121 (1917).

5. S. J. Strickler and R.A. Berg, J. Chem. Phys. 37, 814 (1962).
6. J. B. Birks and D.J. Dyson, Proc. Roy. Soc. A 275, 135 (1963).
7. C. L. Braun, S. Kato and S. Lipsky, J. Chem. Phys. 39, 1645 (1963).
8. I. Tatischeff and R. Klein, in Excited states of biological molecules (ed. Birks, J.B.) Wiley-Interscience, London and New York (1976).
9. J. B. Birks, in Organic molecular photophysics (ed. Birks, J.B.) Vol. 2, 409 (1975).
10. B. Nickel, Chem. Phys. Letters 27, 84 (1974).
11. W. R. Ware and B.A. Baldwin, J. Chem. Phys. 40, 1703 (1964).
12. I. B. Berlman, Handbook of fluorescence spectra of aromatic molecules, Academic Press, New York (1965).
13. J. B. Birks and D. J. S. Birch, Chem. Phys. Letters, 31, 608 (1975).
14. J. P. Dalle and B. Rosenberg, Photochem. Photobiol. 12, 151 (1970).
15. D. J. S. Birch and J. B. Birks, Chem. Phys. Letters, 38, 432 (1976).
16. G. Orlandi and W. Siebrand, Chem. Phys. Letters, 30, 352 (1975).

ELEMENTARY PHOTOCHEMICAL REACTIONS

Albert Weller

Max-Planck-Institut fur biophysikalische Chemie, Abt.

Spektroskopie, D 3400 Gottingen, Germany

(ABSTRACT OF PAPER ONLY)

Excited molecules differ from the same molecules in the ground state with respect to energy (E), electronic structure (ψ) and lifetime (τ) and can, therefore, undergo chemical reactions which do not, or to much lesser extent, occur in the ground state. These photochemical reactions originating from specific electronically excited states often depend critically on the nature of the reacting states and may result in electronically excited products (adiabatic processes) or ground state products (diabatic processes).

Adiabatic processes which occur on excited state potential energy surfaces are reversible insofar as, following product deactivation, very fast reverse reactions on the ground state potential energy surface restore the initial reactants state so that no permanent chemical change occurs. Typical processes of this type are discussed such as (a) proton transfer, (b) electron transfer, (c) H-atom transfer and (d) exciplex formation. The role of spin restrictions in electron (and H-atom) transfer processes is emphasized.

Diabatic processes which include transitions from higher to lower potential energy surfaces are typical for ordinary photochemical reactions such as (1) cis-trans isomerization and (2) photodimerization. The conditions for this type of processes are outlined in terms of appropriate potential energy diagrams.

SYMPOSIUM II

PHOTOREACTIONS IN BIOLOGICAL MACROMOLECULAR
COMPLEXES

PHOTOREACTIONS IN BIOLOGICAL MACROMOLECULAR COMPLEXES

(Introductory Remarks)

Kendric C. Smith

Department of Radiology, Stanford University School of

Medicine, Stanford, California 94305, USA

Most photochemical studies on biological macromolecules have been performed on purified samples of macromolecules. Under these conditions only a relatively small number of photochemical reactions are possible. These include reactions that lead to the production of chain breaks in the polymer, unimolecular alterations in the monomer subunits, and bimolecular reactions involving the subunits.

However, if the macromolecules are not in purified form, but are surrounded by a large number of different types of molecules, as they are in a cell, then the number of different types of bimolecular photochemical reactions that can occur is almost infinite.

While it is much more difficult to work on the photochemistry of complex mixtures of biological compounds than it is on purified molecules, photochemical studies on macromolecular complexes are very important to our understanding of the effects of UV radiation on biological systems. The chemistry and biological effects of such bimolecular heteroaddition reactions involving deoxyribonucleic acid were reviewed recently at an international symposium in Williamsburg, Virginia, and were shown to be of major significance not only to photobiology but also to the fields of aging and carcinogenesis as well.

One fascinating aspect of working with electromagnetic radiation is that it can be used as a two-edged sword. One can study the effects of the radiation on some chemical or biological system in order to understand the response of that system to irradiation. After sufficient knowledge has been gained about the responses of defined systems to irradiation, one can then use radiation as a specific probe to determine structural relationships in other systems.

In this Symposium, Dr. Helene will describe the energy transfer and photochemical reactions that can occur in nucleic acid-protein complexes. As I mentioned before, the photochemistry of the separated molecules differs from that of the complex, and biological systems exist as complexes of macromolecules.

Perhaps the best evidence that a field has achieved a certain maturity is when fundamental observations are converted into methods to solve specific problems. Dr. Schimmel will describe how photochemical techniques can be used to determine the points of close association between enzymes and their substrates, e.g., between aminoacyl transfer RNA synthesases and transfer RNAs.

Dr. Lamola will describe research on the detrimental effects of light upon biomembranes that are relevant to human health. In most cases, these effects are due to photooxidations that result from the photochemical formation of singlet oxygen or of radicals. The effects of different photosensitizers and quenchers will be described.

Thus, we will be exposed in this Symposium to both edges of the photochemical sword; we will hear descriptions of the types of photochemical damage that can occur in macromolecular complexes, and we will hear examples of the use of established photochemical techniques to probe the structural relationships in macromolecular complexes.

K.C. Smith (ed.), Aging Carcinogenesis and Radiation Biology: The Role of Nucleic Acid Addition Reactions, Plenum Press, N.Y. (1976).

EXCITED STATE INTERACTIONS AND PHOTOCHEMICAL REACTIONS IN PROTEIN-NUCLEIC ACID COMPLEXES

Claude HELENE

Centre de Biophysique Moléculaire

45045 Orléans Cedex (France)

INTRODUCTION

The formation of specific protein-nucleic acid complexes is of central importance in molecular biology. Such complexes are involved at every step of genetic expression (transcription, translation) and of its regulation. An investigation of the excited states and of the photochemical behavior of these complexes has different purposes : i) excited state properties may be used to obtain information on the mechanism of complex formation and on the nature of molecular interactions involved in these complexes ; ii) the study of photochemical reactions in protein-nucleic acid complexes should help to understand the action of UV radiations on biological systems. The mechanisms of these photochemical reactions have to be established if one wishes to produce or to remove selectively one particular damage. This requires a knowledge of the excited state properties of the complexes ; iii) the formation of photochemical cross-links between a protein and a nucleic acid in a specific complex will contribute to our understanding of the interactions involved in complex formation by providing information on the regions of the two macromolecules which are in close contact in the complex. The use of photochemical cross-linking reactions should be comparable to that of bifunctional chemical reagents with the advantage that the photochemical reaction will link directly two chemical groups one on each macromolecule.

EXCITED STATE INTERACTIONS AND ENERGY TRANSFER
PROCESSES BETWEEN PROTEIN AND NUCLEIC ACID CONS-
TITUENTS

      The investigation of excited-state interactions and
energy transfer processes in protein-nucleic acid complexes
relies mainly upon the observation of light emission from the
excited states.

      At room temperature only aromatic amino acid side
chains are emitting fluorescence (1). The fluorescence of usual
nucleic acid bases is very weak (quantum yields around 10^{-4})
and will be difficult to detect in the presence of a protein containing aromatic amino acids (2). In some cases (e.g., tRNAs)
the fluorescence of odd bases may be observed. At low temperature (77 K), fluorescence and phosphorescence are emitted by
both aromatic amino acids and nucleic acid bases (1, 3, 4).

1 - Stacking Interactions of Aromatic Amino Acids and Nucleic
        Acid Bases in Frozen Aqueous Solutions

      Interactions between nucleic acid bases and aromatic
amino acids were first observed in frozen aqueous solutions at
77 K (5, 6, 7). Under these experimental conditions the formation
of molecular aggregates induces interactions between solute molecules. The results can be summarized as follows :

      i) tryptophan and nucleic acid bases form electron
donor-acceptor (EDA) complexes in which the indole ring of Trp
is the electron donor and the base is the electron acceptor. These
complexes are characterized by a quenching of both tryptophan
and base fluorescences and the appearance of a new fluorescence
at longer wavelengths with a much lower quantum yield (5, 6).

      ii) tyrosine and pyrimidine bases also form EDA com-
plexes. This is accompanied by a quenching of the fluorescence
of both molecules. In the complexes formed by tyrosine and purine bases, only tyrosine fluorescence is quenched but the purine
fluorescence is not markedly affected (7).

      iii) the fluorescence of nucleic acid bases is not affec-
ted by the presence of equimolar concentration of phenylalanine.
An excited state interaction has been observed only with protonated cytidine under acidic conditions (8).

2 - <u>Energy Transfer Processes in Frozen Solutions</u>

The formation of stacked arrays of molecules in frozen aqueous solutions made it possible to study energy transfer
processes between nucleic acid bases and aromatic amino acids
(9). Mixed aggregates can be formed in which the concentration
of one of the components is small as compared to the other one.
Very efficient energy transfer processes at the triplet level were
observed from bases to tryptophan (10) and from tyrosine to
bases (11), in agreement with triplet state energies which decrease in the order Tyr $>$ C $>$ G $>$ A $>$ T $>$ Trp (12). At the singlet
level, an efficient transfer occured from tyrosine to bases. In
all cases the excitation energy migrated step-by-step from one
excited molecule to its nearest neighbors in the aggregate until
it was trapped by the molecule which had a lower singlet or
triplet state energy. For example, one molecule of tryptophan
was able to trap the <u>triplet</u> state energy of about 200 adenosine
molecules (10). About 70 tyrosine molecules were able to transfer their <u>singlet</u> energy to one base (11).

An efficient energy transfer at the triplet level was
also observed in poly(A)-(Lys-Trp-Lys) complexes (10). The results and the mechanisms were similar to those described for
adenosine-tryptophan aggregates. The triplet excitation energy
migrated from base to base in poly(A) until it was trapped by a
Trp residue inserted between two adjacent bases in the polynucleotide.

3 - <u>Quenching of Aromatic Amino Acid Fluorescence in Oligopeptide-Nucleic Acid Complexes at Room Temperature</u>

At room temperature, excited-state interactions between nucleic acid bases and aromatic amino acids were investigated in oligopeptide-nucleic acid complexes. Oligopeptides of
the general sequence Lys-X-Lys, where X is an aromatic residue,
were shown to interact with nucleic acids at low ionic strength
due mainly to strong electrostatic interactions of lysyl residues
with phosphates. Different physical techniques were used to investigate these interactions including nuclear magnetic resonance (13), circular dichroism (14) and fluorescence (15). The
fluorescence of the aromatic amino acid was always observed
to be quenched in the complexes. A comparison of the results
obtained by the different methods allowed us to draw the following
conclusions :

   i) the tryptophyl residue of the oligopeptide Lys-Trp-Lys form stacked complexes with bases and this stacking interaction is strongly favored in single-stranded as compared to double-stranded nucleic acids (15, 16). Fluorescence quenching is due to the formation of stacked complexes. An electron donor-acceptor interaction is likely to be responsible for this quenching as observed in frozen aqueous solutions (see above).

   ii) the tyrosyl residue of the oligopeptide Lys-Tyr-Lys (and of other related oligopeptides such as Lys-Ala-Tyr-Ala-Lys) is stacked with bases only in single-stranded polynucleotides and in single-stranded regions of nucleic acids (e. g., in tRNAs). However a strong fluorescence quenching is observed with both double-stranded and single-stranded polynucleotides or nucleic acids. Different mechanisms can be proposed to explain this fluorescence quenching : a) stacking with bases does quench tyrosine fluorescence in aggregates (see paragraph 1 above) ; b) hydrogen bonding of the OH group of tyrosine to acceptor groups on the nucleic acid (base, sugar or phosphate) might result in fluorescence quenching due, e. g., to an excited-state proton transfer ; c) an interaction of tyrosine with phosphates during its excited state lifetime might also lead to proton transfer and fluorescence quenching ; phosphate anions do quench tyrosine fluorescence in aqueous solutions although the efficiency of this quenching is less than that of phosphate dianions (17) ; d) singlet energy transfer from tyrosine to bases (which are not fluorescent) should lead to tyrosine fluorescence quenching. This transfer should be efficient as shown from the calculation of critical Förster distances (18) ; e) a conformational change of the peptide due to complex formation might bring the tyrosyl residue close to a quenching group inside the peptide.

   Mechanisms b) and c) were eliminated by investigating the binding of the peptide Lys-Tyr(OMe)-Lys in which the OH group of tyrosine is methylated. Fluorescence quenching is observed upon complex formation with nucleic acids and is as efficient as in the case of Lys-Tyr-Lys, even though neither hydrogen bonding nor proton transfer can occur. Hypothesis e) seems very unlikely since similar results were obtained with Lys-Tyr-Lys, Lys-Ala-Tyr-Ala-Lys and peptides in which both the amino- and carboxyl groups were blocked by acetyl and ethylamide substituents, respectively. The conclusion of these studies is that both stacking and energy transfer are responsible for tyrosine fluorescence quenching in the complexes formed between oligopeptides and single-stranded nucleic acids. Since no stacking was observed with double-stranded DNA, tyrosine fluorescence quenching in this case must be entirely attributed to energy trans- fer to base pairs (F. Toulmé, R. Mayer and C. Hélène, to be published).

iii) the fluorescence of several oligopeptides containing phenylalanine (including Lys-Phe-Lys) is quenched in complexes formed with both single-stranded and double-stranded nucleic acids. Nuclear magnetic resonance studies clearly demonstrate that stacking interactions are much more important in single strands. As in the case of tyrosine, energy transfer to nucleic acid bases is very likely responsible for phenylalanine fluo- rescence quenching (T. Montenay-Garestier and C. Hélène, un- published results).

4 - Interactions Between Nucleic Acid Bases and Disulfides

The possible effect to disulfide bridges of proteins on the excited state properties of nucleic acids has been investiga- ted in a model system : the complexes formed by cystamine and polynucleotides (19). These complexes form at low ionic strength due to electrostatic interaction of the two NH_3^+ groups of cysta- mine with phosphates. The phosphorescence of poly(A) is strong- ly quenched by small amounts of cystamine in low temperature glasses (19) indicating that cystamine acts as a trap for the tri- plet excitation energy migrating amongst adenine bases in poly(A). The most likely mechanism involves an electron transfer from the adenine triplets to the disulfide bridge as already observed with tryptophan and tyrosine (20, 21).

PHOTOCHEMICAL REACTIONS IN PROTEIN-NUCLEIC ACID COMPLEXES

The photochemical behavior of protein-nucleic acid complexes is expected to be different from that of the separated macromolecules. The formation of the complex may alter the mu- tual interactions of reacting groups in each macromolecule and therefore modify the kinetics and quantum yields of photoproduct formation. For example a distortion of the phosphodiester back- bone of a nucleic acid upon binding a protein may modify stacking interactions between bases and thus affect the photodimerization of adjacent pyrimidines. A conformational change in the protein induced by complex formation may modify the environment and the photochemical reactivity of, e.g., tryptophyl residues or

disulfide bridges. However more interesting are new photochemical reactions which will be specific of the particular protein-nucleic acid complex under investigation. The formation of
cross-links between the two macromolecules is a very attractive
method to obtain information on the location of the interacting
parts in the complex (although this method has obviously its own
limitations). The photosensitized splitting of pyrimidine dimers
using tryptophyl residues of proteins as photosensitizers may
prove useful in providing evidence for the involvement of these
residues in complex formation. It should also be kept in mind
that primary photoproducts could act as photosensitizers for other
reactions. This is the case, for example, of N'-formylkynurenine,
a photooxidation product of tryptophan (22).

1 - Photochemical Cross-Linking of Proteins to Nucleic Acids

This subject will be dealt with in detail in this Symposium (see contribution by P. Schimmel) and has been reviewed
recently (23, 24). The formation of cross-links between proteins
and nucleic acids has made it possible to locate the regions of
tRNA molecules which interact with aminoacyl-tRNA synthetases
(23) and the peptides of ribosomal S 4 protein which are in close
contact with ribosomal 16 S RNA (25). The method has also been
recently applied to the identification of those thymine bases of
the lac operator which interact with the lac repressor (26). Using
DNA in which thymine has been replaced by 5-bromouracil, irradiation of the lac repressor-operator specific complex leads to a
cross-linking of the two molecules (27). This cross-linking reaction is due to the photochemical cleavage of the C-Br bond and to
the high reactivity of the uracilyl radicals thus produced.

In most cases however, the chemical nature of the cross-
links has not yet been determined and this must await further
studies. The photochemical formation of covalent bonds between
amino acid side chains and nucleic acid bases has been demonstrated in simple model systems. For example, thymine and uracil react with cysteine to form several photoproducts (29). Thymine can also react with lysine amino groups (30). Purines have
been shown to form adducts at the C(8) position with alcohols
and amines (31). Recently we have shown that carboxylic acids
(side chains of Glu and Asp residues) could photochemically react
with nucleic acid bases, especially thymine and uracil (F. Toulmé and C. Hélène, to be published). All these reactions could be
involved in the formation of photochemical cross-links between
proteins and nucleic acids.

2 - Photosensitized Splitting of Pyrimidine Dimers by Tryptophan-Containing Oligopeptides and Proteins

The splitting of pyrimidine dimers can be photosensitized by several types of molecules. Indole derivatives can act as photosensitizers with respect to both isolated dimers and dimers in DNA. Oligopeptides containing tryptophan residues, such as Lys-Trp-Lys, have been shown to bind strongly to UV irradiated DNA and to photosensitize the splitting of thymine dimers (16, 32). An electron transfer from the indole ring to the dimer was proposed as a very likely mechanism for the photosensitized splitting. This electron transfer requires a close proximity (a stacking interaction) of the indole ring and the pyrimidine dimer.

This reaction was thought of as a possible method to provide evidence for stacking interactions of tryptophyl residues of proteins with nucleic acid bases in protein-nucleic acid complexes especially in complexes involving single-stranded DNA where the presence of pyrimidine dimers is not expected to affect markedly protein binding. If a photosensitized splitting of thymine dimers could be demonstrated in such complexes, this would lend strong support to the involvement of stacking interactions of tryptophyl residue(s) with bases. We chose to study the protein coded by gene 32 of phage T 4 because this protein was known to bind selectively and cooperatively to single-stranded DNA. The fluorescence of the tryptophyl residues of gene 32 protein is quenched upon binding to single-stranded DNA or polynucleotides (33). The presence of pyrimidine dimers does not affect this binding. Upon UV irradiation of the complex formed by gene 32 protein with DNA containing pyrimidine dimers a photosensitized splitting of dimers is observed (33). These two results (fluorescence quenching, photosensitized splitting) are consistent with the involvement of tryptophyl residues of gene 32 protein in stacking interactions with bases.

CONCLUSION

This review shows that many different processes are expected to take place when protein-nucleic acid complexes are excited by UV radiations. Excited-state interactions between the chromophores of the two macromolecules may lead to properties which are characteristic of the complexes. These properties may be used to get some insight into the mechanism of complex formation. They may also contribute to our understanding of the photo-

chemical reactions which take place under UV excitation. The
photochemistry of nucleic acid-protein complexes certainly
plays an important role in the behavior of cells and organisms
submitted to UV radiations. Research in this area will also contribute to a better knowledge of the rules which govern the selective recognition of nucleic acids by proteins.

(1) J. W. Longworth (1971) in "Excited States of Proteins and
    Nucleic Acids", R. F. Steiner and I. Weinryb Ed., Plenum
    Press, pp. 319-484
(2) M. Daniels (1976) in "Photochemistry and Photobiology of
    Nucleic Acids", S. Y. Wang Ed., Academic Press, vol. I,
    pp. 23-108
(3) J. Eisinger and A. A. Lamola (1971) in reference 1 pp. 107-198
(4) C. Hélène (1973) in "Physico-chemical Properties of Nucleic
    Acids", J. Duchesne Ed., Academic Press, vol. I, pp. 119-
    142.
(5) T. Montenay-Garestier and C. Hélène (1968) Nature, 217,
    844-845
(6) T. Montenay-Garestier and C. Hélène (1971) Biochemistry,
    10, 300-306
(7) C. Hélène, T. Montenay-Garestier and J. L. Dimicoli (1971)
    Biochim. Biophys. Acta, 254, 349-365
(8) T. Montenay-Garestier and C. Hélène (1973) J. Chim. Phys.
    70, 1385-1390
(9) T. Montenay-Garestier and C. Hélène (1973) J. Chim. Phys.
    70, 1391-1399
(10) C. Hélène (1973) Photochem. Photobiol., 18, 255-262
(11) T. Montenay-Garestier (1976) in "Excited States of Biological
    Molecules", J. B. Birks Ed., Academic Press, pp. 207-216
(12) C. Hélène (1976) in reference 11, pp. 151-166
(13) J. L. Dimicoli and C. Hélène (1974) Biochemistry, 13, 714-
    723 and 724-730
(14) M. Durand, J. C. Maurizot, H. N. Borazan and C. Hélène
    (1975) Biochemistry, 14, 563-570
(15) F. Brun, J. J. Toulmé and C. Hélène (1975) Biochemistry,
    14, 558-563
(16) J. J. Toulmé, M. Charlier and C. Hélène (1974) Proc. Nat.
    Acad. Sci. USA, 71, 3185-3188
(17) J. Feitelson (1964) J. Phys. Chem., 68, 391-397
(18) T. Montenay-Garestier (1975) Photochem. Photobiol., 22, 3-6
(19) T. Montenay-Garestier, F. Brun and C. Hélène (1976)
    Photochem. Photobiol., 23, 87-91

(20) J. Feitelson and E. Hayon (1973) Photochem. Photobiol.,
     17, 265-274
(21) D. V. Bent and E. Hayon (1975) J. Am. Chem. Soc., 97,
     2612-2619
(22) P. Walrant, R. Santus and M. Charlier (1976) Photochem.
     Photobiol., 24, 13-19
(23) P. R. Schimmel, G. P. Budzik, S. S. M. Lam and H. J. P.
     Schoemaker (1976) in "Aging, Carcinogenesis and Radiation
     Biology", K. C. Smith Ed., Plenum Press, pp. 123-148
(24) K. C. Smith (1976) in "Photochemistry and Photobiology of
     Nucleic Acids", S. Y. Wang Ed., Academic Press, Vol. 2
(25) B. Ehresmann, J. Reinbolt and J. P. Ebel (1975) FEBS
     Letters, 58, 106-111
(26) W. Gilbert (1976) Abstracts Meeting on Molecular Aspects
     of Lac Operon Control, Cold Spring Harbor Laboratory, p. 16
(27) S. Y. Liu and A. D. Riggs (1974) Proc. Nat. Acad. Sci. USA,
     71, 947-951
(28) C. Hélène (1976) in reference 23, pp. 149-163
(29) A. J. Varghese (1976) in reference 23, pp. 207-223
(30) M. D. Shetlar, H. N. Schott, H. G. Martinson and E. T. Liu,
     (1975) Biochem. Biophys. Res. Comm., 66, 88-93
(31) D. Elad (1976) in reference 23, pp. 243-260
(32) M. Charlier and C. Hélène (1975) Photochem. Photobiol.,
     21, 31-37
(33) C. Hélène, F. Toulmé, M. Charlier and M. Yaniv (1976)
     Biochem. Biophys. Res. Comm., 71, 91-98.

STRUCTURAL RELATIONSHIPS IN MACROMOLECULAR COMPLEXES DETERMINED

BY PHOTOCHEMICAL CROSSLINKING

Paul R. Schimmel and Vincent T. Yue

Massachusetts Institute of Technology

Cambridge, Massachusetts 02139 USA

SUMMARY

Structural relationships in protein-nucleic acid complexes
and in a protein-nucleotide complex have been investigated by
photochemical crosslinking. Crosslinking is achieved by direct
irradiation of the complexes with a mercury lamp that gives predominantly 253.7 nm radiation. Six complexes of aminoacyl transfer
RNA synthetases with specific transfer RNAs have been photo-joined.
The areas on the nucleic acid involved in crosslinking have been
determined. When combined with a model for the three dimensional
transfer RNA structure, these data give a picture of how the enzymes
orient on the tRNAs. In addition, photo-crosslinking of ATP to
synthetases has been achieved. (This nucleotide is used by all
aminoacyl tRNA synthetases to drive the aminoacylation reaction.)
This photo-reaction was studied in some detail with Ile-tRNA
synthetase from E. coli B. It was found that the entire ATP molecule (base and phosphate moieties) is stably attached to the enzyme.
Moreover, crosslinkage occurs at a unique, specific site. The
peptide fragment to which the ATP is photo-joined has been isolated
and sequenced, thus localizing in the primary structure the position
of some of the residues that line the ATP receptor site.

INTRODUCTION

Macromolecular complexes are involved in most of the essential
biochemical processes. Some of the well studied examples include
repressor-operator, antigen-antibody, DNA polymerase-DNA, and ribosomal proteins-ribosomal RNA complexes, as well as the many examples
of small ligands such as ATP combined with various proteins. To

41

gain a fundamental understanding of biochemical events much attention has been directed at structural characterization of and mechanism of specificity of formation of these macromolecular complexes.

In this regard, photochemical crosslinking is a powerful
approach toward elucidating structural features of large complexes.
This is particularly true for studies of protein-nucleotide or
protein-nucleic acid complexes, because there appears to be a broad
spectrum of photo-reactions that can occur between amino acid side
chains and nucleotide bases (1-3). Thus, by directly irradiating
complexes at 254 nm, for example, it is possible to link together
the macromolecules (4). Examples where this has been successfully
accomplished include the photo-induced joining of DNA polymerase
to DNA (5), RNA polymerase to DNA (6,7), aminoacyl tRNA synthetases
to tRNAs (8-10), bacteriophage fd gene 5 protein to DNA (11), and
ribosomal proteins to ribosomal RNA (12).

One of the best studied systems that has utilized the photochemical approach is the aminoacyl tRNA synthetases. These enzymes
catalyze the first steps of protein synthesis whereby amino acids
are attached to their cognate tRNA chains (13, 14). (Corresponding
to each amino acid there is a separate aminoacyl tRNA synthetase
and one or more specific tRNAs.) The enzymes all utilize ATP as
the energy source to drive the aminoacylation reaction whereby an
ester linkage is established between a tRNA terminal hydroxyl and
the amino acid. A key question is that of the mechanism whereby
the enzymes are able to recognize distinct tRNA species; this is
a critical event since the attachment of an amino acid to the wrong
tRNA species could lead to an error in protein synthesis. One of
the major contributions of the photo-crosslinking approach is that
for several synthetase-tRNA complexes it has afforded a picture of
the spatial location of some of the enzyme-tRNA contact sites on
the three dimensional tRNA structure (8-10).

Another question that has been attacked is the location of
the ATP site in the primary structure of a synthetase. As discussed
below, this is of interest from several standpoints.

In the discussion that follows, a brief synopsis is given of
the many photo-crosslinking studies of synthetase-tRNA complexes.
This work has been extensively reviewed elsewhere (4, 8-10, 15).
More attention is directed at the detailed characterization of a
photo-crosslinked synthetase-ATP complex. This work has features
of special significance that have not hitherto been reported.

PHOTO-CROSSLINKED SYNTHETASE-tRNA COMPLEXES

A number of synthetase-tRNA complexes have been irradiated
at 254 nm with the result that stable covalent links form between

the enzymes and tRNAs (8-10). Moreover, crosslinking is specific
under the conditions used; that is, when an enzyme is irradiated
with a non-cognate tRNA that doesn't significantly bind, no cross-
linking is observed (9). Other data also point to the specific
nature of the crosslinking reaction. Therefore, there is reason to
believe that the crosslinks that do form represent true enzyme-tRNA
contact sites.

By use of special procedures it has been possible to work out
the sections on the tRNAs that are involved in the crosslinking
reaction (8-10). Because a three dimensional structural model of
transfer RNA is available, the crosslinking information has made it
possible to pinpoint the location of certain of the enzyme-nucleic
acid contact points. This in turn has permitted conclusions to be
drawn about the structural organization of these important macromolecular complexes (8-10).

PHOTO-CROSSLINKED SYNTHETASE-ATP COMPLEXES

Background

As mentioned above, the aminoacyl tRNA synthetases use ATP to
drive the aminoacylation reaction. Thus, each individual synthetase
has the common feature of an ATP receptor site. An ideal way to
approach questions concerning the location and nature of this site
on the various enzymes is, of course, to link bound ATP to its site
and then analyze the resulting product. For this purpose, photo-
affinity labels of nucleotide analogs, such as those described by
Haley and Hoffman (16) and by Cooperman and coworkers (17, 18),
are extremely valuable. These labels rely on the formation of
nitrene or carbene intermediates upon irradiation of the bound
nucleotide analog. These extremely reactive intermediates can then
couple to common groups on a protein (See Ref. 19 for a general
discussion).

An alternate approach is to irradiate directly synthetase-ATP
complexes, without employing one of the useful photo-affinity
analogs. Results of Steinmaus, Rosenthal, and Elad with model
systems suggest that 254 nm irradiation can give photo-joining of
the C-8 position of purine nucleosides to carbon atoms adjacent to
a basic function such as OH (20). Also, Antonoff and Ferguson have
given data suggesting that cAMP can be directly photo-crosslinked
to cAMP receptor proteins present in a crude extract from testis
and adrenal cortex (21). These encouraging results suggest that
direct irradiation of synthetase-ATP complexes may give the desired
crosslinking.

The direct irradiation of synthetase-ATP complexes affords

another advantage in that it enables us to explore in depth certain
questions that were difficult to answer in the case of the photo-
crosslinked synthetase-tRNA complexes. In particular, in the latter
studies the regions on the tRNAs involved in crosslinking were elucidated, but determination of the crosslinking sites on the synthetases is a far more difficult task which was not carried out.
However, with the much simpler synthetase-ATP system it is easier
to determine the location of a crosslinking site(s) on the enzyme.
This makes it possible to examine carefully whether crosslinking
occurs at a unique, specific peptide or whether multiple peptide
units are involved.

With these considerations in mind, we attempted photo-
crosslinking of synthetase-ATP complexes. The attempt was successful and thus enabled us to explore the crosslinking reaction in
depth.

Characteristics of the Crosslinking Reaction

Photo-induced joining of ATP was achieved in three synthetase
systems - isoleucyl, tyrosyl, and valyl tRNA synthetases from
E. coli B. In each instance crosslinking was accomplished by
irradiation with a 15-watt low pressure mercury lamp equipped with
a 2 mm thick Vycor filter. This setup gives predominantly 253.7 nm
irradiation with short wavelength, far uv irradiation cut off by the
filter. The crosslinking was assayed by measuring the amount of
radiolabeled ATP that becomes precipitable by 5% trichloroacetic
acid (TCA) as a result of irradiation in the presence of synthetase.
Control experiments indicated that acid precipitation of ATP results
from stable linkage to the synthetases.

A detailed investigation was made of the Ile-tRNA synthetase-
ATP system. Results of a study of the dosage dependence of the
crosslinking reaction are given in Figure 1. The figure gives
percent crosslinking versus the effective irradiation dose, corrected for screening effects as described by Johns (22). It is
seen that crosslinking increases in a dose dependent fashion until
a maximum of about 0.15 mole of ATP is crosslinked per mole of
enzyme.

The experiment shown in Figure 1 was done with $[8-^{14}C]$ATP.
The question can be raised as to whether the entire ATP molecule
is incorporated or whether only a fragment containing the labeled
portion is actually crosslinked. To check on this issue, irradiations were also done with ^{14}C uniformly labeled ATP ($[^{14}C(U)]$ATP),
$[\alpha-^{32}P]$ ATP, and $[\gamma-^{32}P]$ ATP. The results are tabulated in Table I.
The table shows that the extent of crosslinking monitored with each
labeled ATP is the same. This suggests that the intact molecule

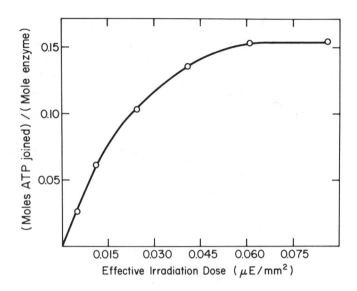

Fig. 1 : Moles of ATP joined per mole of enzyme versus the effective
irradiation dose in microeinsteins/mm^2.

itself is incorporated, and not simply a fragment.

Specificity of Crosslinking

The major question is whether the observed crosslinking occurs
at the specific ATP binding site on the enzyme, or whether linkage
is non-specific. One approach to this problem is to repeat the
experiment with another nucleotide and the enzyme, or with ATP and
another, non-specific protein. Results of these kinds of experiments are given in Table II. Here it is shown that irradiation of
synthetase-AMP solutions gives no detectable crosslinking. (Although
AMP binds to the synthetase, its association is weaker.) This
experiment strongly suggests that the crosslinkage observed with
ATP is not due to a non-specific photo-reaction associated with
the adenine ring. In addition, there is no detectable joining of
serum albumin to ATP. This again indicates that the synthetase-ATP

TABLE I

Photocrosslinking yields for Ile-tRNA synthetase-
labeled ATP mixtures

Compound	moles ATP joined/mole Ile-tRNA synthetase
$[^{14}C(U)]$ATP	0.142
$[\alpha-^{32}P]$ATP	0.163
$[\gamma-^{32}P]$ATP	0.148

$[^{14}C(U)]$ATP / $[\alpha-^{32}P]$ATP / $[\gamma-^{32}P]$ATP = 0.96 : 1.10 : 1.00
The solutions contained in 100 µl : 50 mM Tris-HCl (pH 8),
10 µM Ile-tRNA synthetase, 0.25 mM ATP, and 5 mM $MgCl_2$.
Effective irradiation dose was 43 nanoeinsteins/mm^2 over a
period of 40 min. at 0-4°C.

linkage is specific.

     Another ambiguity arises because irradiation of the synthetase
alone gives rise to inactive enzyme. This is presumably due to
photo-modification of one or more groups that either directly or
indirectly affect the catalytic site. This photo-inactivation also
occurs in the presence of ATP - to an extent considerably greater
than the extent of crosslinking. Therefore, even though cross-
linking appears to be specific, the serious question arises as to
whether joining actually occurs to the active enzyme or to a special
site(s) on the inactive molecule.

     To check this out, two kinds of experiments were designed. One
of these is described here, although the other gave results that
lead to the same conclusion. The synthetase alone was irradiated
with varying dosages to achieve different degrees of catalytic
inactivation. ATP was then added, irradiation was continued, and
the degree of crosslinkage was determined. The idea is to see
whether or not the crosslinking progressively decreases as more
and more of the enzyme is pre-inactivated before addition of the
ATP.

TABLE II

Photo-induced joining of ATP			
Protein	Nucleotide	TCA precipitable CPM	Moles joined per mole enzyme
IleRS	$[8-^{14}C]$ATP	1148	0.15
IleRS	$[^{14}C(U)]$ATP	59	0.0
Serum Albumin	$[8-^{14}C]$ATP	20	0.0
(IleRS	$[8-^{14}C]$ATP)*	37	0.0

All irradiations were carried out to give an effective dosage of at least 32 nanoeinsteins/mm^2 (over a period of 30-40 min), until maximal yields were obtained. Reaction mixtures contained (in 100 µl) about 10 µM protein, 200 µM mononucleotide, 5mM $MgCl_2$, and 50 mM Tris-HCl (pH 8.0).

* No irradiation.

The results are shown in Table III. The table shows that as more of the enzyme is pre-inactivated, less ATP can subsequently be photo-joined. In fact, when the crosslinking yield is normalized by setting the maximal amount equal to 100%, the percent crosslinking correlates well with the fractional amount of active enzyme that is present. This experiment, and the results of another one to be described elsewhere (23), give good evidence that crosslinking specifically occurs to active enzyme.

Together with the data in Table II, we can be reasonably assured that the reaction occurs at the ATP binding site.

Isolation of and Sequence of Labeled Peptide

Having reasonable assurance that crosslinking occurs at the ATP site, the next step was to determine whether one or more peptides are involved in the photo-labeling. The irradiated $[\alpha-^{32}P]$ATP-

TABLE III

___

Effect of Pre-Irradiation on Crosslinking Yields

___

Effective Pre-Irradiation Dose (nanoeinsteins/mm^2)	% Activity	% of Maximal Crosslinking
0	100	100
1.7	91	89
26	21	23
*51	15	15
*51	14	$\leq$14
102	5	$\leq$14

___

Solutions initially contained in 50 μl: 50 mM Tris-HCl (pH 8), 10 μM Ile-tRNA synthetase, and 5 mM $MgCl_2$. After irradiation with varying dosages, lyophilized $[^{14}C(U)]$ATP was added to give a final concentration of 0.25 mM. An irradiation dose of 35 nanoeinsteins/mm^2 over a period of 40 min. was then given.

* Separate experiments.

synthetase complex was removed from free $[\alpha-^{32}P]$ATP and subjected to digestion with protease type VI. The resulting peptides were subjected to two dimensional chromatography and visualized by ninhydrin spray. The location of possible labeled peptides was accomplished by autoradiography.

The results of these experiments showed clearly that although ninhydrin positive material is spread over a large area on the two dimensional chromatogram, the radioactivity is concentrated in one spot. This suggests that the ATP photo-reaction involves only a single peptide. This was confirmed by eluting the labeled material from the chromatogram and performing a N-terminal analysis. This

gave a single amino acid - alanine. Therefore, there is little
doubt that a single peptide photo-reacts at the ATP binding site.

Unfortunately, the peptide isolated from the protease type VI
digestion described above is too large (approximately 20 residues)
for convenient determination of its amino acid sequence by manual
methods. However, it was found that a small peptide can be obtained
by digesting the crosslinked synthetase-ATP complex with a combination of protease type VI, trypsin, and pepsin.

Using the micro Dansyl-Edman procedure discribed by Fleichman
(24), the sequence of the labeled peptide released by these proteases
was determined. The tentative sequence obtained is that of a
hexapeptide and is

Lys-Val-Ala-Gly-Asx-X

where X is a residue which has not been identified because it
doesn't appear to correspond to any of the natural amino acids.
In fact, it may be a modified form of the residue which actually
crosslinks to ATP.

However, it is difficult to be certain which residue is cross-
linked because it appears that the acid hydrolysis conditions in
one step of the Dansyl-Edman procedure are sufficiently harsh to
break the ATP-peptide bond. Therefore, at this point we do not
know the nature of the ATP-peptide crosslink.

Although a complete amino acid sequence has not been determined
for any aminoacyl tRNA synthetase, Kula has determined the sequence
of some cysteine containing peptides of Ile-tRNA synthetase from
E. coli MRE 600 (25). One of these peptides has the sequence

Cys-Val-Ser-<u>Asx-Val-Ala-Gly-Asx</u>-Gly-Glx-Lys

The underlined portion has a sequence identical to the one obtained
by us for the labeled peptide, except that the first residue is Lys
in our peptide versus Asx in Kula's peptide. However, these two
residues are related by a single base change in their triplet codons
so that the difference may only be due to the fact that enzyme from
different strains of E. coli were used in each case. In any event,
the results obtained here pinpoint the location in the primary
structure of a peptide that lines the ATP binding site.

CONCLUSIONS

The results with the synthetase-ATP complex described above
indicate that direct irradiation of the natural complex, without
the introduction of an affinity analog, gives specific crosslinking

to the ATP site. The advantage of this approach is that natural
materials are used. This means that one can not only avoid the
labor of synthesis of an appropriate analog, but equally or more
important, the possibility is eliminated of introducing distorted
or artifactual complexes arising from the extraneous groups commonly
built into an analog.

The photo-chemical coupling of pyrimidine nucleotide inhibitors
to ribonuclease A has been achieved by Sperling and Havron (26).
These investigators also used natural materials as opposed to
pyrimidine nucleotide analogs. Crosslinkage was achieved making
use of acetone as a sensitizer and irradiating at λ > 300 nm. In
this way potentially harmful photo-chemical activation of absorbing
groups on the protein was minimized. In their system, as in ours,
crosslinking appears to occur to a single peptide.

The results of the above described investigation of a
synthetase-ATP complex pinpoint the location in the primary structure of the ATP binding site of Ile-tRNA synthetase. When the
three dimensional structure of this enzyme is known, the spatial
location of the ATP site can thus be identified from the information obtained by the photo-labeling. In other studies described
elsewhere, we have used direct photo-chemical crosslinking to
localize the synthetase binding sites on the known three dimensional structure of transfer RNA (8-10). Thus, direct irradiation
of macromolecular complexes has provided unique information on the
structural and architectural features of these rather complicated
systems.

Since ATP is a common substrate for the various aminoacyl tRNA
synthetases, it is conceivable that each has closely similar peptides lining this part of the molecule. Using the photo-crosslinking
approach, it should be possible to photo-label a peptide from the
ATP site of each enzyme. Comparison of the sequences of these
peptides will give a simple and direct test for homologies in this
part of the molecule. Here again, the photo-chemical approach
provides a direct route to an important structural question.

At this time one of the major needs is for more intimate
knowledge of the kinds of peptide-nucleotide adducts that can
form. A number of interesting results have been obtained with
model systems which suggest that there is a broad spectrum of
possibilities for crosslinking of purines and pyrimidines to
amino acids (1-3). However, far more work is needed in this
area. Once a more detailed picture of the types of linkages
that can form in natural materials becomes available, it will put
the field on a sounder chemical foundation. In the meantime,
however, the wealth of information that can be derived from direct
photo-chemical crosslinking will continue to play a major role in
expanding our understanding of macromolecular systems.

REFERENCES

1. Smith, K.C. (1975) in "Photochemistry and Photobiology of
     Nucleic Acids" S.Y. Wang, (Ed.), Academic Press, New York.
2. Aging, Carcinogenesis, and Radiation Biology (1976) Smith, K.C.
     (Ed.), Plenum Press, New York.
3. Smith, K.C. (1969) Biochem. Biophys. Res. Comm. 34, 354-357.
4. Schimmel, P.R., Budzik, G.P., Lam, S.S.M., and Shoemaker, H.J.P.
     in Aging, Carcinogenesis and Radiation Biology (1976)
     Smith, K.C. (Ed.), Plenum Press, New York.
5. Markovitz, A. (1972) Biochem. Biophys. Acta 281, 522-534.
6. Weintraub, H. (1973) Cold Spring Harbor Symp. Quant.
     Biol. 38, 247-256.
7. Strniste, G.F., and Smith, D.A. (1974) Biochemistry 13, 485-493.
8. Shoemaker, H.J.P., and Schimmel, P.R. (1974) J. Mol. Biol.
     84, 503-513.
9. Schoemaker, H.J.P., Budzik, G.P., Giegé, R.C., and Schimmel, P.R.
     (1975) J. Biol. Chem. 250, 4440-4444.
10. Budzik, G.P., Lam, S.S.M., Shoemaker, H.J.P., and Schimmel, P.R.
     (1975) J. Biol. Chem. 250, 4433-4439.
11. Anderson, E., Nakashima, Y., and Konigsberg, W. (1975) Nucleic
     Acid Res. 2, 361-371.
12. Gorelic, L. (1975) Biochemistry 14 ,4627-4633.
13. Novelli, G.D. (1967) Annu. Rev. Biochem. 36, 449-484.
14. Schimmel, P.R. (1973) Accounts of Chem. Res. 6, 299-305.
15. Schimmel, P.R. (1975) Proceedings of the Tenth FEBS Meeting,
     219-232.
16. Haley, B.E., and Hoffman, J.F. (1974) Proc. Nat. Acad.
     Sci. USA 71, 3367-3371.
17. Brunswick, D.J., and Cooperman, B.S. (1971) Proc. Nat. Acad.
     Sci. USA 68, 1801-1804.
18. Brunswick, D.J., and Cooperman, B.S. (1973) Biochemistry
     12, 4074-4078.
19. Knowles, J. (1972) Accounts of Chem. Res. 5, 155-160.
20. Steinmaus, H., Rosenthal, I., and Elad, D. (1969) J. Am.
     Chem. Soc. 91, 4921-4923.
21. Antonoff, R.S., and Ferguson, J.J.,Jr. (1974) J. Biol.
     Chem. 249, 3319-3321.
22. Johns, H.E. (1968) Photochem, Photobiol. 8, 547-563.
23. Yue, V.T., and Schimmel, P.R., in preparation.
24. Fleichman, J.B. (1973) Immunochem. 10, 401-407.
25. Kula, M.R. (1973) FEBS Letters 35, 299-302.
26. Sperling, J., and Havron, A. (1976) Biochemistry 15, 1489-1495.

PHOTODEGRADATION OF BIOMEMBRANES

A. A. Lamola

Bell Laboratories

Murray Hill, New Jersey USA 07974

INTRODUCTION

That membrane damage plays a role in the photoinactivation of cells especially in the presence of photodynamic sensitizers has been recognized for several decades [1,2]. But compared to other sites of photodamage, biomembrane damage has been given much less attention. I have chosen to address the subject of biomembrane photodegradation for three reasons: (1) A number of studies in the field have been carried out since the last Congress which are important to review. (2) Some key questions have been formulated for which there are no answers presently and I hope to stimulate additional work on them. (3) The destructive effect of light upon biomembranes is relevant to human health in that the phototoxic effects of a variety of compounds may be chiefly a result of such damage.

That the photodynamic disruption of lysosomes may be the basis for the phototoxic effect of a variety of exogenous chemicals on human skin was pointed out several years ago [3,4,5]. Allison, et al., found that the active compounds appear to concentrate in the lysosomes [3,4]. If they cause photolysis of the lysosomes, then the assured result is lysis of the cell due to released lytic enzymes.

The photohemolysis of red cells _in vitro_, a model system which has received a great deal of attention since the pioneering work of Blum [1] and Cook [6], has found an _in vivo_ connection with a side effect of the phototherapy for neonatal hyperbilirubinemia. During therapy some infants have suffered a hemolytic anemia associated

53

with severe skin pigmentation ("bronze baby syndrome") [7]. It
has been suggested that bilirubin present in the red cell membranes
of the infants acts as a sensitizer for the photochemically caused
increase in cell fragility [7,8].

Several groups [9], with the most detailed work by Konrad and
coworkers [10], have pointed out that the photobiological effect
that is the basis for the heightened erythemal response to light
of patients with erythropoietic protoporphyria is the protoporphyrin-
sensitized photolysis of the cells that comprise the walls of the
capillaries at the interface of the dermis and epidermis. Recently
Mukai and Goldstein [11] have shown that malondialdehyde, a product
of the oxidation of polyunsaturated fatty acids (PUFA), is able to
cause mutations in the Ames salmonella test system. This result
adds membrane photooxidation to the list of etiologic possibilities
for sunlight-induced human cancers.

The functions of biomembranes are manifold: The most elementary function is to contain the cell contents; they act to
sequester certain cell components in organelles such as the mitochondria, chloroplasts, and lysosomes; they maintain or produce
concentration gradients of ions and metabolites for energy transduction and storage; they act as reactive surfaces; they are
involved in cell recognition. One can define biomembranes photodegradation in a number of ways: the impairment of some function
of the membrane; the observation of some adverse structural change;
the observation of the loss of a component or the appearance of a
photoproduct. As in most areas of photobiology the emphasis of an
individual worker depends upon hisposition in the spectrum between
spectroscopy and photomedicine. In general, however, a common goal
in photobiology is to relate photobiological effect to photochemical process. As expected for any stressed complicated system
there are difficulties in deriving conclusions about early stages
in the photodegradation of membranes from measurements on the end
photobiological result. Experiments with simpler but relevant
models is a common approach in attempts to understand primary
photochemistry in photobiology. For this purpose it is important
to keep in mind certain aspects of membrane structure and dynamics
that may strongly distort reactivities and kinetics compared with
what is found in perported model systems.

PHOTOOXIDATION AND BIOMEMBRANE PHOTODEGRADATION

From what is in the literature to date I feel free to make the
statement that the photodegradation of biomembranes is virtually
synonomous with light-induced oxidation (uptake of molecular
oxygen) of membrane components. This statement is based on the
fact that photooxidation products have been observed, protection by

antioxidants has generally been observed, and there is generally a large reduction in the rate of the degradative process being assayed when molecular oxygen is excluded. Neither does this imply that damage other than oxidative damage is not important, nor does it imply that the primary photochemical process(es) necessarily involve molecular oxygen.

Lipid components of membranes are easily oxidized in the presence of light and molecular oxygen and appropriate sensitizing or otherwise initiating conditions. A free radical photooxidation pathway [12,13] can obtain easily because of the facile abstracti- bility of the allylic hydrogen atoms of the polyunsaturated fatty acid side chains of the phospholipids and the allylic hydrogens of cholesterol [12,13]. A variety of free radical species generated by photochemical means and molecules in excited states can abstract these labile hydrogens. The relatively stable lipid radical that is formed reacts easily with molecular oxygen to form a hydroperoxy radical which, after it abstracts a hydrogen atom from some source, becomes the first relatively stable oxidation product, an allylic hydroperoxide. The source of the hydroperoxy hydrogen atom is crucial. If it is from some membrane component like the naturally occurring antioxidant α-tocopherol (vitamin E) only the original lipid molecule and the tocapheral have been oxidized. But the source can be another lipid molecule that itself then combines with oxygen. This cycle of reactions can continue in a chain with a great number of turnovers. In this way one initiation event can cause the oxidation of a large number of lipid (or other) mole- cules in the membrane. The hydroperoxides may undergo oxygen- oxygen cleavage under certain conditions to give one or two free radical species and initiate other chains [14]. This chain auto- oxidation of unsaturated lipids has been recognized for a long time [13] as the basis for rancification of fats and oils and the curing of old type oil paints. It may well be of central importance in the photodegradation of biomembranes.

The unsaturated lipids react with singlet oxygen via the ene- reaction to form hydroperoxides directly in one step [15,16]. By this photochemical process one lipid is oxidized per photon absorbed if the hydroperoxide product does not decompose.

Virtually no analyses have been made of the products of photo- oxidation of unsaturated phospholipids. The usual assay for phos- pholipid oxidation is a facile colorimetric determination of malondialdehyde [12]. This approach is not very satisfactory because malondialdehyde is the product of multiple oxidations of a polyunsaturated fatty acid and represents late steps in phos- pholipid oxidation.

The hydroperoxides derived from the primary oxidation of cholesterol have been isolated and are well characterized [16].

Whereas oxidation via singlet oxygen gives virtually exclusively
the 5α-hydroperoxide (Δ^6), free radical oxidation gives the 7α-
and 7β-hydroperoxides (Δ^5) [17,18,19]. It appears that the isolation of cholesterol-5α-hydroperoxide from a photooxidized membrane
is the only method existent by which the involvement of singlet
oxygen in the primary oxidation can be conclusively demonstrated
(see below).

The lipofuscin pigments associated with aging of tissues can
be produced in oxidized membranes. Some of these pigments are
fluorescent and this fluorescence has been used as an assay for
membrane oxidation [12]. It appears that these pigments are formed
by condensation of carbonyl oxidation products of lipids with amino
groups of various membrane components [12].

Destruction of amino acid side chains occur when membranes are
oxidized under a variety of conditions including photolytic conditions. However, the oxidation products have not been identified
in any case. For the protoporphyrin-sensitized photohemolysis of
red cells, a photodynamic process, the usual photooxidizable side
chains are found to be lost from membrane proteins, namely thyl,
methionyl, histidyl, tyrosyl, and tryptophanyl [20,21]. Other
membrane associated materials, such as α-tocopherol, are also
oxidized.

A few data are available concerning the reactivities of
membrane components towards oxidative reactions. The unsaturated
fatty acids and cholesterol react with singlet oxygen in organic
solvents such as pyridine with comparable rates, $10^5-10^6 M^{-1}s^{-1}$,
with the reactivity increasing with the number of allylic hydrogen
atoms in the substrate [22]. Rate constants of reactions of the
reactive amino acids with singlet oxygen in water solution are on
the order of $10^6-10^7 M^{-1}s^{-1}$ [23]. α-Tocopherol quenches singlet
oxygen at a rate of $10^7-10^8 M^{-1}s^{-1}$ depending on the polarity of the
solvent [24]. Almost no data concerning the relative reactivities
of the various membrane components in free radical type oxidation
processes are available. For the lipids it is generally observed
that oxidative rates increase with the degree of unsaturation but
this does not hold for all conditions [14]. It appears, however,
that by the time oxidative damage in membranes causes lysis,
oxidative loss of both lipid and protein components can be demonstrated.

With respect to functional degradation, activities of ATPase,
acetylcholinesterase, and cytochrome-a_3 have been found to be
destroyed under photooxidation conditions [20,25,26]. Sugar and
other molecular transport systems as well as ion pumps are
deactivated [27,28]. However, lysis, or the failure of the cell or
vesicle to hold its contents, remains the main functional property
assayed in membrane photodegradation studies.

Light absorption by pure lipids with non-conjugated double bonds is important only at wavelengths lower than 250 nm. Light of wavelengths lower than about 300 nm can be absorbed by proteins, and nucleotides in the membranes and by other components with benzene ring structures such as α-tocopherol. Absorption of light of wavelengths above 300 nm must be by prosthetic groups and pigments naturally present in the membrane or by exogenous dyes and sensitizers which one way or another become associated with the membrane. Since photoinitiation of radical chain oxidation may be an important process, low efficiency photoprocesses and absorbing species present only as traces cannot be discounted.

PHOTOHEMOLYSIS OF RED CELLS

The photohemolysis of red cells in the absence of added sensitizers has been investigated in some detail [6,29,30]. Most of the work employed light in the ultraviolet region below 300 nm (UV-C, UV-B). Cook and coworkers [29] demonstrated that the photo-hemolysis is of the Colloid-osmotic type and due to an increase in the passive permeability of the membrane to cations rather than solely the destruction of activated ion pumps. In the work of Cook et al. the uv-dose was given in a short time and the resultant hemolysis, which took place after a lag time, was allowed to pro-ceed in the dark. The hemolysis curves (hemolysis vs. time) were typically of some high kinetic "order". The "rate" for the hemo-lysis, defined as the reciprocal of the time required after irradiation to reach 50% hemolysis, was found to be proportional to the square of the light dose. The action spectrum determined on this basis has a peak near 280 nm and a minimum near 265 nm, resembling protein absorption. Thiol reagents protect the red cells as does α-tocopherol as shown by Roshchupkin, et al. [30]. The latter workers suggest that the lag time in hemolysis may be due to inhibition by naturally present α-tocopherol. These authors also report data on the oxygen dependence of the photohemolysis which indicates that oxidative damage is of paramount importance in all the studies made on the system as was already indicated by the protective action of -SH reagents. What is the important primary photochemistry remains an interesting question. It may involve membrane protein. One candidate reaction would be the well known electron ejection from tryptophan. However, one cannot dismiss photochemistry in the cytosol as an important oxidation initiating event. For example, Salhaney and coworkers [31] have recently demonstrated that oxyhemoglobin decomposes to methemoglobin and superoxide ion when irradiated with light in the uv-range. How the key primary photochemistry is tied into the hemolysis event is, of course, an open question, except that oxidative damage is involved somewhere in the scheme. Additional comments about the mechanism of hemolysis are given below.

Before proceeding to sensitized photohemolysis a few words about the kinetics of photohemolysis are in order. Obviously several processes occur before the actual rupture of the cell, and there is probably more than one pathway which leads to hemolysis. Typically the hemolysis curve shows a lag time, during which no hemolysis is observed, followed by the onset of observable hemolysis and completion within a relatively short time, i.e. the hemolysis following an S-shaped curve. It is not possible to define the kinetics of hemolysis in terms of a single parameter. To be accurate the lag-time, the "order" (shape of the curve once hemolysis starts) and the associated "rate constant" must be specified. In other words it is important to follow the complete hemolysis curve and not rely on, for example, the time required for 50% hemolysis as a measure of the hemolysis rate. An even more dangerous "short-cut" is to take the extent of hemolysis at some arbitrary time as a measure of the hemolysis "rate". Unfortunately, most information in the literature on the effect of various potential inhibitors of photohemolysis is based upon such incomplete single time assays. It is usually not possible to discern, for example, whether the inhibitor affected the lag-time or the kinetics of hemolysis once begun or both.

"Normal" red cells are photohemolyzed by near ultraviolet and visible light but the dose requirements is very much higher than for ultraviolet light in the range below about 300 nm [29]. However, red cell photohemolysis can be sensitized by a variety of compounds which absorb light in the near ultraviolet and visible ranges [32]. Among the effective sensitizers are methylene blue, acridine orange, anthracene, porphyrins, chlorophyll, sulfanilimide, and chlorpromazine. All of these sensitizers of red cell photohemolysis are known phototoxic agents. The mechanism of photohemolysis has not been investigated in all cases. In those that it has the photodynamic (photosensitized oxidation) mechanism has been demonstrated.

The photohemolysis of red cells sensitized by protoporphyrin (PP) has been studied most extensively [20,25,33-36] because in several porphyrias, diseases of abnormal porphyrin metabolism, the red cells contain photosensitizing porphyrins. In erythropoietic protoporphyria (EPP), a fraction of the red cells contain very large amounts of protoporphyrin (metal free) and show brilliant fluorescence [37]; all the red cells have abnormal photosensitivity. The action spectrum for photohemolysis matches the absorption of protoporphyrin, molecular oxygen is required, and the hemolysis is inhibited by free radical inhibitors. Polyunsaturated fatty acids and amino acid side chains are oxidized as well as glutathione. A number of enzyme activities are destroyed concomitantly. Malondialdehyde and hydrogen peroxide are produced, and membrane proteins become cross-linked [38]. Incorporation of β-carotene into the membrane affords moderate protection [33,39].

Two questions about the photohemolysis of EPP cells have received most attention: (1) What change(s) in the membrane cause the hemolysis? (2) What is the primary photochemistry that leads to the subsequent hemolysis?

The photohemolysis of EPP is of the colloid osmotic type, that is, the passive permeability of the membrane to K^+ and Na^+ is increased [40]. This means that damage other than that to ion pumps must be involved. Oxidative cleavage of unsaturated fatty acid side chains would give products analogous to lysolicithins. However, the hemolytic competance of such fatty acid oxidation products has not been tested. Work in our laboratory has demonstrated that cholesterol-5α-hydroperoxide is formed upon irradiation of red cell ghosts containing protoporphyrin [41]. If about 5% of red cell membrane cholesterol is replaced by this hydroperoxide or by the 7α-hydroperoxide, the cell becomes progressively more osmotically fragile and hemolyses [39,41]. The hemolysis induced in this way shows many characteristics identical to the photohemolysis of EPP cells. However, it has been shown that these cholesterol hydroperoxides do not directly disturb the membrane integrity by reason of their structures. These compounds cause hemolysis because they undergo catalyzed decomposition which initiates free radical autooxidation processes in the membrane [39]. A recent suggestion is that photohemolysis is the result of a direct photooxidative cross-linking of membrane proteins by routes which do not require lipid oxidation [38].

With respect to the elucidation of the primary photochemistry associated with photohemolysis it should be recognized at the start that different primary photochemistry can result in similar if not identical later stages in membrane destruction. For example, incorporation of either cholesterol-5α-hydroperoxide, a singlet oxygen product, or cholesterol-7α-hydroperoxide, a free-radical oxidation product, into red cell membranes appear to lead to hemolysis in the same way [39].

Normal red cells incubated in a dilute solution of protoporphyrin and then washed, are found to be photosensitive [20,25,39]. The photohemolysis of these "pseudo-EPP" red cells was found to be virtually identical to the photohemolysis of EPP red cells on the basis of several criteria. On this account pseudo-EPP red cells have been used by several workers as models for EPP red cells [20,25]. However, one study reported that various potential protective agents gave different results for the two kinds of cells [25]. For example, while β-carotene and ascorbic acid both afforded similar moderate protection for both kinds of sensitized cells, cysteamine afforded striking protection for EPP cells but actually increased the photohemolysis rate of the pseudo-EPP cells. The basis for the differences in inhibitor action may be provided

by the more recent observations that EPP cells and pseudo-EPP cells
differ drastically with respect to the binding site of the photosensitizer protoporphyrin and with respect to the distribution of
the protoporphyrin in the cell population [39,42,43]. In EPP cells
more than 99% of the protoporphyrin is bound to hemoglobin in the
cell interior [43]. The binding site on the hemoglobin is unknown
but it is not the heme binding site. In vivo there is a rapid
diffusion of this protoporphyrin out of the cell to the plasma which
is efficiently cleared in the liver [42]. The half life of protoporphyrin in the cells is on the order of one day so that only the
youngest cells have the majority of the protoporphyrin and the
oldest cells have close to normal concentrations [42]. When red
cells are incubated with aqueous solutions of protoporphyrin there
is rapid takeup of the porphyrin by the cell membranes [20,39].
This membrane associated protoporphyrin is slowly taken up by the
hemoglobin (1/e rate $\sim$ 1 day at 37 C) so that for a few hours after
the initial incorporation most of the protoporphyrin is located in
the membranes of these pseudo-EPP cells [39]. The distribution of
protoporphyrin among pseudo-EPP cells is, of course, homogeneous.
The simplest measure of the location of the protoporphyrin is by
spectrofluorometry. The maximum of the main peak in the fluorescence
spectrum of hemoglobin bound protoporphyrin is 624 nm [43], while
that of membrane bound protoporphyrin is 635 nm [39,44]. Through
a series of careful kinetic studies it has been shown that the
hemoglobin-bound protoporphyrin in EPP cells is the primary sensitizer for their photohemolysis, and that, for fresh pseudo-EPP
cells it is, of course, the membrane-bound protoporphyrin that
sensitizes the photohemolysis [39]. It has also been shown that
10^{-6}M hemoglobin-protoporphyrin contained in dipalmitoyllecithin
vesicles can sensitize the photooxidation of diphenylisobenzofuran
in the vesicle wall [45].

It goes without saying that proper interpretation and
modelling of phototoxic phenomena require knowledge about the
binding site(s) of the sensitizer. Unfortunately, little is known
about the location of photodynamic sensitizers in cells.

A familiar tactic in the elucidation of primary photochemical
processes is to examine the action of various potential inhibitors
(quenchers) which are designed to act at specific stages in the
sequence of intermediate steps comprising the overall photochemical
process. The interpretation of the effects of potential inhibitors
in a system as complicated as red cell photohemolysis must be made
with caution. The location of the inhibitor relative to the
sensitizer is, of course, crucial. Dynamical aspects, for example,
limited motion of molecules in membranes, may also preclude clear
interpretation. In the absence of good kinetic criteria specificity in reactivity may still allow conclusions based on inhibition studies. Such an outlook has been developed with respect

to the use of purportedly specific quenchers of singlet oxygen.
Unfortunately, to this author's knowledge, there is no compound in
use which solely on the basis of its ability to quench can safely
differentiate between the "free-radical" (type 1) and single
oxygen (type 2) mechanisms for photodynamic action in complicated
systems. β-Carotene, an excellent quencher of singlet oxygen, can
act as a free radical inhibitor in the red cell membrane [39].
α-Tocopherol, the well known naturally occurring radical inhibitor,
has recently been shown to be a good singlet oxygen quencher [24].
Dabco and diphenylisobenzofuran react with singlet oxygen and
radicals. The oxidation product of diphenylisobenzofuran, o-
dibenzoylbenzene, is commonly taken to be indicative of singlet
oxygen. However, it is likely that various oxidation modes would
give this product. It seems that cholesterol hydroperoxidation
does provide a "clean" method to detect singlet oxygen in a membrane
system. The 5α-hydroperoxide of cholesterol appears to be produced
only by way of singlet oxygen [16,17]. However, the technical
problems associated with the assay of the cholesterol hydroperoxides
are formidible.

Besides a better understanding of the actual chemistry which
occurs (i.e. product determination) in biomembranes, mechanistic
interpretations would benefit a good deal from an understanding of
the location and motion of sensitizers and quenchers within the
membranes.

REFERENCES

1. O. Raab, Z. Biol., 39, 524 (1900).
2. H. F. Blum, Photodynamic Action and Diseases Caused by Light,
    Reinhold Publishing Co., New York, 1941.
3. A. C. Allison and M. R. Young, Life Sciences, 3, 1407 (1964).
4. A. C. Allison, J. A. Magnus, and M. R. Young, Nature, 209,
    874 (1966).
5. T. F. Slater and P. A. Riley, Nature, 209, 151 (1966).
6. J. S. Cook, J. Cell. Comp. Physiol., 47, 55 (1956).
7. G. B. Odell, R. S. Brown, and A. E. Kopelman, J. Pediatrics,
    81, 473 (1972).
8. A. W. Girotti, Biochem., 14, 3777 (1975).
9. M. A. Pathak and J. H. Epstein: Normal and Abnormal Reactions
    of Man to Light, in T. B. Fitzpatrick, et. al. "Dermatology
    in General Medicine", McGraw-Hill Book Co., New York (1971).
10. K. Konrad, H. Hönegsmann, F. Gschnait, and K. Wolff, J. Invest.
    Dermatology, 65, 300 (1975).
11. F. H. Mukai and B. D. Goldstein, Science, 191, 868 (1976).
12. J. F. Mead, Free Radical Mechanisms of Lipid Damage and
    Consequences for Cellular Membranes in W. A. Pyrov, ed., "Free
    Radicals in Biology", Vol. 1, Academic Press, New York (1976).

13. N. Uri, in "Autooxidation and Antiooxidants", ed. by
    W. O. Lundberg, Vol. 1, Wiley (Interscience), New York (1961).
14. A. L. Tappel, in W. O. Lundberg (ed.), "Autooxidation and
    Antiooxidants", Vol. 1, Wiley (Interscience), New York (1961).
15. H. R. Rawls and W. van Santen, J. Am. Oil Chem. Soc., 47, 121
    (1970); Am. N.Y. Acad. Sci., 171, 135 (1970).
16. C. S. Foote, Photosensitized Oxidation and Singlet Oxygen, in
    W. A. Pyrov, ed., "Free Radicals in Biology", Vol. 2,
    Academic Press, New York (1976).
17. G. O. Schenck, K. Gollnick, and O. A. Neumüller, Justig
    Liebigs Ann. Chem., 603, 46 (1957); G. O. Schenck and
    O. A. Neumüller, ibid., 618, 194 (1958).
18. M. J. Kulig and L. L. Smith, J. Org. Chem., 38, 3639 (1973).
19. L. L. Smith and F. L. Hill, J. Chromatogr., 66, 101 (1972);
    L. L. Smith, J. I. Teng, M. J. Kulig and F. L. Hill, J. Org.
    Chem., 38, 1763 (1973).
20. B. D. Goldstein and L. C. Harber, J. Chin. Invest., 51, 892
    (1972).
21. A. A. Schothorst, et al., Chin. Chim. Acta., 39, 161 (1972).
22. F. H. Doleiden, S. R. Fahrenholtz, A. A. Lamola, and
    A. M. Trozzolo, Photochem. Photobiol., 20, 519 (1974).
23. I. B. C. Matheson, et al., Photochem. Photobiol., 21, 165
    (1975).
24. S. R. Fahrenholtz, F. H. Doleiden, A. M. Trozzolo and
    A. A. Lamola, Photochem. Photobiol., 20, 505 (1974);
    B. Stevens, R. D. Small, and S. R. Perez, ibid., 20, 511
    (1974); C. S. Foote, T.-Y. Ching, and G. G. Geller, ibid.,
    20, 515 (1974).
25. A. A. Schothorst, et al., Chin. Chim. Acta., 33, 207 (1971).
26. B. Epel, in A. C. Giese (ed.), "Photophysiology", Vol. 8,
    Academic Press, New York, 1973, p. 209.
27. L. R. Barran, et al., Biochem. Biophys. Res. Comm., 56, 522
    (1974).
28. G. D. Sprott, et al., Photochem. Photobiol., 24, 21 (1976).
29. J. S. Cook, Photopathology of the Erythrocyte Membrane, in
    B. F. Trump and A. Arstila (eds.), "Pathobiology of Cell
    Membranes", Vol. 1, Academic Press, New York, 1975, and
    references therein.
30. D. I. Roshchupkin, Photochem. Photobiol., 21, 63 (1975).
31. J. Salhaney, et al., private communication.
32. For example, G. Kahn and B. Fleischaker, J. Invest. Derm.,
    56, 85 (1971).
33. A. A. Schothorst, et al., Chin. Chim. Acta., 28, 41 (1970).
34. A. A. Schothorst, et al., Chin. Chim. Acta., 39, 161 (1972).
35. J. Hsu, B. D. Goldstein, and L. C. Harber, Photochem. Photo-
    biol., 13, 67 (1971).
36. A. F. de Goey, P. H. Ververgaert, and J. van Steveninck,
    Chin. Chim. Acta., 62, 287 (1975).
37. I. A. Magnus, et al., Lancit, 2, 448 (1961).

38. A. F. de Goey and J. van Steveninck, reported at this Congress,
     p. 111.
39. A. A. Lamola, T. Yamane and F. H. Doleiden, submitted to
     Photochem. Photobiol.
40. A. S. Fleischer, et al., J. Invest. Dermatol., 46, 505 (1966).
41. A. A. Lamola, T. Yamane, and A. M. Trozzolo, Science, 179,
     1131 (1973).
42. S. Piomelli, A. A. Lamola, M. B. Poh-Fitzpatrick, et al.,
     J. Clin. Invest., 56, 1519 (1975).
43. A. A. Lamola, S. Piomelli, M. B. Poh-Fitzpatrick, et al.,
     J. Clin. Invest., 56, 1528 (1975).
44. A. A. Lamola and M. B. Poh-Fitzpatrick, unpublished data.
45. A. A. Lamola and A. M. Mattucci, unpublished data.

SYMPOSIUM III

PHOTOMOVEMENT IN MICROORGANISMS

PHOTOMOVEMENT IN MICROORGANISMS - INTRODUCTION TO THE
SYMPOSIUM

Wolfgang Haupt

University of Erlangen-Nürnberg

Institute of Botany, D-8520 Erlangen, Germany

The locomotion of motile organisms can be influenced
or controlled by light in different ways:
1) The stationary light intensity can simply control the
linear velocity of movement: photokinesis.
2) A sudden change in light intensity can elicit a transient change in speed and/or direction of movement: photophobic response ("photophobism" or "photophobotaxis" in
older literature).
3) The direction of light can control the direction of
movement: phototaxis ("phototopotaxis" in older literature).

If we work with a sample of microorganisms in an
inhomogeneously illuminated environment, each of these
types of response can result in a pattern of distribution
of the organisms, and without further investigation it is
sometimes difficult or even impossible to decide which of
the three responses we are dealing with. In such cases,
it might be justified to use the purely descriptive term
photoaccumulation; but one always should realize that
this is a preliminary term of ignorance, which will be
replaced as soon as analysis of that special case has
proceeded sufficiently.

On the other hand, there seems to be no difficulty
to theoretically distinguish between photokinesis, photophobic response, and phototaxis. This becomes evident if
we consider the light signals effective in behavior. In
photokinesis, only the intensity of light at any moment
has to be measured by the organism. This is a common

problem in photobiology, and photobiologists are faced
with it in nearly all kinds of responses. More complicated is the signal in photophobic responses: The organism has to perceive a change of light intensity as a
function of time. This implies an important difference:
photokinesis is a steady-state response to a stationary
and permanent signal, but photophobic response is a
transient response to a single non-stationary signal,
hence a pre-steady-state response. Finally, in phototaxis the signal to be perceived is the light direction;
hence, this signal is vectorial.

As a consequence: When investigating and analyzing
photomovement responses, first of all we have to discover
the structures and pigments responsible for the perception of light and of its proper parameters, viz. intensity
or change of intensity or direction. Though much progress
has been made in recent time, there are still fundamental
question marks, as will become obvious in this symposium.
This is true especially since there are by no means a
master structure, a master pigment or a master equipment
in all organisms for this task, but several completely
different possibilities have been evolved in different
taxonomic groups.

The types of photomovement differ not only in the
kind of signal at the begin of the reaction chain, but
also in the kind of response. This will become obvious
especially if we try to introduce and to define the terms
"positive" and "negative". There is general agreement in
phototaxis: The response is called positive if the organism is approaching the light source, but negative if
movement is directed away from the light source. The
precision of those orientations can vary continuously
from random orientation to strict positive or negative
phototaxis. Hence, we may call phototaxis a quantitative
response.

Similarly, not too serious difficulties arise with
photokinesis. Most scientists agree to call the photokinesis positive if the velocity in light is increased
over the dark value, but negative if the organism is
slower in light than in dark. Typically, we are dealing
with the velocity beeing a steady function of the intensity, hence photokinesis also is a quantitative response.

This is, however, completely different with photophobic responses. It is true, there are many types and
mechanisms of response in different organisms. But a given organism, as a rule, is able to perform only one well

defined pattern of behavior as a photophobic response;
this pattern is fixed by the morphology and the physiology of the organism and cannot be changed fundamentally.
Especially, there is no possibility to distinguish between a "positive" and "negative" response; if, e.g.,
the flagellate Euglena turns toward its dorsal side after
a photophobic signal, we never can expect a turn towards
the ventral side after an opposite signal (light-on versus
light-off). Instead: if a signal is effective at all, the
only possible program is switched on. Hence, the photophobic response is to be considered as a qualitative response. It is obvious that, under these circumstances,
photophobic responses cannot be characterized as positive
or negative. But another distinction has to be made: The
effective photophobic signal can be a decrease or an increase in intensity, depending on the taxonomy of the
organism, on the environment, and on the overall light
intensity. Since we should not use "positive" versus
"negative" (as has been done in earlier time), we may
characterize the signal as "light-off" and "light-on",
or even better as "step-down" and "step-up". By this, we
characterize the signal rather than the response.

If we try to analyze a reaction chain of a photomovement response in an organism, we have to consider not
only the perception of the signal as an input and the final response as an output, but also the black box in between, which may be called "signal transduction" or "sensory transduction". This contains the question: how can
the signal - after being perceived - process the energy
flow in such a way as to result in the observed motor
response?

Perception of the signal, sensory transduction, and
the response proper have been investigated with different
success in different organisms. Obviously, best progress
has been made in Euglena among the eucaryotes and in the
blue-green alga Phormidium among the procaryotes. Nevertheless, even in these outstanding organisms, much information is still lacking, and much work is still to be
done. But at least some speculations are possible to arrive at an understanding of the reaction chains. These
speculations will be an important part of the lectures,
besides the well established experimental results.

Once a complete reaction chain in photomovement would
have been disentangled in one of these organisms, it would
be tempting to take this organism as a model for all
others. But incomplete as our knowledge is, one important
fact has become obvious in recent research: There is an

amazing diversity between taxonomic groups concerning
photomovement. This is true for the structures and pigments of signal perception in all three types of photomovement; this is true for the mechanisms at least of the
photophobic and phototactic responses; and consequently,
differences are suggested also within the black boxes.
To make the story as complicated as possible, even in a
given organism the reaction chains of photokinesis, photophobic response and phototaxis are partially or completely different from each other. It is therefore very
difficult or even impossible to extrapolate from one organism to the other. This will become evident, e.g., by
referring to the flagellate Chlamydomonas the photomovement of which cannot be explained on the basis of Euglena
at all.

Keeping this diversity in mind, it is fascinating to
find one general result: Whatsoever the type and mechanism
of response, photomovement in photosynthetic organisms
finally results in a pattern of spatial distribution which
guarantees optimal use of light as an energy source. This
means: To meet the same requirement, different organisms
have developed completely different strategies which will
be reported at the begin of our symposium. This symposium,
then, will be centered on the one hand to showing the
frontier of research in analyzing favorable reaction
chains in photomovement. On the other hand, the speakers
will point to the great diversity among organisms and
among responses. As a result, we should be able to put
forward some key questions left open for future research,
and to discuss promising experimental approaches.

PHOTOMOVEMENT IN MICROORGANISMS: STRATEGIES OF RESPONSE

Mary Ella Feinleib

Biology Dept., Tufts University

Medford, Mass. 02155 U.S.A.

Photomovement strategy in microorganisms displays a marvelous
diversity of evolutionary solutions to the same problem: how to
move into an optimally-illuminated region of the environment.
The aim of the present lecture is to describe some of these response
strategies and to point out key questions left open for future
investigation. The response types fall into three major categories,
defined by Professor Haupt in the preceding lecture: photokinesis,
phobic response (or photophobism) and phototaxis.

PHOTOKINESIS

It is doubtful that any microorganism uses photokinesis
per se as a major strategy for moving into the light. In order for
photoaccumulation to occur solely via this type of response,
organisms should move more slowly in the light than in the dark;
i.e., they should demonstrate "negative photokinesis" (1). In
general, however, the opposite is true - except at very high
intensity, organisms move faster in the light than in the dark,
and some are motile only in the light. In the extreme case of an
organism that requires light for movement, positive photokinesis
is simply a prerequisite for carrying out photophobic or phototactic maneuvers.

PHOTOPHOBIC RESPONSE

In photosynthetic bacteria, photophobic response is the major
photomovement strategy. The effectiveness of this strategy can
clearly be seen in the corkscrew-shaped sulfur bacterium, Thiospi-

<u>rillum</u>. At one pole, the cell has a bundle of flagella which
rotates as a unit. When the flagellar bundle extends out from
the pole, it functions as a "push propeller," causing the cell to
move forward (and to rotate about its long axis.) When <u>Thiospi-</u>
<u>rillum</u> encounters a sudden decrease in light intensity, the
flagellar bundle snaps over like an umbrella and reverses its
direction of rotation. This causes the cell to reverse its
direction of swimming, with the flagellar bundle now acting like
a "pull-propeller" (2,3,4). If each cell in a population reverses
swimming direction when it crosses the border from light to dark,
the result is an accumulation of cells in the illuminated region.

　　　　Photophobism encompasses a whole class of strategies,
with the pattern of movement differing from one genus to the
next. In each case, the direction of response is strictly
programmed with respect to the morphology of the organism and is
not related to the direction of the stimulus light. Four examples
of photophobic strategies are schematized in Figure 1. In each
case, the stimulus is a single sudden change in light intensity
(step-up or step-down). Reversal of movement in parallel with
the original path is characteristic of phobic responses in photosynthetic bacteria and blue-green algae. <u>Rhodospirillum</u>, a

TYPES of PHOTOPHOBIC RESPONSE

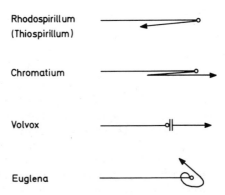

<u>Figure 1</u>. Schematic representation of four types of photophobic
response. In each case, direction of translational movement is
indicated by an arrow and position of cell (or colony) at start
of response is indicated by a circle. Double vertical lines
for <u>Volvox</u> indicate transient cessation of movement. (After
Haupt, unpublished.)

bipolarly flagellated bacterium, reverses its direction of
movement and then continues to swim in the new direction.
(Thiospirillum responds the same way under favorable conditions.)
In the bacterium Chromatium, which is unipolarly flagellated,
photophobic reversal of movement is followed within seconds by
resumption of movement in the original direction (4). Green
flagellated algae respond quite differently. In Volvox there
is no direction change. Photophobic response consists simply of
a transient cessation of movement, after which movement is resumed
in the original direction (5). Euglena, by contrast, executes a
turn toward its dorsal (stigma-containing) side, effected by
a sharp outward beat of the flagellum in the "ventral" direction.
This maneuver results in a randomly-directed change in the cell's
forward "heading".

Despite their diversity, all photophobic responses demand
that the organism acquire the same kind of information about
the stimulus; namely, it must detect a change of light intensity
as a function of time (dI/dt). Fundamental questions facing the
investigator are:
1. What kind of signal is generated by the temporal change
in light intensity?
2. How does this signal trigger phobic response?
3. What ultrastructural change brings about the final change
in movement?
In procaryotes, these questions have been partly answered. A
sudden change in light intensity causes a change in photosynthetic
electron flow (6), which, in turn, triggers a change in locomotion.
Advances have recently been made in our understanding of flagellar
movement in bacteria, at least in chemotactic forms (7). Unlike
a eucaryotic flagellum, a bacterial flagellum (or flagellar bundle)
rotates as a rigid helix without wave propagation along its length.
The flagellar drive mechanism may be viewed as a rotary motor
that changes its direction of rotation in response to information
transmitted from the receptor. The intervening steps between
electron-flow signal and flagellar-motion response are yet to
be elucidated.

For eucaryotes, all we have to date are some interesting
models. Piccinni and Omodeo (8) propose that flagellates detect
temporal change in intensity by means of an intensity "comparator."
The signals generated by light are presumed to resemble receptor
potentials in animals. The first signal, before the intensity
change, is transmitted via a synapse-like delay-device, while the
new signal travels by a direct route so that both reach the
"comparator" at the same time. This scheme poses the challenge
of identifying a cell structure that might perform the functions
attributed to the "comparator." Theories concerning the ultrastructural changes underlying flagellar reorientation in Euglena
(8,9) also remain to be tested.

PHOTOTAXIS

The characteristic behavior of motile algae is phototaxis. Phototactic strategies are as varied as the algae that perform them. Three types of positive phototaxis are schematized in Figure 2. The diatom <u>Navicula</u> displays what is probably a rudimentary form of phototaxis, found also in blue-green algae of the family <u>Oscillatoriaceae</u> (1). In diffuse light, the organism normally oscillates back and forth, parallel to its long axis, without showing a preferred direction. When a unilateral

TYPES of PHOTOTACTIC RESPONSE

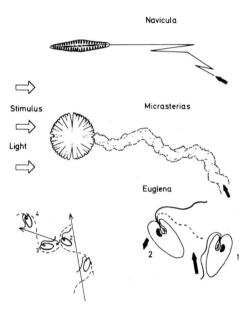

Figure 2. Schematic representation of three types of phototactic orientation. In each case, arrow at right shows direction of movement at onset of stimulation. <u>Navicula</u>: Typical phototactic movement path; cell shown already oriented. <u>Micrasterias</u>: Cell shown already oriented, with slime trail behind it. <u>Euglena</u>: <u>Right</u>. Two positions of cell during swimming along helical path. Thickening at flagellar base: paraflagellar body. Adjacent dark patch: stigma. Position 2. Dashed-line flagellum: orientation before (phobic) response. Solid-line flagellum: reorientation during response. Arrow at position 2: Direction of turn during response. <u>Left</u>: 1,2: Axis of swimming path before response, with two cell positions. 3,4: Axis of path after response, with two further positions of same cell. See text for explanation.

stimulus light is introduced (at any angle other than 90° with
respect to the cell's long axis), movements toward the light
source are prolonged, while movements away from it are abbreviated.
The critical feature of this strategy is that Navicula moves
toward a light source without steering. Any orientation change
occurs by chance alone. For this strategy, the organism has only
to determine in which direction along the stimulus-beam axis
the light is travelling. Recent work by Wenderoth with
Pleurosigma (10) indicates that the reaction time depends upon
the difference in light intensity impinging on the two poles of
the cell.

More "sophisticated" phototactic strategies are found in
desmids such as Micrasterias (11) in flagellates such as Euglena,
and in some blue-green algae of the family Noctocaceae (1).
Under unilateral illumination, these organisms actively reorient
or "steer" until (in positive phototaxis) they are moving toward
the light source. Two strikingly different examples of phototactic steering are seen in Micrasterias and Euglena.

Micrasterias steers slowly and smoothly into the light beam
until its longitudinal axis is parallel with the beam. This
desmid typically creeps over the substratum, oscillating slightly
from side to side. As it moves, the cell leaves behind it a
mucilage trace. Studies of stained traces laid down during
phototactic response (11) reveal that the oscillations in the
trace decrease in amplitude as stimulus intensity is increased.
Cells that do not oscillate also become oriented, indicating
that oscillation is not required for phototaxis. Surprisingly,
Micrasterias can even stand on edge and turn its "face" toward
the light source while remaining otherwise motionless.

Euglena steers toward a light source via a series of "trial-
and-error" adjustments (12,13). Normally, the organism swims
along a helical path while also rotating about its longitudinal
axis. If a stimulus beam is introduced perpendicular to the cell's
long axis, the stigma intermittently comes between the light
source and the photoreceptor (paraflagellar body). When the
photoreceptor is shaded, it receives a "step-down" stimulus, and
the cell responds by deviating toward its "dorsal" side, just
as in the photophobic response described earlier. As a result,
the swimming-path axis swings over slightly in the direction of
the stimulus beam (Figure 2, lower left). Euglena continues to
perform these small phobic responses until it is heading toward
the light source. In that position, the cell no longer receives
periodic-shading stimuli.

Despite the differences in their steering strategies,
Micrasterias and Euglena require the same information about the
stimulus light; they both must detect the direction of the light

source. There are two fundamental mechanisms by which light
direction can be perceived. An organism may compare: 1) light
absorbed at two photoreceptive sites at <u>one instant</u> or 2) light
absorbed at one photoreceptive site at <u>two instants</u>, ordinarily
by taking two intensity "readings" at two positions with respect
to the light source.

Since <u>Micrasterias</u> can turn toward the light while remaining
otherwise motionless, it seems unlikely that the cell takes two
intensity readings at two positions. The slow and steady steering
of this organism seems explicable only in terms of a one-instant
mechanism. We have no idea, however, where the two photoreceptive
sites are, or how the absorption difference between them is detected.
Periodic shading in <u>Euglena</u> is a well-established case of a two-
instant mechanism (14). The cell compares light absorbed by the
photoreceptor at two successive instants: without and with shading
by the stigma. The information needed for directional perception
in <u>Euglena</u> thus reduces to that needed for photophobic response:
detection of a temporal change in light intensity.

A two-instant strategy seems to be a practical evolutionary
choice for a small, fast-moving organism such as a flagellate,
and is known to occur in <u>Volvox</u> (5,15) and in <u>Gyrodinium</u> (16).
Even in these cases (including <u>Euglena</u>), we have yet to discover
how the error signal is fed back into the steering apparatus
(see model by Diehn, 17) and how the final steering maneuver is
effected on the ultrastructural level. In other flagellated
algae, we do not know what phototactic strategies are used. The
familiar biflagellate <u>Chlamydomonas</u> can serve as a case study of
questions that remain open, of approaches toward answering them
and of the difficulties involved.

1. <u>Does Chlamydomonas detect light direction by a two-instant
strategy</u>? Our main approach to this question (18,19) has been
to determine whether the cell can become oriented in response to
a single light flash - too short to allow the cell to compare
flash intensity at two different positions. If all the cells
became oriented following such a short flash, one could conclude
that a one-instant mechanism operates.

Studies on single cells were conducted using the video
microscope system shown in Figure 3 (18). The cells were exposed
to a continuous blue (442 nm) laser light of high intensity. This
stimulus elicted strong negative phototaxis in all the cells.
After 10s, a single 6 μs flash from a stroboscopic lamp (used with
a blue filter) was superimposed on the laser beam, at 90° to that
beam. The laser stimulus continued for 10s after the flash. Cell
movements were recorded on video tape and were analyzed by
individual frames. About half the cells responded to a single
flash as shown in Figure 4. Cell A shows a transient deflection

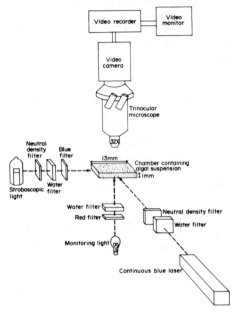

Figure 3.

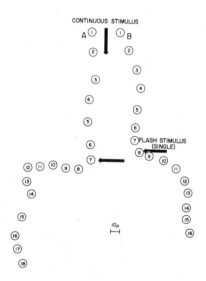

Figure 4A (left). Tracing of negative-turn response elicited by a single flash. Cell positions are shown at 0.1s intervals and are numbered consecutively in direction of movement. 4B (right). Same as 4A, but positive turn response.

away from the flash source ("negative turn response"); cell B shows
a deflection toward the source ("positive turn reponse"). We have
evidence that a turn response represents the initial stage of
phototactic orientation. Most importantly, there was a statistically significant predominance of negative turns, as might be
expected in phototactic orientation to a high-intensity stimulus.

These results confirm those of our studies on cell populations
(19): orientation can occur in response to a short flash. We
cannot conclude, however, that Chlamydomonas necessarily perceives
light direction by a one-instant mechanism. The fact that only
half the cells showed a turn response suggests that the only ones
to respond were those in which the photoreceptor happened to be
in a "correct" position with respect to the flash source at
the instant of the flash. This interpretation is consistent
with a two-instant model in which the cell compares light absorbed
before and during the flash, i.e. the turn response may be a
phobic reaction to a step-up stimulus. On this model, the occurrence
and the direction of a turn response would depend on the orientation
of the cell about its longitudinal axis during the flash. Since
the direction of a phobic response is fixed with respect to the
cell's morphology, the orientation of a negative-turn cell during
the flash should differ by 180° from that of a positive-turn
cell. In the case of a continuous stimulus, the cell's orientation
at the onset of stimulation should be correlated with the lag
time of the phototactic response.

In order to test these hypotheses directly, one should know
the precise orientation of the cell about its longitudinal axis.
We have found a promising technique for conducting these studies.
Between crossed polarizers, the stigma appears as a bright spot
against the darker background of the cell. This provides a
convenient marker for following the cell's rotation, and one
which may well play a role in phototaxis. It is difficult,
however, to record on video tape stigmas of moving cells under
conditions appropriate for studying phototaxis. An alternative
method has been suggested by rereading the older literature.
Ludwig (20) argued that a cell which rotates about its longitudinal
axis while swimming must also follow a helical path of given
handedness, with the same side of the cell constantly directed
toward the axis of the helix. Our video tapes show that
C.reinhardtii indeed rotates (despite Ringo's contentions to the
contrary, 21) and also follows a helical path, which looks
sinusoidal when projected in two dimensions. If the predicted
constancy obtains, we should be able to establish which side of
the cell faces the axis of the helix. The cell's exact orientation
during a flash can then be deduced by noting the phase of the
sinusoidal path at that instant.

2. Is the stigma essential for phototactic steering in

Chlamydomonas? Over twenty years ago, it was reported that stigmaless mutants of Chlamydomonas are phototactic, but that their response is less pronounced than that of the wild-type algae (22). In those studies, movement of a cell population was observed with the naked eye. Since that time it has generally been assumed that the stigma aids in phototactic navigation, but is not essential for it. We are making video recordings of photomovement in several so-called "eyespotless" strains of Chlamydomonas (provided by Dr. R. Smyth, Boulder, Colo., 23). A morphological difference between mutant and wild type can readily be seen with a light microscope. In some mutant strains (eg. 624 ey (+)) only ca. 5% of the cells have an obvious stigma and the stigma looks smaller and less-heavily pigmented than in the wild type. In other mutants (eg. 627 ey (-)) a stigma is visible in as many as 50% of the cells and looks similar to that of the wild type. The mutants are now being examined with an electron microscope (Leipold, Erlangen, Germany). An important companion study would be a determination of pigment content, particularly of the carotenoids.

Our stigma mutants show both positive and negative phototaxis, but their behavior is strikingly different from that of the wild type. Figure 5 A, B and C are tracings of swimming tracks typical, respectively, of wild type, mutant 627 ey (-) (50% stigmas) and mutant 624 ey (+) (ca. 5% stigmas), during phototactic response to continuous blue light of high intensity. Each tracing was begun 30 seconds after introduction of the stimulus and shows cell position at 1/5s intervals for a total of three seconds. All three algal strains show a preferred direction away from the light source, but there is marked difference in the directness of their swimming paths. (Swimming rate does not appear to differ very much among the strains.) To obtain an index for directness of swimming path, we divide the length of the straight line between the cell's initial position and its final position by the actual path length covered in three seconds. Pathlength is measured with a map measurer. (A perfectly straight path would have an index of one.) According to our preliminary results, the directness-of-path index is 0.94 in the wild type, 0.74 in mutant 627 (50% stigmas) and 0.54 in mutant 624 (5% stigmas).[1] It would be important to know whether mutant cells that show a relatively straight path (upper two tracks, Fig. 5B) in fact possess a normal stigma. (Interpretation of the phototactic data is complicated by the fact that some of the mutants (eg. 624 ey) grow more slowly than the wild type, and may lack cell components

---

1. The wide variation in directness of path is also interesting in connection with our earlier study of the relationship between stimulus intensity and net response of a cell population (24). We then found that directness of path was the most important factor in determining the magnitude of the net response.

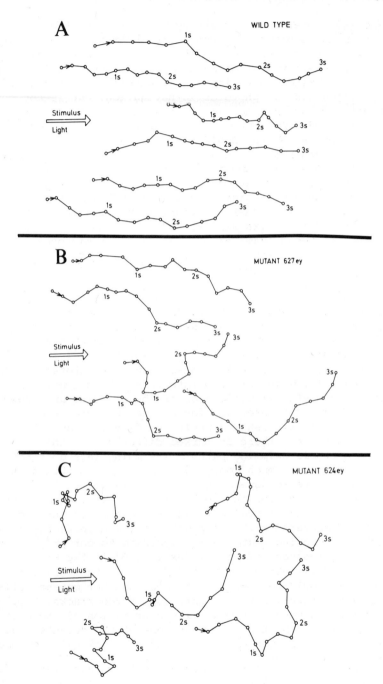

Figure 5. Swimming-track tracings from videomicroscopy of three Chlamydomonas strains during high-intensity unilateral illumination, Cell positions: every 1/5s, from 30s-33s after light on.

other than the stigma, that are also important for phototaxis.)

Despite the differences in phototactic behavior, 100 percent of the cells in all strains examined show a distinct stop response upon introduction of a high-intensity stimulus light. (24)   These results suggest that stop response and steering are, at least in part, separable phenomena. The same conclusion was drawn from earlier observations (Feinleib, unpublished) - when the direction of a continuous stimulus light is changed by 90°, wild-type Chlamydomonas can reorient without performing a stop response.

Our mutant studies to date are consistent with the hypothesis that the stigma acts as a fine adjustment for phototactic steering in Chlamydomonas, perhaps as an accessory shading device. This raises the question - 3. If the stigma of Chlamydomonas functions as a shading device, what does it shade? i.e. where is the photo-receptor for phototaxis located? In this organism, there is no paraflagellar body, and the stigma is relatively far from the flagellar bases (25). Peripheral to the stigma, however, is a thickened region of the plasmalemma. It has been postulated (26,8) that this area is the site of the photoreceptor and is periodically shaded by the stigma. To test this hypothesis, it would be convenient to have a mutant that lacks the plasmalemma thickening. Isolation of such a mutant should be an interesting challenge for Chlamydomonas geneticists.

4. What controls the "switch" between positive and negative phototaxis? Chlamydomonas orients equally well toward or away from a stimulus light (24), suggesting that positive and negative phototaxis share a common steering mechanism. The nature of the positive/negative switching device is unknown. The sign of phototactic orientation is influenced by many factors, including light intensity (27, 28, 24), illumination prior to testing (29, 30, 31, 32) and ionic composition of the medium (33, 34). The direction of net response in a cell population may also change during the course of a single several-minute exposure to a stimulus light (31, 35, 36).

Recent work (37, 38, 39) demonstrates that yet another factor can "throw the switch" in short-term reversals of phototactic direction; namely, contact with a physical barrier. Under laboratory conditions, the barriers are usually the walls of the experimental chamber. According to Riedl (39), in cultures with mixed positive and negative phototaxis, almost 100% of the cells swimming toward the light source reverse direction within two seconds of reaching the end-wall, whereas only about 50% of the cells swimming away from the light show a similar reversal at the far wall; the other 50% remain at the wall. In view of these observations, investigators studying phototaxis in population systems should be particularly careful in interpreting phototactic reversals recorded during the course of an experiment.

5. <u>How are the final steering maneuvers effected</u>? In <u>Chlamydomonas</u>, as in many other phototactic organisms, even the phenomenology of steering remains largely unknown. A useful next step would be high-speed cinematography of flagellar movements during phototactic orientation and during positive/ negative reversal. Once we have a detailed description of the steering maneuvers, we can ask more specific questions about the underlying mechanisms.

REFERENCES

1. Nultsch, W. 1975. Phototaxis and photokinesis. In: "Primitive Sensory and Communications Systems: the Taxes and Tropisms of Micro-organisms and Cells" (M.J. Carlisle, ed.) London-New York, Academic Press: 29-90.

2. Buder, J. 1915. Zur Kenntnis des <u>Thiospirillum jenense</u> und seiner Reaktionen auf Lichtreize. <u>Jb. wiss. Bot.</u> 56: 529-584.

3. Pfenning, N. 1968. <u>Thiospirillum jenense</u>. Lokomotion und phototaktisches Verhalten. Inst. für den wissentschaftlichen Film. Göttingen.

4. Nultsch, W. 1970. Photomotion in microorganisms and its interaction with photosynthesis. In: "Photobiology of Micro-organisms." (P. Halldal, ed.) London, Wiley: 213-252.

5. Huth, K. 1970. Bewegung und Orientierung bei <u>Volvox aureus</u>. I. Mechanismus der phototaktischen Reaktion. <u>Z. Pflanzen-physiol.</u> 62: 436-450.

6. Häder, D-P. 1976. These proceedings.

7. Berg, H.C. and R.A. Anderson. 1973. Bacteria swim by rotating their flagellar filaments. <u>Nature.</u> 245: 380-382.

8. Piccinni, E. and P. Omodeo. 1975. Photoreceptors and phototactic programs in protista. <u>Boll Zool</u>. 42: 57-79.

9. Jahn, T. L. and E.C. Bovee. 1972. Locomotive and motile responses in <u>Euglena</u>. In: "The biology of <u>Euglena</u>" (E.E. Buetow, ed.). New York-London, Acad. Press: 45-108.

10. Nultsch, W. (personal communication.)

11. Neuscheler, W. 1967. Bewegung und Orientierung bei <u>Micrasterias denticulata</u>. Breb. im Licht. I. Zur Bewegungs - und Orientierungsweise. <u>Z. Pflanzenphysiol</u>. 57 (1): 46-59.

12. Jennings, H.S. 1906. "Behavior of the Lower Organisms." New York, Columbia University Press.

13. Mast, S.O. 1911. "Light and the Behavior of Organisms." New York, Wiley.

14. Haupt, W. 1959. Die Phototaxis der Algen. In: "Handbuch der Pflanzenphysiologie." XVII (1). (W. Ruhland, ed.) Berlin-Heidelberg-New York. Springer Verlag.

15. Schletz, K. 1976. Phototaxis bei <u>Volvox</u> - Pigmentsysteme

der Lichtrichtungsperzeption. Z. Pflanzenphysiol. 77: 189-211.

16. Hand, W.G. and J. Schmidt. 1975. Phototactic orientation by the marine dinoflagellate Gyrodinium dorsum Kofoid. II. Flagellar activity and phototactic response. J. Protozool. 22: 494-498.

17 Diehn, B. 1973. Phototaxis and sensory transduction in Euglena. Science. 181: 1009-1015.

18. Boscov, J.S. 1974. Responses of Chlamydomonas to single flashes of light. M.S. dissertation. Tufts University. Medford, Mass., U.S.A.

19. Feinleib, M.E. 1975. Phototactic response of Chlamydomonas to flashes of light. I. Response of cell populations. Photochem. Photobiol. 21: 351-354.

20. Ludwig, W. 1930. Untersuchungen über die Schraubenbahnen niederer Organismen. Z. Vergleich. Physiol. 9: 734-801.

21. Ringo, D.L. 1967. Flagellar motion and fine structure of the flagellar apparatus in Chlamydomonas. J. Cell. Bio. 33: 543-571.

22. Hartshorne, J.N. 1953. The function of the eyespot in Chlamydomonas. New Phytol. 52. 292-297.

23. Smith, R.D., G.W. Martinek and W.T. Ebersold. 1975. Linkage of six genes in Chlamydomonas reinhardtii and the construction of linkage test strains. J. Bacteriol. 124: 1615-1617.

24. Feinleib, M.E. and G.M. Curry. 1971a. The relationship between stimulus intensity and oriented phototactic response (topotaxis) in Chlamydomonas. Physiol. Plant. 25: 346-352.

25. _____. 1971b. The nature of the photoreceptor in phototaxis. In: "Handbook of Sensory Physiology. I. Principles of Receptor Physiology" (W. Loewenstein, ed.) Berlin, Springer Verlag.

26. Walne, P.L. and H.J. Arnott. 1967. The comparative ultrastructure and possible function of eyespots: Euglena granulata and Chlamydomonas eugametos. Planta. 77. 325-353.

27. Famintzin, A. 1967. Die Wirkung des Lichtes auf Algen und einige ihnen nahe verwandte Organismen. Jb. wiss. Bot. 6: 1-48.

28. Buder, J. 1917. Zur Kenntnis der phototaktischen Richtungsbewegungen. Jb. wiss. Bot. 58: 105-220.

29. Strasburger, E. 1878. Wirkung des Lichtes und der Wärme auf Schwärmsporen. Jen. Zeitschr. f. Naturwiss. 12.

30. Halldal, P. 1960. Action spectra of induced phototactic response changes in Platymonas. Physiol. Plant. 13: 726-735.

31. Feinleib, M.E. 1965. Studies on phototaxis in Chlamydomonas reinhardtii. Ph.D. dissertation. Harvard University, Cambridge, Mass., U.S.A.

32. Nultsch, W. 1975. Effect of external factors on phototaxis of Chlamydomonas reinhardtii. I. Light. Arch. Mikrobiol. 179: 215-216.

33. Halldal, P. 1957. Importance of calcium and magnesium ions in phototaxis of motile green algae. Nature. 179: 215-216.

34. _____. 1959. Factors affecting light response in phototactic algae. Physiol. Plant. 12: 742-752.

35.   Mayer, A.M.   1968.   Chlamydomonas: adaptation phenomena
in phototaxis.   Nature.   217:   875-876.
36.   Nultsch, W., G. Throm and I.v. Rimscha.   1971.
Phototaktische Untersuchungen an Chlamydomonas reinhardtii
Dangeard in homokontinuierlicher Kultur.   Arch. Mikrobiol.   80.
351-360.
37.   Stavis, R.   1974a.   Phototaxis in Chlamydomonas: a
sensory receptor system.   Ph.D. dissertation.   Yeshiva University,
New York, N.Y., U.S.A.
38.   _____.   1974b.   The effect of azide on phototaxis in
Chlamydomonas reinhardtii.   Proc. Nat. Acad. Sci.   U.S.A.   71 (5):
1824-1827.
39.   Riedl, G.   1976.   Zulassungsarbeit, Botan. Inst. der
Universitat Erlangen-Nürnberg.   In preparation.

PIGMENTS INVOLVED IN THE PHOTOMOTION

OF MICROORGANISMS

R. BENSASSON

ER 98, Labóratoire de Chimie Physique
Université de Paris, 91405 ORSAY

Some microorganisms contain specific pigments which capture the energy of light to trigger and modify their locomotion. In these organisms, which are mostly photosynthetic, three different behavior patterns of photomotion can occur: photokinesis, for which the linear velocity of motion is a function of the light intensity, photophobic (or photophobotactic) response, which is a transient change in motion resulting from a temporal change in light intensity, phototaxis (or phototopotaxis) in which the orientation of motion is controlled by the direction of light.

Although our communication was supposed to deal with pigments involved in phototaxis, we shall also include data dealing with pigments involved in photophobic response, photokinesis and even phototropism.

We shall (1) report the different pigments involved in photomotion and compare the properties of carotenoids and flavins for the blue light effects requiring a yellow pigment (2) try to find some correlation between the class of microorganism, the type of photoresponse and the pigment involved, (3) emphasize the implication of the pigment in the metabolism of the microorganism, (4) stress the way by which photomodification of certain pigments might change the properties of the membranes where they are inserted.

The different pigments involved

Table I shows for the main prokaryotes and eukaryotes, the pigments involved for each type of photomotion. These pigments will now be considered successively:

1) The yellow pigments

a)Action spectra.
In phototaxis most of the action spectra, are in the
visible range with maximum activity between 450 and 500 nm. This
has favoured the view that carotenoids or carotenoproteins are the
photoreceptors, (*Porphyridium cruentum*, *Chlamydomonas reinhardtii*
and most flagellates; Nultsch *et al* 1971, Bûnning and
Schneiderhöhn 1956, Halldal 1963, Forward 1973). In some species
however, as *Euglena gracilis*, *Nitzschia communis*, *Phormidium
autumnale...*, a second active band is found between 350 and 400nm.
These action spectra coincide with the absorption maximum of
flavins or flavoproteins. In a third case, the maximum activity
in the 450-500 nm region is accompanied by a second region around
280-290 nm as in the case of *Gymnodium splendens*, *Platymonas
subcodiformis*. These action spectra are similar to that of
Halobacterium halobium and would favor a carotenoprotein.

Yellow pigments with maximum activity around 450-500 nm
are also found as photoreceptors for phobic response (*Nitzschia
communis*) or for photokinesis (*Eudorina*, *Pandorina and Chlamydomonas*).

b) Identity of the yellow pigments.
As already reported above, candidates are carotenoids,
carotenoproteins, flavins or flavoproteins:

α) Flavins-flavoproteins. In the case of *Euglena gracilis*
the work of Benedetti and Checcucci (1975) and Checcucci *et al.*
(1976) has shown that the paraflagellar body exhibits flavin like
fluorescence and that the action spectra for phototaxis coincides
with a flavin spectrum. This result could be generalised to other
action spectra of photomotion with a near UV band accompanying a
visible band at 450-500 nm. Similarly, phototropic and light
growth response of Phycomices might be due to riboflavin (Delbrück
et al, 1976). The possible involvement of flavins in several
photobiological processes has been the reason for many
investigations of their photochemical properties (for a review
see Song, Moore and Sun, 1972). Flavins have a very rich
photochemistry, in contrast to carotenoids, due to their $n\pi^*$
states which are responsible for their important singlet to
triplet transition probability. Another interesting property of
flavins and flavoproteins might be their ability to undergo a
reduction to a half reduced state (1 e⁻ equivalent) and a fully
reduced state (2 e⁻ equivalent). Thus a flavoprotein might
participate in the electron transport by shuttling back and forth
between its different redox states. These properties could be
satisfactory for the photoreceptor of photomotion.

β) Carotenoids-carotenoproteins. Only cis carotenoids which
have an additional subsidiary band at 340-360 nm, accompanying the
main visible absorption band, positioned between 450 and 500 nm,
might be candidates for action spectra showing absorptions at
370 and 450 nm. But cis carotenoids are absent in microorganisms.

Therefore the work of Hager (1970), which has shown the appearance
of a new band at 350-360 nm for some trans carotenoids in ethanol
water mixtures, is a crucial argument in favour of carotenoids for
photoactic and phototropic action spectra showing peaks around 370
and 450 nm. But Song and Moore (1974) have since shown that the
350-360 nm band found by Hager for lycopene in ethanol-water
mixture, does not appear for β trans carotene and that it is due to
two stacked lycopenes. This means that if a dimeric model of carotenoids could be the photoreceptor, the action spectra would only
show the 360 nm band. All these findings cast doubt on carotenoids
as photoreceptors, in the case of an action spectrum with the two
bands 370 and 450 nm. But are carotenoids still possible photoreceptors for other cases? Certainly when they act as accessory
pigments as in the case of *Rhodospirillum rubrum* where they accompany
chlorophyll. But in absence of the red absorption due to chlorophyll
carotenoids should not be considered as possible photoreceptors
involving singlet excited states because of their short life time
(Song and Moore, 1974) or involving triplets because of their
extremely low triplet energy levels (below 94 kJmole^{-1} level of O_2
($^1\Delta$g) Bensasson, Land and Maudinas, 1976) and because of their
extremely low singlet to triplet quantum efficiency (below 0.002,
Bensasson, Dawe, Land, Long unpublished). Moreover their quantum
yield of photoisomerisation is negligible and only takes place in
presence of a photosensitizer. Thus, the energy absorbed by carotenoids is rapidly degraded into thermal energies when they do not
act as accessory pigment. However, carotenoids have the ability to
isomerize with a high yield as in the chromophore moiety of the
carotenoprotein bacteriorhodopsin. That is why a carotenoprotein
might be a possible candidate for the cases where the action
spectrum has an absorption at 400-500 nm and also around 280 nm
where aromatic aminoacids should absorb (*Platymonas subcordiformis,*
Haldall, 1961). Nevertheless, even in that case, flavoproteins might
still be candidates, the absence of the 370 nm band being due to
the screening of an inactive pigment (see ref. in Haldall, 1967).
It should be mentioned that few proteins-polyene complexes have been
found (Nishimura and Takamatsu 1957, Thompson 1975).

A property of carotenoids which deserves some attention is that
they might be expected to undergo charge transfer reactions and
could be involved in electron transfer reactions due to their
extended conjugation (Platt, 1959).

In short, carotenoids show few properties that would make them
satisfactory candidates for the yellow pigment inducing photomotion.

2) Chlorophylls
The phobic response of all prokaryotes (*Purple bacteria
blue green algae, Halobacterium halobium...*) and of some
eukaryotes (like *Navicula peregrina and Porphyridium cruentum*) is
stimulated by the photosynthetic pigments. The action spectrum
shows chlorophyll a sometimes accompanied by biliproteins as in
the case of *Phormidium*, among *blue-green algae* or *Porphyridium
cruentum*, among eukaryotic microorganisms.

The action spectra of photokinesis of *Purple bacteria, blue green algae, flagellates, diatoms and red algae* indicate for a correlation with photosystem I (*Phormidium* type) or with both photosystems (*Anabaena* type). In the case of *Porphyridium cruentum* a phycobilin B phycoerythrin accompanies chloropyll a.

Although as reported earlier, most photoactic action spectra have pointed to yellow pigments, in the case of *Anabaena variabilis* (Nultsch, unpublished) and in the case of the *Desmid Micrasterias* (Neuscheler, 1967) the action spectrum points to chlorophyll a.

3) Other pigments.
Two other kinds of pigments are associated with photomotion.

a. Biliproteins are suggested in some microorganisms as photoreceptors for phototaxis. The involvement of chlorophyll being excluded, as red light is quite ineffective. Those microorganisms are *Prorocentrum micans* whose action spectrum (Haldall, 1958) shows a maximum at 570 nm (C Phycoerythrin), *Phormidium uncinatum and autumnale* whose action spectrum favours (Nultsch, 1974) the biliproteins-C-Phycoerythrin and C-phycocyanin (maximum around 560 nm and shoulder at 615 nm) accompanying (370 and 500 nm peaks) and *Cryptomonas* (Watanabe and Furnya, 1974) whose action specturm implicates Phycoerythrin and perhaps some other blue receptor.

Phycobiliproteins are very active in photosynthesis and the energy transfer of those compounds to chlorophyll has been experimentally proven (Duysens, 1951). That is why their role as accessory pigments when they accompany chlorophyll for phototaxis of *Oscillatoria mougeottii* or *Anabaena variabilis* can be well understood. However their role in absence of a red light effect will not be understood until more studies on their photochemistry have been carried out.

b. In *Dictyostellium discoideum*, the photoreceptor for photo-taxis appears to be a ferroheme protein oxidized to the ferri form by the action of light (Poff and Butler, 1974).

Correlation between the class of microorganism, the type of photomotion and photopigment involved.

The classification of microorganisms reported in Table I allows the following generalisations: for prokariotic cells, which are the less complex microorganisms the pigments involved in photomotion are the pigments of the photosynthetic process by which sunlight is converted into the chemical energy required for metabolism. For eukariotic cells, the pigments involved in photomotion are usually specific. Also of interest in the case of eukariotic cells is the fact that those pigments are organised in a specialised structure. In this very general scheme there are some exceptions: for instance, *diatoms* (less evoluted

than other eukaryotes as they have no distinct photoreceptor
structure) can have as chromophores for motion the pigments of
the photosynthetic process.

Photomotion in relation to metabolism.

The action of photosynthetic inhibitors, uncouplers, redox
systems and of ATP has been investigated. The main results
obtained are the following:

in photokinesis of *blue-green algae*, movement appears to
result from an additional supply of ATP by photophosphorylation
either cyclic or non cyclic (Nultsch, 1974).

In photophobic responses of *blue green algae* and of
Rhodospirillum rubrum movements are triggered by the photosynthetic electron flow (Häder and Nultsch, 1973 and Häder, 1974, 1975).

In photokinesis or in phobic responses of *flagellates* and in
phototaxis of most organisms, there is no relationship between
photomotion and photosynthesis since only blue light is effective.
The mechanism might be a flavoprotein photomodification inducing
a change in the ATP production. Another type of mechanism is
found in *Halobacterium halobium* whose purple membrane contains
a pigment, bacteriorhodopsin, a retinal-protein conplex whose
absorption spectrum coincides with the action spectrum of the
positive phobic response (Hildebrand and Dencher, 1974). This
membrane builds up an electrochemical gradient and functions as a
light driven proton pump used by the cell for ATP synthesis,
(Danon and Stoeckenius, 1974).

In general, photomotion enables the microorganisms to swim
towards light conditions which are favourable for energy conversion.

Photomotion and ion movements.

It has been shown that light alters the intracellular level of
cations as Mg^{++}, Ca^{++}, K^+ in *Platymonas* (Haldall, 1960). Several
findings on the control of ciliary reversal by internal calcium
concentrations in the case of *Paramecium* offer a stimulating model
for the phobic response of photosynthetic microorganisms (Naith
and Kaneko, 1972, Eckert, 1972). Moreover, the effect of cations
on motility of microorganisms is reminiscent of the effect of
cations in determining ATPase activity and muscle contraction
(Needham, 1952); it is also reminiscent of the presence of a high
concentration of Na-K activated ATPase in the outer segment of the
red cell (Bonting, 1966) and of the ion fluxes observed during
photostimulation of visual photoreceptors (Hagins, 1965). It is
also reminiscent of modulation of membrane properties by phytochrome (Borthwick, 1972). Thus the modification of a membrane
permeability to ions by a change of conformation of a photoreceptor inserted in the membrane could be an interesting model for many
cases of photomotion. Future findings on the molecular mechanism
of photomotion, vision, muscular contraction and photomorphogenesis
should stimulate one another.

PROKARYOTIC CELLS

	PHOTOMOTION	PIGMENT	REFERENCE

K = photokinesis
P = photophobic response
T = phototaxis

Y.P. = yellow pigment
CAR. = carotenoprotein
FLA. = flavoprotein

BACTERIA

	PHOTOMOTION	PIGMENT	REFERENCE
Rhodospirillum rubrum	K	(bacteriochlorophyll a	Throm, 1968
	P	(+ spirilloxanthin	Clayton, 1953
Halobacterium halobium	P	bacteriorhodopsin	Hildebrand and Dencher, 1975

CYANOPHYTA
(blue green algae)

	PHOTOMOTION	PIGMENT	REFERENCE
Phormidium	K	(chlorophyll a	Nultsch and Hellman, 1972
	P	(+ C phycocyanin	Nultsch, 1962
	T	Y.P. (FLA.) + biliprotein	Nultsch, 1961
Anabaena variabilis	K	(chlorophyll a	Nultsch and Hellman, 1972
	T	(+ C phycocyanin	Nultsch, 1975

N.B. The action spectra for the photomotion of each microorganism and sometimes the *in vivo* absorption spectra of the microorganism are shown in the corresponding references of Table I. The absorption spectra of the pigments can be found for carotenoids in Vetter *et al.*,1971, for flavins, flavoproteins, and cytochromes in Wainio,1970, for chlorophylls in Goedher,1966, for phycobilins in Rüdiger, 1975.

EUKARYOTIC CELLS

ORGANISM	PHOTOMOTION	PIGMENT	REFERENCE
CONTOPHORA			
Cryptomonas	T	C. phycoerythrin	Watanabe and Furuya, 1974
Peridinium	T	Y.P. (CAR. ?)	Haldall, 1958
Gonyaulax	T	Y.P. (CAR. ?)	Haldall, 1958
Prorocentrum	T	C. phycoerythrin hydroxyechinenone ? FLA. complexe ?	Haldall, 1958 Krinsky and Goldsmith, 1960 Edmondson and Tollin, 1971
Gymnodium splendum	P and T	CAR. ?	Forward, 1974
Navicula pereprina	P	chlorophyll a	Wenderoth, 1975
Nitzschia communis	K	chlorophyll a + fucoxanthin Y.P. (FLA.)	Nultsch, 1971 Nultsch, 1971

EUKARYOTIC CELLS

ORGANISM	PHOTOMOTION	PIGMENT	REFERENCE
CHLOROPHYTA			
Euglena gracilis	T K P	FLA. β carotene chlorophyll b FLA.	Checcucci et al., 1976 Wolken and Shin, 1958 Nultsch, 1975 (Diehn, 1969 (Froehlich and Diehm, 1974
Chlamydomonas reinhardii	T	Y.P. CAR. (or FLA. ?)	Nultsch et al., 1971
Micrasterias	T	chlorophyll	Neuscheler, 1967
Platymonas subcordiformis	T	Y.P. CAR. (or FLA. ?)	Haldall, 1961
Dunaliella	T	Y.P. CAR. (or FLA. ?)	Haldall, 1958
Stephanoptera	T	Y.P. CAR. (or FLA. ?)	Haldall, 1958
Dictyostelium	T	hemoprotein	Poff and Butler, 1974.

REFERENCES

BENEDETTI P.A. and A. CHECCUCCI (1975) Plant. Sci. Letters 4, 47-51.
BENSASSON R.V., E.J. LAND and B. MAUDINAS (1976) Photochem. Photobiol. 23, 189-193.
BENSASSON R.V., E.A. DAWE, E.J. LAND and D.A. LONG. Unpublished.
BONTING S.L. (1966) Ophtalmologica Basel 152, 527-529.
BORTHWICK   (1972) Phytochrome Proceedings of a Symposium Eretria Greece, Sept. 1971. K. Mitrakos and W. Shropshire Jr. Eds., p. 27-44, Academic Press New York.
CHECCUCCI A., G. COLOMBETTI, R. FERRARA and F. LENCI (1976) Photochem. Photobiol. 23, 51-54.
DANON A. and W. STOECKENIUS (1974) Proc. Natl. Acad. Sci. U.S.A. 71, 1234-1238
DELBRUCK M., A. KATZIR and D. PRESTI (1976) Proc. Natl. Acad. Sci. U.S.A. 73, 1969-1973.
DUYSENS L.N.M. (1951) Nature 168, 548-550.
ECKERT R. (1972) Science 176, 473-481.
EDMONTON D.E. and G. TOLLIN (1971) Biochemistry 10, 113-124.
FORWARD R. (1974) J. Protozool. 21, 312-315.
GOEDHER J.C. (1966) in The Chlorophylls. L.P. Vernon and G.R. Seely eds. p. 147-184. Academic Press, New York.
HADER D.P. and W. NULTSCH (1973) Photochem. Photobiol. 18,311-317.
HADER D.P. (1974) Arch. Mikrobiol. 96, 255-266.
HADER D.P. (1975) Arch. Mikrobiol. 103, 169-174.
HAGER A. (1970) Planta 91, 38-53.
HAGINS W.A. (1965) Cold Spring Harb. Symp. Quant. Biol.30,403-418.
HALDALL P. (1958) Physiol. Plant. 11, 118-153.
HALDALL P. (1960) Physiol. Plant. 13, 726-735.
HALDALL P. (1961) Physiol. Plant. 14, 133-140.
HALDALL P. (1967) Photochem. Photobiol. 6, 445-460.
KRINSKY N.I. and T.H. GOLDSMITH (1960) Arch. Biochem. Biophys. 91 271-279.
NAITOH Y. and H. KANEKO (1972) Science 176, 473-481.
NEEDHAM D.M. (1952) Adv. Enzymol. 13, 151-197.
NEUSCHELER W.Z. (1967) Pflanzenphysiol. 57, 151-172.
NISHIMURA M. and TAKAMATSU K. (1957) Nature 180, 699-701.
NULTSCH W., G. THROM and I.V. RIMSCHA (1971) Arch. Mikrobiol. 80, 351-369.
NULTSCH W. (1974) in "Algal Physiology and Biochemistry" (W.C.P. STEWART, ed.) p. 864-893, Blackwell Scientific Publications, Oxford.
NULTSCH W. (1975) in "Primitive Sensory and Communication Systems" (M.J. CARLILE ed.) Academic Press New York p. 29-90.
NULTSCH W., G. THROM and I.V. RIMSCHA (1971) Arch. Mikrobiol. 80, 351-369.

PLATT J.R. (1959) Science 129, 372.
POFF K.L. and W.L. BUTLER (1974) Photochem. Photobiol. 20,241-244.
RUDIGER W. (1975)Ber. Dtsh. Bot. Ges. 88, 125-139.
SONG P.S., T.A. MOORE and M. SUN (1972) in "The Chemistry of Plant
 Pigments" C.O. CHICHESTER ed.) Proc. Am. Chem. Soc. Symposium
 Chicago 1970. Academic Press, New York p. 33-74.
SONG P.S. and T.A. MOORE (1974) Photochem. Photobiol. 19, 435-441.
THOMSON A.J. (1975) Nature 254, 178-179.
VETTER W., G. ENGLERT, N. RIGASSI and V. SCHWEITER (1971) in
 "Carotenoids" (O. ISLER, ed.) p. 189-202, Birhäuser Verlag. Basel
 und Stuttgart.
WATANABE M. and M. FURUYA (1974) Plant. Cell. Physiol. 15, 413-
 420.
WAINIO W.W. (1970) in "The Mammalian Mitochondrial Respiratory
 Chain" Academic Press. New York.
WENDEROTH K. (1975) Thesis Marburg.

SPECULATIONS ABOUT SENSORY TRANSDUCTION

Donat-P. Häder

Department of Botany, University of Marburg

D-3550 Marburg, Lahnberge, West-Germany

SUMMARY: Light-induced reactions in microorganisms may be regarded as being composed of three successive steps in a reaction chain: stimulus perception, sensory transduction and visible response of the motor apparatus. The transmission of data may be brought about by a mechanical, chemical or electrical signal. In photokinesis in prokaryotes ATP serves as chemical transmitter. For sensory transduction in the phototaxis in Euglena several models have been developed, the more recent of which make use of electrical potentials. Phobic responses in blue-green algae also seem to be mediated by light-induced changes in the membrane potential, which can be followed by external and internal electrodes.

THE REACTION CHAIN IN PHOTOMOVEMENT OF MICROORGANISMS

Any of the three light reactions performed by microorganisms, i.e. phobic reaction, phototaxis and photokinesis (see introductory lecture by Prof. Haupt) may be regarded as being composed of three subsequent steps in a reaction chain. 1. The stimulating signal, which may be either light intensity, light direction or change in intensity of light must be perceived by the photopigments,which may be localized in a specialized photoreceptor or scattered in different cell organelles, e.g. on the photosynthetic thylacoids. 2. This signal, once it has been perceived, has to be processed into a transportable form and then transmitted to the effector. That part of the reaction chain is called sensory trans-

95

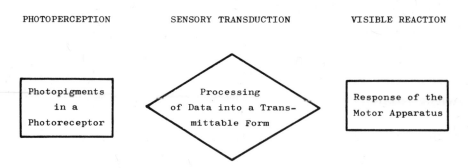

Fig. 1. Reaction chain of photomotion in microorganisms

duction. 3. Data travelling along this "telephone line"
influence the activity of the motor apparatus, which
produces the visible reaction of the organism (Fig. 1).
The concept of three successive steps has already been
postulated by Massart (1).

Two well-known examples might clarify this concept:
In the phototaxis of Euglena light perception is performed by a photoreceptor pigment, which is probably a
flavin (2,3,4) localized in the paraflagellar body. The
reaction of the motor apparatus consists in an erection
of the locomotor flagellum from its normal position
parallel to the ventral surface. Even though the photoreceptor is localized quite near the effector the primary photoevent must be translated into a signal which
influences the motor apparatus. In blue-green algae the
photoreceptor pigments are localized on the photosynthetic thylacoids. The motor apparatus has not yet been
identified though it is probably situated at the lateral
cell wall. Again we are faced with the problem of sensory transduction.

The photopigments involved in the photoreactions
of microorganisms have been studied quite extensively
in a number of organisms during the last decades and
a lot of information is available on the motor response
at least in the flagellated forms. But our knowledge on
sensory transduction is quite limited, therefore much
of this lecture will be highly hypothetical.

POSSIBLE MECHANISMS FOR SENSORY TRANSDUCTION

The transmission of data could be brought about by

1. production of a chemical transmitter
   a. propagated by undirected diffusion or
   b. propagated by specialized organelles (e.g. micro-
      tubules)
2. mechanical stimulation of the motor apparatus
3. propagation of an electrical signal, travelling
   along a membrane system
4. a structure with piezoelectric properties.

Several of these possibilities seem to be utilized
by different microorganisms.

SENSORY TRANSDUCTION BY MEANS OF A CHEMICAL TRANSMITTER

In photokinesis many, especially prokaryotic, organisms seem to make use of a chemical transmitter. In
photosynthetic bacteria and blue-green algae light perception by the photosynthetic pigments results in an
increased ATP production. This high-energy compound is
transported to the motor apparatus — whatever that is
in cyanophyceae — presumably by simple diffusion, since
it takes about 10 minutes until the organism has adjusted its speed to a new light intensity. The fact,
that photokinesis is inhibited by uncouplers of photophosphorylation, such as DNP, NaN3, atebrine and CCCP
(5), must be regarded as evidence for the energetic
coupling. Furthermore externally applied ATP increases
the speed of Rhodospirillum rubrum by the factor of
three, though only after a lag phase of several hours
(6).

MODELS OF SENSORY TRANSDUCTION IN EUGLENA

Several models have been proposed for the sensory
transduction in Euglena. An early model developed by
Wolken (7) describes stigma and flagellum as primitive
eye and elementary nervous system. But today it is widely accepted that the stigma is not the photoreceptor in
Euglena.

It has been found that absorbed light does not
serve as energy source of phototactic orientation by
some energy conversion mechanism, since the dependence
of reaction on light intensity follows the Weber law
and not a saturation curve (8). This indicates that
light perception only triggers the steering mechanism.
Therefore Tollin (8) proposed that the photoreceptor
triggers the ATP flux to the locomotor flagellum, since
phototaxis in Euglena — but not in all flagellates —
seems to be energized by photosynthesis: Inhibitors of

photosynthesis, e.g. DCMU (9) and uncouplers as CCP (8)
block phototaxis. Furthermore dark-bleached Euglena
have been reported to show only weak phototaxis, though
normal motility (8). But this model only accounts for
changes in the flagellar beat frequency, but not for a
directional orientation toward the light source.

Jahn and Bovee (10) offered a different model
based on the piezoelectric properties of the structures
involved. Light perception by the paraflagellar body
(PFB) conducts electrons along a chain of redox components which ultimately stimulate the microtubules of the
flagellum. Since the structure of the axoneme is quasi-
cristalline, the authors postulate piezoelectric properties, which have been found in other biological
structures. An electrical wave travelling along the flagellum would cause a mechanical wave. This is further
accompanied by waves of ATPase activity, stimulated by
cation shifts (11). It has been further proposed that
the paraflagellar rod, a cristalline structure, which
has been found in many euglenoid flagellates between
PFB and axoneme, may be the structural component of sensory transduction (12).

The model developed by Piccinni and Omodeo (13)

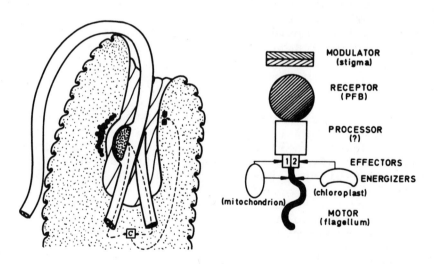

Fig.2   Fig.3

Fig. 2. Model of sensory transduction in Euglena (modified after Piccinni and Omodeo, 13)
Fig. 3. Schematic representation of the reaction chain
in Euglena (after Diehn, 17)

makes use of the propagation of electrical signals and
processing of data (fig. 2). The photoreceptor emits a
signal which reaches a comparator at the base of the
flagella by travelling along the basal part of the
longer flagellum. In order to measure a change in light
intensity a delayed signal is needed to be compared
with the present light intensity. The authors propose
a signal transmission via a synaptic type junction between PFB and the non-emergent flagellum. The information derived from this comparison is transmitted to the
effector, which may be localized in the wall of the re-
servoir next to the locomotor flagellum. This hypothesis
cannot be generalized for other organisms, since e.g.
Phacus and Euglena pisciformis have been reported to
lack the short flagellum (14).

In the ciliate Paramecium Eckert (15) and Machemer
(16) demonstrated that frequency and direction of ciliar
beating is controlled by membrane potential changes.
Similarly in flagellates the primary photoevent could
cause a change in the membrane potential, which triggers
the motor response. But light-induced membrane potential
changes in Euglena have not yet been published.

On the basis of light-induced potential changes
Diehn (17) devised a computer simulation of phototaxis
(Fig. 3). During lateral illumination the stigma modulates the photon flux to the receptor (PFB). From this
signal the information is derived by a processor of unknown identity, which triggers the effector at the flagellum base.

Up to now there is no generally accepted model for
the sensory transduction in photomovement of flagellates,
but it is interesting to note that all modern theories
have one element in common: the production and light-
induced change of bioelectrical membrane potentials.
Froehlich and Diehn (18) measured light-induced potential changes in artificial polarized bilayer lipid membranes containing a flavin, which they regarded to
simulate the in vivo photoprocesses in Euglena.

SENSORY TRANSDUCTION IN PHORMIDIUM UNCINATUM

For phobic reactions in the blue-green alga Phormidium uncinatum a model has been devised which also
is based on light-induced potential changes. According
to the Mitchell chemiosmotic hypothesis the photosynthetic electron and proton transports produce a bio-

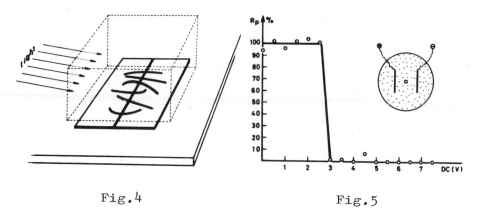

Fig.4   Fig.5

Fig. 4. Apparatus to measure potential differences between front and rear end of blue-green algae trichoms
Fig. 5. Inhibition of photophobic accumulation in a
light trap by an external DC field

electric potential across the thylacoid membrane. Since
in blue-green algae the thylacoids are not enclosed by
a chloroplast envelope, but connected with the cell membrane, this potential is superimposed to the normal
cell potential. Phobic reactions occur, when an organism
enters an area of lower light intensity. The decreased
photosynthetic electron transport causes a drop in the
membrane potential in the darkened front cells of a
trichom. This potential difference can be measured by
two external platinium electrodes, which are bridged
by parallel preorientated trichoms (fig. 4). The sum
potential of the population follows the light-dark regime (90 seconds dark, 420 seconds white or monochromatic light) with oscillations of about 5 - 10 mV. The
amplitude of the potential changes depends on light intensity and the action spectrum indicates the participation of both photosystem I and II pigments.

Recently we have been successful in penetrating
a single cell with a 0.1 μ tip diameter glass electrode.
The resting potential is about -35 mV. A light-dark
regime causes oscillatory potential changes of about
2 mV.

To prove, whether these light- induced potential
changes are responsible for phobic reactions in Phormidium uncinatum, we applied an external electric DC

field of 30 mm length to a suspension of organisms in a
semisolid agar (fig. 5). Upon irradiation of a light
trap from below the algae accumulate in the illuminated
area by repetitive phobic reactions (19,20). Accumula-
tions were unaffected by voltages up to 2.5 V. Higher
DC potentials completely inhibited phobic reactions.
The mean trichom length in the suspension was 96.97 µ;
thus a 3.0 V DC field induces a potential of about
9.7 mV from front to rear end. This potential, which inhibits phobic orientation, has about the same magnitude
as the externally measurable potential change (see
above). Thus it seems probable that the externally
applied electric field cancels the endogeneous potential
difference in a trichom and thus abolishes the phobic
reaction.

During the last few years evidence has been piling
up that also in the non-photosynthetic bacterium Halobacterium halobium phobic reactions are mediated by
electrochemical potential changes; though the production of the proton gradient is brought about by a different mechanism which may be similar to the primary
photoprocess in vision (21,22).

REFERENCES

1. Massart, J. (1902) Biol Zentralbl., 12, 9 - 23
2. Benedetti, P. A., Checcucci, A. (1975) Plant Sci.
   Let., 4, 47 - 51
3. Checcucci, A., Colombetti, G., Ferrarra, R., Lenci,
   F. (1975) Photochem. Photobiol., 21, 51 - 54
4. Mikolajczyk, E., Diehn, B. (1975) Photochem. Photobiol., 22, 269 - 271
5. Nultsch, W. (1974) Abh. Marburger Gelehrten Ges.
   Jg. 1972, Nr. 2, 143 - 213
6. Throm, G. (1968) Arch. Prot., 110, 313 - 371
7. Wolken, J. J. (1961) Euglena - An experimental organism for biochemical and biophysical studies.
8. Tollin, G. (1973) In: Pérez-Miravete (ed.), Behaviour of Micro-organisms. London, New York, Plenum
   Press
9. Pederson, T. A., Kirk, A., Bassham, J. A. (1966)
   BBA 112, 189
10. Jahn, Th. L., Bovee, E. C. (1968) In: Buetow, D. E.
    (ed.) The Biology of Euglena. Acad. Press New York,
    London
11. Bovee, E. C., Jahn, Th. L. (1972) J. Theor. Biol.,
    35, 259 - 276
12. Kivic, A. P., Vesk, M. (1972) Planta, 105, 1 - 14

13. Piccinni, E., Omodeo, P. (1975) Boll. Zool., <u>42</u>, 57 - 79
14. Haye, A. (1930) Arch. Prot., <u>70</u>, 1 - 86
15. Eckert, R. (1972) Science, <u>176</u>, 473 - 481
16. Machemer, H. (1974) J. Comp. Physiol., <u>92</u>, 293 - 316
17. Diehn, B. (1973) Science, <u>181</u>, 1009 - 1015
18. Froehlich, O., Diehn, B. (1974) Nature, <u>248</u>, 802 - 804
19. Häder, D.-P., Nultsch, W. (1973) Photochem. Photobiol., <u>18</u>, 311 - 317
20. Häder, D.-P. (1976) Z. Pflanzenphysiol., <u>78</u>, 173 - 176
21. Sherman, W. V., Caplan, S. R. (1975) Nature, <u>258</u>, 766 - 768
22. Krinsky, N. J. (1974) Photochem. Photobiol., <u>20</u>, 532 - 535

PANEL DISCUSSION ON PHOTOMOVEMENT

Moderator : W.NULTSCH (Lehrstuhl für Botanik,
            Fachbereich Biologie, Lahnberge, D-3550
            Marburg, West Germany)

Panel   : R.B.BENSASSON, W.R.BRIGGS, M.E.FEINLEIB,
            D.-P.HÄDER, W.HAUPT, F.LENCI, P.-S.SONG

NULTSCH: As shown by the speakers of this afternoon,
three types of photomovement exist in microorganisms:
phototaxis, photophobic responses and photokinesis. Irrespective of the reaction type the reaction chain can
be subdivided into the following parts: stimulus perception by the photoreceptor, transformation of the absorbed energy into another form of energy, its transduction to the effector, and the motor response. Following this chain we start with the discussion of the
photoreceptor problem in phototaxis. Since in most
organisms light of shorter wavelengths is phototacti-
cally active, yellow pigments come into question as
photoreceptor pigments, and mainly carotenoids and
flavins are discussed as possible candidates.

BENSASSON: The following observations favor carotenoids:
1. Some phototactic organisms lack the near UV peak in
their action spectrum; thus the action spectrum better
fits with carotenoid absorption. 2. A colourless
Nitzschia does not show phototactic reactions, while a
carotenoid-containing form does. 3. It is a well known
fact, that carotenoids may act as accessory pigments
in photosynthesis. 4. The absence of phototactic activity of longer wavelengths is an argument against a

possible energy transfer to other pigments. 5. The par-
tecipation of bacteriorhodopsin and rhodopsin in the
phobic response of halobacterium halobium and the vision process in animals are well-known examples of carotenoid action.

SONG: These observations are certainly valid points in
favoring carotenoids over other pigments such as
flavins. However: there are also difficulties with C_{40}-
type carotenoids as photoreceptor: regarding observations N°1 above it should be mentioned that screening
due to the absorbance of the near UV in most organisms
increase inversely with wavelength, so that it is easy
to obtain a distorted or weak near UV-effect in the
action spectra. As to N°2, carotenoids may well aid the
photoresponsive mechanism by protecting the photoreceptor (eg: flavins, chlorophylls, biliproteins, etc) from
the singlet oxygen attack. Furthermore, the shading
effect of stigma carotenoids is essential in the photomotion of certain organisms, while the primary photoreceptor itself can be a non-carotenoid pigment.
Observations N°3 and 4 require some elaboration: in
spite of extremely short lifetime of the S_1 state
(less than 10^{-13} s), carotenoids are able to dispose
the excitation energy efficiently to the acceptor molecules such as chlorophyll in chloroplasts. We have recently shown that a special molecular topology of carotenoids (peridinins) and chlorophylls is necessary
to overcome short lifetime (τ_f) and accomplish the energy transfer. Thus, it is conceivable that the primary
photoreceptor is non-carotenoid, absorbing at wavelengths longer than that of carotenoids, so that the
energy transfer from the former to the latter can account for the appearance of carotenoid-type action
spectra. The direct excitation of the primary photoreceptor and recording of the corresponding photoresponse
action spectrum can be obscured by its low concentration
and screening pigments. Thus, it would be of interest
to employ the tunable dye laser to record the action
spectrum of a photoresponse at wavelengths longer than
the carotenoid-type action spectrum. It is doubtful
that the short lifetime of carotenoids can be overcome

without suitable energy donor (carotenoid)-acceptor
interactions.

Although bacteriorhodopsin and rhodopsin are photoreceptors of carotenoid origin, their electronic polarization of the excited state is highly anisotropic (e.
g., strong permanent dipole moment and pK change in the
excited state) due to the asymmetric π-electron system
and protonation of the Schiff base. Such a static electronic polarization of the excited state is essential
for an effective coupling between the electronic excitation of the chromophore and molecular relaxation (e.
g., conformational change) of apoproteins or membranes.
Most C_{40}- and higher carotenoids lack the structural
feature for this type of electronic polarization.

HÄDER: Since phototaxis incorporates the detection of
direction of light, a difference in absorption must be
measured. This can be brought about by the introduction
of a screening pigment. Thus, even if carotenoids do
not actually play a role as photoreceptor molecules in
certain organisms, they are involved in the action of
photoperception.

BENSASSON: Is the colourless Nitzschia a mutant?

NULTSCH: No, it is a heterotrophic wild type, which has
been isolated from Fucus.

BENSASSON: Then the loss of phototactic activity is due
to the absence of carotenoids.

FEINLEIB: There are three different roles carotenoids
can play in phototactic responses: 1. They may act as
photoreceptor per se. 2. They may act as shading device
3. They may act as protecting agents.

NULTSCH: In most organisms in which photophobic responses are coupled to photosynthesis, carotenoids are active energy transfer to chlorophyll a.

SONG: The reason why carotenoids cannot act as primary
photoreceptors is the lack of an asymmetric electronic
polarization of the excited state which is needed for
the energy transfer. One possible way carotenoids can
act as the primary photoreceptor is through photoisome-

rization, as in rhodopsin. However, both photoisome-
rization paths via S_1 and T_1 states of carotenoids are
not likely to be efficient because of the short lifetime (due to radiationless transitions other than isomerization) and low intersystems crossing efficiency,
respectively: recent tunable dye laser experiments by
Prof.Delbrück strongly implicate flavin as the phototropic photoreceptor of Phycomyces, showing that the
action spectrum has a weak peak at 595 nm corresponding
to the singlet-triplet absorption band of flavin. Since
the quantum efficiency of most photoresponses approaches
100%, it is expected that a flavin photoreceptor in vivo is nonfluorescent, with $\tau_f \leq 0.5$ ns and k (primary
photoprocess) 10^{10} s^{-1}. Flavins can readily meet these
requirements.

BRIGGS (CARNEGIE INST. WASHINGTON, STANFORD, U.S.A.):
Munoz and Butler demonstrated flavin-mediated photoreduction of cytochrome (b-type) in Neurospora mycelium.
The action spectrum has a broad UV band and a band
showing fine structure in the blue. We have isolated a
membrane fraction which displays the same light-induced
cytochrome reduction. But the important question remains: does this pigment system have anything to do with
photoresponses in Neurospora such as carotenogenesis
or suppression of circadian rhythm which can be seen
in the conidium formation in the timex mutant. We have
obtained a Neurospora strain containing both timex and
poky mutations. The poky mutation is known to have a
deficiency in b-type cytochromes, and we wanted to determine, first whether it was also deficient in the
flavin-mediated photoreduction system, and, second,
whether its photoresponses were impaired. Preliminary
experiments showed that the poky-timex-mutant is deficient in both mitochondrial and extramitochondrial
b-type cytochromes. Membrane preparations from poky-
timex showed less than 10% photoreduction of b-type
cytochrome as wild type and roughly 20 times as much
light was required to suppress circadian rhythmicy in
poky-timex as in timex alone. Furthermore the carotenogenesis in the poky-timex is impaired in photosensitivity.
These results suggest that the flavin-b-type cytochrome

is involved in the Neurospora photoresponses. The photoreduction itself, observed only when the system is
poised within a narrow redox potential range, is a convenient assay for the pigment system, but is not necessarily a part of the normal photoresponse chain. The
experiments described, were done by Mr Robert Brain.

LENCI (C.N.R. - BIOPHYSICS LAB., PISA, ITALY): Among
microorganisms exhibiting photoorientation by means of
a "two-instant-mechanism", Euglena gracilis is an example in which it has been ascertained that the pigments
of the shading device (carotenoids) are different from
those of the photoreceptor (flavins). This has recently been confirmed in our Laboratory recording "in vivo"
absorption spectrum of the stigma and fluorescence
spectrum of the paraflagellar body. In principle therefore, action spectra for stigma-mediated responses
should be proportional to the product of the absorption
spectra of carotenoids and of flavins, whilst action
spectra of stigma-independent responses (e.g., accumulation elicited by short-lasting flashes, negative phototaxis of stigma-less mutants) should reflect only and
simply flavin spectral properties. As a matter of fact,
action spectra for different motile photoresponses are
all the same and proportional to the absorption spectrum of flavins/flavoproteins. Two question are now to
be answered: why are action spectra of stigma-mediated
responses not affected by the spectral properties of
this organelle and which is the role of the shading device in the photobehaviour regulation of Euglena. The
first question can be answered, taking into account
that phototaxis is a photo-sensing-type process (light
variations trigger photomotile responses), on the basis
of some calculation on the relationship between carotenoid content of the stigma and intensity of transmitted light. The relationship between stigma pigmentation degree,which is related to the metabolism of the
cell, and overall photobehaviour can then be found considering the sign of the response: actually when the
stigma becomes absorptionless, i.e. when its efficiency
is lost, the cell exhibits only negative phototaxis,
even at the lowest light intensities.

HÄDER: We have to accept that there are different photoreceptor pigments being used by different microorganisms
in their various photoreactions. We should now concentrate on finding, what the primary photoreactions are
in the photoreceptor molecule, since this is the starting of the reaction chain, which carries out sensory
transduction, i.e. the connection between photoreceptor
and the motor apparatus in the organism.

NULTSCH: At the end of the discussion it should be mentioned that other pigments are also active as photoreceptors in phototaxis, such as biliproteins in Phormidium and biliproteins as well as chlorophyll a in Anabaena variabilis, as observed in our laboratory very
recently. Obviously different mechanisms of phototaxis
are developed in microorganisms, using different photoreceptor pigments.

SYMPOSIUM IV

PHOTOSYNTHESIS

PRESENT STATUS OF THE O_2 EVOLUTION MODEL

Bessel Kok and Bruno Velthuys

Martin Marietta Laboratories
1450 South Rolling Road
Baltimore, Maryland 21227

According to the generally accepted model of O_2 evolution (1, 2), the O_2 centers, each associated with a photosystem II unit, act independently of each other. Driven by 4 successive photoacts, the O_2 enzyme cycles through 5 oxidation states:

$$S_O \xrightarrow{h\nu} S_1 \underset{k_d}{\overset{h\nu}{\underset{\longleftarrow}{\longrightarrow}}} S_2 \underset{k_d'}{\overset{h\nu}{\underset{\longleftarrow}{\longrightarrow}}} S_3 \xrightarrow{h\nu} S_4 \xrightarrow{H_2O} S_O + O_2$$

In dark reactions, states S_3 and S_2 decay to state S_1 which is stable. State S_O is also stable; however, it is formed only from S_4. The damping of the oscillation of the O_2 flashyield is due to a) "misses" - randomly occurring failures of the photochemical charge separation and b) "double hits" -- the occurrence of two charge separations in a single flash.

In this paper we discuss some recent observations; in general these have revealed additional complicating phenomena, without significantly changing the original hypotheses.

Whole Algae vs. Chloroplasts

It has been noticed recurrently (3-5) that whole algae, presumably optimal material, compared to isolated chloroplasts, show poorer oscillations, characterized by more misses ($\sim 20\%$ vs. 5-10%). With (6) we argue that this "paradox" rests on the peculiar conditions necessary to observe the O_2 flashyield oscillation: close vicinity of a reducing electrode and a long dark period that is only briefly interrupted by a weak illumination

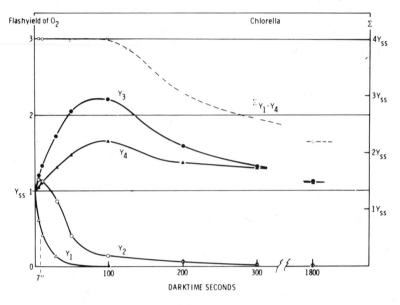

Fig. 1

that is below the compensation point of respiration. In the reducing intracellular environment the system II acceptor pools (Q and A) are not kept fully oxidized which causes misses and "rapid" deactivation: a ($\sim 1''$) backreaction between Q^- and S_n (7).

Figure 1 illustrates these phenomena in whole algae. Flash yield sequences (20 flashes, spacing 1 sec) were measured after various dark times following a previous sequence. The sum of the first 4 flashyields (an index of the photochemical efficiency) declines, due to a gradual reduction of the electron acceptor pools. Y_3 ($\sim [S_1]$) only rises to a modest level of $2.2Y_{ss}$ and then declines, along with Y_4. Deactivation is fast, Y_1 (i.e. S_3) decays with a halftime of ~ 7 sec; together with a high number of misses this accounts for a rapid damping of the yield oscillation.

In our opinion apparent anomalies which have been observed in whole algae (3-5, 8) can be explained adequately, i.e., without additional assumptions, by redox changes of the acceptor pools and deactivation loss with its associated change of the S state distribution.

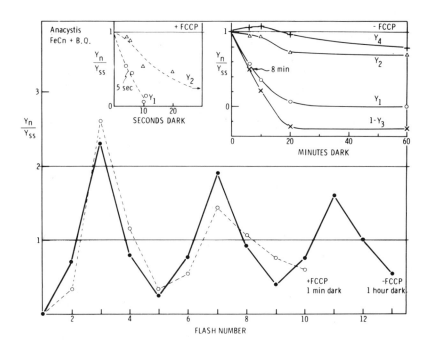

Fig. 2

One might predict that most any treatment which inhibits respiratory O_2 uptake can improve the yield oscillation. This has indeed been reported and is again illustrated in Fig. 2. In this case, Anacystis cells were used which, without additions, more or less followed the pattern shown in Fig. 1. In the presence of oxidized benzoquinone which penetrates the algae and stops their respiration, the oxygen flash yield sequence shows a deep, sustained oscillation and very few misses (5-6%). The average flash yield remains constant throughout the sequence. Apparently Q and its secondary acceptors remain fully oxidized. As shown in the right hand insert in Fig. 2 deactivation is very slow; the half-life of Y_1 is 8 min. and Y_2 shows hardly any decay.

As will be discussed below, this constancy of Y_2 does not necessarily imply stability of S_2 in dark; it might be due, at least in part, to another (double hit) phenomenon. Our data, however, do not support the assumption (9) that S_2 is made in dark by oxidation of S_1.

Deactivation

Figure 2 includes another experiment (open circles) made with the same sample, but after 1.6 μM FCCP was added to the medium. FCCP greatly accelerates the loss of S_3 and S_2 in darkness (10, 11). In this case, the decay of Y_1 became 100 times faster (8 min to 5 sec., c.f. inserts Fig. 2). Consequently, a dark period of 1 min. sufficed for complete deactivation and the observation of a new flash yield oscillation. This oscillation is somewhat more damped than the original one. However, because of the 1 sec spacing between flashes, the rapid deactivation added ($\leq$ 10%) deactivation loss to the still very few misses. This experiment thus implies that a 100-fold acceleration of deactivation causes no significant increase of the number of misses. This is unlike the conditions described in the previous section, where misses and deactivation rate seemed to be related. We conclude, with (11), that FCCP induces a "short-circuit", which does not involve Q.

In the presence of FCCP, under conditions as used in Fig. 2 we have made "preflash" experiments using dark periods of only one minute before and after one (or three) preflash(es). The subsequent flash sequence showed a normal increase (or decrease) of the ratio Y_3/Y_4 (indicative of the ratio $[S_1]$ / $[S_0]$. From this observation we conclude that the rate of deactivation does not (grossly) affect the final distribution of the states.

Except for cooling to low temperatures (77°K) we have not found a way to stop deactivation -- especially the loss of S_3. Presumably, we deal with a reductive process but even under quite strongly oxidizing conditions, which should remove endogenous electron donors (but unfortunately also tend to damage the system), the decay of $Y_1([S_3])$ is rapid. Thus, the donor(s) involved in deactivation either have a high midpoint potential or are inaccessible.

We have also considered another possible explanation (which would be in conflict with the assumption of non-cooperation between O_2 centers): that deactivation is due to a transfer of charges between centers, leading to a slow evolution of oxygen in dark. Recently, we have been able to perform experiments with sufficient sensitivity to observe the very small, slow O_2 production which should accompany such an event. So far, we have not detected any O_2 formation during deactivation.

Losses Due To Reductants Associated With the O_2 System

After a treatment of the chloroplasts with a low concentration of hydroxylamine the yield pattern tends to show a delay of two numbers

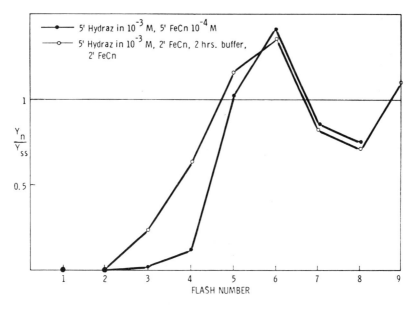

Fig. 3

$(Y_5$ high rather than $Y_3)$ (12). Presumably, the first two flashes in a sequence oxidize H. A. (or another reductant formed by it), rather than water. We have observed that this effect is not specific for H. A. and that the two equivalent reduction is preferred but not absolutely.

Figure 3 illustrates the effect of a treatment with hydrazine. As was observed with H. A., with increasing hydrazine "doses" (concentration and/or exposure time) the maximum yield which is normally at the third flash gradually shifts to the 4th, 5th, 6th, etc; which reflects the incorporation of 1, 2, 3, etc. reducing equivalents per O_2 center. High doses also affect Y_{ss}. In darkness, recovery to a normal flash pattern is incomplete even after long washings with buffer containing FeCn, or benzoquinone (Fig. 3 open circles). Under reducing conditions other agents, e.g. pyocyanin, induce similar effects, including a loss of Y_{ss}. Partial recovery of Y_{ss} could be obtained, requiring illumination followed by a dark time - somewhat analogous to the photoactivation of the Mn catalyst (13).

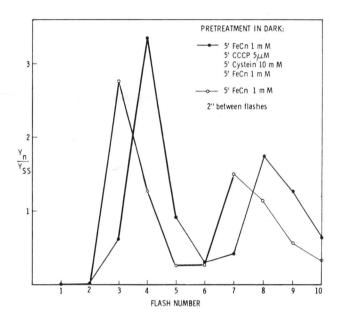

Fig. 4

Figure 4 illustrates a seemingly similar, but in reality very different "dark loss". In this case a one equivalent reducing compound is involved, which reacts only slowly (~ 1") with S_2 and/or S_3. As a result, the flash yield maximum is displaced one step: from Y_3 to Y_4 - but only if the spacing between the flashes is long ($\geq$ 1 sec). With a brief flash spacing the two patterns looked rather similar, because in this case the loss is made in the course of a number of flashes. The reductant is readily and irreversibly induced by a low concentration of FCCP and, as was reported earlier (14), also by reduced DPIP. In fact FCCP and DPIPH only accelerate its formation; given enough time, the donor is formed under all conditions we have investigated. Fig. 5 demonstrates the occurence of this event even in the presence of a high concentration of FeCn: during a long dark time (and measured with 2" spacing between flashes) Y_4 rises and Y_3 declines. The kinetic behavior of this one equivalent electron donor strongly suggests its identity with the precursor of (slow) EPR signal II (15).

In preflash experiments we allowed the system to deactivate, gave 1 (or 3) flash(es) and then exposed the chloroplasts to DPIPH (and a FeCn wash). The subsequently measured flash pattern still reflected the preflash(es) but with a delay of one flash: 1 preflash enhanced the ratio Y_4/Y_5, 3 preflashes decreased it. This again shows that the "one equivalent" reductant and the S enzyme are separate compounds.

Interconversion Of S_0 And S_1

One of the original assumptions was that states S_0 and S_1 are stable in dark. Later, however, it was observed (3) that the Y_3/Y_4 ratio often tends to lose its dependence upon preillumination and approach the "normal" value, presumably reflecting an S_1/S_0 ratio of 3. In the presence of FeCn a complete conversion to S_1 appears to take place (16). In our laboratory this change of the Y_3/Y_4 ratio has been a variable phenomenon. In the absence of FeCn after a preillumination with 3 flashes the ratio can remain low for a long time (2) or increase significantly.

In the preceding section we discussed various interfering phenomena which so far have prevented an unambiguous observation of a dark conversion $S_1 \longrightarrow S_0$. The evidence for the dark $S_0 \longrightarrow S_1$ is stronger, but is far from conclusive:

Expt. Fig. 5 shows O_2 flash yield patterns which were observed after various dark times following 3 preflashes given to a 5' deactivated system. In the absence of FeCn Y_3 became about equal to Y_4. In the presence of FeCn the time courses of Y_3 and Y_4 were distinctly biphasic: Y_4 reached a maximum after about

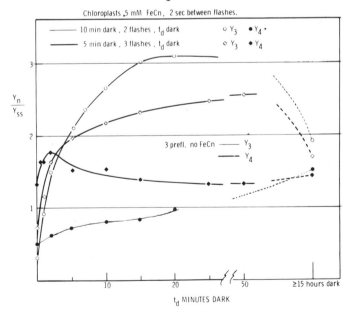

Fig. 5

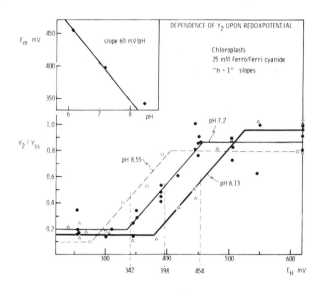

Fig. 6

3 min dark, and then declined to a low value (~1.3) with a half-time of about 10 min. Simultaneously Y_3 increased in an anti-parallel manner. It thus seems that we view the oxidation $S_0 \longrightarrow S_1$, driven by ferricyanide. If so, however, this conversion is incomplete: Fig. 5 shows a parallel experiment in which 2 rather than 3 preflashes were given. In this case, S_1/S_0 will be high after deactivation, without the necessity of a $S_0 \longrightarrow S_1$ dark conversion. In fact Y_3/Y_4 becomes higher than after 3 preflashes. Therefore, there is no true equilibration, which is inconsistent with the assumption of a dark $S_0 \longrightarrow S_1$ conversion.

Double Hits

Earlier data (17) appeared to show that also the transition $S_1 \longrightarrow S_2$ can be induced in darkness: incubation of a totally de-activated sample of chloroplasts with ferricyanide caused a significant increase of Y_2. Recently, we have been able to obtain quite high values of Y_2 ($\geq Y_{ss}$) by either increasing [FeCn] or changing the salt composition of the medium. Fig. 6 shows the dependence of the increase of Y_2 upon the redox potential of the medium. At pH 7, the increase was half maximal at ~400 mV. The titration curves showed best fit with a one equivalent slope and the E_m value changed about 60 mV per pH unit. These data may imply that a semiquinone intermediate in the oxidized state causes a high value of Y_2.

Additional observations which concerned the "activation deficit" in weak light and the effects of flash intensity and flash length revealed that the high Y_2 value was due to a large number of double hits in the first flash and not a dark conversion $S_1 \longrightarrow S_2$. As will be argued elsewhere in more detail, in this double hit event another electron acceptor ("C") in addition to Q becomes reduced. Fig. 6 then pertains to the midpoint potential of this additional system II acceptor.

ACKNOWLEDGEMENTS

This work was supported by the National Science Foundation, Grant No. PCM74-20736 A01 and the Energy Research and Development Administration Contract E(11-1)-3326.

REFERENCES

1. Kok, B., Forbush, B. and McGloin, M. (1970) Photochem. Photobiol. 11, 457-475.
2. Forbush, B., Kok, B. and McGloin, M. (1971) Photochem. Photobiol. 14, 307-321.
3. Joliot, P., Joliot, A., Bouges, B. and Barbieri, G. (1971) Photochem. Photobiol. 14, 287-305.
4. Lavorel, J. (1976) J. Theor. Biol. 57, 171-185.
5. Lavorel, J. (1976) FEBS Lett. 66, 164-167.
6. Ley, A., Babcock, G. and Sauer, K. (1975) Biochem. Biophys. Acta 387, 379-387.
7. Radmer, R. and Kok, B. (1973) Biochem. Biophys. Acta 314, 28-41.
8. Lavorel, J. and Lemasson, C. (1976) Biochem. Biophys. Acta 430, 501-516.
9. Greenbaum, E. and Mauzerall, D.C. (1976) Photochem. Photobiol. 23, 369-372.
10. Renger, G. (1972) Biochim. Biophys. Acta 256, 428-439.
11. Renger, G., Bouges-Bocquet, B. and Delosme, R. (1973) Biochim. Biophys. Acta 292, 796-807.
12. Bouges, B. (1971) Biochim. Biophys. Acta 234, 103-112.
13. Cheniae, G. M. and Martin, I. F. (1969) Plant Physiol. 44, 351-360.
14. Velthuys, B. R. and Visser, J. W. M. (1975) FEBS Lett. 55, 109-112.
15. Babcock, G. T. and Sauer, K. (1973) Biochim, Biophys. Acta 325, 483-503.
16. Bouges-Bocquet, B. (1973) Biochim. Biophys. Acta 292, 772-785.
17. Kok, B., Radmer, R. and Fowler, C. F. (1974) in Proc. 3rd. Int. Congr. Photosynth., Rehovot (Avron, M., ed) pp. 485-496, Elsevier, Amsterdam.

LOW TEMPERATURE REACTIONS IN PHOTOSYNTHESIS

J. Amesz

Department of Biophysics, Huygens Laboratory

University of Leiden, The Netherlands

SUMMARY

A brief discussion is given of the photochemical reactions of systems 1 and 2 at low temperatures. In addition to the primary charge separation these include also some secondary electron transfer reactions, in particular on the oxidizing side of system 2. Measurements of so-called electrochromic absorption changes in chloroplasts at -30 to -50 $^{\circ}$C indicate that in the presence of an electron donor and acceptor for system 1 high membrane potentials across the thylakoid lamellae are generated upon illumination, due to a low permeability of the membrane. Similar results were obtained with bacterial chromatophores. Analysis of the absorption changes indicates an electrochromic rather than a conformational change of the bacterial pigments.

INTRODUCTION

Photosynthesis is the result of various physical and chemical processes. Of these, the so-called "light steps" consist of light absorption and transfer of excitation energy to the reaction centers and the photochemical transfer of an electron from a donor (chlorophyll or bacteriochlorophyll) to an acceptor molecule; the "dark steps" include secondary electron transport between components of the electron transport chain, and various enzymatic and diffusion processes. Most of these dark steps have a rather high activation energy and, moreover, depend on the presence of liquid water, and therefore it is not surprizing that photosynthesis is effectively inhibited by lowering the temperature. However, the photochemical reactions proper are in principle insensitive to temperature lower-

ing, and it has been observed that even some secondary electron
transport reactions may occur at very low temperatures. Although low
temperature studies of photosynthesis were initiated several years
ago already,in general it may be stated that detailed information
about the properties of photosynthetic material at low temperature
has become available only recently. In the following we shall discuss some of these studies, including recent results from our
laboratory.

PHOTOSYSTEM 2

There is good evidence now that the primary photoreaction of
system 2 is similar to that of photosystem 1 and of purple bacteria,
and consists of the transfer of an electron from an excited chlorophyll a dimer (P680) to an acceptor. Part of the evidence comes
from experiments at low temperature. Upon illumination of chloroplasts with a short flash of light an ESR signal was observed (1,2)
with a decay time of a few ms at 10 - 100 K that could be ascribed
to oxidized P680 (P^+680). The shape and g value of the signal were
similar to that of P^+700, the primary electron donor of system 1,
suggesting that the signal was likewise due to a dimeric chlorophyll
cation radical. Similar results were obtained by optical measurements at 680 nm and in the near-infrared region (3,4). Together,
these experiments provide convincing evidence that P680 is photooxidized reversibly at low temperature in chloroplasts.

A convenient way to study the reduction of the primary electron
acceptor at low temperature is by measurement of the absorbance
changes of C-550 (5). These changes are thought to be due to
an "indicator pigment", perhaps phaeophytin (6), the absorption
bands of which shift in response to the reduction of the acceptor.
This acceptor is probably plastoquinone (6), which is reduced to the
semiquinone anion upon illumination. Studies of the response of
C-550 induced by a flash revealed similar kinetics as of P680: a
rapid rise followed by a decay with a half-time of a few ms (7).
However, the decay did not proceed to completion, indicating that
a fraction (about 20 %) of the acceptor remains reduced after a
flash. This indicates that at low temperature after a flash in most
of the reaction centers a back reaction between P^+680 and reduced
acceptor (Q^-) occurs, but that in some centers P^+680 oxidizes a
secondary donor D, giving a stable charge separation:

$$D\ P680\ Q \underset{}{\overset{h\nu}{\rightleftarrows}} D\ P^+680\ Q^- \longrightarrow D^+\ P680\ Q^-$$

Upon repeated flash illumination an increasing number of reaction
centers are converted into the condition D^+ P680 Q^-. In part of the
reaction centers this condition is stable for at least several hours
at 90 K (2).

The scheme given in the previous paragraph is too simple to quantitatively describe all observations, including those on chlorophyll fluorescence (2,8,9). Photooxidation of cytochrome b_{559} at low temperature indicates that this substance acts as a secondary electron donor, but this appears to be the case only in half of the reaction centers (2). Presumably, in the remaining part an as yet unknown substance is oxidized. If the chloroplast suspension is brought to a high redox potential before cooling, both donors are apparently oxidized already in the dark, and a chlorophyll dimer, different from P680, now seems to act as secondary donor (1,2).

In the temperature region between about -40 and -100 $^{\circ}$C the photochemical properties of chloroplasts appear to be even more complicated than at still lower temperatures, and show a dependence on the number of flashes given as preillumination before cooling (10-12). This dependence is probably related to the number of positive charges in the electron pathway from water to system 2 (the so-called S-state, see ref. 13). Cytochrome b_{559} photooxidation was observed after one or two flashes preillumination, when the reaction centers had been converted to states S_2 or S_3, but not in dark-adapted chloroplasts (state S_1).

PHOTOSYSTEM 1

The properties of photosystem 1 at low temperature are in one respect more simple than those of system 2: electron transport from secondary donors to the primary electron donor P700 is inhibited already at relatively high temperature (14). Photooxidation of P700 at low temperature in isolated chloroplasts and intact cells was observed several years ago already and more recent investigations have also provided information about the properties of the electron acceptor(s) of system 1. Even more than for system 2, ESR measurements had an important role in these studies.

The photooxidation of P700 has been studied by measurements of absorbance changes, the most prominent effect being a light-induced absorbance decrease near 700 nm, and of the well-known ESR signal of P^+700, due to the formation of a chlorophyll a dimer cation radical. In addition, Malkin and Bearden (15) observed another light-induced signal at low temperature. This signal, which is only measured with sufficient sensitivity at temperatures below about 25 K could be assigned to a reduced iron-sulphur protein, "bound ferredoxin". Comparative studies of the kinetics of P700 and bound ferredoxin showed a stoichiometry at temperatures between 10 and 200 K and indicated that the only dark processes that occurred are back reactions between these compounds (2,16,17). Three different components were observed in the decay after illumination, with half-times of about 25 ms, 300 ms and 30 s (2). In addition to this, with decreasing temperature an increasing fraction of reaction cen-

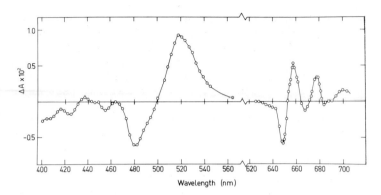

Fig. 1. Electrochromic absorbance changes in spinach chloroplasts at -35 °C with 53 % glycol, 0.17 mM PMS and 3.3 mM ascorbate (21).

ters, up to 80 % at 20 K showed an irreversible photoconversion. This phenomenon was observed not only in isolated spinach chloroplasts but also in intact cells of various algae (16), indicating that it reflects a fundamental property of system 1, the significance of which is not understood yet. However, others reported a reversible photooxidation of P700 near 10 K which did not involve ferredoxin but another, as yet unidentified compound (18,19). The identity of the primary electron acceptor thus remains an open question, and further experiments will be needed to reconcile these contrasting observations.

ELECTROCHROMIC EFFECTS

A class of light-induced absorbance changes which do not specifically reflect the redox state of an electron carrier are the so-called electrochromic shifts, which are thought to be due to the generation of a membrane potential across the photosynthetic lamellae (20). The acceptors and donors of photosystems 1 and 2 are probably located at opposite sides of the membrane. Both photosystems thus act as an electrogenic pump upon illumination, which transfers electrons from the inside to the outside of the thylakoid. Ionic movements in the aqueous phases rapidly "delocalize" the charge that is transferred, at least at not too low temperature. The most conspicuous electrochromic change is the well-known increase in absorbance near 518 nm. This change is mainly due to a shift of the absorbancy of a carotenoid, probably β-carotene. As illustrated by Fig. 1, chlorophylls a and b do also display electrochromic shifts upon illumination, most notably in the red region of the spectrum.

The electrochromic shifts do also occur at sub-zero temperature. At 100 K, however, only a system 2-driven 518 change was observed(21).

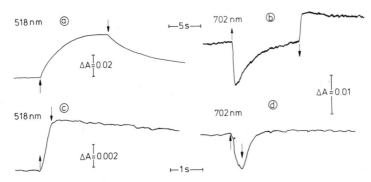

Fig. 2. Kinetics of light-induced absorbance changes of chloroplasts at -45 °C in the presence of glycol and PMS-ascorbate. Light on and off as indicated by arrows (21).

At this temperature delocalization of the field presumably does not occur, and the absence of a system 1-driven change may suggest a different location of the pigment with respect to the reaction centers of the two photosystems. At -50 °C, in a liquid medium, the charge separation in both photosystems was found to contribute equally to the electrochromic effect at 518 nm (14). The system 1-induced shift showed the same kinetics as P700, and both decayed in a few hundred ms, apparently by a back reaction between P^+700 and the reduced electron acceptor.

An up to 20-fold stimulation of the absorbance increase at 518 nm was observed after addition of an artificial electron donor and acceptor for system 1, and the kinetics of the electrochromic shift and of P700 now were quite different (Fig. 2). Of a number of donors tested, reduced N-methylphenazonium methosulphate (PMS) proved to be the most effective one; its oxidized form, and methylviologen were effective electron acceptors (22). The size of the 518 nm change suggests the generation of membrane potentials of up to 450 mV in these conditions. Although the rates of electron transport are at least an order of magnitude lower than at room temperature, they are apparently large enough to create high membrane potentials because of the strongly reduced permeability of the membrane, which is also reflected in the slow rate of decay upon darkening. The difference spectrum of Fig. 1 was obtained in these conditions.

Fig. 3 shows an experiment in which a shoft flash of light was given with or without a background of light. It can be seen that with increasing membrane potential an increasing proportion of the additional potential generated by the flash dissipates rapidly. This may be due to a stimulation by the electric field of the back reaction between P^+700 and the reduced acceptor. Experiments with flashes of different intensities suggest that the efficiency of the

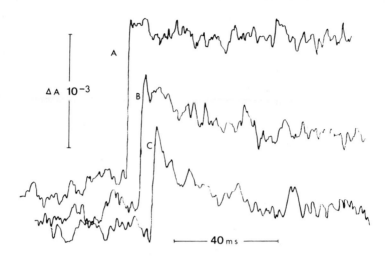

Fig. 3. Flash-induced absorbance changes at 518 nm at -40 °C with glycol and PMS-ascorbate, without (A) or with background light to give a membrane potential of about 100 (B) and 200 mV (C).

primary electron transfer is not significantly affected by the membrane potential.

Photosynthetic bacteria and bacterial chromatophores display absorbance shifts of carotenoids and bacteriochlorophyll that are in many respects similar to those in higher plants. However, analysis of the carotenoid changes at room temperature indicated that these were caused by a relatively large red shift (about 10 nm) of the absorption bands, too large to be explained by an electrochromic shift (22;cf. 23). Moreover, the shift appeared to be a step-wise phenomenon suggesting a conformational rather than an electrochromic effect.

Recent experiments in our laboratory (24) show that, like with chloroplasts, the pigment shifts in chromatophores of Rhodopseudomonas sphaeroides could be stimulated strongly (by about 10-fold) at -30 to -40 °C by addition of a donor-acceptor system. Fig. 4 shows the difference spectrum of the stimulated absorbance changes, with bandshifts of carotenoid and bacteriochlorophyll. However, comparison of the stimulated carotenoid spectrum with that caused by a single charge separation per reaction center, showed that the first spectrum was not only larger, but also shifted by about 1 nm (Fig. 4). Also, the kinetics of the absorbance changes near the isosbestic wavelengths were more complicated than those near the maxima and minima of the difference spectrum; e.g. at 516 nm they showed an increase followed by a decrease upon illumination as compared to a monotonous increase at 524 nm and could be described by a gradual

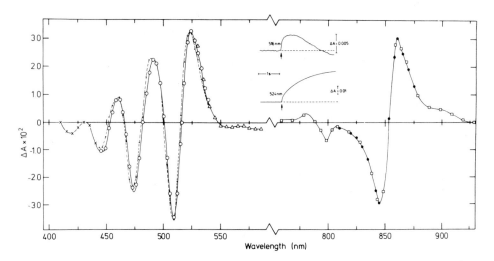

Fig. 4. Absorbance difference spectra of chromatophores upon illumination at -30 °C. Solid line: PMS and ascorbate added. The insert
shows the kinetics at 518 and 524 nm. Broken line (enlarged): no
additions.

bandshift by about 2 nm towards the red. This indicates that, at
least at low temperature, the bandshifts, like those in chloroplasts,
are electrochromic, and correspond to a shift by about 0.2 nm per
electron transferred per reaction center. If this is true, the bands
that shift in response to the electric field are located near 452,
481 and 515 nm, which is at longer wavelength than the bands in the
overall absorption spectrum; presumably they are also narrower.

    The investigation was supported by the Netherlands Foundation
for Chemical Research (SON), financed by the Netherlands Organization for the Advancement of Pure Research (ZWO). Thanks are due to
Mr. J.M. Glasbergen for aid with some of the measurements.

REFERENCES

1. Malkin, R. and Bearden, A.J. (1975) Biochim. Biophys. Acta 396,
   250-259.
2. Visser, J.W.M. (1975) Thesis, University of Leiden.
3. Floyd, R.A., Chance, B. and DeVault, D. (1971) Biochim. Biophys.
   Acta 226, 103-112.
4. Mathis, P. and Vermeglio, A. (1975) Biochim. Biophys. Acta 396,
   371-381.
5. Butler, W.L. (1973) Acc. Chem. Res. 6, 177-184.

6. Van Gorkom, H.J. (1976) Thesis, University of Leiden.
7. Mathis, P. and Vermeglio, A. (1974) Biochim. Biophys. Acta 368, 130-134.
8. Mathis, P., Michel-Villaz, M. and Vermeglio, A. (1974) Biochem. Biophys. Res. Comm. 56, 682-688.
9. Den Haan, G.Λ. (1976) Thesis, University of Leiden (in prep.).
10. Joliot, P. and Joliot, A. (1973) Biochim. Biophys. Acta 305, 302-316.
11. Vermeglio, A. and Mathis, P. (1973) Biochim. Biophys. Acta 314, 57-65.
12. Amesz, J., Pulles, M.P.J. and Velthuys, B.R. (1973) Biochim. Biophys. Acta 325, 472-482.
13. Kok, B., Forbush, B. and McGloin, M. (1970) Photochem. Photobiol. 11, 457-475.
14. Amesz, J. and de Grooth, B.G. (1975) Biochim. Biophys. Acta 376, 298-307.
15. Malkin, R. and Bearden, A.J. (1971) Proc. Natl. Acad. Sci. US 68, 16-19.
16. Visser, J.W.M., Rijgersberg, C.P. and Amesz, J. (1974) Biochim. Biophys. Acta 368, 235-246.
17. Ke, B., Sugahara, K., Shaw, E.R., Hansen, R.E., Hamilton, W.D. and Beinert, H. (1974) Biochim. Biophys. Acta 368, 401-408.
18. McIntosh, A.R. and Bolton, J.R. (1976) Biochim. Biophys. Acta 430, 555-559.
19. Evans, E.H., Cammack, R. and Evans, M.C.W. (1976) Biochem. Biophys. Res. Comm. 68, 1212-1218.
20. Witt, H.T. (1971) Q. Rev. Biophys. 4, 365-477.
21. Vermeglio, A. and Mathis, P. (1974) Biochim. Biophys. Acta 368, 9-17.
22. Amesz, J. and de Grooth, B.G. (1976) Biochim. Biophys. Acta 440, 301-313.
23. Amesz, J. and Vredenberg, W.J. (1966) in: Currents in Photosynthesis (Thomas, J.B. and Goedheer, J.C., eds.) pp. 75-81, Donker, Rotterdam.
24. Crofts, A.R., Prince, R.C., Holmes, N.G. and Crowther, D. (1975) in: Proc. 3rd Int. Congr. Photosynth. Res. (Avron, M., ed.) pp. 1131-1146, Elsevier, Amsterdam.
25. De Grooth, B.G. and Amesz, J., unpublished observations.

CONTROL OF THE ELECTRON TRANSFER BY THE TRANSMEMBRANE ELECTRIC FIELD AND STRUCTURE OF SYSTEM II CENTERS

P. Joliot and A. Joliot

Institut de Biologie Physico-Chimique

13, rue Pierre et Marie Curie-Paris(5e)France

I. INTRODUCTION

In applying the chemiosmotic theory to photosynthetic organisms, Mitchell (1) assumed that the donors and acceptors of both photoreactions are respectively located on the inner and outer surfaces of the thylakoid membrane. As a result, the photochemical charge separation should induce the formation of an electrostatic potential across this membrane. The existence of such an electric field has been clearly demonstrated by the extensive studies of Witt and coworkers. The transmembrane electric field shifts the absorption spectrum of the pigments included in the membrane, particularly the carotenoids (electrochromic effect). Junge and Witt (2) showed that the amplitude of the potential across the membrane is proportional to a 515nm absorption increase and a 480nm absorption decrease.

Furthermore, it is known that the membrane potential controls the electron transfer in Photosystem II centers. The study of both the formation of the membrane potential and of the control of the rate of the electron transfer by the electric field can provide new information on the structure of Photosystem II. The following scheme includes some recent results concerning System II centers (rate constants measured at 2°, A. Joliot, unpublished results).

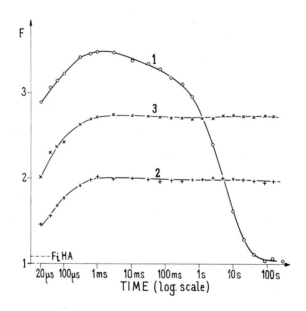

Chla$_{II}$ and Q are respectively the primary donor and acceptor. The primary photochemical step induces the
transfer of one electron from Chla$_{II}$ to Q :
Chl-Q $\rightarrow$ Chl$^+$-Q$^-$. The photochemical centers involving either the primary donor or acceptor in the oxidized form
quench the fluorescence (3,4). According to den Haan et
al. (5), the subsequent transfer of the positive charge
from the Chlorophyll to the secondary donor (here called Y) occurs in less than 1µs (t^1/$_2$ $\simeq$ 30ns according to

Fig.1.   Fluorescence yield in the presence of 20µM DCMU
as a function of dark time following a brief saturating
flash. Time scale is logarithmic. 1, Control. 2, 10mM
hydroxylamine incubated 20 minutes. 3, 100mM hydroxylamine incubated 20 minutes. F = 1 corresponds to the initial fluorescence level measured in the presence of DCMU
before the actinic flash. The dashed line F$_i$HA is the
same measurement after addition of hydroxylamine.

the data of Mauzerall (6). To explain the biphasic fluorescence rise (see Fig.1.Curve 1), A. Joliot (7) assumed the existence of at least one more step ($t^1/2 \simeq 70\mu s$
at 2°) from Y to Z, the final donor involved in water
splitting.

Hydroxylamine is known to inactivate the secondary
donors Y and Z by removing the associated Manganese (8).
According to den Haan et al. (5), Chl^+ is then reduced
by an auxiliary donor D, in a time much slower than the
Y to Chl transfer. The time constant for this reaction
is about 100µs at 2° (Fig.1. Curve 2). The low efficiency of Photoreaction II measured in the presence of hydroxylamine is due to the competition between the
charges stabilization $Chl^+-D \rightarrow Chl-D^+$ and the back reaction $Chl^+-Q^- \rightarrow Chl-Q$ which, according to Lavorel (9),
gives rise to a strong luminescence emission. In the
presence of 100mM hydroxylamine (Fig.1. Curve 3), a faster reduction of Chl^+ occurs, which increases the overall efficiency of Photoreaction II. Hydroxylamine is
probably able to reduce Chl^+ directly.

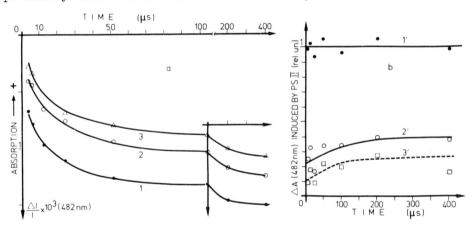

Fig.2a. 482nm absorption change induced by a single
short saturating flash(3µs). Chlorella. Curves 1 and 2:
dark adapted algae. 1,Control 2=2mM hydroxylamine, incubation > 15min. 3=2mM hydroxylamine+20µM DCMU, algae
have been preilluminated to irreversibly block System II
centers(26)and then dark adapted for more than 2min.

Fig.2b. 482nm absorption change induced by photoreaction
II. 1'=Control (1-3 Fig.2a) ; 2'=2mM hydroxylamine (2-3
Fig.2a); 3'=100mM hydroxylamine : the corresponding data
are not shown on Fig.2a. Absorption change for the control (Curve 1') is normalized to 1.

II. SYSTEM II PHOTOINDUCED ELECTRIC FIELD

Using a spectrophotometric method described previously (10,11), P. Joliot and A. Joliot (12) have measured the electric field generated by Photoreaction II
at 515nm and 482nm before and after action of hydroxylamine. Fig.2a shows the 482nm absorption decrease induced by a 3μs saturating flash. At this wawelength,
the absorption change linked to the carotenoid triplet
effect can be neglected. According to Cox and Delosme
(13), one observes a biphasic decrease either when both
photosystems are active (Curve 1) or only when Photosystem I is functional, Photoreaction II having been blocked by hydroxylamine + DCMU. The biphasic character of
the absorption decrease at 482nm induced by Photoreaction I is discussed in refs. 11 and 13.

We wish to consider here only the absorption
change induced by Photoreaction II (Fig.2b), which can
be calculated by subtracting curve 3 from curve 1 in
Fig.2a. This absorption change (Curve 1', Fig.2b) appears in less than 3μs (time resolution of our method)
and no significant change is observed between 3μs and
400μs. Thus, the electric field generated by Photoreaction II reaches a steady-state value in less than 3μs.
In the presence of 2mM hydroxylamine alone (Curve 2',
Fig.2b), one observes a smaller amplitude of the System
II induced field. In a time range (<10μs) much shorter
than the half-time of the back reaction ($\simeq$100μs(9)),
the electric field is only about 25% of the control :
we thus conclude that the electric field associated with
the primary Photoreaction II $Chl-Q \rightarrow Chl^{+}-Q^{-}$ is about
four times smaller than that observed when the positive
charge is stabilized on the secondary donors. In the
presence of 100mM hydroxylamine, the amplitude of the
Photosystem II induced field at times shorter than 10μs
is only about 10% of the control. As shown in Fig.3,
the absorption change measured at 515nm presents similar properties.

In Fig.4 are compared the amplitude of the fluorescence and of the 515nm absorption increase measured
209μs after a saturating flash i.e. after completion
of charge stabilization. In the presence of 100mM hydroxylamine, the efficiency of Photoreaction II is close to
its maximum value (78%) while the photoinduced electric
field is only 8% of the control. This value varies from
one algal culture to another but remains always smaller
than 20% of the control. Possible interpretations of
these results will be discussed below.

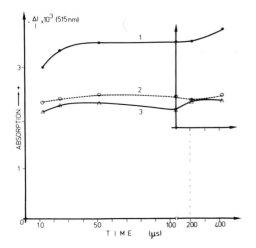
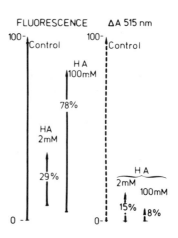

Fig.3.   515nm absorption change induced by a single short
saturating flash. Curves 1,2,3. Same conditions as in
Fig.2a. To minimize the contribution of the carotenoid
triplet, measurements have been performed at times longer than 12µs.

Fig.4.   Fluorescence and 515nm absorption increase induced by photoreaction II, measured 209µs after a saturating flash. Incubation time = 2mM hydroxylamine (HA) :
15min; 100mM HA 10min. Fluorescence is measured on dark
adapted algae in the presence of 20µM DCMU.

III. CONTROL OF THE RATE OF THE ELECTRON TRANSFER BY
      THE ELECTRIC FIELD

      A first example is the modulation of the rate of
the System II back reaction by the electric field. Several authors (14,15,16) have established that a membrane
potential generated by ion gradients increases the luminescence intensity, which depends on the rate of the
back reaction occuring in Photosystem II centers. Using
Chlorella in the presence of DCMU, we studied (17) the
effect of the photoinduced electric field on the rate
of the back reaction $Z^+YChlQ^- \rightarrow ZYChlQ$. We showed that :
1) the electric field generated by Photoreaction I modulates the back reaction occuring in System II centers ;
thus, the electric field appears to be delocalized along
the thylakoid membrane, a conclusion which is in agreement with Mitchell's theory ;
2) The Photosystem II centers are not equally sensitive
to the field : the more sensitive the centers to the

field, the faster the back reaction. This unequal sensitivity of the centers to the field probably reflects a variability in their structural properties.

The field effect can be interpreted in two ways which are not mutually exclusive :

a) a change in the equilibrium constants on the donor side which increases the concentration of Chl^+ ;

b) a direct effect on the rate of the charge recombination between Chl^+ and Q^-.

A second example of the control of the rate of electron transfer by the electric field is given by the experiments of Diner and Joliot (18) : these authors showed that after saturating flashes, there exists a component of the fluorescence relaxation correlated with the decay of the 515nm absorption change. The electric field induces a non-photochemically active form of the centers which is not due to a reduction of Q. One of the main conclusions of this work is that Photosystem II centers have widely different sensitivity to the field, which indicates, as above, a structural heterogeneity among the centers.

By studying at low temperature, the rate of the back reaction $Z^+YChlQ^- \rightarrow ZYChlQ$, A. Joliot (19) also demonstrated a heterogeneity of System II centers. In the presence of DCMU, the rate of the back reaction is markedly temperature dependent down to $-40°$. Below $-40°$, this rate becomes practically temperature independent, suggesting that a tunneling process is involved ; in these conditions, the half times of the back reaction are spread over a very large time range (100ms to hours) If each center is presumed to undergo a first order back reaction, then the rate constant for different centers differ by several orders of magnitude. Mathis and Vermeglio (20) studied spectrophotometrically the back reaction $Chl^+Q^- \rightarrow Chl\ Q$ at $-196°$ and showed that there kinetics were close to an exponential function with a time constant of about 3ms. To take into account these different results, we are led to the conclusion that the structure of the Chl-Q complex is relatively constant ; the heterogeneity of the centers with respect to the rate of the back reaction could arise from a variability in the distance between the primary and the secondary donors.

IV. MODELS

Experiments described in Figs. 2 and 3 show that after destruction of the secondary donors by hydroxylamine, the field induced by Photoreaction II is only a small fraction of that measured in the control. This result implies that under these conditions both Chl and Q are located close to the outer face of the membrane. By completely different experiments, Babcock and Sauer (21) came to the same conclusion. Blankenship and Sauer (22) demonstrated that after Tris-treatment which is known to release the Mn associated with the secondary donors (23), Mn^{++} remains trapped within the internal phase of the thylakoid. This experiment shows that the secondary donors are located on the inner face of the membrane which is highly impermeable to Mn^{++} ions. Nevertheless, under these conditions, externaly added Mn^{++} is able to act as a very efficient electron donor to Chl^+ (21,24) which thus must be located close to the outer surface of the membrane. While both our results and those of Babcock and Sauer lead to the conclusion that after destruction of the secondary donors the chlorophyll is located close to the outer surface, no direct information is available concerning its position under normal conditions. The small increase of the Photosystem II induced signal observed in the first 200µs in the presence of hydroxylamine (Fig.2b, Curves 2' and 3') suggests that the auxiliary donor D is closer to the inner face of the membrane than Chl ; this agrees with Babcock and Sauer's conclusion that one of the donor sites is only accessible to lipophylic donors.

The following possible models for System II centers take into account the different experimental data (Fig.5)
①) In a first model, the photoactive Chl is always located close to the outer surface of the thylakoid. In this model, the formation of the membrane potential should follow the non photochemical step $Chl^+Y \rightarrow Y^+Chl$, which is known to be very fast (likely < 30ns from Mauzerall's experiments (6) and difficult to observe by conventional spectrophotometric methods. The fast rise of the 515nm absorption change (total rise < 80ns) observed by Wolff et al. (25) is not inconsistent with this model.
②) In a second model, Chl is normally located on the inner face of the membrane and, after inactivation of the secondary donors, migrates towards the outer face.
③) A third model, intermediate between the two preceeding ones, takes into account the heterogeneity of the

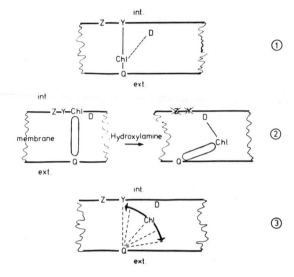

Fig.5.   Possible models for Photosystem II centers.

System II centers which we noted in Section III. From measurement of the rate of the back reaction, we concluded that the distance between Chl and Q is constant while the distance between Chl and the secondary donor Y is variable. The orientation of the Chl-Q complex might vary from one center to another. It is possible that the conformation in which Chl is close to the outer face of the thylakoid membrane (as in model 2) is favored when the secondary donors are inactivated.

REFERENCES

1) Mitchell,P. (1966) Biol. Rev. 445.
2) Junge,W. and Witt,H.T. (1968) Z.Naturforsch, 23b, 244.
3) Duysens, L.N.M. and Sweers,H.E. in Studies on Micro-algae and Photosynthetic Bacteria. Univ. of Tokyo press. (1963) p.353.
4) Butler,W.L., Visser,J.W.M. and Simons,H.L. (1973) Biochim. Biophys. Acta. 292, 140.
5) Den Haan,G.A., Duysens,L.N.M. and Egberts,D.J.N. (1974) Biochim. Biophys. Acta. 368, 409.
6) Mauzerall,D. (1972) Proc. Natl. Acad. Sci. US. 69, 1358.
7) Joliot,A. (1975) in Proc. 3rd. Cong. Photosynthesis. Rehovot, 1974) (M.Avron ed.) p. 315. Elsevier Sci. Publ. Co.

8) Cheniae, G.M. and Martin,I.F. (1970) Biochim. Biophys. Acta. 197, 219.
9) Lavorel,J. (1973) Biochim. Biophys. Acta. 325, 213.
10) Joliot,P. and Delosme,R. (1974) Biochim. Biophys.
    Acta. 357, 267.
11) Joliot,P., Delosme,R. and Joliot,A. (1976) Biochim.
    Biophys. Acta. In press.
12) Joliot,P. and Joliot,A. (1976) C.R. Acad. Sci.
    Paris. In press.
13) Cox,R. and Delosme,R. (1976) C.R. Acad. Sci. Paris.
    282, 775.
14) Miles,C.D. and Jagendorf,A.T. (1969) Arch. Biochem.
    Biophys. 129, 711.
15) Barber,J. and Kraan,G.P.B. (1970) Biochim. Biophys.
    Acta. 197, 49.
16) Wraight,C.A. and Crofts,A.R. (1971) Eur. J. Biochem.
    19, 386.
17) Joliot,P. and Joliot,A. (1975) in Proc. 3rd. Cong.
    Photosynthesis. Rehovot, 1974 (M. Avron ed.) p. 25.
    Elsevier, Sci. Publ. Co.
18) Diner,B. and Joliot,P. (1976) Biochim. Biophys.
    Acta. 423, 479.
19) Joliot,A. (1974) Biochim. Biophys. Acta. 357, 439.
20) Mathis,P. and Vermeglio,A. (1975) Biochim. Biophys.
    Acta. 369, 371.
21) Babcock,G.T. and Sauer,K. (1975) Biochim. Biophys.
    Acta. 396, 48.
22) Blankenship,R. and Sauer,K. (1974) Biochim. Biophys.
    Acta. 357, 257.
23) Cheniae,G.M. and Martin,I.F. (1966) in Energy Conver-
    sion by the Photosynthetic apparatus. Brookhaven.
    Symp. in Biology. 19, 406.
24) Lozier,R., Baginsky,M. and Butler,W.L. (1971) Photo-
    chem. Photobiol. 14, 323.
25) Wolff,C., Büchwald,H.E., Rüppel,H., Witt,K. and
    Witt,H.T. (1969) Z. Naturforsch. 24b, 1038.
26) Bennoun,P. (1970) Biochim. Biophys. Acta. 216, 357.

MODULATING EFFECTS ON THE DELAYED LUMINESCENCE

FROM PHOTOSYSTEM II OF PHOTOSYNTHESIS

Shmuel Malkin

Biochemistry Department, The Weizmann Institute of

Science, Rehovot, Israel

INTRODUCTION

A glance at the literature will reveal quite a few papers which report that delayed luminescence from isolated chloroplasts varies in a parallel way to the variations in the fluorescence yield (1-3). On the other hand several papers (4-8) report variation (an increase) of the delayed luminescence with the development of the "high energy state" of photophosphorylation (pH gradient and membrane potential). The impression from these (4-6) reports is however the absence of a clear-cut correlation to the fluorescence. Most of the above reports give data on delayed luminescence measurement at a particular point of time of its decay (i.e. at around 1 msec.). It is important to be able to scan the dependence of delayed luminescence on the fluorescence yield and other parameters over a large proportion of its decay curve, and to establish the conditions for simple relations between the delayed luminescence, fluorescence yield and other parameters. This is one of the purposes of this report.

Crofts et al. (9) summarized a simple-minded theoretical treatment in which the delayed luminescence depends on several factors: The concentration of the luminescence precursors, the emission yield (monitored by the prompt fluorescence yield), and the membrane electric potential and pH gradients. The dependence on the electric potential gradient was particularly proved by salt injection experiments which induce diffusion potential and enhance the delayed luminescence (7). According to the theoretical treatment all the other factors besides concentrations influence either the rate parameters (viz. energy of activation) of the reaction which generates the chlorophyll excited state, or the yield of emission from

139

the excited state. One may refer to them as <u>modulating</u> factors.

   The fluorescence yield changes indicate the reduction of Q
($Q \to Q^-$), the primary electron acceptor of photosystem II (10).
According to Lavorel (11) Q^- is one of the precursors which enter
into a recombination reaction with the primary electron donor to
regenerate the chlorophyll excited state. This assumption will be
considered as an open question for discussion in this report. All
other factors being constant, the dependence of the delayed luminescence on the fluorescence yield should be of the following form:

$$(1) \quad L = \phi_F \cdot (\phi_F - \phi_{FO}) \cdot K$$

where $\phi_{\bar{F}}$ is the fluorescence yield (at the same time where L is
measured) ϕ_{FO} is the fluorescence yield of a dark-adapted sample
(constant fluorescence) $\phi_F - \phi_{FO}$ is the momentary value of the variable fluorescence. K contains all the other influencing parameters.
ϕ_F reflects the emission yield factor and $\phi_F - \phi_{FO}$ the extent of Q
reduction. An unknown contribution to ϕ_F and ϕ_{FO} must be substracted, which is a non-relevant background fluorescence ("dead" fluorescence).

   According to eqn. 1, L must be related quadratically to ϕ_F.
This relation should be particularly strongly expressed in the
experiment, since in many samples and for a considerable range of
time $\phi_{FO} << \phi_F$ and hence $L \simeq K\phi_F^2$.

   In this paper however we repeat the basic observation of
Wraight (2) that L is linearly proportional to ϕ_F. This must be
interpreted in either one of 2 ways: (i) <u>that Q^- is not a lumines-</u>
<u>cence precursor</u>, and that the delayed luminescence changes are
caused only by changes in the emission yield, thus influenced only
indirectly by the $Q \to Q^-$ reaction. The factor $\phi_F - \phi_{FO}$ in eqn (1)
must then be dropped; (ii) that Q^- is a luminescence precursor but
the <u>macroscopic</u> changes of the emission yield are not expressed in
the delayed luminescence intensity. This is because after the
recombination reaction the delayed luminescence is generated in a
particular photosynthetic unit having a <u>microscopic</u> fluorescence
yield corresponding to ϕ_0 (open unit). A corollary of this result
is that photosynthetic units are independent with respect of energy-
transfer between them (2). To distinguish between the two possibilities, conditions which change ϕ_F without affecting the extent of
Q must be found. A parallel change of the luminescence will then
favor assumption (i).

EXPERIMENTAL

Apparatus: The basic apparatus was a conventional phosphoroscope equipped with two detectors, one for measuring luminescence
during the off period of the exciting light and the second for
measuring prompt fluorescence during the on period of the exciting
light. The two detectors were connected each separately to two
oscilloscopes. Other details of the apparatus are shown in Fig. 1.

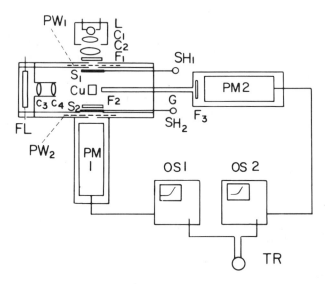

Fig 1. Sketch of the apparatus for measuring delayed luminescence
and fluorescence:
L light source; C_1-C_4 condensers; F_1-F_3 filters; PW_1 and PW_2 phosphoroscope wheels; S_1 and S_2 sliding shutters; SH_1 and SH_2 handles
to manipulate the shutters; CU cuvette; FL flash lamp as an additional light source for preillumination; PM_1 and PM_2 photomultipliers for delayed luminescence and prompt fluorescence, respectively;
OS 1 and OS 2 - oscilloscopes, to monitor PM 1 and PM 2, respectively; TR - common triggering source for OS 1 and OS 2; G light
guide from CU to PM 2.

Materials: Whole chloroplasts were prepared from spinach, as
described previously (12), following essentially the method of
Stokes and Walker (13). In most experiments the chloroplasts were
osmotically shocked by diluting them into water followed by the
addition of a reaction mixture. No essential difference was found
between the behaviour of whole or broken chloroplasts. Very dilute
suspensions were used (∿1-2 μgr. chlorophyll per ml.).

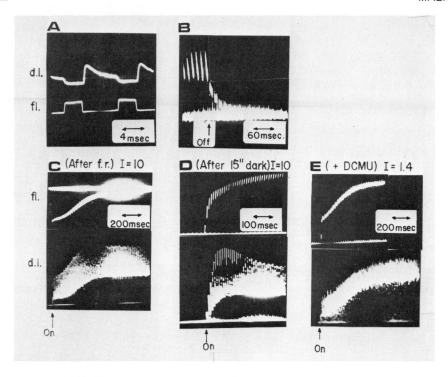

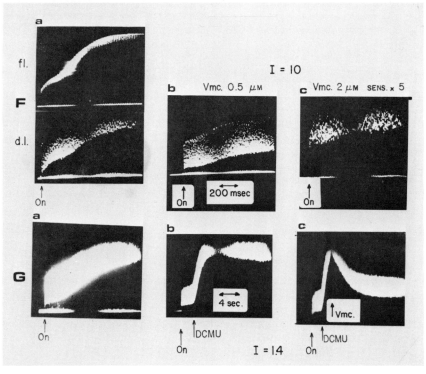

Fig. 2. Delayed luminescence (d.l.) and fluorescence (fl.) patterns
for chloroplasts in 0.1M KCl.

A.   Parallel measurement of prompt and delayed luminescence. Bottom
     trace-fluorescence (signifying also the on period of the phospho-
     roscope). Upper trace - delayed luminescence. Actinic light -
     480-620nm filter (10n.Einst./cm^{2}sec average incident intensity);
     steady-state conditions.

B.   Isolating the slow component of delayed luminescence dacay.
     The fast component is recorded by the (almost) vertical lines.
     At the arrow the shutter was closed and the decay of the slow
     component is isolated. Actinic light - 480-620 nm (2.5n.Einst/
     cm^{2}sec); steady-state conditions.

C.   Induction pattern for the fluorescence (top) and the delayed
     luminescence (bottom) - after far-red (725nm-1.8n Einst./cm^{2}sec
     for 1 min.). Actinic light 480-620 nm (10 n Einst./cm^{2}sec).

D.   The same as C, but after a dark period (15") elapsing from a
     previous exposure to actinic light.

E.   The same as C, but in presence of 5x10^{-6}M DCMU. Actinic light
     554nm (1.4 n Einst./cm^{2}sec).

F.   The effect of valinomycine on the induction curves. Conditions
     as in C. (a) Control. (b) With valinomycine (0.5µM). (c) With
     valinomycine (2µM). The fluorescence curve shown in the top
     of (a) is the same for (b) and (c).

G.   (a-c) The effect of addition of valinomycine. Conditions as
     in E.

RESULTS AND DISCUSSION

General Characteristics of the Delayed Luminescence Trace

Fig. 2A shows a selection of two phosphoroscope cycles with
fluorescence and delayed luminescence recorded on the same oscilloscope (a dual beam mode). It shows the time relationships of the
on period (at which fluorescence was excited and recorded), the off
period and the time window (actually during most of the off period)
at which delayed luminescence was recorded. This figure was taken
at the steady state achieved by the actinic light. However, the
essential form of the delayed luminescence trace is not much different if measured at other regions of time elapsing from the start
of the exposure. It is composed of relatively fast decaying com-
ponent ($\tau_{1/2} \sim$ 2 msec) followed by a slow component ($\tau_{1/2} \sim$ 50 msec).
The slow component cannot be defined with confidence from the phosphoroscope traces, since the measurement period was terminated too
early. However, if the actinic light was terminated by closing an
additional shutter the slow component was isolated (Fig. 2B).

This profile of the delayed luminescence was noted before by
Bertch et al for algae but not in isolated chloroplast where the
fast phase was almost absent (14). Itoh and Murata (15) observed
the same two components in isolated chloroplasts and also noted
that the two components behaved in a different way as a function of
the exposure time.

To observe slow phenomena and at the same time keep track of the
delayed luminescence during its entire decay time the intensity of
the oscilloscope ray was adjusted so that only the traces from
the delayed luminescence proper remained on the screen while the
on and off (almost vertical) traces were not. One thus obtains an
envelope of many traces (which are indistinguishable at very long
time scale) of which the bottom will be taken as representing the
slow component. This is an approximation of course, but not a bad
one; by defining the slow component by use of a shutter the results
closely extrapolated to the bottom of the envelope, with a difference which was almost constant during the induction period.

The fast delayed luminescence component behaved in a completely
different way than the slow component. The results showed that one
should focus attention to the slow component, since this component
shows distinct and clear relation to the fluorescence.

Induction Pattern

The experiment in which the fluorescence yield changes considerably is the transition from dark to light of initially dark
adapted isolated chloroplast (or after far-red preillumination.
The fluorescence yield changes from ϕ_{F_0} to ϕ_{Fmax} in a characteristic

way (fluorescence induction - cf. ref. 10). These are the best
conditions to observe the relation between delayed luminescence
and the fluorescence yield.

Fig. 2C shows that in such experiment the bottom and the top
of the delayed luminescence envelope increase almost in parallel.
The fast component alone (top envelope minus bottom envelope) increases very little (by~ 20%) compared to the increase in the bottom envelope (by ~600%). This experiment was performed after far-
red (photosystem I light) preillumination.

In this case there is a very good correlation of the luminescence and the fluorescence yield changes - both delayed luminescence and the fluorescence show characteristic inflection at the
beginning of the induction and increase in the same fashion.

If instead of far-red preillumination a very long dark period
was interposed (>~2 min.) from a previous actinic exposure to the
start of the irradiation(and the measurement) the results were
essentially the same. If however only a short dark period was
interposed (e.g. after 15" only, cf. fig. 2D), the delayed luminescence induction became more complicated and the correlation to the
fluorescence was lost or became much less obvious: The bottom of
the delayed luminescence envelope increased faster than the fluorescence. The top initially increased within the same time
range as the bottom but then decayed to a lower steady-state value.
This characteristic type of induction was noted by Itoh and Murata
(15). The decrease of the top is within the same time range as the
increase of the fluorescence and deserves further study (cf. also
Ref.3). When DCMU was added the fluorescence and luminescence were
correlated even if a short dark period (e.g. 10-20") was interposed.
(Fig. 2E).

One may conclude that under proper conditions (far-red preillumination, or sufficient dark adaptation) the slow component of
the delayed luminescence is quite perfectly correlated with the
fluorescence yield. It was found also that the light intensity
should not be too strong to observe such correlation ($<$~10 n
Einsteins/cm^{2}sec for 550 nm incident light). At strong intensities
the correlation was lost. However only a few experiments with high
light intensity were done to be reported in any detail.

The Effect of Uncouplers

Under experimental conditions where no electron transport cofactors are added one does not expect that the "high-energy state"
of photophosphorylation will be developed to any great extent,
especially so in the presence of DCMU. It was therefore a surprise
that the following tested compounds inhibited the extent of the
(slow component) delayed luminescence rise. These compounds were:
gramicidine, nigericine and valinomycine (in the μM range). Figs.2F
a-c show the effect of increasing valinomycine concentration. The

normalized shapes of the luminescence induction are roughly not
changed (the inflection and the subsequent rise preceding the inflection disappeared in the presence of the antibiotics. The fluorescence induction itself was not influenced at all. Figs. 2G a-c
demonstrate another kind of experiment to show the effect of valinomycine in the presence of DCMU. Upon addition of valinomycine
the delayed luminescence decreased considerably. Two kinds of
control experiments were done (Figs. 2G a,b), one shows the normal
induction pattern without DCMU and the second demonstrates the
effect of DCMU in raising quickly the level of the delayed luminescence. It also shows that the steady-state level with DCMU is
reasonably constant as a control for Fig.2G c, but it has a tendency
to decrease slowly. Indeed after very long periods ($\sim$1') a low
steady-state of delayed luminescence is achieved with DCMU. This
must be considered as a secondary effect.*

Formulation of the Results

The correlation between the variable part of fluorescence
during its induction and the variable part of the slow component of
the delayed luminescence is very nearly linear (other unknown
parameters might distort the results somewhat). This is shown by
Fig.3. The effect of valinomycine is to decrease the amplitude of
change but only relatively little the shape of the luminescence
induction. Hence one can approximately write:

(2) $L = A + B \, (\phi_F - \phi_{Fo})$

A and B are constants during any single experiment, ϕ_F and L are
functions of the time t. B is the amplitude which is influence by
the antibiotics.

The effect of valinomycine depends on the presence of K^+ in
the reaction medium, as fig. 4 shows. Only B is affected, but
not A.

The Effect of the Reaction Medium

The effect of cations on the fluorescence parameters, especially on ϕ_{Fmax} is well documented (16-20). The effect is mainly on
the rate constants for transfer and deactivation of the electronic

---

* In view of this experiment the sometimes conflicting reports
about DCMU action might somewhat be understood. It is very probable that its primary effect as an electron-transfer inhibitor does
not affect directly the delayed luminescence. (The increase of
delayed luminescence in Fig. 2G b is caused by the increase in ϕ_F).
It could affect delayed luminescence at conditions where the
"high-energy state" is developed, by inhibiting electron transfer
and hence also abolishing the "high energy state".

excitation, and not on the level of Q reduction. This effect is
used to introduce changes in ϕ_F independently of the changes in Q^-.

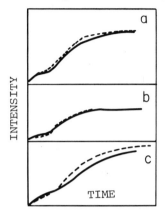

Fig. 3. Direct comparison of the (slow component) delayed luminescence and fluorescence, the extent of which were normalized to the
same scale. (____) delayed luminescence (----) fluorescence.
   a. whole chloroplasts, suspended in the grinding medium, actinic
      light 480-620 nm 10 nEins./cm^{2}sec.
   b. Osmotically shocked chloroplasts, suspended in 0.1M KCl,
      actinic light 480-620 nm 5 nEins./cm^{2}sec.
   c. The same experiment of fig. 2Fa.
In all the experiments far-red preillumination was given (cf.fig.2C).

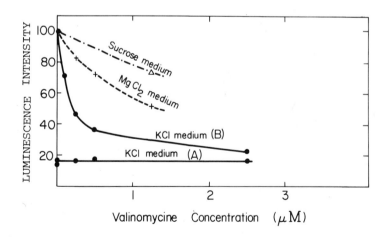

Fig. 4. Concentration dependence of the valinomycin effect in
different media.
KCl medium = 0.1M KCl, 5mM Tricine pH 7.6. MgCl$_2$ medium = 2.5 mM
MgCl$_2$ 25 mM Tricine pH 7.6, 0.1M Sucrose Sucrose medium = 0.25 M
Sucrose only.   The lower KCl medium curve is for the initial part
of the delayed luminescence A, the others are for the steady-state
extent of the variable part (B).

The results are summarized in Table I. One can see that changes in ϕ_{Fmax} are paralleled by comparable changes in the total extent of L. This is what is expected if it is the fluorescence yield proper which affects the luminescence, but not the level of Q^-. In the last case the changes in the medium should not produce any changes in the luminescence, except by chance.

Table I: Effect of the Reaction Medium on the Fluorescence (Fl.) and the "Slow" Component of the Delayed Luminescence (d.l.)

Medium	d.l. (mv)	Ratios	Fl. Changes (mv)	Variable Fl. Ratios
Sucrose 0.2 M	25	2.1	160 → 590	1.95
+ KCl 20 mM	12	1	170 → 390	1
+ MgCl$_2$ 6 mM	19	1.6	170 → 520	1.6

The fact that delayed luminescence is modulated by the variable fluorescence $(\phi_F - \phi_{FO})$ shows that most of the constant fluorescence ϕ_{FO} is "dead" and not contributed from active photosystem II units. This is in accord with an old conclusion by Clayton (1). The true fluorescence yield of photosystem II is mostly the variable fluorescence.

Dark Resting Potential

According to the results expressed by eqn. 2 B is constant during one experiment but is modified by valinomycine. It is conceivable that valinomycine, by allowing a current of cations, (K^+), across the thylakoid membrane, changes the value of a given electric potential. Hence a simple minded interpretation would be that B is a function of a membrane potential. Since it is not a function of the time it shows that no extra potential is developed in the light but that this potential is created by dark processes.

Assuming that this potential ΔV affects the delayed luminescence in the same way as suggested by Crofts et al (9), namely:

$$(3) \quad L \propto \exp \frac{F \Delta V}{RT}$$

one can roughly estimate a lower limit for this potential from the data in fig. 2F a-c, assuming that in fig. 2Fc the potential is

completely abolished. The value obtained is close to 100 mV. This
value should not be taken too seriously in view of the somewhat
arbitrary assumptions used in deriving eqn.(3) (e.g. that the re-
combining species for the delayed luminescence are situated on the
two opposite faces of the membrane), but is very suggestive.

The effect of a dark resting potential which is abolished by
ionophores can explain part of the results obtained before (4) with
regard to the delayed luminescence induction at different conditions,
which was attributed solely to a development of light induced membrane potential and pH gradient. These assumptions must now be
rechecked, since they do not follow unumbigously from the results.

The effect of valinomycine could be explained alternatively in
terms of the model in which the extent of Q reduction influences
directly the delayed luminescence - a model which was unfavored
above. According to this explanation the potential is due to the
local polarisation of the reaction centers by the charge separation
photoreaction. (Following the model of Witt et al (21)). However,
this potential should collapse by ion movement and more rapidly so
in the presence of the ionophores. This is obviously not the case.

The Component A of Delayed Luminescence

The ratio A/B is of the order 1/60 - 1/20. A is not influenced
by the ionophores, hence it behaves as an additional delayed luminescence of different mechanism, superimposed on the delayed luminescence expressed by B. Nevertheless one could think that A represents the level of delayed luminescence for a zero membrane potential. In this case, however, it should depend on the fluorescence
yield.
However, since A is very small, no absolute experimental elimination of the possibility that it represents an instrumental
artifact was yet made (e.g. - delayed luminescence from the filter,
etc.). This still awaits an answer.

The Dependence of A and B on the Light Intensity

The verification of the relation of delayed luminescence and
$\phi_F - \phi_{Fo}$ was tried also with respect to the steady-state values at
different light intensities (Fig. 5). Here however the correlation
collapsed since the delayed light intensities fell down at low light
intensities long before the fall down of the fluorescence yield.
This could be interpreted in terms of changes in A and B as a function of the light intensity. This is quite understandable since
these factors should also contain, for example, the concentrations
of the precursors which are functions of the intensity. Interestingly B tends toward saturation at the higher intensities while
A does not.

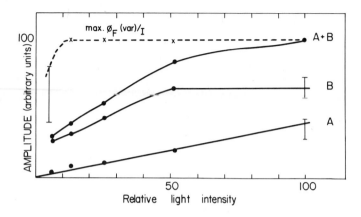

Fig. 5. The effect of light intensity on the parameters of delayed
luminescence and on the fluorescence yield.
The delayed luminescence is expressed as $A+B \cdot (\phi_{Fmax} - \phi_{Fo})$. Chloroplasts
suspended directly in 0.1M KCl. The fluorescence yield remained
constant down to about 13% of the maximal light intensity and decreased at lower intensities; (due to an error in the measurement
it is impossible to ascertain the extent of this decrease from the
experimental records for the point of 6% intensity). 100% intensity
$\sim$ 10n Eins./cm^{2}sec.

The Relation of the "Fast" and "Slow" Delayed Luminescence
Components

In a first glance the "fast" and the "slow" delayed luminescence
seem to represent different processes superimposed on the top of each
other. This is corroborated by the different behavior of these
components during the induction, and especially the approximately
constant value of the "fast" phase during the induction. One can
however devise kinetic schemes by which these two phases will represent the same mechanism for generating delayed luminescence with
two consecutive steps. However, the apparent lack of any control
of ϕ_F on the "fast" phase is not easily explained.

CONCLUSIONS

The "slow" phase ($\sim$ 50 m.sec.) component of delayed luminescence
varies in a characteristic way during irradiation, suggesting the
control of the fluorescence yield, but not of Q^-. The effect of
ionophores suggests the existence of a resting membrane potential
which affects steady-state value of the delayed luminescence.

REFERENCES

1. Clayton, R.K. (1969) Biophys. J. 9. 61-76
2. Wraight, C.A. (1972) Biochim. Biophys. Acta 283. 247-258
3. Mar, T. Brenber, J. and Roy, G. (1975) Biochim. Biophys. Acta
   376. 345-353
4. Wraight, C.A. and Crofts, A.R. (1971) Eur. J. Biochemistry 19.
   386-397
5. Evans, E.H. and Crofts, A.R. (1973) Biochim. Biophys. Acta 292
   130-139
6. Itoh, S., Murata, N. and Takamiya (1971) Biochim.Biophys.Acta
   245. 121-128
7. Barber, J. (1972) Biochim. Biophys. Acta 275. 105-116
8. Newman, J., Barber,J. and Gregory, P. (1973) Plant Physiol. 51
   1069-1073
9. Crofts, A.R., Wraight, C.A. and Fleischman, D.E. (1971) FEBS
   Lett. 15. 89-100
10. Malkin, S. and Kok, B . (1966) Biochim. Biophys. Acta 126.
    413-432
11. Lavorel, J. (1975) Luminescence in Bioenergetics of Photosyn-
    thesis (Govindjee, Ed.) pp. 223-317, Academic Press
12. Barber, J., Telfer, A. and Nicolson, J. (1974) Biochim. Biophys.
    Acta 357. 161-165
13. Stokes, D.M. and Walker, D.A. (1971) Plant Physiol. 48. 163-165
14. Bertch, W.F., West, J. and Hill, R. (1969) Biochim. Biophys.
    Acta 172. 525-538
15. Itoh, S. and Murata, N. (1975) in: Proc. 3rd Int. Cong. Photosyn.
    (Avron M., Ed.) Vol.I pp.115-126, Elsevier, Amsterdam
16. Murata, N. (1971) Biochim. Biophys. Acta 226. 422-432
17. Homann, P. (1969) Plant Physiol. 44. 932-936
18. Barber, J., Telfer, A., Mills, J. and Nicolson, J. (1975) in
    Proc. 3rd Int.Cong. Photosyn. (Avron M., Ed.) Vol.I, pp.53-63,
    Elsevier, Amsterdam
19. Wydrzynski,T., Gross, E.L. and Gouindjee (1975) Biochim. Biophys.
    Acta 376. 151-161
20. Malkin, S. and Siderer, Y. (1974) Biochim. Biophys. Acta 368
    422-431
21. Witt, H.T. (1975) in Excited States of Biological Molecules
    (Birks, J.B., Ed.), pp.245-261, Willey-Interscience, London
    & New York

- - -

Acknowledgement:
A significant portion of this work was carried out in the Botany
Department of the Imperial College, London. I wish to thank
Dr. J. Barber for his interest and help in this work, and Mrs. J.
Nicolson for the technical assistance.
This work was supported by an EMBO fellowship.

IN VITRO PHOTOSYNTHESIS

DAVID ALAN WALKER

THE UNIVERSITY OF SHEFFIELD
DEPARTMENT OF BOTANY
SHEFFIELD S10 2TN (UNITED KINGDOM)

Light-dependent O_2 evolution by isolated chloroplasts was first observed by Haberlandt in 1888 (1) and confirmed by Ewart in 1896 (2) but for all practical purposes the quantified study of in vitro photosynthesis started with the experiments of Robert Hill (3, 4). Following his development of a spectroscopic method for measuring O_2 (5), he prepared chloroplasts by crushing leaves in a solution containing 0.29 M sucrose and 0.033 M orthophosphate (3, 4, 6). He illuminated these with myoglobin in the presence of an aqueous extract derived from an acetone leaf powder. In his own words "It was a very thrilling moment when I saw the spectrum of oxymyoglobin. Then later on a sad disappointment: the presence or absence of CO_2 made no difference. This was really lucky, however, because if CO_2 had 'worked' I might well have got no further" (7). Because his chloroplasts did not "work" with CO_2, Hill pursued the contribution made to his reaction by the acetone powder and showed that the oxygen produced corresponded to the reduction of a hydrogen or electron acceptor. In so doing he laid the foundations on which much of our present understanding of photosynthetic electron transport rests. It is not the aim of this article to attempt to evaluate the contribution made to photosynthesis by the army of workers who have followed Hill's example and started their laboratory day by grinding leaves in a variety of media. All it seeks to do is to take a very circumscribed and personal view of the emergence of what, for want of a better term, can be described as "in vitro photosynthesis" and to indicate the relevance of studies in this field to photosynthesis as a whole. "In vitro photosynthesis" implied the utilisation of isolated chloroplasts (or systems derived from chloroplasts) which will "work" with CO_2 in the sense that they can utilise this compound as an oxidant. It will be suggested

that, except in a somewhat limited way, chloroplasts cannot
support the entire photosynthetic process unaided.

Implicit in the classic overall equation, which emerged from
the work of de Saussure and others (8-10), was the notion that
photosynthesis involved photodecomposition of CO_2 i.e.

$$CO_2 \;+\; H_2O \longrightarrow [CH_2O] \;+\; O_2$$

Although this possibility attracted serious attention for many
years (see e.g. 11) Hill's reaction appeared to confirm Van Niels
proposal (12, 13) that light energy is used to bring about the
photolysis of water.   The equation then became

$$CO_2 \;+\; 2H_2\underline{O} \longrightarrow [CH_2\underline{O}] \;+\; H_2O \;+\; \underline{O}_2$$

However, Hill's work also strengthened the view, see e.g. Kny (14)
that the chloroplast was "not perhaps a complete photosynthetic
system in itself" (Hill, 4).   The first attempts to demonstrate
$^{14}CO_2$ fixation by isolated chloroplasts (15) pointed to the same
conclusion.

In 1954 the whole field of in vitro photosynthesis was
revolutionised by the major contributions of Arnon & Whatley and
their colleagues in Berkeley, California (see e.g. 16-22).   It
had already been shown that chloroplasts could reduce NADP (23-
25 and cf. 6) and it now became clear that isolated chloroplasts
could also generate ATP (from ADP and Pi) and reduce CO_2 to the
level of carbohydrate.   Although, at first, the rates of CO_2
fixation were low (16, 21, 26-30) and although the associated
release of O_2 necessitated the employment of anaerobic conditions
for its demonstration (31) there was no real doubt that in vitro
photosynthesis had been achieved.   "When the conversion of $^{14}CO_2$
by isolated chloroplasts to sugars and starch was confirmed and
extended by other laboratories the capacity of chloroplasts to
carry on complete extra-cellular photosynthesis was no longer
open to question" (22).

The failure of the first chloroplast preparations to fix CO_2
at other than modest rates is almost certainly related to the
procedures which were used in their isolation "Because of the
earlier negative results, special experimental safeguards were
deemed necessary to establish that chloroplasts alone, without
other organelles or enzyme systems and with light as the only
energy source were capable of a total synthesis of carbohydrates
from CO_2.   The chloroplasts were washed and, to eliminate a
possible source of chemical energy and metabolites, their
isolation was performed, not as formerly in isotonic sugar
solutions (3, 4) but in isotonic sodium chloride" (22).   Although
saline media yield chloroplasts capable of high rates of electron

transport and photophosphorylation they have rarely permitted the isolation of chloroplasts capable of even modestly high rates of CO_2 assimilation (see e.g. 32, 36, 47).   Although never certainly established, this is probably attributable to ion penetration, osmotic swelling, and consequent damage (36, 77).   Certainly the reintroduction of sugar media was followed by a marked improvement in rate (33) and this was clearly correlated with retention of the chloroplast envelopes (34).   [It may be noted that the terms "intact" or "whole" indicate the presence of the limiting envelopes only when applied to chloroplasts prepared in sugar media after 1964 (see e.g. Hall - 35).   Before 1964 the same descriptions frequently meant little more than the fact that the chloroplast still appeared as a discrete body when viewed under the light microscope, i.e. it had not yet fragmented].

     The method developed in 1964 also incorporated brief blending, rapid centrifugation and no washing (33).   Again, the relative contribution and importance of some of these procedures is arguable (see e.g. 36, 37) and there is little doubt that the choice of sorbitol (as a relatively inert osmoticum) does not lead to more active chloroplasts than a number of alternative sugar alcohols or sugars (including Hill's original choice of sucrose).   [Some of the details which become incorporated into other procedures have perplexed subsequent investigators and it may be worth recording that isoascorbate was used because it was available in the United Kingdom at that time as the sodium salt, whereas its isomer was not.   Similarly Mg was chelated with EDTA to avoid precipitation in the presence of orthophosphate and this seemingly illogical procedure was then persisted with for the simple reason that in the early experiments inclusion of the chelate gave better results than the omission of the Mg (38]].

     During the period 1964-1966 the rates recorded for chloroplasts prepared in sugar-orthophosphate mixtures gradually rose until a value just over 100 μmoles.mg^{-1}chlorophyll.hr^{-1} was recorded for the first time (39).   The increase was not a result of further modifications in isolation but followed inevitably from a gradual switch from peas to market spinach and finally to good field grown spinach as this became available.   In order to obtain the best rates it was normally necessary to add a cycle intermediate (such as ribose-5-phosphate) though in the better preparations these additives shortened the initial induction period without appreciable effect on the rate (40, 41).

     In 1966, Jensen & Bassham (42) recorded rates of more than 100 (μmoles.mg^{-1}chlorophyll.hr^{-1}) for the first time in mixtures containing no added cycle intermediates.   This was a notable achievement but it is important to appreciate that it did not derive from an improvement in the procedures used for isolation but rather from an improvement in the conditions of assay.

Table 1

Effect on photosynthetic rate of additives in preparative
and assay media

Exp.	Prep. Medium	Assay Medium	Rate
1a	Jensen & Bassham (42)	Jensen & Bassham (42)	46.5
1b	Kalberer et al (36)	Kalberer et al (36)	51.0
2a	Jensen & Bassham (42)	Jensen & Bassham (42)	38.2
2b	Kalberer et al (36)	Kalberer et al (36)	43.8
3a	Kalberer et al (36)	Kalberer et al (36)	52.8
3b	Kalberer et al (36)	Kalberer et al (36) minus PPi	27.5

Notes.   The data is from Kalberer et al (36) and the rates are
modest compared with those reported by Jensen & Bassham (42) and
those achieved in much contemporary work based on Jensen & Bassham
procedures but similar comparisons (see e.g. 43) also point to the
conclusion that for a given sample of spinach, with a sugar or
sugar alcohol as the osmoticum (33, 37) there is usually little or
no advantage to be derived from the inclusion of further additives
to the preparative medium such as $NaNO_3$ (compare Exps. 1a & 1b,
2a & 2b above), whereas, unless the experimenter is particularly
fortunate in the concentration of Pi which he chooses (the optimum
is sharp and variable - see e.g. ref. 45) the rate will be improved
by the inclusion of inorganic pyrophosphate (compare 3a & 3b).

Indeed as subsequent work showed (36, 38) the Jensen & Bassham
isolation procedures differed in no major particular from those
previously employed and the inclusion of compounds such as KNO_3
(see e.g. Table 1) did not materially affect the rates.   What
was of crucial importance was the introduction of inorganic
pyrophosphate (PPi) into the assay mixtures.   In our own
laboratory we have used a variety of media for different purposes
and, in order to facilitate comparison with earlier work have
often continued to use additives of uncertain usefulness.   We
have, however, recorded rates of CO_2-dependent O_2 evolution in
excess of 100 µmoles.mg $^{-1}$chlorophyll.hr^{-1} from chloroplasts prepared in media containing only sorbitol and buffer. Similarly, high
rates can be observed in assay media containing only sorbitol,
buffer, bicarbonate and an optimal concentration of orthophosphate.
Although sorbitol is not superior to other sugar alcohols (36, 37)
it is metabolically somewhat inert and since its introduction (33)
it has become time-honoured in its use.   In our hands, we have
found that inorganic pyrophosphate is to be preferred in the
grinding medium as well as in the assay medium (43) but there is
no substitute for "good" plant material.   Heber (personal
communication) has probably achieved rates in excess of 300 more
consistently than anyone else in the field but not even he can
prepare active preparations from inferior spinach.   It may be

noted that if spinach leaves are illuminated by the same quality
and intensity of light used in the experiments with isolated
chloroplasts and particularly if stringent precautions are taken
to maintain the leaf temperature (as well as the air temperature)
at 20° their net photosynthetic rate in air is near to 100 (68)
and there can be little doubt that in this regard the performance
of isolated chloroplasts is inferior to those within the parent
tissue.   It is true, of course, that chloroplasts are normally
provided with saturating CO_2 but saturation is reached at a low
level (68) and, in augmented CO_2, we have been unable to increase
net photosynthesis by spinach leaves at 20° beyond the 300-400
rates achieved with chloroplasts in Heber's laboratory] .

     What PPi does to intact chloroplasts is still by no means
certainly established but there is no doubt that it moderates the
effect of Pi (37, 44, 45) and in this regard it can even be argued
that it acts as a substitute for the cytoplasm (46).   By and
large, the isolated chloroplast is an orthophosphate consuming
organelle.   Except on rare occasions (48, 49), and then for
reasons still unresolved, isolated chloroplasts do not produce
free sugars during photosynthesis, they produce sugar phosphates
and phosphoglycerate (20, 27, 29, 39, 42, 47, 50).   [A great many
compounds including glycollate (20) can be produced under certain
conditions but under optimal conditions the major products are
dihydroxyacetone phosphate and hexose monophosphates] .   Accordingly,
the chloroplast must be provided with Pi as well as with CO_2 and
if Pi is withheld, photosynthesis soon falls to a low rate which
presumably reflects the internal recycling of Pi which is released
during starch synthesis (52).   In the short term (Fig. 1) there
is a stoichiometry of 3 molecules of O_2 evolved for each molecule
of Pi added after Pi depletion (51) and the ratio of P^{32} to C^{14}
incorporated in steady-state photosynthesis is in the region of
1 to 6 (53) reflecting the fact that hexose monophosphates are
major products and that other sugar phosphates, PGA and starch
collectively add up to a similar ratio.   For these reasons Pi
must be supplied and if the investigator is fortunate enough to
find the exceedingly sharp optimum (often in the range of 0.25 to
0.5 mM) he will be rewarded by a fast rate after a short lag.
However, the optimum is both sharp and variable (depending on the
parent tissue and its pretreatment) and if exceeded there is a
progressive lengthening of the lag and inhibition of the final
rate (34, 45).

     The subject of induction is too large to be pursued here at
length but it has been extensively reviewed (47, 55, 56) and there
is considerable evidence to support the early proposition by
Osterhout and Haas (57) that the initial lag is associated with
the light activation of catalysts and the building up of intermediates from a low level in the dark to the higher level demanded
by steady-state photosynthesis.   Phosphate is believed to extend

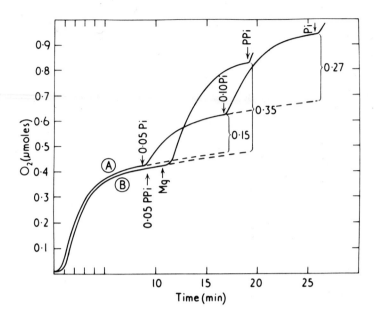

<u>Figure 1.</u>   The isolated chloroplast as a phosphate consuming
organelle.   Two reaction mixtures containing intact spinach
chloroplasts were illuminated in the absence of exogenous Pi.
After a characteristic induction lag CO_2-dependent O_2 evolution
commenced at a rate of about 45 µmoles.mg $^{-1}$chlorophyll.hr^{-1}(the rate
under optimal conditions was 108) but fell to about one tenth of
this value (after about 8-9 minutes) as endogenous Pi was depleted.
The residual rate may then reflect internal recycling of Pi on
starch synthesis.   At this stage the addition of 0.05 µmoles of
Pi to A yielded almost exactly 0.15 µmoles of O_2 before the rate
again declined.   When twice as much Pi (0.10 µmole) was then
added an approx. 1:3 stoichiometric relationship between Pi added
and O_2 evolved (0.27 µmoles) was again observed.   In B the
addition of 0.05 µmoles of PPi was without effect until the further
addition of Mg allowed external hydrolysis.   The O_2 evolved (0.35
µmoles) was then approximately twice that initiated by equimolar
Pi.   [Previously unpublished but, apart from initial omission of
Mg, procedures essentially the same as those described by Cockburn
<u>et</u> <u>al</u>, 51].

induction because it facilitates export of triose phosphate (60)
via Heldt's phosphate translocator (61).   Excess Pi therefore
inhibits because it exaggerates normal export (56, 58, 59) and it
was for this reason that good rates were only originally achieved
in the presence of added metabolites (39) or PPi (42).

As implied above, the action of PPi is complex.   It does not
appear to penetrate the intact spinach chloroplast at anything
faster than a very slow rate but in the presence of exogenous Mg
it can undergo external hydrolysis catalysed by PPiase released
from ruptured plastids (37, 44).   In this way it can function as
a "Pi-stat".   The rate of Pi formation does not become excessive
because Mg-PPi is the substrate and anionic Pi is inhibitory.   In
addition, however, PPi materially reverses Pi inhibition,presumably
by interfering with the action of the Pi-translocator (54, 58).

If these facts are borne in mind it can be seen that the use
of Pi in the first isolation of chloroplasts capable of "good" rates
of photosynthesis was, in its own way, as fortuitous as the failure
of Hill's chloroplasts to fix CO_2.   If these chloroplasts had
"worked" as well as those assayed in PPi the delicate inter-
relationship between Pi and metabolite transport would not have
attracted the early attention that it did.

At this stage we may therefore be permitted to recognise
metabolite transport as one area in which advance has followed the
availability of chloroplasts capable of active CO_2 assimilation.
A general picture of metabolite transport is given in Figure 2 but
the principal traffic may be regarded as an influx of CO_2 and Pi
and a corresponding efflux of triose phosphate.   It would also

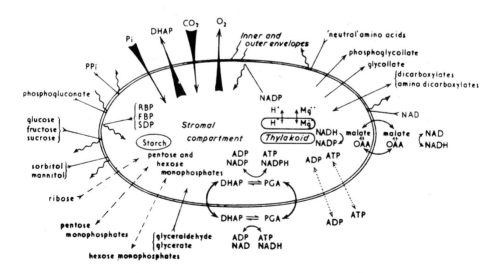

Figure 2.   Summary of movements between the chloroplast and its
cytoplasmic environment.   The principal imports and exports (Pi,
DHAP, CO_2 and O_2) are indicated by bold arrows and the slowest
movements (for example, ADP and ATP, by broken arrows).   Curved
arrows indicate the operation of shuttle mechanisms.

follow that the ability of the chloroplast to "carry on complete
extra-cellular photosynthesis" is, after all, still open to question.
Thus, when added together, the combined reactions of photophosphorylation and the reductive pentose phosphate pathway do not add
up to the simplified overall equation for photosynthesis because
they yield triose phosphate as a product (Scheme 1 - SUM).   Agreement can be achieved (Scheme 1 - Grand Total) but only by introducing an additional reaction (Scheme 1 - Hypothetical reaction) in
which triose phosphate is converted to a free sugar.   Within the
chloroplast, something akin to this is achieved as triose phosphate
is converted into starch

$$n \ (\text{triose-P}) + n \ H_2O \longrightarrow \text{starch} + nPi$$

but normally only up to about 10% of the total photosynthetic
product accumulates as starch.   However, if triose-P is normally
exported (58, 59, 61, 62) and sucrose is normally synthesised within
the cytoplasm (63) the chloroplast will depend very largely on
events (Fig. 3) within the cytoplasm for continued photosynthesis
(46).   As already noted, photosynthesis by isolated chloroplasts
soon falls to a low rate in the absence of exogenous Pi.   In vivo
there now seems little doubt that this Pi will be maintained in the
cytoplasm by sucrose synthesis from triose phosphate (46, 52, 59)

$$4 \ \text{triose phosphate} + 4H_2O \longrightarrow \text{sucrose} + 4Pi$$

Certainly if cytoplasmic Pi is sequestered by mannose feeding
(which yields non-metabolised mannose phosphate in species such as
spinach-beet) photosynthesis is slowed and more of the photosynthetic product is retained in the chloroplast as starch (46, 64).
Similarly, with isolated chloroplasts, both the rate of photosynthesis and the distribution of products between the stroma and the
media is profoundly influenced by the exogenous Pi concentration
(45, 46, 54).

     In recent years, in vitro photosynthesis has developed in a
new direction following improvements to the reconstituted chloroplast system first devised by Whatley et al (65).   At first the
improved reconstituted system would only evolve O_2 in the presence
of PGA and ATP (66) but gradually it became possible to extend the
reaction sequences involved (67, 68).   Thus Bassham et al have
achieved CO_2 fixation with PGA as substrate (69) and Walker &
Slabas have reported O_2 evolution with triose phosphate (70).   In
both instances the reactions involved embrace the entire reductive
pentose phosphate pathway.   Independence of all but catalytic
substrate is more difficult to demonstrate, presumably because key
products are more easily built up within the confines of the intact
plastid than in the much larger volume (x 500 to 1000) of the
average reaction mixture.   Nevertheless autocatalysis has been
reported (68) in mixtures containing above normal stromal protein.

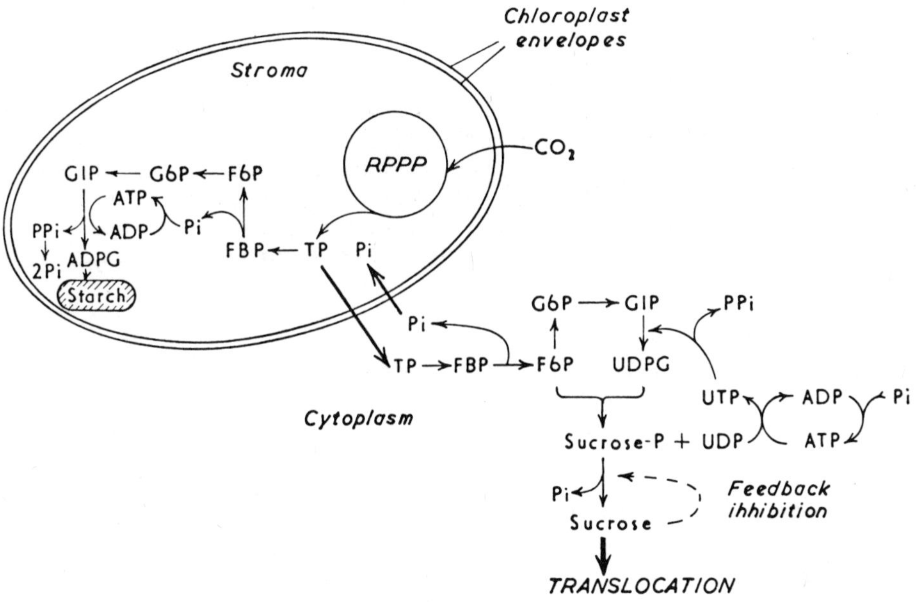

Chloroplast envelopes

Stroma

RPPP

CO₂

GIP ← G6P ← F6P

ATP

PPi   ADP   Pi

2Pi   ADPG   FBP ← TP   Pi

Starch

Pi ←

TP → FBP

Cytoplasm

G6P → GIP   PPi

TP → FBP → F6P   UDPG

UTP   ADP   Pi

ATP

Sucrose-P + UDP

Pi   Feedback inhibition

Sucrose

TRANSLOCATION

Figure 3.   Dependence of photosynthesis on cytoplasmic events.
In order to photosynthesise at high rates the chloroplast requires
a continuous supply of orthophosphate.   Some orthophosphate is
made available within the stromal compartment as a consequence of
starch synthesis.   The remainder is supplied from cytoplasm and
the figure illustrates one example of Pi recycling in which triose
phosphate is exported to the cytoplasm via Heldt's translocator and
Pi is returned following its release in sucrose synthesis.   Feedback inhibition of photosynthetic carbon assimilation could arise
as illustrated by inhibition of sucrose phosphatase (79). Alternatively, some sequestration of cytoplasmic orthophosphate could result
from the formation of hexose monophosphates.   In either event
photosynthesis would be influenced by the [organic phosphate]/[Pi]
ratio within the cytoplasm.

In mixtures containing no excess stromal protein other than ferredoxin, the present reconstituted systems are self-sufficient with
regard to NADPH and ATP and will fix CO2 and evolve O2 at least as
well as the parent tissue.

The furthest extension of in vitro photosynthesis which can be
foreseen at this time is reconstitution from purified fractions and
a step in this direction has already been taken (71).   Such
reconstitution would, by analogy with organic chemistry, add the
dimension of synthesis to what has so far been pursued as an

analytical investigation.

        Again, the usefulness of the reconstituted system
may be questioned.   At this moment its value appears to be centred
on regulation and indeed its gradual extension of the system to
include more and more reactions has only become possible with a
fuller understanding of the control mechanisms involved.   For
example, work with the reconstituted system has contributed to
present views concerning Mg activation of RBP carboxylase and
FBPase (72, 73).

        As for many other "control mechanisms" it is not always easy
to see what benefit the organism may derive from their operation,
although this may become more evident as work proceeds.   This is
particularly true of ADP-inhibition of PGA reduction induced by
consumption of ATP in the reaction catalysed by ribulose-5-
phosphate kinase (73-75).   This will tend to ensure that the
pentose monophosphate and triose phosphate pools within the stroma
will be relatively small, whereas the pool of PGA will be relatively
large.   A high triose phosphate would facilitate export and if
ribulose bisphosphate were produced more rapidly than it could be
carboxylated (e.g. in a CO_2-limited situation) some brake on its
formation might alleviate excessive photorespiratory loss.

        The reconstituted system also has the virtue that it permits
the investigator to reach into the chloroplast or at least to work
with a system which is not too far removed from the intact plastid
but which is without the barriers normally presented by the intact
envelopes.   It may also be prepared from species which have not
so far yielded functional intact chloroplasts (76).   In short, its
future as an experimental tool would seem to be assured.

<u>Scheme 1</u>

<u>Photophosphorylation</u>

$$9 \text{ ADP} + 9 \text{ Pi} \longrightarrow 9 \text{ ATP} + 9 \text{ H}_2\text{O}$$

$$6 \text{ NADP} + 6 \text{ H}_2\text{O} \longrightarrow 6 \text{ NADPH} + 6 \text{ H}^+ + 3 \text{ O}_2$$

<u>Reductive Pentose Phosphate Pathway</u>

$$3 \text{ CO}_2 + 3 \text{ ribulose 1,5-bisphosphate} + 3 \text{ H}_2\text{O}$$
$$\longrightarrow 6 \text{ 3-phosphoglycerate}$$

$$6 \text{ 3-phosphoglycerate} + 6 \text{ ATP}$$
$$\longrightarrow 6 \text{ glycerate 1,3-bisphosphate} + 6 \text{ ADP}$$

6 glycerate 1,3-bisphosphate + 6 NADPH + 6 H$^+$

$\longrightarrow$ 6 glyceraldehyde 3-phosphate + 6 NADP + 6 Pi

2 glyceraldehyde 3-phosphate

$\longrightarrow$ 2 dihydroxyacetonephosphate

1 glyceraldehyde 3-phosphate + 1 dihydroxyacetonephosphate

$\longrightarrow$ 1 fructose 1,6-bisphosphate

1 fructose 1,6-bisphosphate + 1 H$_2$O

$\longrightarrow$ fructose 6-phosphate + Pi

1 fructose 6-phosphate + 1 glyceraldehyde 3-phosphate

$\longrightarrow$ 1 xylulose 5-phosphate + 1 erythrose 4-phosphate

1 erythrose 4-phosphate + 1 dihydroxyacetonephosphate

$\longrightarrow$ 1 sedoheptulose 1,7-bisphosphate

1 sedoheptulose 1,7-bisphosphate + 1 H$_2$O

$\longrightarrow$ 1 sedoheptulose 7-phosphate + Pi

1 sedoheptulose 7-phosphate + 1 glyceraldehyde 3-phosphate

$\longrightarrow$ 1 xylulose 5-phosphate + 1 ribose 5-phosphate

1 ribose 5-phosphate $\longrightarrow$ 1 ribulose 5-phosphate

2 xylulose 5-phosphate $\longrightarrow$ 2 ribulose 5-phosphate

3 ribulose 5-phosphate + 3 ATP

$\longrightarrow$ 3 ribulose 1,5-bisphosphate + 3 ADP

SUM 3 CO$_2$ + 2 H$_2$O + 1 Pi $\longrightarrow$ 1 glyceraldehyde 3-phosphate + 3 O$_2$

Additional
hypothetical   1 glyceraldehyde 3-phosphate + 1 H$_2$O
reaction
$\longrightarrow$ 1 glyceraldehyde + 1 Pi

Grand   3 CO$_2$ + 3 H$_2$O $\longrightarrow$ C$_3$H$_6$O$_3$ + 3 O$_2$
Total
    or   CO$_2$ + H$_2$O $\longrightarrow$ $\boxed{CH_2O}$ + O$_2$

Scheme 1. The combined reactions of photophosphorylation (ATP +
NADPH generation) and the reductive pentose phosphate pathway.
It will be seen that the "SUM" of the above reactions can only be
made equal to the correct "Grand Total" (i.e. the classic equation
for photosynthesis as written in equation 1) if the triose phosphate
product is hydrolysed. The additional hypothetical hydrolysis is

achieved within the chloroplast by the conversion of triose phosphate to polysaccharide but as Fig. 1 shows this internal re- cycling of Pi does not normally proceed at a fast enough rate to satisfy maximal photosynthesis.   If it is accepted that triose phosphate is normally exported from the chloroplast in exchange for cytoplasmic Pi (and that sucrose synthesis is an event which principally occurs outside the chloroplast) then it follows that the chloroplast cannot support the entire photosynthetic process unaided.

REFERENCES

(1)   HABERLANDT, G. (1888)   Flora. 71, 291-309.
(2)   EWART, A. J. (1896)   J. Linnean Soc. London (Bot) 31,
          364-461, 554-576
(3)   HILL, R. (1937)   Nature. 139, 881-882.
(4)   HILL, R. (1939)   Proc. Roy. Soc. Lond. B, 127, 192-210.
(5)   HILL, R. (1936)   Proc. Roy. Soc. Lond. B, 120, 472-483.
(6)   HILL, R. (1965)   Essays Biochem. 1, 121-151.
(7)   HILL, R. (1975)   Ann. Rev. Plant Physiol. 26, 1-11.
(8)   de SAUSSURE, T. (1804)   Recherches Chimiques Sur la
          Vegetation (V. Nyon, Paris).
(9)   von BAEYER, A. (1864)   Ber. Deut. Chem. Ges. 3, 63-75.
(10)   WILLSTÄTER, R. & STOLL, A. (1918)   Untersuchungen Über die
          Assimilation der Kohlensäure.   Springer, Berlin.
(11)   WARBURG, O., KRIPPAHL, G. & LEHMAN, A. (1969)   Amer. J. Bot.
          56, 961-971.
(12)   van NIEL, C. B. (1931)   Arch. Mikrobiol. Z. 3, 1-112.
(13)   van NIEL, C. B. (1941)   Advan. Enzymol. 1, 263-328.
(14)   KNY, L. (1897)   Ber. Deut. Bot. Ges. 15, 388-403.
(15)   BROWN, A. H. & FRANCK, J. (1948)   Arch. Biochem. 16, 55-60.
(16)   ARNON, D. I., ALLEN, M. B. & WHATLEY, F. R. (1954)   Nature.
          174, 394-396.
(17)   ARNON, D. I., WHATLEY, F. R. & ALLEN, M. B. (1954)   J. Amer.
          Chem. Soc. 76, 6324-6329.
(18)   ARNON, D. I. (1958)   In: The photochemical Apparatus - its
          structure and function.   (eds. R. C. Fuller, J. A.
          Bergeron, L. G. Augenstine, M. E. Koshland, H. J. Curtis)
          Brookhaven Symp. in Biol. No.11, pp.181-235.
          Brookhaven Nat. Lab., Upton, N.Y.
(19)   ARNON, D. I. (1959)   Nature. 184, 10-21.
(20)   ARNON, D. I. (1961)   In: Light and Life.   (eds. W. D.
          McElroy & B. Glass)   pp. 489-566.   Johns Hopkins Press,
          Baltimore.
(21)   ARNON, D. I. (1967)   Physiol. Rev. 47, 317-358.
(22)   ARNON, D. I. (1971)   Proc. Nat. Acad. Sci. 68, 2883-2892.
(23)   VISHNIAC, W. & OCHOA, S. (1951)   Nature. 167, 768-769.
(24)   TOLMACH, L. J. (1951)   Nature. 167, 946-948.
(25)   ARNON, D. I. (1951)   Nature. 167, 1008-1010.
(26)   ARNON, D. I. (1955)   Science. 122, 9-16.

(27) GIBBS, M. & CYNKIN, M. A. (1958)   Nature.   182, 1241-1242.
(28) GIBBS, M. & CALO, N. (1959)   Plant Physiol. 34, 318-323.
(29) HAVIR, E. A. & GIBBS, M. (1963)   J. Biol. Chem. 238, 3183-
         3187.
(30) LOSADA, M., TREBST, A. V. & ARNON, D. I. (1960)   J. Biol.
         Chem. 235, 832-839.
(31) ALLEN, M. B., ARNON, D. I., CAPINDALE, J. B., WHATLEY, F. R.
         & DURHAM, L. J. (1955)   J. Amer. Chem. Soc. 77, 4149-
         4155.
(32) GIBBS, M., LATZKO, E., EVERSON, R. G. & COCKBURN, W. (1967)
         Proc. Intern. Minerals Chem. Symp. Chicago, 1966 (eds.
         A. San Pietro, F. A. Greer & T. J. Army) pp. 111-130.
         Academic Press, New York, London.
(33) WALKER, D. A. (1964)   Biochem. J. 92, 22-23.
(34) WALKER, D. A. (1965)   Plant Physiol. 40, 1157-1161.
(35) HALL, D. O. (1972)   Nature New Biol. 235, 125-126.
(36) KALBERER, P. P., BUCHANAN, B. B. & ARNON, D. I. (1967)
         Proc. Nat. Acad. Sci. (Wash.) 57, 1542-1549.
(37) WALKER, D. A. (1971)   In: Methods in Enzymology. (ed.
         A. San Pietro) 23, 211-220.   Academic Press, London &
         New York.
(38) AVRON, M. & GIBBS, M. (1974)   Plant Physiol. 53, 140-143.
(39) BUCKE, C., WALKER, D. A. & BALDRY, C. W. (1966)   Biochem. J.
         101, 636-641.
(40) COCKBURN, W., BALDRY, C. W. & WALKER, D. A. (1967)   Biochim.
         Biophys. Acta. 143, 603-613.
(41) JENSEN, R. G. & BASSHAM, J. A. (1968)   Biochim. Biophys.
         Acta. 153, 219-226.
(42) JENSEN, R. G. & BASSHAM, J. A. (1966)   Proc. Nat. Acad. Sci.
         56, 1095-1101.
(43) COCKBURN, W., WALKER, D. A. & BALDRY, C. W. (1968)   Plant
         Physiol. 43, 1415-1418.
(44) SCHWENN, J. D., LILLEY, R. McC. & WALKER, D. A. (1973)
         Biochim. Biophys. Acta. 325, 586-595.
(45) LILLEY, R. McC., SCHWENN, J. D. & WALKER, D. A. (1973)
         Biochim. Biophys. Acta. 325, 596-604.
(46) HEROLD, A. & WALKER, D. A. (1976)   In: Handbook on
         Transport across Membranes. (eds. G. Giebisch,
         D. C. Tosteson & H. H. Ussing)   Springer-Verlag,
         Heidelberg, New York. (in press)
(47) GIBBS, M. (1971)   In: Structure and Function of
         Chloroplasts. (ed. M. Gibbs) pp. 169-214.   Springer-
         Verlag, Berlin, Heidelberg, New York.
(48) EVERSON, R. G., COCKBURN, W. & GIBBS, M. (1967)   Plant
         Physiol. 42, 840-844.
(49) MIYACHI, S. & HOGETSU, D. (1970)   Plant Cell Physiol. 11,
         927-936.
(50) WALKER, D. A. (1967)   Proc. NATO Adv. Study Inst.
         Aberystwyth, 1965.   (ed. T. W. Goodwin) Vol.2, 53-69
         Academic Press, New York.

(51) COCKBURN, W., BALDRY, C. W. & WALKER, D. A. (1967)
     Biochim. Biophys. Acta. 131, 594-596.
(52) WALKER, D. A. (1976)   In: Current Topics in Cellular
     Regulation.   (eds. B. L. Horecker & E. Stadtman)
     Vol. 11.   (in press)
(53) BALDRY, C. W., BUCKE, C. & WALKER, D. A. (1966)   Nature,
     210, 793-796.
(54) COCKBURN, W., BALDRY, C. W. & WALKER, D. A. (1967)   Biochim.
     Biophys. Acta.   143, 614-624.
(55) WALKER, D. A. (1973)   New Phytol.   72, 209-235.
(56) WALKER, D. A. (1976)   In: The Intact Chloroplast.   (ed.
     J. Barber) Chapter 7, 235-278. Elsevier, Amsterdam.
(57) OSTERHOUT, W. J. V. & HAAS, A. R. C. (1918)   J. Gen. Physiol.
     I, 1-17.
(58) WALKER, D. A. (1974)   Med. Tech. Publ. Int. Rev. Sci.
     Biochem. Ser. 1. (ed. D. H. Northcote)   Vol.11, 1-49
     Butterworths, London.
(59) WALKER, D. A. (1976)   In: Encyclopedia of Plant Physiology.
     (eds. A. Pirson & M. Zimmermann)   Vol. 3.   Springer-
     Verlag, Berlin, Heidelberg, New York.   (in press)
(60) WALKER, D. A. & CROFTS, A. R. (1970)   Ann. Rev. Biochem.
     39, 389-428.
(61) HELDT, H. W. & RAPLEY, L. (1970)   FEBS Lett.   10, 143-148.
(62) HEBER, U. (1974)   Ann. Rev. Plant Physiol.   25, 393-421.
(63) BIRD, I. F., CORNELIUS, M. J., KEYS, A. J. & WHITTINGHAM, C.P.
     (1974)   Phytochemistry.   13, 59-64.
(64) CHEN-SHE, S. H., LEWIS, D. H. & WALKER, D. A. (1975)   New
     Phytol.   74, 383-392.
(65) WHATLEY, F. R., ALLEN, M. B., ROSENBERG, L. L., CAPINDALE,
     J. B. & ARNON, D. I. (1956)   Biochim. Biophys. Acta.
     20, 462-468.
(66) STOKES, D. M. & WALKER, D. A. (1971)   Plant Physiol.   48,
     163-165.
(67) WALKER, D. A., McCORMICK, A. V. & STOKES, D. M. (1971)
     Nature.   233, 346-347.
(68) WALKER, D. A. & LILLEY, R. McC. (1974)   Plant Physiol. 54,
     950-952.
(69) BASSHAM, J. A., LEVINE, G. & FORGER, J. III. (1974)   Plant
     Sci. Lett.   2, 15-21.
(70) WALKER, D. A. & SLABAS, A. R. (1976)   Plant Physiol. 57,
     203-208.
(71) SLABAS, A. R. & WALKER, D. A. (1976)   Arch. Biochem. Biophys.
     (in press)
(72) LILLEY, R. McC., HOLBOROW, K. & WALKER, D. A. (1974)   New
     Phytol.   73, 657-662.
(73) WALKER, D. A., SLABAS, A. R. & FITZGERALD, M. P. (1976)
     Biochim. Biophys. Acta. 440, 147-162.
(74) SLABAS, A. R. & WALKER, D. A. (1976)   Biochim. Biophys. Acta.
     430, 154-164.

(75)   SLABAS, A. R. & WALKER, D. A. (1976)   Biochem. J.   154, 185-
           192.
(76)   DELANEY, M. E. & WALKER, D. A. (1976)   Plant Sci. Lett.
           (in press)
(77)   KAHN, A. & von WETTSTEIN, D. (1961)   J. Ultrastruct. Res.
           5, 557-574.
(78)   LILLEY, R. McC. & WALKER, D. A. (1975)   Plant Physiol.   55,
           1087-1092.
(79)   HAWKER, J. S. (1967)   Biochem. J.   102, 401-406.

SYMPOSIUM V

COMPARATIVE EFFECTS OF EXCITING AND
IONIZING RADIATIONS

QUANTITATIVE MUTAGENESIS BY CHEMICALS AND BY RADIATIONS:

PREREQUISITES FOR THE ESTABLISHMENT OF RAD-EQUIVALENCES

R. LATARJET

Institut du Radium

26, rue d'Ulm, 75005 PARIS, France

INTRODUCTION

Owing to the horrors of Hiroshima and Nagasaki, and because of the foreseeable and inevitable development of nuclear energy for peaceful purposes, as early as 1950 ionizing radiations were considered a risk requiring international legislation. A special commission, I.C.R.P. was constituted within the United Nations, and, after an excellent study, enacted certain rules which defined the permissible doses of radiation for most circumstances of utilization and exposure. These rules were considered as being far from perfect, but it was wisely decided not to await their perfection, and to modify the rules according to improvements in our understanding of fundamental problems. For twenty years, these rules, regularly updated, have been observed practically universally, and have formed the basis of national legislations.

No similar guidelines exist in the field of chemical mutagenic pollutants. However this form of pollution is quantitatively much more important than pollution by radiations. Coal and oil--fired power-stations, industrial and domestic combustions, the automobile and tobacco, liberate chemicals which have adverse effects upon genetic material. If world authorities have never been really concerned with this form of pollution, it is because it has so far remained diffuse and sly. No spectacular accident of large magnitude has ever ignited the relatively unimaginative interna-

tional opinion. In the field of protection, chemical pollution
lags 25 years behind radiation pollution, and clearly this de-
fault must end.

Chemical pollution must be submitted to a rigorous international legislation. We need to decide the permissible maximal
dose for the main representatives of this pollution. It would be
too long and too tedious a task to follow, for each of them, all
the steps which were necessary to bring about today's radiation
legislation. However a somewhat shorter and more general route
may exist, that of "rad-equivalents".

The lesions produced in the genetic material by chemical mutagens, on the one hand, and radiations, on the other, are very
similar. In both cases, they are either lesions in DNA (chain
breaks, base lesions which produce a change in the shape of the
double helix, interchain cross-links) or changes in the bonds
between this DNA and the proteins which surround it. The lesions
are sufficiently similar to elicit, in both cases, the activity
of the same repair systems (which act either by resealing breaks,
by opening cross-links, by excision of the damaged bases, by genetic recombination, or, finally, by an inducible repair mechanism). The similarity between chemical and radiation induced mutagenesis can be demonstrated by checking (usually on microorganisms, but also on mammalian cells in vitro) that a strain which
is hyper-sensitive to radiation because it lacks some repair system, is also hyper-sensitive to most chemical mutagens.

These similarities between the lesions suggest that one can
establish an equivalence between the "dose" of a chemical and a
dose of radiation,on the basis of the effects produced on some
biological systems of reference. Once such equivalence has been
established, one could extrapolate the rules of radiation protection to protection against that chemical. Is this principle applicable, and under which conditions? What prerequisites must be ful
filled? The goal of this paper is to answer these questions. Obviously the problem appears difficult and uncertain. But such a
study is waranted by the important consequences of the resolution
of this problem.

There is a real and urgent need for rad-equivalents.

THE QUALITATIVE ASPECT: ANALOGIES AND GENERALIZATION

    The equivalence between two quantities comprises a qualitati-
ve aspect and a quantitative aspect. The qualitative aspect comes
first. Analogies between the two quantities must be recognized
as sufficient. These analogies cannot be total, otherwise the two
quantities would be merged into a single one. The analogies must
be sufficient for one to be able to consider the two quantities
as "interchangeable" with respect to some important property,
given sufficiently general circumstances. Only later can one approach the quantitative aspect by attempting to express the equivalence by a number. The meaning and scope of that number will depend upon the generality of the circumstances in which the equivalence can be applied.

    As an example, let us consider the first law of thermodyna-
mics. First, we have acknowledged some deep analogies between heat
and work: work generates heat; heat can be converted into work. We
have then deemed it permissible to measure an equivalence between
the units expressing the amount of these two quantities: thus one
calorie is equivalent to 4,18 joules.

    This value of equivalence has a very wide bearing because it
depends neither on the mode nor on the direction of the transformation of one of the quantities into the other, nor even on the
system in which it takes place. The equivalence does not negate
the profound differences that exist between work and heat, and
which are dealt with by the second principle of thermodynamics
(Carnot). Charles Fabry used to say that "the layman knows that re-
winding a watch is not the same thing as placing it in the fire".

    In our present case, we acknowledge that the effects of ra-
diations and of the chemicals under consideration have enough
similarities in the field of mutagenesis, as far as the lesions of
the genetic material and the final effects (mutations) are concerned, as to allow us to admit a principle of equivalence and to look
for the values of that equivalence particular to each mutagen.

    But we do not know whether, for each chemical, the value, as
it is measured,has a degree of generality, i.e. of independence
from the experimental circumstances, sufficient to endow it with
real meaning. The principle of rad-equivalences is valid, but we
cannot tell yet whether it is applicable in practice.

THE DOSE OF A CHEMICAL MUTAGEN VERSUS THE DOSE OF RADIATION

Since we have to establish a quantitative equivalence between a dose of a chemical mutagen and a dose of radiation, we must first define unambiguously what these doses are.

A dose of ionizing radiation D, expressed in rads, is an amount of radiant energy absorbed by the unit of mass of the irradiated object. It is a mass density of absorbed energy. This dose D, is itself the product of the mass density of absorbed energy flux Φ and the time of exposure t

$$(1) \quad D = \Phi \times t$$

A profound analogy exists between the flux of radiation and the concentration of a chemical. At a given moment, the flux of radiation represents the density of interacting elements (photons, electrons,. .) in the vicinity of the biological target under consideration. At a given moment, the chemical concentration represents the density of interacting elements (active molecules)in the vicinity of the same target. This "true" concentration is always unknown within a complex object such as a cell. But providing one remains below saturation, it is proportional to the concentration C within the surrounding medium. Thus, the chemical dose takes the form :

$$(2) \quad D = C \times t$$

which is analogous to (1). Equivalence between the doses is nothing more than the equivalence between the flux of the radiation and the concentration of the chemical. In other words :

-rads correspond to the product of a concentration and a time

-a concentration corresponds to rads per unit of time.

For example, we can say that a concentration x of a chemical in air is equivalent to y millirads per week; or that in order to deliver one rad equivalent one must maintain the concentration x for z weeks ($z = 10^3/y$).

This is valid when c and t are well defined, for example in the case of a long term pollution of air or water (which are the most important cases). But this is not always the case, for example when one deals with a short-lived accidental contamination. It should be noted that the dose D, as defined above, represents, together with a proportionnality factor linked to absorption, the amount of the chemical which interacts with the object. That amount can often be determined, without C and t being known; for ex-

ample when an individual accidentally ingests or inhales a mutagen,
the amount which has been absorbed can usually be determined[+].

THE RECIPROCITY BETWEEN CONCENTRATION AND TIME EXPOSURE

Let us come back to the general case where one compares a dose   c x t of a chemical mutagen with a dose of Φ x t of radiation. Such defined doses have a meaning only if, when the two factors vary inversely such that their product remains constant, the
biological effect which ensues remains approximately the same.
For radiations, this is called "reciprocity" between flux and time.

Since the end of the 19th century, many studies have dealt
with this problem of reciprocity. They first started in
the field of the blackening of the photographic image, to yield
what has been called "Schwartzwhild's law". Then, at the beginn-
ng of this century, they dealt with radiobiology in order to formulate an approach to the radiotherapy of cancer. As a general rule, reciprocity is roughly fulfilled within a limited interval of
flux and time. The simpler the irradiated system, the more extended this interval. On each side of it, failures of reciprocity are
observed. When the flux becomes too weak and exposure time too
long, restauration phenomena, in their broadest sense, have time
to come into play and decrease the efficiency of the dose. Conversely, when the flux becomes very high and the exposure time very
brief, saturation effects may show up which may also decrease the
efficiency of the dose since several interactions with the same
receptor may occur even though a single one would suffice to produce the same effect. In spite of these failures of reciprocity
the dose, as defined by the product Φ x t, remains useful in the
treatment of radiobiological problems.

Thus it appears necessary to carry out preliminary studies to
ensure that the same principles hold for chemical mutagenesis. One
should choose a system and a chemical, and study the effects of
that chemical on the system, using equal doses but varying C and t
inversely; by these means, we can ascertain whether the effect remains constant within a sufficient interval of C and t. Here also
can we surmise that, at low concentrations, restauration phenomena
can take place, and that, at high concentrations, saturation ef-

---

[+] This point should be developed further, but this would bring
    us too far from the general character of this paper.

fects will show up, both decreasing the efficiency of the dose.

We already possess the results of two experiments of this type. One, direct, i.e., carried out with this purpose, is due to
Moustacchi et at. (3). They used formadehyde treatment of yeasts.
The other, indirect, is due to Maltoni (2), who used vinylchloride
on mice. In both cases, an approximate reciprocity was observed
within a rather wide interval of concentration and time.

THE SPECTRUM OF EQUIVALENCES FOR A GIVEN CHEMICAL

If we accept that an equivalence for a chemical mutagen can
be found, we should note that this equivalence will depend on the
biological system utilized and, probably, on the effect chosen;
this is true for a variety of reasons, the most evident of which
is that the absorption and the possible metabolism of the chemical
depend upon the nature of the system and its actual physiological
conditions. For example, in an isolated cell, absorption is governed by the permeability of the cell membrane, which itself varies
with the growth conditions and the phases of the mitotic cycle.
These factors do not necessarily influence the effects of the radiation in the same direction and with the same amplitude.

Let us consider a mutagen and let us modify the parameters:
nature and physiological state of the system, biological effect,
animal or plant species used, etc... and let us determine the rad-
equivalence in each case. The equivalence will fluctuate within
an interval, or, if one prefers, a spectrum of values which is
more or less wide. The notion of rad-equivalence attached to the
substance has a general meaning and scope only if this spectrum is
narrow enough to have a "centre of gravity" which is close enough
to the individual values. If so, the equivalence of the center
of gravity can be attached to the substance. If, conversely, the
dispersion is too wide one should reject an equivalence of a general character and restrict oneself to a few cases of particular
importance.

Before looking closer at the difficulties which will inevitably be
encountered, let me underline three points which should be kept in
mind :
a)   The differences in behaviour between chemicals and radiations
     should not hinder us if we keep ourselves fully aware of them.
     As already mentionned, the obvious differences between heat
     and work have not prevented the establishment of the first
     law of thermodynamics which has been so useful.

b)   One can only compare two quantities numerically if they are
      interchangeable with respect to the effect under considera-
      tion. They can remain interchangeable, when the effect they
      produce is the same, even if the mechanisms are different.
      This is particularly true for the lethal effect, when we are
      interested in death, whatever its mechanism. One can numer-
      ically compare the lethal effectiveness of a mustard gas and
      of an atom bomb [+].

c)   But this is not true when the difference in mechanisms leads
      to a difference in the consequences of the effect. This is
      particularly true if one considers mutations as a whole, be-
      cause that whole is very heterogeneous. For example, let us
      suppose that the mutagenic effect of the chemical upon the
      chosen system principally proceeds through small deletions,
      whereas x-rays principally act on the same system through
      chromosomal aberrations. The consequences of the deletions
      are likely to be more severe than those of abberrations. If
      equivalence is measured on the basis of total mutations, the
      genetic consequences of the exposure to the chemical will be
      underestimated if compared to those of x-rays. In this case
      "total mutations" does not constitute the interchangeable
      quantity. The equivalence should be based on the number of
      recessive lethals which are the indicators of small deletions.

      Let us now envisage the main circumstances which may compli-
cate a solution of the problem.

      1°) Same biological system, same effect, same conditions, but
dissimilar dose-effect curves.
      This case is not rare. For example the curve may be exponent-
ial with x-rays and sigmoid with the chemical. In this case, the
equivalence varies with dose. Certain authors hastily conclude
that the principle of equivalence should be rejected in such cases.
Of course, it is no longer simple, but I think that it may still
furlfill our expectation, although in a more elaborate way or in
a less general manner.

      Let us draw both curves on the same figure and choose the

---

(+) In this paper the lethal effect is considered for mutagens on-
    ly, in parallel with their mutagenic, and occasionally carcino-
    genic effects.

abscissae in such a way that the linear parts of the curves are
parallel (this is always possible) - fig. 1 - Facing 1 rad on the
abscissae is the chemical dose D_1. We can already say that for
all doses of the linear part xx of the curve, the equivalence is
the same. The dose $D_1 \times \frac{Oa}{Ob}$ corresponds to rad 1. For a smaller
dose, for example that which corresponds to a point A on the curve
- which can be that of a particularly interesting pollution -,
the radiation equivalent is immediately given by the curves; it
is the abscissa of point B, that is 0,5 rad in the present drawing.

Note : The initial plateau of a chemical curve suggests the
existence of a threshold, i.e., of a dose Do below which no effect
is produced. Such is the case when the chemical naturally exists
at low concentration Co in the system (formol for ex. is a normal
metabolite in most cells). If the doses D are not expressed as
C x t but as (C-Co) t, the two curves in fig. 1 come closer, and
the correcting factor $\frac{Oa}{Ob}$ practically disappears.

2$^{\bullet}$) Same system, same effect, different conditions.
It may happen that the sensitivities of the system to radiation and to the chemical vary in the opposite direction when certain conditions of the system are changed. Then the values of
equivalence have a tendency to diverge. This situation was observed by Moustacchi et al., when comparing the lethal effects of
formaldehyde and γ-rays on yeast either growing or resting. When
one passes from stationary phase to exponential growth, the sensitivity to the chemical increases 5,4 fold, whereas the sensitivity
to radiation decreases 42 fold. Thus the ratio between the two
sensitivities varies by a factor of 227 (1).

Other factors may intervene. The same authors have observed
that haploid and diploid yeasts are equally sensitive to formaldehyde whereas haploids are more radiation sensitive than diploids.

If one envisages all the possible variables and the other
effects to be considered (mutagenic), one realizes how wide the
equivalence spectrum can be for the cells of a single species.
But, here again, if we stick to the practical point of view, i.e.,
that of natural population in a dynamic equilibrium, the situation is simpler. Any pollution will selectively attack the cells
when they are in the phase of greatest sensitivity - this phase
in the cycle being the same or not for the two agents. In the
preceeding case, if we limit ourselves to the sensitive phases,
the spectrum of equivalence values shrinks from a ratio of 250 to

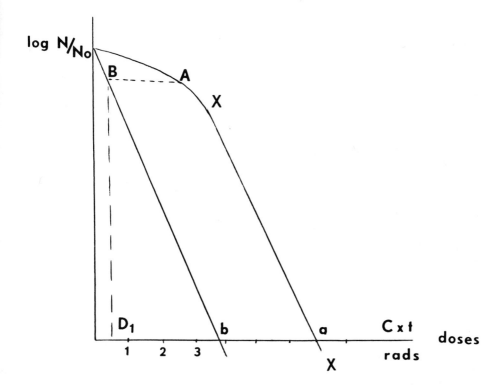

Figure 1.

a ratio of 4.

3*) Same effect, same conditions, different species.
The sensitivity to the mutagenic effect of radiations varies
from one animal or plant species to another. The sensitivity to
the effect of chemical mutagens also varies. If these two variations were parallel, the rad-equivalence of a given compound would
be the same for all species. Evidently this question is of particular interest with respect to the extrapolation of experimental
results to Man.

Recently, the sensitivity of a species to the mutagenic effect of radiations has been described as an increasing linear
function of the amount of DNA present in the haploid genome (4) -
a function which is represented by the ABCW curve, according to
the name of the authors. Does a similar relationship exist for
chemical mutagens, and are the corresponding curves parallel to
the ABCW one ? A positive answer has just been given to this
question in the case of ethyl methane sulfonate (5), a result obviously favourable to the generalization of the rad-equivalent
concept.

However, the ABCW relationship is surprising. In order to
interpret it, one most admit either a) that the entire genome
participates in the mutation of a locus; or b) that when the amount of DNA increases, the number of genes remains the same, but
their individual (radiosensitive) size increases accordingly.
These two assumptions are not easy to accept; some well established facts argue against them. Thus, the ABCW relationship has
been criticized (6), and the criticisms are also valid for the
ethyl methane sulfonate curve. Therefore, it is premature to
accept the interspecific generalization of a rad-equivalence.
The question remains open.

4*) So what ?
We are in a dilemma.
On the one hand, mutagenic chemical pollution cannot remain
in its present anarchic situation. Society and the responsible
authorities need a quantitative definition of what is permissible
and what is not in the most important circumstances. As already
pointed out, comparisons with pollution by radiations offer an
efficient way to proceed with this problem. For example we would
wish to quantitatively compare chemical pollution by coal and oil
fired power-stations and radiation pollution by the nuclear power

plants, producing the same amounts of energy. We would like to
have a similar comparison between urban pollution by polycyclic
hydrocarbons and radiation pollution with regard to carcinogenic
effects in Man.

On the other hand, rad-equivalence comprises uncertainties
and gaps, which were pointed out at an early stage (7, 9, 10),
and which raise such criticisms as to place the principle itself
in doubt.   Certain geneticists, such as F. Sobels (1) and C.
Auerbach (8) already go as far as to reject it. What should we
decide ?

Let us return to the case of radiations as it was around
1950. If, at the time one had respected the scruples of rigour-
ous scientists, if one had waited until certain fundamental problems in radio-genetics (in Man particularly) had been solved in
a satisfactory manner, we would still be waiting. The anarchy
which pervades chemical pollution would still pervade pollution
by radiations; and one trembles to think what nuclear technology
would be to-day if international norms had not been fixed 25 years
ago on some basis considered as sensible and provisionally satisfactory, although imperfect.

Let us note that this imperfection promptly motivated new
research in order to improve the situation; the imperfections
have been a powerful incentive in radiobiology. To a large extent
the situation has been improved because the problem was accepted.
I think that we must follow that example for the rad-equivalents.

As already mentioned, it is not possible to envisage any general solution now. Will it be found some day ?   It is too early
to tell. But one can already tackle the problem in a limited fashion, either according to favourable circumstances - as already
done by Hahn (11) and Nauman (1) - or by choosing some particularly significant systems. Let me give two examples :

1˙ Formaldehyde is one of the mutagens most generally produced by combustions; in addition, it pollutes the atmosphere in
the vicinity of the sources of natural gas. In such a region in
the south-west of France, dosages of formaldehyde are regularly
performed. In some villages, they have revealed a permanent pollu-
tiontion of the order of $20\mu g\ m^{-3}$, with short lived peaks which
occasionally reach $100\mu\ g\ m^{-3}$. What equivalence can we give ?
The results obtained by Moustacchi and her collaborators (1, 3)
allow us to give a "limited" answer. In the lethal effect of this

compound on yeast and on Escherichia coli, these authors have
found that the centre of gravity of the spectrum of values is :

1 millimole per liter x one minute ∿ 14 rads

On this basis, 20 µg m^{-3} maintained for one year is equivalent to 5,25 rads; a value which can reasonably be considered valid
with respect to the lethal effect upon that part of the microflora
which yeast and E. coli represent. This value cannot, for the
time being, be extrapolated either to other effects or to other
species. But, concerning an important effect, it gives an order
of magnitude which cannot be ignored.

2˙) As regards carcinogenic pollution, it is possible to
quantitatively determine the in vitro cancerization of human cells
in culture, either by representative hydrocarbons (benpyrene) and
by the X or γ radiations used as references. This experiment
should yield results within about two years. It would thus be
possible to establish a "limited" equivalence (human cells in vitro) of great interest.

On the basis of these types of results, I believe that such
studies would proliferate, and would receive substantial support,
because no responsible agency could ignore the significance and
the consequence of such results. The limitations on what one
can hope to obtain, and the validity of what can be kept, will
gradually become clear. I think that, considering these positive
perspectives, the obstacles raised by the care for scientific ri-
gour - a care which every one will take as well as possible -
will not impede this great endevour whose goal is to bring order
into the field of mutagenic pollution.

This work is part of a large programme carried out at the Institut
du Radium with the financial support of "Electricité de France". I
thank Dr. John A.Lewis for his help in translating the text into
English, and Dr. Ethel Moustacchi for her useful comments.

REFERENCES

1. First European Symposium on Rad-equivalence (Institut du Radium, Orsay, May 1976). Published by EURATOM (in press).
2. Maltoni, C., Excerpta Medica Int. Congress Series, 1975, No. 375, p. 216.
3. Chanet, R., Izard, C., Houstacchi, E., Mut. Res., 1976, 35, 29.
4. Abrahamson, S., Bender, M. A., Conger, A. D., Wolff, S., Nature, 1973, 245, 460.
5. Heddle, J. A., Athanasiou, K., Nature, 1975, 258, 359.
6. Schalet, A. P., Sankaranarayanan, K., Mut. Res., 1976, 35, 341.
7. Bridges, B. A., Mut. Res., 1974, 26, 335.
8. Auerbach, C., Mut. Res., 1975, 33, 3.
9. Bridges, B. A., Environ. Health Perspect., 1973, 6, 221.
10. Crow, F., Environ. Health Perspect., 1973, 6, 1.
11. Hahn, G. M., Rad. Res., 1975, 64, 533.

A COMPARATIVE STUDY OF SEVERAL SURVIVAL RESPONSES OF MAMMALIAN CELLS AFTER UV AND AFTER X-IRRADIATION

G.M. Hahn, G.C. Li and J. Braun*

Department of Radiology, Stanford University School of

Medicine, Stanford, California 94305, USA

SUMMARY

Dose response curves were obtained from Chinese hamster cells irradiated in different growth stages and under different nutritional conditions with either X-ray or ultraviolet light. All X-ray survival curves were asymptotically exponential, while the shape of the UV-curves was bi-phasic. Survival of X-irradiated cells was much less influenced by cultural factors than survival of cells exposed to UV. Recovery of cells from potentially lethal damage had a $T\frac{1}{2}$ of 1-2 h after exposure to the ionizing irradiation, while it took considerably longer to recover 50% of the UV induced damage. Similarly "fixation" of X-ray damage was relatively rapid ($T\frac{1}{2} \simeq 30$ min) compared to "fixation" of UV damage ($T\frac{1}{2} \simeq 3.5$ h). Finally, exposure of cells to $43°C$ for 1 h immediately after X-irradiation resulted in synergistic cell killing and subsequent complete inhibition of recovery from potentially lethal damage. On the other hand similar heat treatment of UV-exposed cells resulted only in additive cell killing and did not inhibit recovery.

INTRODUCTION

The survival of cells exposed to irradiation (either UV or X-ray) is determined essentially by two factors: 1) the induction of lesions which have the potential of being lethal and 2) the subsequent repair (or lack of repair) of these lesions. To assay the first of these, survival experiments must be performed which

---

*Current address: Harvard University School of Medicine, Boston, Massachusetts, USA.

minimize the occurance of repair. Obviously this requires at
least some knowledge about the repair processes themselves. To
study these in bacterial systems, repair deficient mutants have
proven extremely useful. A complicated series of repair "pathways"
have been identified and elucidated (1). However, in the case of
mammalian cells, only very few repair deficient mutants have been
isolated (2, 3). Therefore, no comparable genetic technology has
developed for the analysis of repair phenomena in these vastly more
complicated cells. Indirect approaches are necessary to attempt to
gain the type of information provided by the mutants in the bacterial strains.

   One such indirect approach might be to modify external parameters such as nutritional factors, pH, temperature etc., and then
to examine the effects of such modifications on the shape of survival curves and on recovery kinetics. Perhaps it might be possible
to thereby cause mammalian cells to phenotypically mimic the genetic
differences found in bacteria. As a start in the direction of such
an undertaking we have collected a variety of responses of Chinese
hamster cells to X-ray and to ultraviolet (UV). In this paper we
contrast the responses of these cells to the two types of irradia-

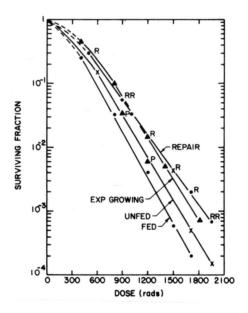

   Fig. 1.   Survival of HA1 cells exposed to graded doses of
X-ray: effects of pre- and post-irradiation conditions. P, exponentially growing cells incubated in HBSS for 3 h prior to X-ray;
R, 24 h recovery; RR, 48 h recovery. Survival curves after recovery intervals (24 h, 48 h) from unfed or plateau phase cultures
are indistinguishable.

tion. These responses show interesting differences in the sensitivity of survival to various environmental changes. Cells irradiated
with UV are most affected by nutritional differences; cells irradiated with X-ray show greater temperature effects. In addition
the repair of damage and the "fixation" of lesions (conversion of
potentially lethal to lethal lesions) follow different kinetics.

MATERIALS AND METHODS

Cells and culture conditions have been described in detail (4).
Density inhibited plateau phase cells ($\sim 10^6$ cells/cm^2) were maintained by daily medium exchange (5). X-irradiation was at 85 KeV,
9.6 mA, approximately 130 rads/min with 0.05 cm Al filtration. UV
irradiations were carried out in open petri dishes with medium removed; to insure uniform dosimetry (750 ergs/mm^2/min) cells were
scraped from the edge of the dishes and the cleared areas sterilized
with alcohol swabbing after irradiation (6). Drug exposures were
according to the protocol of Ray et al. (7), while labeling of cells
with BUdr and irradiation with visible light followed the outline
of Yang et al (8). Finally hyperthermia experiments were done
under precise control of temperature and pH as described by Hahn
(9).

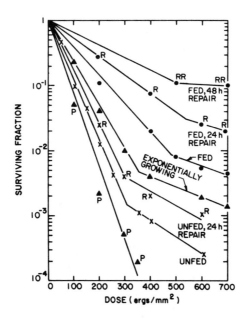

Fig. 2.   Survival of exponentially growing and plateau phase
HA1 cells exposed to graded doses of UV: effects of pre- and post
irradiation conditions. P, exponentially growing cells incubated
in HBSS 3 h prior to UV; R, 24 h recovery; RR, 48 h recovery.

RESULTS

Comparison of survival of cells exposed to X-ray or UV.

The dose responses of Chinese hamster cells exposed either to X-irradiation or to UV are shown in Fig. 1 and 2 respectively. Examined are: effects of cell density and nutritive state at time of irradiation, and effects of recovery of irradiated cells after exposure. X-ray survivals at doses beyond the shoulder are all log-linear; UV curves are biphasic, but without shoulder. In general, X-ray survival is relatively insensitive to environmental manipulations; UV survival, on the other hand, is strongly in-fluenced both by the post- and pre-irradiation milieu.

Recovery and Fixation Kinetics at 37°C.

Density inhibited cells were irradiated with 100 ergs/mm^2 (UV) or 2000 rads (X-ray). They were then overlaid with HBSS, trypsin-ized and replated at the times indicated, to assay survival as shown in Fig. 3. After UV recovery proceeds relatively slowly; beyond

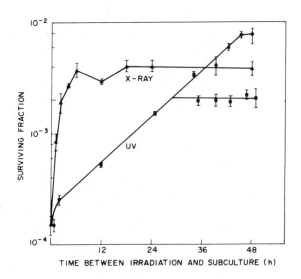

Fig. 3.   Recovery of plateau phase cells after exposure to 1500 rads of X-irradiation or to 100 ergs/mm^2 of UV irradiation. Twenty-four hours after UV irradiation, the number of cells per dish starts decreasing. The upper curve assumes that all cells, UV-inactivated or not, have equal probability of lysing. The lower curve assumes that cell lysis occurs only among the UV-inactivated cells. There is little or no cell lysis after X-irradiation, hence only one curve is shown.

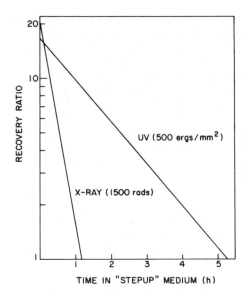

Fig. 4. Fixation of potentially lethal lesions after X- and UV irradiation. Cells were grown in medium containing 11% serum to plateau phase and then irradiated with either 1835 red X-ray or 1000 ergs/mm² UV. They were then stimulated for periods up to 3 h by being placed in "step up" medium containing 15% serum. Recovery ratio is the ratio of surviving fraction of unstimulated cells to those stimulated for the indicated period. The lines drawn are least square fits to data points from four or more experiments. Presence or absence of cycloheximide (20 µg/ml) or puromycin (50 µg/ml) did not affect rate of fixation.

24 h the survival of cells is difficult to assay because many cells begin to lyse. If it is assumed that cell lysis (resulting from a membrane defect) is not related to reproductive capacity (a function of the integrity of DNA) then recovery continues until about 48 h. On the other hand if only those cells lyse that are already reproductively dead, then the survival remains constant beyond 24 h. Possibly, actual recovery does go on beyond 24 h, but at a reduced rate. After X-irradiation, recovery is much more rapid and is complete 6 h after exposure. Lysis represents no problem; the cell number remains unchanged for up to 48 h. Fixation of damage after UV appears also to be slower than after X-irradiation. This is shown in Fig. 4. In these experiments, cells were grown in medium containing 4% fetal calf serum to a stabilized plateau density of about 4×10^5 cells/cm². The cells were then stimulated into proliferation by being "stepped up" as a result of exchange of the 4% medium to medium containing 15% serum. At

various times after "step up", the medium was removed, the cells were rinsed twice with HBSS and then overlaid with HBSS for the recovery period. In Fig. 4, the abscissa is time in the medium containing the high serum concentration, the ordinate survival. Thus, 0 time corresponds to complete recovery without any stimulation interval. The damage is completely "fixed" when extending the time in "step up" medium (before recovery) no longer reduces survival. After UV irradiation this was about 5 h; after X-irradiation only about 1 h. The presence of inhibitors of protein synthesis (cycloheximide or puromycin) in the stimulation medium did not seem to affect the rate of fixation (6).

Effect of 43° Temperature on Survival.

Figures 5 and 6 show the effects on the radiation response of plateau phase cells of 60 min of exposure of cells to 43°.   In Fig. 6 survival values were normalized with respect to the cell

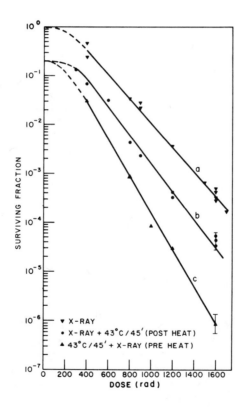

Fig. 5.   Effect of sequential X-ray and hyperthermia (43°C, 1 h) on survival of HA1 cells. a: Controls (X-ray only), b: X-ray followed by hyperthermia, c: hyperthermia followed immediately by X-ray.

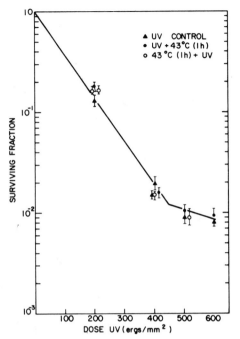

Fig. 6.   Effects of prior or subsequent hyperthermia (43°C, 1 h) on UV response. These results were normalized to controls exposed to 43°C for 1 h (but not irradiated).

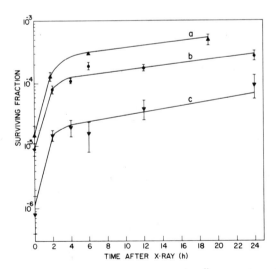

Fig. 7.   Effects of hyperthermia (43°C, 0-45 min) prior to X-ray (1600 rads) on recovery from potentially lethal damage. a: 15 min; b: 30 min, c: 45 min of hyperthermia data from Ref. 10.

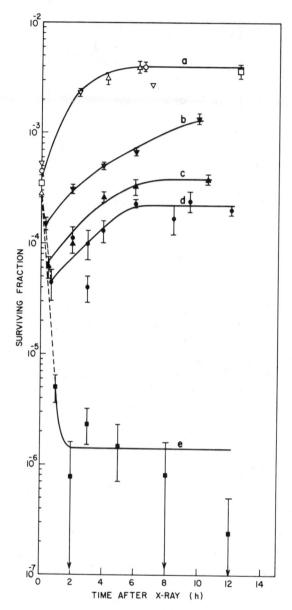

Fig. 8.   Effects of hyperthermia (0-1 h), after X-ray (1600 rads) on recovery from potentially lethal damage.   a: No heat;   b: 15 min;   c: 30 min;   d: 45 min;   e: 60 min.

killing caused by 43° exposure, while Fig. 5 shows actual survival
values. Heating either prior to or after X-irradiation modified
the X-ray survival curve. There was no sensitization by hyperthermia to UV-irradiation or vice versa.

Effect of 43° Temperature on Recovery From Potentially Lethal Damage.

    Experiments similar to those depicted by the results of Fig. 3
were performed, except that cells were exposed to 43°C for 1 h
prior to or after irradiation, and then permitted to recover at 37°
following the combined treatments. Results for hyperthermia and
X-irradiation are shown in Fig. 7 and 8. In the experiments depicted in both these figures, the combination treatments resulted
in synergistic cell killing if the cells were explanted immediately
after the combined treatments. Heating for 1 h following X-irradiation prevented subsequent recovery; inverting the order modified
the rate of recovery by introduction of a slow component but did
not reduce its magnitude. Hyperthermia and UV, in either order,
did not significantly modify either the rate or the magnitude of
the recovery from UV irradiation (Fig. 9).

DISCUSSION

    The results show some marked differences between the response
of Chinese hamster cells to the ionizing X-irradiation when compared
to the response to the non-ionizing UV radiation. These are summarized as follows:

    1. Environmental factors are much more important in deter-
mining the survival of cells to UV than to X-ray.
    2. Recovery from damage occurs after either type of irra-
diation. However, the X-ray induced lesions are repaired with a
time constant $(T_{\frac{1}{2}})$ of about 1-2 h; after UV irradiation, depending
upon the interpretation of the data, this repair has a $T_{\frac{1}{2}}$ of 12-24
h. Fixation of UV-induced damage may also proceed at a slower rate.
    3. A sequential exposure of cells to X-ray and to hyperthermia
with treatments given in either order results in synergistic cell
inactivation; after sequential treatments of UV and hyperthermia
no synergy is observed.
    4. A 1 h exposure of 43°C hyperthermia inhibits repair of
potentially lethal damage, provided X-irradiation precedes the heat
exposure. The same hyperthermic treatment either before or after
UV irradiation has little if any effect on the recovery process.

    The first two results may be related to the report by Regan
and Setlow (11) of two types of repair of DNA molecules in human
cells. After X-irradiation they find a repair process with a $T_{\frac{1}{2}}$
of 1 h; this repair involves insertion of 3-4 nucleotides per
damaged site. Repair after UV irradiation proceeds much more slowly ($T_{\frac{1}{2}}$ of $\sim$9 h) and the sizes of the patches require about 100

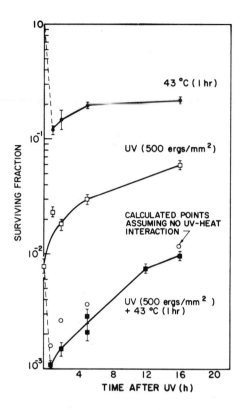

Fig. 9.   Effect of hyperthermia (43°C; 1 h) on recovery of
potentially lethal lesions in UV-irradiated HA1 cells. Top curve:
heat only; center curve: UV only; bottom curve: UV followed (or
preceded) by hyperthermia. The open circles are calculated points
obtained by multiplying the appropriate points on the heat only
and UV only curves. These points do not differ significantly from
the bottom curve, indicating an absence of heat and UV interaction.

nucleotides. It is not unreasonable to suppose that this latter
mechanism requires considerably more energy than the former, and
hence the rate and magnitude of UV repair would be more affected by
cultural conditions. However, the tacit assumption that "repair
of potentially lethal damage" is directly related to repair of DNA,
is today still an unproven hypothesis and hence arguments such as
the one just presented may justly be viewed with some skepticism.

The last two findings regarding the interaction of X-ray and

hyperthermia (and the lack of such interaction between UV and
hyperthermia) may yield more information regarding hyperthermic
potentiation of X-ray damage than about the X-ray and UV effects
themselves. The fact that heating prior to X-irradiation does not
inhibit recovery indicates that heat does not inhibit the repair
enzymes. Heat after X-irradiation may very well modify some of the
X-ray induced lesions so that repair becomes impossible. X-ray
induced lethality may be largely related to DNA strand breakage.
One can readily hypothesize that hyperthermia facilitates chemical
changes at broken ends and that these changes then prevent rejoining
of the broken strands. Cell killing by UV, on the other hand, is
mostly a result of the induction of pyrimidine dimers. There is
no reason to suspect that a relatively short heat exposure (1 h)
would modify these stable chemical configurations.

As a final comment we point out that similar recovery phenomena
also occur after exposure of cells to various drugs (6). The
kinetics of recovery vary from X-ray like (e.g., bleomycin or
nitrogen mustard) to mixed or almost UV like (methyl methane sulfonate or 5-bromodeoxyuridine and visible light).   While such
classification is useful, care must be taken not to push the analogy
too far. Until for each agent the molecular basis of recovery has
been established, kinetic correlations can only be the basis for
the formation of working hypotheses regarding the nature of the
target and the mechanisms of its repair.

REFERENCES

1. Smith, K.C., Youngs, D.A., and Van der Schueren, E., (1975)
    in Molecular Mechanisms for Repair of DNA (P.C. Hanawalt and
    R.B. Setlow, eds.), Part B, pp. 443-451, Plenum Press, N.Y.

2. Cleaver, J.E., (1973) Cancer Res. $\underline{33}$, 362-369.

3. Epstein, J., Williams, J.R., and Little, J.B., (1973) Proc.
    Nat. Acad. Sci. U.S.A. $\underline{70}$, 977.

4. Yang, S.-J., Hahn, G.M., and Bagshaw, M.A., (1966) Exp. Cell
    Res. $\underline{42}$, 130-135.

5. Hahn, G.M., Stewart, J.R., Yang, S.-J., and Parker, V., (1968)
    Exp. Cell Res. $\underline{49}$, 285-292.

6. Hahn, G.M., (1975) Radiat. Res. $\underline{64}$, 533-545.

7. Ray, G.R., Hahn, G.M., Bagshaw, M.A., and Kurkjian, S., (1973)
    Cancer Chemother. Rpts. Part I, $\underline{57}$, 473-475.

8. Yang, S.-J., Hahn, G.M., and Van Kersen-Bax, I., (1970)
    Photochem. Photobiol. $\underline{11}$, 131-136.

9.   Hahn, G.M., (1974) Cancer Res. <u>34</u>, 3117-3123.

10. Li, G.C., Evans, R.G., and Hahn, G.M., (in press) Radiat. Res.

11. Regan, J.D., and Setlow, R.B., (1974) Cancer Res. <u>34</u>, 3718.

IONIZING AND ULTRAVIOLET RADIATIONS : GENETIC EFFECTS AND REPAIR

IN YEAST

E. MOUSTACCHI, R. CHANET and M. HEUDE

Institut du Radium, Biologie, Bâtiment 110

91405 - ORSAY, FRANCE

Yeast is one of the classic eucaryotic multicellular organisms studied by radiation biologists. Well defined by genetic methods, Saccharomyces cerevisiae also provides a powerful tool in recent environmental mutagenesis studies.

Some of the early notions concerning the influence of ploidy on the radiosensitivity (1), the role of the cell position in the cycle (2-3) and the existence of DNA repair mechanisms (4) were based on the analysis of radiation induced killing in yeast.

I. The number of genes which govern the response to radiations

In order to unravel the mechanisms involved in the control of the response to radiations, the modern contribution consisted as a first step in isolating radiation-sensitive mutants (5-12). Such mutants, selected on the basis of their radiosensitivity for reproduction death, occupy at least 30 distinct genetic loci (12). Their characterisation included delineation of their effect on cross-sensitivity to different types of radiations and chemicals (13-17) or elevated temperature (18-19), their spontaneous (20) and radiations (8-10, 21-23) or chemicals (24-25) induced nuclear and mitochondrial (8, 26) mutations frequencies, their sensitivity to inter and intragenic recombination induction (27-28), the variations in radiosensitivity in synchronized cell cycle (29-31).

The symbol rad is given to such radio-sensitive mutants. The genetic loci which primarly confer UV sensitivity are numbered from 1 to 50 where rad50 to upwards designate genes which principally affect sensitivity to ionizing radiations (33). Among the 30 rad loci mentioned, 13 govern specifically the UV response, 5 affect principally the X-rays response and 12 control both the UV and X-rays sensitivities. Several of these genes modify the abili-

197

ty of the homozygous rad diploid to undergo normal meiosis and the
ability to the spores to germinate (32).

Other mutants selected for their reduced capacity for UV mutagenesis (rev$_1$ to rev$_3$ (34) and umr$_1$ to umr$_3$ (35))or for γ rays
and UV induced recombination (rec$_1$ to rec$_5$ (36)) have been identified.

As discussed by Haynes (37), the existence of so many genes
which interfere with the response to irradiation raises important
questions regarding the multiplicity of factors involved in the
control of recovery or radioresistance in wild type eucaryotic
cells. It has been shown that DNA repair inhibition is certainly
involved for some of the mutants but it is far from excluded that
part of these gene products affect radiosensitivity in other ways
not yet defined.

II. Genes interactions

The first step of characterization of the different genes
being sufficiently documented, yeast radiobiologists have been
faced with the determination of the interactions among these genes. The aim of such studies is to establish the different pathways involved and their interelation with respect to recognition
and reshaping of damages induced in DNA by ultraviolet and ionizing radiations. A preliminary approach consisted in comparing the
UV and X-rays response of single mutants to double or multiple mutant strains (14, 68-70) with respect to cell killing or more recently for induced mutagenesis (23). Three situations were found :

1) Epistasis which implies that the genes products mediate
consecutive steps in the same pathway or constitute a part of a
multimeric enzymatic complex. For example, such a response has
been found for rad$_1$, rad$_2$, rad$_3$, rad$_4$ (70) and rad$_{22}$ (23). These
genes govern specifically the UV-sensitivity, in all cases the
mutant strain demonstrate a high sensitivity to UV-induction of
nuclear mutations (8-11, 22-23) and recombination (8, 10, 27), as
compared to the wild type strain. This phenotype is reminiscent of
E.coli uvrA. As shown later these genes belong to the excision
repair pathway.

2) Additivity which indicates that the genes products act on
different substrates in independant pathways. Such an interaction
has been found between rad$_1$ or rad$_2$ and rad$_6$ (68) ; this last mutant is sensitive to both X-rays and ultraviolet radiations induced killing whereas it is particularly resistant to mutations induction (23). This phenotype reminds the Escherichia coli rec$^-$ or
lex$^-$ mutants. The rad$_6$ function is also necessary for the completion of meiosis and for X-rays induced mitotic crossing over.

3) Synergism which indicates that the genes control different steps in two pathways which compete for the same substrate. This situation occurs for rad_{18} (which is sensitive to both UV and X-rays) which is synergistic with both rad_1 or rad_2 and rad_{52} (36). This last mutant is specifically X-rays sensitive (7-9), the mutation eliminates the resistance of budding yeasts cells and removes the shoulder on the survival curves of homozygous diploid cells (38).

Due to the great number of genes involved in the yeast response to radiations only a small number of relevant combinations of double mutants have been investigated up to now. The work accomplished up to now includes 12 genes and tentative schemes of repair pathways have been proposed (68,37). It appears that in addition to photoreactivation (a phr^- mutant has been isolated (39)), there exist at least three UV-repair pathways and two X-rays pathways. Two of them appear to be common to both UV and X-rays. The phenotypes of rad_6, rev and umr mutants support the existence of branched UV mutation pathways (23). The RAD6 dependant pathways is likely to be error-prone (23).

However until a definitive molecular characterization of the steps is provided one should keep in mind reservations concerning pathways analysis based on phenotypes of single and multiple mutants. For instance due to a given mutation, gaps may be generated by defects in the filling steps of the excision-resynthesis repair, this may in turn offer unusual substrates for enzymes not normally acting in wild type cells. Moreover, the regulatory function of enzymes may be abnormal in the mutants providing opportunities for interactions not normally occuring in wild type cells. Finally it should be reminded that different alleles of a given rad gene may exhibit different properties (40-41). At the moment certain of the distinct pathways for repair have been characterized at the macromolecular level in yeast.

III. Pathways biochemically characterized

1) The excision-repair of UV-induced pyrimidine dimers :

From the kinetics of pyrimidine dimers induction by 254 nm light it can be estimated that 1 erg/mm^2 of UV light applied to yeast cells will generate 22 (42) to 24 (43) pyrimidine dimers in each genome. Wild type haploid yeast cells will tolerate around 300 erg/mm^2 (6000 dimers at least per genome) without showing any detectable ill effects. The existence of very sensitive rad mutants indicates that the resistance of wild type cells to these potentially lethal lesions is due to metabolic repair functions of the cells.

However, one should be aware of course that the preponderant finding of pyrimidine dimers at "biological" UV doses cannot be a proof of their being the unique and most important lesions. Evaluation of the role of the pyrimidine dimers in UV-lethality for a given organism is generally based on determination of photoreactivable sector —a 100 % photoreactivation argues in favour of the pyrimidine dimers as being the potentially lethal lesion-, and on the number of dimers being lethal in a mutant deprived of any UV dark repair mechanism —in principle one should find 1 dimer per lethal hit for such a mutant-. For Saccharomyces cerevisiae the photoreactivable sector varies between 40 to 66 % according to the strain. This suggests that pyrimidine dimers certainly play an important role in lethality but unless some unknown factor(s) reduces the photolyase efficiency (chromosomal structure for instance) other lesions may be involved. It may be noticed that the non-photoreactivable damages are still subject to repair by dark liquid holding since the survival of cells is enhanced when visible light is applied to irradiated cells prior to liquid holding as compared to liquid holding alone or to photoreactivation alone (44). Directly relevant to the problem is the observation showing that in a $\underline{rad}_1$ $\underline{rad}_{18}$ $\underline{rad}_{57}$ multiple mutant a LD_{37} of 0.06 ergs/mm^2 is found (68). This implies, on the basis of 22 dimers formed per erg/mm^2, 1 to 2 dimers per lethal hit. This data suggest that nearly all of the lethality due to UV can be accounted by the formation of pyrimidine dimers in DNA. Moreover it shows that the products of these three genes are located on the three main pathways responsable of the wild type resistance to UV (68).

It has been shown that one of these pathways is the excision-resynthesis process. Mutants $\underline{rad}_{1-1}$ (45), $\underline{rad}_{1-3}$ (44), $\underline{rad}_{1-2}$ (45) and $\underline{rad}_2$ (46) fail to remove dimers from their DNA in conditions where the wild type can. It appears that the incision step is blocked at least for $\underline{rad}_{1-3}$ from indirect evidence (44). For a repair proficient strain virtually all the dimers induced in the nuclear DNA are removed on dark incubation (71). However, like in bacteria and in mammalian cells the excision system is saturated at relatively low doses of UV (47). The excision process is dependant upon the cellular energy metabolism. It requires the integrity of the mitochondrial system since a respiratory deficient cytoplasmic mutant is more UV sensitive than the corresponding wild type ; it does not show a recovery of survival on dark liquid holding and it demonstrates a reduced amount of excision of pyrimidine dimers (48).

Very little is known about the degradative and resynthesis steps in yeast. The degradation of nuclear DNA after a UV treatment is limited to a few percent (10 to 15 %) for mitotic cells (49-50). Whereas it is more extensive for meiotic cells (up to 40 %) (51). In this last case the recombination frequency is reduced and the meiotic segregation of chromosomes is prevented. From the response to dark liquid holding followed by a visible light treatment it is likely that a few of the $\underline{rad}$ mutants (rad_{50}, rad_{15}) may be involved in the gap filling steps (52) but this remains to be proved by a biochemical approach.

It is worth to mention that pyrimidine dimers when induced
in mitochondrial DNA are photoreactivatable (53) but are not subject to a controlled excision-repair neither at high (53) nor at
low (71) UV doses. A dose-dependant degradation of this DNA is
observed on dark incubation of these cells (S. Hixon and E.
Moustacchi, submitted).

2) Repair of double-strand breaks :

The induction frequency of double-strand breaks induced by γ-rays
at biologically relevant doses has been measured in wild type
yeast cells by the reduction of molecular weight of DNA in neutral
sucrose gradients (54-55). The induction frequency 0.58 x 10^{-10}
breaks per Dalton-Krad is within the range of values found for
bacteria and mammalian cells (0.4 to 2.7 x 10^{-10}).

The repair of double-strand breaks reflected by the reappea-
rance of large molecular weight DNA on incubation of irradiated
cells has been demonstrated in logarithmically growing haploid
(54) and diploid (55) wild type yeast cells. For doses up to 100
krad the wild type can tolerate 34 to 38 double-strand breaks. The
rad_{52} mutant strain shows an increase in sensitivity to ionizing
radiations due predominantly to an increase in dominant lethality
(38-54) ; the diploids and triploids homozygous for rad_{52} are more
radiosensitive than rad_{52} haploid cells. On post-irradiation incubation this mutant is uncapable to rejoin the double-strand
breaks. The number of double-strand breaks per lethal event is
found to be close to 1 in the rad_{52} mutant (55) which suggests
that this type of lesion if unrepaired is a major class of lethal
event in yeast.

The rad_{52} mutation is also important because it reduces the
X-ray induction of intragenic recombination (56) (rad_{52} is alle-
lic to rec2, Mortimer personal communication), and demonstrates a
high mutator activity (20). It is not yet clear how these processes are related to the presence of unrepaired double-strand
breaks intermediates.

3) Repair of single-strand breaks :

As expected ionizing radiations induce single-strand breaks in
yeast DNA as seen from the reduction of the molecular weight in
alkaline sucrose gradients. Ninety percent of the breaks were
found to be repaired within one hour by yeast protoplasts (Cox and
Haynes, personal communication). Intact cells may be capable to
perform complete repair. One X-ray sensitive mutant rad_{50} was
found to be defective in this type of repair. Improved methods
are now available (57) which may allow in next future a screening
of different mutants likely to be affected in the repair of single-strand breaks.

For the moment there is no direct demonstration of post-replication repair in yeast. Several genetic studies argue in favor

of the existence of such a mechanism in this organism (27-28, 68).
The availability of respiratory competent strains incorporating
thymidine-5'-monophosphate specifically into their DNA (58), the
progresses accomplished in the technics of "heavy-light" labeling
of DNA (59-60) and for the inhibition of nucleases activities
opens the door to such a biochemical study. Since there have been
no study on the genetic control of post-replication repair in eucaryotic systems and that it has not yet been possible to relate
such repair to other radiation-induced genetic events, such an
analysis in the yeast system may be extremely rewarding.

The actual state of knowledge in the field of yeast repair
pathways is summarized in fig. 1.

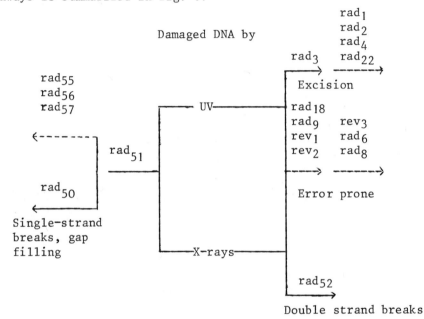

Fig. 1 : UV and X-rays pathways for Saccharomyces cerevisiae based
         on data refered in the text with the exception of rad_{55},
         rad_{56} and rad_{57} (Nakai, personal communication). Broken
         lines indicate an undetermined number of steps.

IV. Variations of radiosensitivity in synchronized cell cycle

In an early study of radiation effects in yeast (2), it was
observed that dividing cultures were more radioresistant than stationary cultures. This observation was later shown to be the result of a much greater resistance to X-rays in budding as compared
to interdivisional cells (3). Recently it has been demonstrated
that this radioresistance is correlated with the DNA synthetic
phase of the cell cycle.

For both UV and γ-rays the maximal radioresistance is found in G2 phase whereas the higher radiosensitivity occurs in G1 and early S period (29, 61-63). The pattern of fluctuations is modified or even suppressed by liquid holding of the cells after irradiation (62), by the addition of caffeine (63) or by specific mutations in the repair pathways (30). It should be noticed that none of the actually identified rad mutants demonstrate a clearcut enhancement in γ-rays sensitivity in G1 phase as compared to the wild type cells. Slightly more pyrimidine dimers are found to be induced at the most UV sensitive stage of the cycle (64). However these differences are not sufficient to account for the amount of variations in radiosensitivity. The excision-resynthesis type of repair appears to act very efficiently and constantly throughout the cell cycle from the comparison of wild type to rad_{1-3} synchronized cells (30). On the contrary the repair mechanism involved in intragenic recombination (blocked in rec_5 mutant for example) would mainly act during S and G2 phases (30). The functional integrity of the RAD50 and RAD51 genes is also required for the end of S and G2 radioresistance (R. Chanet, unpublished). One of the main conclusion of these experiments is that an efficient repair process takes place in G2 phase. The proximity of homologous chromatides kept together by the centromere may favor such a process (31).

Yeast cultures progressing from the exponential to the stationary phase of growth also show changes in cell sensitivity to heat stock at 52° (18) and to chemicals such as formaldehyde (65), ethylmethane sulfonate, nitrous acid and mitomycin C (66). A number of the rad mutants demonstrate also a higher sensitivity to certain of these agents (17, 19, 24). This suggests that there are common steps in the repair of radiations and heat or chemicals induced DNA lesions. However in contrast to sensitivity to radiations yeast cells in exponential phase (or in end of S and G2 phases for synchronized cultures) were found to be most sensitive to cell death after treatments with heat or chemicals. It is likely that these last agents may induce more DNA lesions at certain stages in the cycle (67). Repair processes certainly play a role but in that case it may be a minor factor.

References

1. Latarjet, R., and Ephrussi, B., (1949) C.R. Acad. Sci. Paris, 229, 306-308.
2. Lacassagne, A., and Holweck, F., (1930) C.R. Soc. Biol. Paris, 104, 1221-1223.
3. Beam, C.A., Mortimer, R.K., Wolfe, G.R., and Tobias, C.A., (1954) Arch. Biochem. Biophys., 49, 110-122.
4. Patrick, M.H., Haynes, R.H., and Uretz, R.B., (1964) Radiation Res., 21, 144-163.
5. Nakai, S., and Matsumoto, S. (1967) Mutation Res., 4, 129-136.

6. Snow, R., (1967) J. Bacteriol., 94, 571–575.
7. Cox, B.S., and Parry, J.M., (1968) Mutation Res., 6, 37–55.
8. Moustacchi, E., (1969) Mutation Res., 7, 171–185.
9. Resnick, M.A., (1969) Genetics, 62, 519–531.
10. Zakharov, I.A., Kozina, T.N., and Federova, I.V., (1970) Mutation Res., 9, 31–39.
11. Averbeck, D., Laskowski, W., Eckhardt, F., and Lehmann–Brauns, E., (1970) Mol. Gen. Genetics, 107, 117–127.
12. Game, J.C., and Cox, B., (1971) Mutation Res., 12, 328–331.
13. Brendel, M., Khan, N.A., and Haynes, R.H., (1970) Mol. Gen. Genetics, 106, 289–295.
14. Brendel, M., and Haynes, R.H., (1973) Mol. Gen. Genetics, 125, 197–216.
15. Waters, R., and Parry, J.M., (1973) Mol. Gen. Genetics, 124, 135–143.
16. Laskowski, W., and Lehmann–Brauns, E., (1973) Biophysika, 10, 51–59.
17. Zimmerman, F.K., (1968) Mol. Gen. Genetics, 102, 247–256.
18. Schenberg–Frascino, A., and Moustacchi, E., (1972) Mol. Gen. Genetics, 115, 243–257.
19. Evans, W.E., and Parry, J.M., (1972) Mol. Gen. Genetics, 118, 261–271.
20. Von Borstel, R.C., Cain, K.T., and Steinberg, C.M. (1971), Genetics, 69, 17–27.
21. Resnick, M.A., (1969) Mutation Res., 7, 315–332.
22. Lawrence, C.W., Stewart, J.W., Sherman, F., and Christensen, R., (1974) J. Mol. Biol., 85, 137–162.
23. Lawrence, C.W., and Christensen, R., (1976) Genetics, 82, 207–232.
24. Prakash, L., (1974) Genetics, 78, 1101–1118.
25. Prakash, L., (1976) Genetics, 83, 285–301.
26. Moustacchi, E., (1971) Mol. Gen. Genetics, 114, 50–58.
27. Hunnable, E.G., and Cox, B., (1971) Mutation Res., 13, 297–308.
28. Kowalski, S., and Laskowski, W., (1975) Mol. Gen. Genetics, 136, 75–86.
29. Chanet, R., Williamson, D.H., and Moustacchi, E., (1973) Biochim. Biophys. Acta, 324, 290–299.
30. Chanet, R., Heude, M., and Moustacchi, E., (1974) Mol. Gen. Genetics, 132, 33–40.
31. Fabre, F., (1973) Radiation Res., 56, 528–539.
32. Game, J.C., and Mortimer, R.K., (1974) Mutation Res., 24, 281–292.
33. Game, J.C., (1974) in : "Molecular Mechanisms for repair DNA" Ed. by P.C. Hanawalt and R.B. Setlow, Plenum Press, New York, Part B (1975) pp. 541–544.
34. Lemontt, J.F., (1971) Genetics, 68, 21–33.
35. Lemontt, J.F., (1974) Genetics Suppl., 73, 153–159.
36. Rodarte–Ramon, U.S., and Mortimer, R.K., (1972) Radiation Res.,

$\underline{49}$, 133–147.
37. Haynes, R.H., (1975) in : "Molecular Mechanisms for repair of DNA" Ed. by P.C. Hanawalt and R.B. Setlow, Plenum Press, New York, Part B., pp. 529–540.
38. Ho, K.S., and Mortimer, R.K., (1973) Mutation Res., $\underline{20}$, 45–51.
39. Resnick, M.A., (1969) Photochem. Photobiol., $\underline{9}$, 307–312.
40. Waters, R., and Parry, J.M., (1973) Mol. Gen. Genetics, $\underline{124}$, 145–156.
41. Parry, J.M., and Parry, E.M., (1973) Radiation Res., $\underline{55}$, 29–40.
42. Fath, W.W., and Brendel, M., (1975) Z. Naturforsch., $\underline{30c}$, 811–817.
43. Unrau, P., Wheatcroft, R., Cox, B., and Olive, R., (1973) Biochim. Biophys. Acta, $\underline{312}$, 626–632.
44. Waters, R., and Moustacchi, E., (1974) Biochim. Biophys. Acta, $\underline{353}$, 407–419.
45. Unrau, P., Wheatcroft, R., and Cox, B., (1971) Mol. Gen. Genetics, $\underline{113}$, 359–362.
46. Resnick, M.A., and Setlow, J.K., (1972) J. Bacteriol., $\underline{109}$, 979–986.
47. Waters, R., and Moustacchi, E., (1975) J. Bacteriol., $\underline{121}$, 901–906.
48. Waters, R., and Moustacchi, E., (1975) Photochem. Photobiol., $\underline{21}$, 441–444.
49. Hatzfeld, J., (1973) Biochim. Biophys. Acta, $\underline{299}$, 43–53.
50. Wilmore, P.J., and Parry, J.M., (1976) Mol. Gen. Genetics, $\underline{145}$, 287–291.
51. Salts, Y., Simchen, G., and Pinon, R., (1976) Mol. Gen. Genetics, $\underline{146}$, 55–59.
52. Parry, J.M., and Parry, E.M., (1969) Mutation Res., $\underline{8}$, 545–556.
53. Waters, R., and Moustacchi, E., (1974) Biochim. Biophys. Acta $\underline{366}$, 241–250.
54. Ho, K., (1975) Mutation Res., $\underline{30}$, 327–334.
55. Resnick, M.A., and Martin, P., (1976) Mol. Gen. Genetics, $\underline{143}$, 119–129.
56. Resnick, M.A., (1975) in : "Molecular Mechanisms for repair of DNA" Ed. by P.C. Hanawalt and R.B. Setlow, Plenum Press, New York, Part B., pp. 549–556.
57. Forte, M.A., and Fangman, W.L., (1976) Cell, $\underline{8}$, 425–431.
58. Wickner, R.B., (1975) in : "Methods in Cell Biology" Ed. by D.M. Prescott, Academic Press, New York, Vol. XI, pp. 295–302.
59. Williamson, D.H., and Fennell, D.J., (1974) Mol. Gen. Genetics $\underline{131}$, 193–207.
60. Leff, J., and Lam, K.B., (1976) J. Bacteriol., $\underline{127}$, 354–361.
61. Perper, T., (1975) Radiation Res., $\underline{63}$, 97–118.
62. Chanet, R., and Heude, M., (1974) Mol. Gen. Genetics, $\underline{131}$, 21–27.

63. Nunes de Languth, E., and Beam, C.A., (1973) Radiation Res., 53, 226–234.
64. Chanet, R., Waters, R., and Moustacchi, E., (1975) Int. J. Rad. Biol., 27, 481–485.
65. Chanet, R., Izard, C., and Moustacchi, E., (1975) Mutation Res., 33, 179–186.
66. Parry, J.M., Davies, J.P., and Evans, W.E., (1976) Mol. Gen. Genetics, 146, 27–35.
67. Howell-Saxton, E., Smith, D.C., Zamenhoff, S., and Zamenhoff, P.J., (1974) Mutation Res., 24, 227–237.
68. Cox, B.S., and Game, J., (1974) Mutation Res., 26, 257–264.
69. Game, J.C., and Cox, B.S., (1973) Mutation Res., 20, 35–44.
70. Game, J.C., and Cox, B.S., (1972) Mutation Res., 16, 353–362.
71. Prakash, L., (1975) J. Mol. Biol., 98, 781–795.

ATAXIA TELANGIECTASIA: AN INHERITED HUMAN DISEASE INVOLVING RADIOSENSITIVITY, MALIGNANCY AND DEFECTIVE DNA REPAIR

M.C. Paterson, B.P. Smith, P.A. Knight and A.K. Anderson

Biology and Health Physics Division, Atomic Energy of Canada Limited, Chalk River Nuclear Laboratories, Chalk River, Ontario, KOJ 1J0, Canada

SUMMARY

A unique feature of ataxia telangiectasia (AT), a hereditary multisystem disease, is extreme sensitivity to ionizing radiation, observed both clinically and in cell culture. Hence, we have measured the DNA repair capabilities of ten diploid fibroblast strains derived from unrelated AT donors, following anoxic ^{60}Co γ-irradiation. Compared to two control strains, six of the ten mutant strains are markedly deficient in γ-induced repair replication. Two defective strains were defined further. While capable of rejoining single-strand breaks normally, both are impaired in the initial incision step in excision repair of base defects, assayed as γ-modified sites sensitive to a *Micrococcus luteus* endonuclease activity. Cell fusion studies assign three repair-deficient strains to two complementation groups; this result, coupled with a normal repair-replication ability in four of the ten AT strains, indicates genetic heterogeneity in the disease. AT strains appear otherwise normal, including their ability to repair UV damage. Aside from providing molecular insight into this complex disorder, our findings characterize AT as a γ-ray analogue of the UV-sensitive skin disease, xeroderma pigmentosum. Moreover, since AT patients are cancer-prone, faulty DNA repair is implicated in neoplastic transformation. Finally, given that (i) impaired embryonic differentiation best explains the clinical features of AT and (ii) defective DNA repair is of etiological relevance, we are led to conclude that DNA damage can lead to congenital malformations; thus, enzymatic DNA repair processes play a vital role in normal neonatal development.

INTRODUCTION

Ataxia telangiectasia (AT) (Louis-Bar syndrome) is an autosomal recessive, multisystem disorder in man clinically exhibiting
progressive cerebellar ataxia, oculocutaneous telangiectasis and
recurrent sinopulmonary infection (1-3). The infectious complication is associated with an immunodeficient state (4) often involving both immune systems, that is, humoral (e.g. diminished-to-
absent levels of immunoglobulins IgA, IgE) and cellular (e.g.
thymic hypoplasia, delayed skin-test reactivity) (3-5). Accessory
manifestations typically include increased incidence of lympho-
reticular malignancy (2-4), bronchiectasis (1,2), markedly elevated
concentrations of serum-α-fetoprotein (6), and spontaneous chromosomal fragility (7).

Several reports (8-10) dealing with AT patients receiving
standard radiotherapy for tumor treatment illustrate yet another
feature of this complex disease: an unusually severe response to
X-irradiation culminating in death of the patient. Laboratory
studies indicate that radiosensitivity is also displayed at the
cellular level. Ionizing irradiation induces chromosomal aberrations at elevated levels in leukocytes obtained from AT donors
(11); in addition, diploid fibroblasts cultured from afflicted
patients are deficient in their ability to form colonies following
gamma (γ)-irradiation (12). In view of the well-known causal
relationship between defective enzymatic repair of DNA lesions
produced by an external agent and enhanced cell killing by that
agent (13,14), these observations prompted us to determine the DNA
repair properties of AT strains in response to ^{60}Co γ-irradiation.
Our results, summarized here, provide direct biochemical evidence
that many AT strains are, in fact, deficient in enzymatic DNA repair — specifically, an excision-type repair process active on
γ-ray-modified base residues. Moreover, we demonstrate the heterogeneous nature of the genetic defect in the disease, as indicated
by (i) complementation between excision-deficient strains after
cell fusion and (ii) the existence of strains which, although
derived from patients clinically diagnosed as suffering from AT,
have no apparent deficiency in their capacity to repair γ-ray-
damaged DNA.

METHODS

Twelve human diploid strains, two normal and ten AT, were
used: CRL 1141 and CRL 1147, established from healthy volunteers;
and CRL 1312, CRL 1343, CRL 1347, AT3BI, AT4BI, AT5BI, AT7BI,
AT81CTO, AT82CTO, and AT97CTO, derived from AT patients of unrelated kindred. The two normal and the first three AT strains
were obtained from the American Type Culture Collection, Rockville,
Maryland; the next four AT strains were kindly provided by

Dr. A.M.R. Taylor, University of Birmingham; and the last three were a gift of Dr. M. Buchwald, The Hospital for Sick Children, Toronto. The cell culture methodology practiced here has been described previously (15).

In brief, our experimental design for assessing DNA repair was as follows: monolayer cultures in late logarithmic growth phase were γ-irradiated anoxically and subsequently monitored for their ability to execute two enzymatic DNA repair processes, (i) rejoining of single-strand breaks (i.e. scissions directly induced in the phosphodiester backbone of individual DNA strand by γ-rays) and (ii) excision repair of γ-modified base residues, as measured by two complementary techniques, conventional equilibrium density-gradient centrifugation (15,16) and our *in vitro* enzymatic assay (15,17). (By administering the radiation treatment in a nitrogen, rather than an air, atmosphere, the incidence of base defects relative to strand breaks was markedly increased, thereby improving, as will become apparent later, the signal-to-noise ratio of both techniques [15].)

Genetic complementation analysis was performed on three repair-deficient AT strains using a standard somatic cell hybridization technique (18,19). In this analysis, selected strains were first fused in pairs, employing a high titre of inactivated Sendai virus to produce cell populations primarily consisting of multinucleate heterokaryons, each containing numerous intact nuclei and cytoplasm derived from both fusion partners. Following overnight incubation, the fused cultures were irradiated, and assayed for their repair capability, as measured by the equilibrium centrifugation technique.

All experimental protocols and treatment schedules followed here, including those involved in carrying out (i) ^{60}Co γ-irradiation (24-27 krad min^{-1}) under anoxic conditions, (ii) equilibrium density-gradient centrifugation, (iii) the *in vitro* enzymatic assay and (iv) complementation analysis of fused cell cultures, are described in detail elsewhere (15-19) and in general in the figure captions.

RESULTS

Utilizing the equilibrium centrifugation technique, normal and AT strains were initially evaluated for their ability to carry out γ-ray-induced repair replication, since this repair indicator is considered to reflect the total number of nucleotides incorporated into DNA at damaged sites in the course of repairing both strand-breakage and base damage (20,21). The extent of repair replication performed by two normal and six mutant strains after receiving graded doses of anoxic ^{60}Co γ-rays is shown in Fig. 1. The magnitude of repair replication detected in the normal strains,

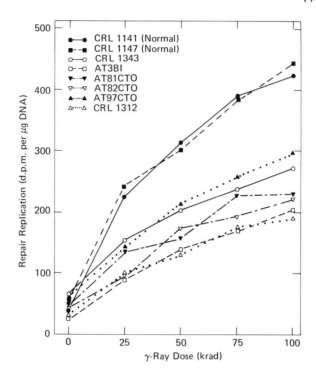

Figure 1. DNA repair replication in indicated normal and AT
strains as a function of anoxic ^{60}Co γ-irradiation. Monolayer
cultures of each strain were first incubated for 2 h in growth
medium (90% modified F12 - 10% calf serum) supplemented with 6.5
µM bromodeoxyuridine and 1 µM fluorodeoxyuridine, irradiated in a
nitrogen gas phase, immediately labelled for 2 h in fresh pre-irradiation medium containing 10 µCi ml^{-1} [*methyl*-^{3}H]thymidine (dThd;
specific activity, 45-53 Ci mmol^{-1}) and 1 mM hydroxyurea and finally
incubated for 1 h in fresh pre-irradiation medium (15). After cell
lysis, the radioactive, density-labelled DNA was purified and
centrifuged to equilibrium in neutral NaI density gradients, thereby physically separating 'light' DNA having undergone repair synthesis from 'heavy' DNA resulting from *de novo* synthesis (15,16).
To improve resolution, 'light' DNA was pooled and recentrifuged as
above. The gradients were then fractionated and the DNA content and
radioactivity in each fraction were determined to measure repair
replication, expressed as specific activity (d.p.m. per µg DNA) in
'light' DNA. Datum points represent the mean of three or more
measurements (s.d. ≤15%). (For experimental details, see refs.
15,17).

CRL 1141 and CRL 1147, rises steadily with increasing dose,
although the rate of increase is not proportional to the dose.

While the general response of the AT strains is also dose-dependent, the amount of repair replication is clearly less in the mutant strains than in the normal strains, ranging from ∿45% in CRL 1312 to ∿70% in AT97CTO at the maximal dose of 100 krad. These findings suggest that the AT strains are impaired in either strand-rejoining or excision repair or perhaps both processes.

To pinpoint the defective repair mechanism in the AT strains, we monitored strand-rejoining and excision repair separately in a normal (CRL 1141) and a mutant (AT3BI) strain, using the *in vitro* enzymatic assay. Fig. 2 shows typical results of this experiment for the two strains following exposure to 50 krad of γ-radiation in a nitrogen atmosphere. As is evident in the left panel, single-strand breaks disappear from the DNA of mutant cells with expo-nential kinetics very similar to those seen for normal cells, indicating the presence of a proficient strand-rejoining process in the AT3BI strain. Two other mutant strains, CRL 1312 and CRL 1343, also reunite strand breaks with normal kinetics (data not shown). These findings are in accord with those of others for different AT strains (12,22).

Gamma-induced endonuclease-sensitive sites, however, disappear from the DNA of AT3BI cells at a rate approximately three times slower than that observed in normal cells (Fig. 2, right panel). For example, whereas the normal strain acts on all except ∿20% of the initial sites within 3 h, the mutant eliminates no more than 50% after 6 h. A second repair replication-deficient AT strain, CRL 1312, also displays a similar impairment in site removal (unpublished data). Inasmuch as (i) the sequence of reactions promoting the rejoining of single-strand breaks seems to mimic the last three steps in the excision-repair process (excision of nucleotides at the damaged site, repair synthesis to fill the re-sulting gap, ligation to restore strand continuity) and is executed by a similar, if not an identical, battery of enzymes, and (ii) this rejoining mechanism appears normal in AT strains, it seems therefore that the kinetics of site removal reflect the initial in-cision step in excision repair of γ-ray-modified base defects. Thus, the most reasonable explanation is that both AT3BI and CRL 1312 cells fail to incise near damaged bases normally and are therefore defective in the first step of excision repair.

In microbial systems the excision-type mechanism mediating the repair of γ-modified base defects seems to parallel the ex-tensively studied process operating on ultraviolet (UV) light-induced cyclobutane pyrimidine dimers (21). It was therefore of interest to follow the metabolic fate of UV-induced DNA damage in AT strains. A proficient excision-repair response was observed in the three strains (AT3BI, CRL 1312, CRL 1343) studied, as demonstrated by normal kinetics for dimer removal (assayed enzy-matically, using a dimer-specific endonuclease purified from

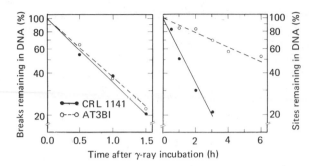

Figure 2. Disappearance of γ radioproducts from the DNA of normal (CRL 1141) and AT (AT3BI) strains following exposure to 50 krad of γ-rays delivered under anoxia. Time courses for the disappearance of single-strand breaks (left panel) and γ-induced endonuclease-sensitive sites (right panel) were constructed as follows: (i) ^{3}H-labelled cells were irradiated and incubated for varying times in parallel with ^{14}C-labelled, unirradiated (control) cells. (ii) The parallel cultures were collected, the cells were mixed, and their DNA's containing [3]H- and ^{14}C-dThd were purified; then one DNA sample was incubated in the presence and a second in the absence of a crude cell-free *M. luteus* extract (essentially fraction II, ref. 31) known to possess endonuclease activity toward γ-modified base defects (32,33). (iii) The number of breaks in the irradiated, relative to the unirradiated, DNA in the two samples was then determined by computer analysis of differences in the velocity sedimentation patterns of denatured DNA in alkaline sucrose gradients (17). The value for the DNA sample incubated alone was taken as the number of clean breaks (actually, clean breaks and alkali-labile lesions [34,35]) while the extra breaks detected in the extract-treated sample became a measure of the number of endonuclease-sensitive sites. Data in each curve were normalized by expressing the yield of radioproducts in incubated cultures as a percentage of that found for the corresponding non-incubated culture (100% ∿0.8 breaks, ∿1.2 sites per 10^7 daltons DNA). Datum points represent the mean of three measurements (s.d. ≤13%). (From Paterson *et al.*, Nature *260*, 444-447 [1976] with permission of Macmillan Journals Ltd.)

Micrococcus luteus as the test enzyme) and normal levels of UV-induced repair replication (15, unpublished data).

Genetic heterogeneity in ataxia telangiectasia is suggested by the finding that four of the ten AT strains under study carry out γ-ray-stimulated repair replication to an extent not significantly different from that attained by the two normal strains (Fig. 3), implying that certain AT strains are not

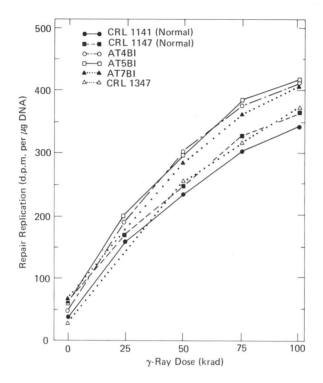

Figure 3. DNA repair replication in indicated normal and AT
strains in response to varying doses of anoxic ^{60}Co γ-rays. The
experimental design was as in Fig. 1. Datum points represent the
mean of three or more measurements (s.d. ≤10%).

deficient in either strand-rejoining or excision repair of γ radioproducts. We have examined one strain, AT4BI, in greater detail
with the aid of the enzymatic assay. Data to be published elsewhere confirm that both repair processes are indeed normal. Thus,
although derived from patients having characteristic clinical
features of the disease, some mutant strains possess no apparent
repair deficiency.

     As a preliminary step in the systematic study of the precise
biochemical defect in the AT strains defective in excision repair,
we have further characterized the mutations in three strains
(AT3BI, CRL 1312, CRL 1343) by performing a complementation analysis
on appropriate fused cultures. As is evident in Fig. 4, the level
of γ-induced repair replication in fused hybrid cultures of AT3BI/
CRL 1343 parentage is appreciably higher than in fused cultures
containing only one cell type (either AT3BI/AT3BI or CRL 1343/
CRL 1343). We are led to conclude that complementation can occur

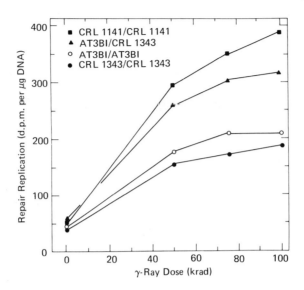

Figure 4.   Dose-dependent DNA repair replication in indicated
fused cultures of normal and AT strains.   Cells of two chosen
strains were artifically hybridized according to Harris and Watkins
(18); thereafter the experimental protocol was essentially as
described in Fig. 1.   Datum points represent the mean of four
measurements (s.d. ≤16%).

between these two strains, resulting in a masking of the biochemical
defect in each; complementation also demonstrates that genetic
heterogeneity can exist, different mutated loci giving rise to the
observed deficiency in repair synthesis.   The failure of the hybrid
AT3BI/CRL 1343 culture to reach a level equal to that of fused
normal cells is an effect ascribed to our fusion conditions which,
despite the use of a high titre of virus, do not produce a culture
containing multinucleate heterokaryons exclusively.   In the second
combination of AT hybrid cultures (CRL 1312/CRL 1343), but not in
the third (AT3BI/CRL 1312), the magnitude of repair replication
approaches that seen in a control (normal cell) fusion (CRL 1141/
CRL 1141) (data to be published elsewhere).   On the basis of these
studies, the three excision repair-deficient strains are assigned
to two complementation groups, AT3BI and CRL 1312 to group A, and
CRL 1343 to group B.

DISCUSSION

The DNA repair properties of diploid strains derived from
patients afflicted with ataxia telangiectasia strongly suggest that

this multisystem disorder is an ionizing-radiation counterpart of XP, the sunlight-sensitive skin disease whose only known biochemical defect is faulty repair of UV-induced DNA damage (13,14). Data presented here indicate that many, but not all, AT strains are incompetent in excision repair of γ radioproducts just as many, but not all, XP strains are defective in excision repair of UV photoproducts. Extensive genetic heterogeneity has been demonstrated in XP; not only do excision-deficient strains fall into no less than five distinct complementation groups (23) but there also exists so-called variant strains (13). These XP strains exhibit normal excision repair of UV lesions but are impaired in a second mechanism, postreplication repair (24). Similarly, our results allocate three excision-deficient AT strains to two complementation groups; moreover, some of the AT strains examined respond to γ-irradiation by executing both single-strand-rejoining and excision repair proficiently. Two of these repair proficient strains, AT4BI and AT5BI, also rejoin double-strand breaks with no apparent difficulty, as shown recently by Lehmann and Stevens (25). On the other hand, both display reduced colony survival when challenged with γ-rays (12; C.F. Arlett, cited in ref. 25). In short, these AT strains closely resemble XP variants and it is therefore tempting to postulate faulty postreplication repair following γ-irradiation as the molecular defect responsible for the radiosensitivity.

Neither the chemical structure of the γ-induced radioproducts subject to excision repair nor the precise molecular defect in the repair-deficient AT strains have been identified as yet. The most reasonable explanation for the abnormal repair properties of these strains, however, is that they lack the full complement of functional enzyme activity catalyzing intrastrand incisions near radiation-damaged base residues and thus fail to initiate excision repair normally. Since (i) ionizing radiation probably damages all four heterocyclic bases in DNA *in vivo* (20,21) and (ii) a crude *M. luteus* extract, undoubtedly containing a multitude of activities, is used as the test enzyme in our studies, the enzymatic assay may be monitoring the combined actions of a number of lesion-recognizing enzymes toward a number of different γ radioproducts. Furthermore, recent reports (26,27) suggest that in the case of at least some radioproducts, strand incision may not necessarily be accomplished by a conventional one-step reaction in which a repair endonuclease cleaves a phosphodiester bond at a damaged site. Rather, two steps may be involved, that is, breakage of a deoxyribose-base bond and release of a damaged base by the action of an N-glycosidase in concert with an endonuclease attack on the resulting unsubstituted deoxyribose residue. Besides distinguishing between these two possible modes of incision, future experimentation into the biochemical basis for excision deficiency in AT strains should help elucidate intermediate steps and mediating enzymes involved in the poorly understood excision-repair mechanism and the exact chemical composition of its presumed substrates, γ-modified residues of the four DNA bases.

It is well-established that AT patients (2-4), like XP patients
(13,14), are predisposed to an elevated incidence of malignancy,
suggesting that DNA damage, if permitted to accumulate due to defective enzymatic repair, can play a role in the induction of cancer.
In XP, a connection between neoplastic transformation and faulty
repair of UV-induced damage is strengthened by the clinical observation that multiple carcinomas are largely confined to sunlight-
exposed regions of the skin (14). A causal relationship between
defective repair and carcinogenesis remains a distinct possibility
in AT, as well. Considerable caution must be exercised in advancing
this postulate, however, as the development of malignancy in AT may
be etiologically related to a defective immune surveillance system
(3,4) and the involvement of a reduced capacity to perform DNA repair may be only secondary.

Assuming faulty DNA repair is etiologically pertinent in AT,
it is intriguing to speculate on the causal relationship between
the molecular defect and the clinical features of the disease.
The pattern of multiple abnormalities typically seen in afflicted
patients, has led others (2,4,28) to implicate disturbed tissue
differentiation during early embryonic development in the pathogenesis of the disorder; in particular, a defective interaction
between the entodermal and mesodermal germ lines. If so, this
implies that unrepaired DNA damage has led to the developmental
malformations and consequently efficient repair of that damage is
of critical importance during early gestation. This is reasonable
since cellular genetic homeostasis, presumably achieved by the
combined actions of enzymatic repair processes, would seem vital
during the first trimester—the crucial period when the embryo
develops from a single cell into a highly differentiated fetus
having many adult characteristics. Involvement of defective DNA
repair processes in other congenital disorders presumed to have a
developmental defect seems a reasonable prognosis.

Aside from offering a plausible explanation for the constel-
lation of clinical abnormalities associated with AT, our studies
uncover a valuable source of mutant strains for elucidating enzymatic repair mechanisms active on the DNA of human systems injured
by γ-rays. Finally, since damage to cellular DNA by countless
external agents, chemical in addition to physical, can be conveniently classified on the basis of the enzymatic repair response
as either ionizing-like or UV-like or both (29,30), AT strains in
concert with XP strains should prove to be a powerful human test
system for evaluating the deleterious consequences of any number
of environmental agents suspected of having mutagenic and
carcinogenic activities.

REFERENCES

1. Boder, E., and Sedgwick, R.P. (1963) Little Club clin. dev.
    Med., *8*, 110-118.
2. Peterson, R.D.A., Kelly, W.D., and Good, R.A. (1964) Lancet,
    1, 1189-1193.
3. Peterson, R.D.A., Cooper, M.D., and Good, R.A. (1966) Am. J.
    Med., *41*, 342-359.
4. McFarlin, D.E., Strober, W., and Waldmann, T.A. (1972) Medicine,
    51, 281-314.
5. Strober, W., Wochner, R.D., Barlow, M.H., McFarlin, D.E., and
    Waldmann, T.A. (1968) J. clin. Invest., *47*, 1905-1915.
6. Waldmann, T.A., and McIntire, K.R. (1972) Lancet, *2*, 1112-1115.
7. Harnden, D.G. (1974) In *Chromosomes and Cancer* (German, J.,
    ed.), pp. 619-636, Wiley, New York.
8. Gotoff, S.P., Amirmokri, E., and Liebner, E.J. (1967) Amer.
    J. Dis. Childh., *114*, 617-625.
9. Morgan, J.L., Holcomb, T.M., and Morrissey, R.W. (1968) Amer.
    J. Dis. Childh., *116*, 557-558.
10. Cunliffe, P.N., Mann, J.R., Cameron, A.H., Roberts, K.D., and
    Ward, H.W.C. (1975) Br. J. Radiol., *48*, 374-376.
11. Higurashi, M., and Conen, P.E. (1973) Cancer, *32*, 380-383.
12. Taylor, A.M.R., Harnden, D.G., Arlett, C.F., Harcourt, S.A.,
    Lehmann, A.R., Stevens, S., and Bridges, B.A. (1975) Nature,
    258, 427-429.
13. Cleaver, J.E., and Bootsma, D. (1975) Annu. Rev. Genet., *9*,
    19-38.
14. Robbins, J.H., Kraemer, K.H., Lutzner, M.A., Festoff, B.W.,
    and Coon, H.G. (1974) Ann. Intern. Med., *80*, 221-248.
15. Paterson, M.C., Smith, B.P., Lohman, P.H.M., Anderson, A.K.,
    and Fishman, L. (1976) Nature, *260*, 444-447.
16. Lohman, P.H.M., Sluyter, M.L., Matthijs, I.A.A., and Kleijer,
    W.J. (1973) Anal. Biochem., *54*, 178-187.
17. Paterson, M.C. Adv. Radiat. Biol., *7*, (in the press).
18. Harris, H., and Watkins, J.F. (1965) Nature, *205*, 640-646.
19. Paterson, M.C., Lohman, P.H.M., Westerveld, A., and Sluyter,
    M.L. (1974) Nature, *248*, 50-52.
20. Setlow, R.B., and Setlow, J.K. (1972) Annu. Rev. Biophys.
    Bioeng., *1*, 293-346.
21. Cerutti, P.A. (1974) Life Sci., *15*, 1567-1575.
22. Vincent Jr., R.A., Sheridan III, R.B., and Huang, P.C. (1975)
    Mutat. Res., *33*, 357-366.
23. Kraemer, K.H., Weerd-Kastelein, E.A. de, Robbins, J.H., Keijzer,
    W., Barrett, S.F., Petinga, R.A., and Bootsma, D. (1975)
    Mutat. Res., *33*, 327-340.
24. Lehmann, A.R., Kirk-Bell, S., Arlett, C.F., Paterson, M.C.,
    Lohman, P.H.M., Weerd-Kastelein, E.A. de, and Bootsma, D.
    (1975) Proc. Nat. Acad. Sci. USA, *72*, 219-223.
25. Lehmann, A.R., and Stevens, S. Biochim. Biophys. Acta,
    (in the press).

26. Lindahl, T. (1974) Proc. Nat. Acad. Sci. USA, *71*, 3649-3653.
27. Friedberg, E.C., Ganesan, A.K., and Minton, K. (1975) J. Virol. *16*, 315-321.
28. Peterson, R.D.A., Cooper, M.D., and Good, R.A. (1965) Am. J. Med., *38*, 579-604.
29. Regan, J.D., and Setlow, R.B. (1974) Cancer Res., *34*, 3318-3325.
30. Setlow, R.B., Faulcon, F.M., and Regan, J.D. (1976) Int. J. Radiat. Biol., *29*, 125-136.
31. Carrier, W.L., and Setlow, R.B. (1970) J. Bacteriol., *102*, 178-186.
32. Paterson, M.C., and Setlow, R.B. (1972) Proc. Nat. Acad. Sci. USA, *69*, 2927-2931.
33. Setlow, R.B., and Carrier, W.L. (1973) Nature, New Biol., *241*, 170-172.
34. Paterson, M.C., Roozen, K.J., and Setlow, R.B. (1973) Int. J. Radiat. Biol., *23*, 495-508.
35. Johansen, I., Brustad, T., and Rupp, W.D. (1975) Proc. Nat. Acad. Sci. USA, *72*, 167-171.

EFFECTS OF VACUUM-UV AND EXCITED GASES ON DNA

H. JUNG

Institut für Biophysik und Strahlenbiologie
Universität Hamburg

W. SONTAG, C. LÜCKE-HUHLE, K.F. WEIBEZAHN & H. DERTINGER

Institut für Strahlenbiologie, Kernforschungszentrum
75 Karlsruhe, Federal Republic of Germany

INTRODUCTION

When dealing with the 'comparative effects of exciting and ionizing radiations' investigations into the biological effects in the intermediate energy range appear to be of special interest as evident from the following arguments:

When exposed to ionizing radiation in the dry state, enzymes, DNA, RNA and viruses containing single-stranded nucleic acid are inactivated according to exponential kinetics. From the 37 per cent dose (D_{37}) the volume of the radiosensitive target may be calculated using the well-known mathematical expressions of target theory. The results obtained generally show a rather good agreement between the molecular weight of the radiosensitive target and the molecular weight of the irradiated macromolecules or that of the single-stranded genome of the irradiated viruses, respectively (see Dertinger and Jung [1], equation 5.5, figure 28, and table 15). These calculations are usually made assuming the mean energy expenditure per primary absorption event to be 60 eV. Consequently, such calculations mean essentially that biomolecules or viruses containing single-stranded nucleic acid are inactivated with a probability or 'killing-efficiency' near to unity after 'receiving' an average amount of energy of 60 eV. The value of 60 eV has been obtained from cloud-chamber photographs by analysing the frequency distribution of ion pairs per ion cluster [2, 3] and, in good agreement, by measuring the energy loss of electrons in thin foils of plastic material [4] or DNA [5]. Therefore, this figure simply reflects the physics of radiation absorption and gives no indication of the energy necessary to damage biological macromolecules.

Since biomolecules are readily inactivated by UV-radiation having quantum energies of about 5 eV, on the molecular level all types of ionizing radiation cause a pronounced 'over-kill'.

To obtain some information about the energy required for damaging biomolecules, we started a series of experiments in which amounts of energy between 4 and 20 eV were transferred to DNA, and the resulting structural and functional changes were analysed. In the energy range under consideration, at least two distinct ways of transferring energy exist: (1) Irradiation with UV light of extremely short wavelength, so-called vacuum-UV, and (2) exposure of the specimens to gases excited to metastable states.

MATERIALS AND METHODS

Monochromatic UV-radiation was obtained from a concave grating monochromator constructed by Berger [6] delivering radiation of quantum energies between 4.9 eV (254 nm) and 21.2 eV (58.4 nm). Details concerning construction, operating procedures and dosimetry have been described elsewhere [6, 7, 8].

Excited gases were generated in a gas-flow system as used earlier by Jung and Kürzinger [9]. In this set-up a constant stream of gas containing a certain fraction of atoms or molecules in metastably excited states passes over the samples exposed in thin layers. The excitation energy of the metastable states is transferred to the biomolecules by so-called 'collisions of the second kind'. The energies of the lowest metastable states amount to 4.3 eV for H_2, 6.17 eV for N_2, 11.56 eV for Ar and 19.81 eV for He [10].

A device that might be considered a modern version of 'Ulbricht's sphere' was used to determine the absorption spectrum and the photoelectric emission of DNA [8]. The method of exposing dry powders of DNA and DNA constituents to excited gases for electron paramagnetic resonance (e.p.r.) investigations has been described by Weibezahn and Dertinger [11, 12], the computer-controlled analysis of e.p.r. spectra by Hartig and Dertinger [13].

The biological techniques, such as DNA extraction from phage ØX174, sample preparation by lyophilization on microscope cover slips, assaying the infectivity of DNA in the spheroplast system, and measuring the number of strand breaks by ultracentrifugation in an alkaline sucrose gradient, have been communicated by Lücke-Huhle and Jung [14, 15].

RESULTS

Fig. 1 shows the absorption cross-section of dry calf thymus DNA as a function of quantum energy. Absorption was measured for native as well as for heat-denatured DNA. Both absorption curves, however, did not show any difference within the limits of experimental error. By a detailed analysis [8] the various maxima observed within the energy range under discussion could be explained to have their origin in specific $\pi\to\pi^*$ and $\sigma\to\sigma^*$ transitions of the base and sugar moieties, respectively. This conclusion holds irrespective of the fact that there are some discrepancies between our measurements and those of Inagaki et al. [16] still awaiting further investigation.

In addition, the photo-electron emission from calf thymus DNA was investigated. From 8 to 13 eV the quantum yield increases by about two orders of magnitude, reaching a constant level between 13 and 22 eV. Further details have been described by Sontag and Weibezahn [8].

When DNA and DNA constituents are exposed to excited gases, the e.p.r. spectra observed are identical for discharge-excited hydrogen, argon, and helium [11, 12]. Taking into account the different energies transferred, these results are somewhat surprising. As shown in Fig. 2 the spectrum obtained after exposing DNA to excited argon is completely different from the spectrum of γ-irradiated DNA. Not only is there a large difference in the central region but also in the low- and high-field directions, where a strong contribution of the eight-line spectrum of the 5,6-dihydro-5-thymyl-radical is observed after irradiation. Apparently these lines are scarcely visible in the Ar*-spectrum.

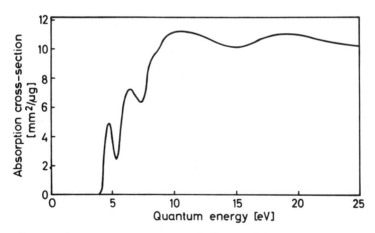

Fig. 1. Absorption cross section of dry calf thymus DNA as function of quantum energy of incident radiation.

From the detailed results described elsewhere [11, 12] it is quite evident that energy transfer from excited gases leads to quite different reactions than does ionizing radiation. The spectra observed after γ-irradiation could be shown to be due to radical-ions resulting from previously ionizing the molecules investigated [13], whereas radicals generated by the attack of excited gases are formed by direct C-H bond dissociation in the bases without preceding ionization [11, 12].

When ØX174-DNA is exposed to vacuum-UV at various wavelengths, the proportion of unbroken molecules decreases and approaches a constant level at long exposure times indicating that part of the DNA molecules are enclosed in small salt crystals that cannot be penetrated by the short-wavelength UV-quanta. After subtracting this constant fraction the experimental values fall on straight lines in semi-log plots. The slope of these lines is characterized by the 37 per cent exposures (in ergs/mm^2), the reciprocals of which represent the cross-sections for the production of single-strand breaks as a function of photon-energy. Using the absorption data from Fig. 1 the quantum yield can be calculated. Fig. 3 shows the quantum yields for the production of single-strand breaks as well as those for the inactivation of the infectivity of the ØX-DNA molecules. With increasing photon energy both cross-sections increase reaching a constant plateau at energies above 12 eV.

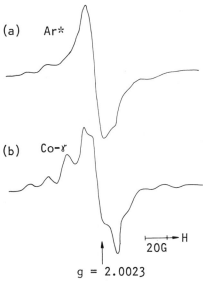

Fig. 2. First derivative of DNA e.p.r. spectra.
(a) Exposure to discharge-excited argon
(b) Gamma irradiation

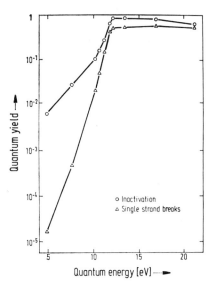

Fig. 3. Quantum yields for strand break formation and inactivation of ØX174-DNA by vacuum-UV as function of photon energy

From these results it becomes obvious that the breakage of the DNA backbone is the relevant type of damage to inactivate ØX174-DNA at quantum energies above 10 eV whereas base damage is the inactivating event predominant at energies below this.

The results obtained after exposing ØX174-DNA to four different excited gases are compiled in table 1. The figures were determined by plotting the percentage of infectious DNA as well as the percentage of unbroken molecules versus duration of the exposure. After subtracting the constant fractions amounting to about 70 per cent under all experimental conditions exponential dose-effect curves were obtained in all cases. The ratio of the two 37 per cent exposure times (t_{37}) determined for each gas gives the fraction of ØX-DNA molecules inactivated by strand breakage. By comparing the relative frequencies of strand breaks (last column of table 1) it becomes obvious that this frequency does not depend on the gas used. Thus, we arrive at the conclusion that (77 ± 10) per cent of the inactivated ØX-DNA molecules carry at least one strand break; within the energy range investigated this figure is independent of the amount of energy transferred.

For comparison, inactivation and strand breakage were determined in ØX174-DNA after γ-irradiation in the dry state [17]. Fig. 4 shows the percentage of infectious ØX-DNA as well as the percentage of unbroken molecules at various γ-ray doses. The D_{37} values calculated from the slopes of the dose-effect curves amount to (658 ± 46) krads for the loss of plaque-forming ability and (749 ± 81) krads for the occurrence of strand breaks, indicating that under the experimental conditions applied (88 ± 16) per cent of the inactivated ØX174-DNA molecules carry at least one strand break. Since sedimentation in an alkaline sucrose gradient yields the breaks formed directly by irradiation as well as those deriving

Table 1. Energy transfer from metastable states of excited gases to ØX174-DNA

Gas	Energy (eV)	Plaque-forming ability t_{37} (sec)	Strand breaks t_{37} (sec)	Percentage of ØX-DNA molecules inactivated by strand breaks
H_2^*	4.3	10.7 ± 1.0	14.3 ± 2.5	75 ± 20
N_2^*	6.2	26.0 ± 1.9	32.2 ± 5.1	81 ± 20
Ar^*	11.6	18.1 ± 3.0	29.4 ± 4.4	62 ± 19
He^*	19.8	5.5 ± 0.3	6.7 ± 0.7	82 ± 13
			Weighted mean:	77 ± 10

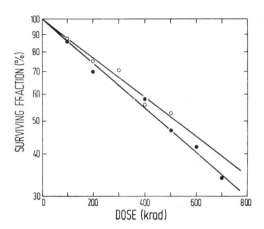

Fig. 4. Inactivation and strand breakage in ØX174-DNA after γ-irradiation in the dry state. ● Loss of plaque-forming ability as tested in a spheroplast system. o Fraction of unbroken molecules as determined by sedimentation in an alkaline sucrose gradient.

from the rupture of 'alkaline-labile bonds' [18] the results have to be corrected for this contribution [17]. Thus, we come to the conclusion that (69 ± 19) per cent of the ØX174-DNA molecules irradiated in the dry state are inactivated by direct strand breakage and only one third by other types of damage including alkaline-labile bonds.

Our findings are summarized in Fig. 5. Here the percentage of molecules inactivated by strand breaks is plotted versus the amount of energy transferred by the absorption of vacuum-UV quanta or by 'collisions of the second kind' from metastably excited gases, respectively. The energy transfer by ^{60}Co γ-radiation represents an average value of a very broad distribution ranging from about 10 to several hundreds of electron volts. This is indicated by the arrows associated with the triangular symbols. The values for vacuum-UV radiation have been obtained from the ratio of the quantum yields for single-strand breaks and inactivation as plotted on Fig. 3, those for excited gases and ^{60}Co-γ have been taken from table 1 and Fig. 4, respectively.

Above 12 eV the fraction of strand breaks is independent of the energy primarily absorbed and, obviously, also independent of the energy transferred be it from UV-quanta, excited gases or γ-radiation. The average value amounts to about 75 per cent. Whereas the values for excited gases do not show any variation over the entire energy range investigated, the curve for vacuum-UV reveals a strong increase between 5 and 12 eV, demonstrating the increasing role of strand breaks in ØX-DNA inactivation by vacuum-UV in this particular energy interval.

To explain the difference between the action of UV quanta and of excited gases, one should keep in mind that the action of a charged particle on a molecule is similar to the action of a photon field having a spectral distribution proportional to $1/\nu$ [19]. It is well known that energy transfer from photons to molecules is governed by selection rules. Thus, when irradiating with UV light only certain classes of electrons within the molecule (σ or π electrons) may take part in dissipating the energy. 'Collisions of the second kind', however, are radiationless processes and, consequently, not governed by these selection rules, so that the energy transferred from the metastably excited gases may be distributed over the electron system as a whole. Since 4.3 eV is too low an energy to ionize the molecule, our results (viz. that the frequency of strand breaks is independent of the energy transferred) lead to the conclusion that excited gases do not cause ionizations, even if the energy (for example, 19·8 eV in the case of excited helium) is well above the ionization level of the macromolecules exposed.

This conclusion is strongly supported by the aforementioned e.p.r. experiments showing that the spectra observed after exposing DNA and DNA constituents to excited gases derive from radicals that

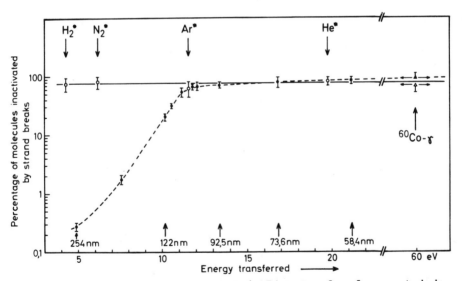

Fig. 5. Percentage of inactivated ØX174-DNA molecules containing at least one strand break. (●) Vacuum-UV. (o) Metastably-excited gases. (▲) Co-γ radiation; breaks deriving from 'alkali-labile bonds' are included. (Δ) Co-γ radiation, after subtracting the contribution of 'alkali-labile bonds'.

are generated by direct C–H bond dissociation in the bases without preceding ionization.

Our present and by no means final model of the reactions induced by energy transfer from vacuum-UV and excited gases to DNA is the following:

At energies between 4 and 7.5 eV, the absorption of DNA has its origin mainly in $\pi \rightarrow \pi^*$ transitions of the bases whereas between 8 and 13 eV the increase of DNA absorption has to be correlated preponderantly to ionization of sugar-phosphate-groups. Ionization of bases becomes noticeable at relatively higher energies. Above quantum energies of about 10 eV strand breaks probably deriving from ionizations are the predominant types of damage whereas base damage caused by excitations are predominant at energies below this.

The excited gases transfer their energy to the electron system of the molecule as a whole leading probably to excited states of strongly collective character. These states dissipate their energy by dissociating aromatic C–H bonds at the base moiety. The type of lesion produced is independent of the energy transferred. From our observation that the fraction of strand breaks among the lesions destroying the plaque-forming activity of ØX-DNA is as high as 77 per cent and is also independent of the energy transferred, we may speculate that the radicals originally located at the base moiety lead to strand breakage. Details of this mechanism are not yet fully understood, but seem to be similar to the reactions postulated in trying to explain the action of UV light on bromouracil-substituted DNA [20, 21, 22].

REFERENCES

[1]   H. Dertinger & H. Jung. Molecular Radiation Biology. Springer-Verlag, New-York-Heidelberg-Berlin, 1970.

[2]   A. Ore & A. Larsen, Radiat. Res. 21, 331 (1964).

[3]   A. Ore, Physica Norwegica 5, 259 (1971).

[4]   A.M. Rauth & J.A. Simpson, Radiat. Res. 22, 643 (1964).

[5]   C.D. Johnson & T.B. Rymer, Nature 213, 1045 (1967).

[6]   K.U. Berger, Z. Naturforsch. 24b, 722 (1969).

[7]   W. Sontag & H. Dertinger, Int. J. Radiat. Biol. 27, 543 (1975).

[8]   W. Sontag & K.F. Weibezahn, Radiat. Envir. Biophys. 12, 169 (1975).

[9]   H. Jung & K. Kürzinger, Radiat. Res. 36, 369 (1968).

[10] B. Brocklehurst, Radiat. Res. Rev. 1, 223 (1968).

[11]   K.F. Weibezahn & H. Dertinger, Int. J. Radiat. Biol. 23, 271 (1973).

[12]   K.F. Weibezahn & H. Dertinger, Int. J. Radiat. Biol. 23, 447 (1973).

[13]   G. Hartig & H. Dertinger, Int. J. Radiat. Biol. 20, 577 (1971).

[14]   C. Lücke-Huhle & H. Jung, Int. J. Radiat. Biol. 24, 221 (1973).

[15]   C. Lücke-Huhle & H. Jung, Int. J. Radiat. Biol. 25, 339 (1974).

[16]   T. Inagaki, R.N. Hamm, E.T. Arakawa & L.R. Painter, J. Chem. Phys. 61, 4246 (1974).

[17]   C. Lücke-Huhle, A. Seiter & H. Jung, Z. Naturforsch. 29c, 98 (1974).

[18]   C. Lücke-Huhle, M. Pech & H. Jung, Radiat. Envir. Biophys. 11, 165 (1974).

[19]   R.L. Platzman, Vortex 23, 372 (1962).

[20]   G. Hotz & H. Reuschl, Molec. Gen. Genet. 99, 5 (1967).

[21]   W. Köhnlein & F. Hutchinson, Radiat. Res. 39, 745 (1969).

[22]   M.L. Dodson, R. Hewitt & M. Mandel, Photochem. Photobiol. 16, 15 (1972).

SYMPOSIUM VI

PHOTOSENSITIZED REACTIONS OF NUCLEIC
ACIDS AND PROTEINS

PHOTOSENSITIZATION IN BIOLOGICAL SYSTEMS

John D. Spikes

Department of Biology, University of Utah

Salt Lake City, Utah 84112   U.S.A.

Protoplasm, in general, is relatively insensitive to direct effects of visible and near ultraviolet light since most molecules of biological importance do not absorb appreciably in this wavelength range. However, in the presence of an appropriate photosensitizer, cells, organisms and many kinds of biomolecules are damaged and destroyed by light. We usually date the start of the study of photosensitization as an area of photobiology from the observation by Oscar Raab, a medical student in Munich, that acridines sensitize paramecia to killing by light (1). However, on this occasion, we should take note of the earlier work of Arturo Marccaci, a physiologist in Perugia, who reported in 1888 that low concentrations of quinine, which had no effect in the dark on grain germination, the development of amphibian eggs, etc. produced severe damage in the light (2). Photosensitized cells are killed by selective photochemical damage to certain cell organelles, with effects on the organelle resulting from selective alteration of macromolecules in the organelle; alteration of the macromolecule results in turn from selective damage to certain of its subunits. Thus, studies of photosensitization in biology range from an examination of the photophysics and photochemistry of sensitizer molecules to the physiology and pathology of the photosensitized injury and killing of mammals. Several review papers on photosensitization in biological systems are listed in the references (3-9).

The light energy involved in most photosensitized reactions is absorbed by the photosensitizer molecule rather than by the biological system affected. Light of wavelengths greater than approximately 320 nm is generally used in photosensitization studies, since a number of biologically important molecules

231

including nucleic acids and proteins absorb at shorter wavelengths. To function as a photosensitizer, a molecule must be
capable of more than merely absorbing light of the wavelength
involved; in general, it must have the capability to be excited
by light into a long-lived energy-rich form, the triplet-state.
Many kinds of synthetic and natural molecules can act as photosensitizers.

The majority of the photosensitized reactions in biological
systems that have been studied involve molecular oxygen, i.e.,
they are sensitized photooxidation processes. These reactions
in biological systems are often termed "photodynamic action" or
"photodynamic" reactions. The sensitizer molecule is typically
neither consumed nor destroyed in these reactions, but is used
repeatedly in a somewhat "catalytic" fashion. The excited
sensitizer in photooxidation reactions can react directly with
the substrate (the molecule being oxidized) by an electron or
hydrogen transfer process; the resulting semi-reduced substrate
can then react with oxygen giving an oxidized form of the substrate. Alternatively, the excited sensitizer can react with
molecular oxygen by an energy transfer process giving an excited
singlet form of oxygen which then oxidizes the substrate. The
relative participation of these different pathways depends on
the sensitizer, the substrate, and the reaction conditions. In
contrast, some photosensitizers do not require molecular oxygen
for their action, and may be consumed in the sensitized reaction.

Our symposium today examines the current status of our
knowledge of the photosensitized reactions of nucleic acids and
proteins. Significant advances have been made in this area of
photobiology since the last international Congress. The first
paper will explore the role of molecular oxygen in photosensitized reactions of proteins and nucleic acids. The last three
papers are concerned primarily with photosensitized reactions of
nucleic acids and their components in which the sensitizer is
consumed by bonding covalently with the substrate on irradiation.
The best understood reactions of this type are those sensitized
by furocoumarins such as 5-methoxypsoralen. These sensitizers
show promise in the photochemotherapeutic treatment of proliferative diseases of the skin, such as psoriasis.

It is altogether fitting that this symposium emphasizes the
photochemistry and photobiology of the skin-sensitizing psoralens,
since much of the fundamental research in this area of photobiology has been carried out in our host country, Italy.

REFERENCES

1. Raab, O. (1900) Z. Biol. 39, 524-546.
2. Marccaci, A. (1888) Arch. Ital. Biol. 9, 2-4.
3. Spikes, J.D. "Photosensitization", Chap. 4 in The Science of
    Photobiology (Smith, K.C., ed.), Plenum Press, New York
    (in press). A general review ranging from molecules to
    organisms.
4. Foote, C.S. "Photosensitized oxidation and singlet oxygen:
    Consequences in biological systems", pp. 85-133 in Free
    Radicals in Biology, Vol. II (Pryor, W.A., ed.), Academic
    Press, New York (1976). A general review with quantitative
    considerations of fundamental mechanisms.
5. Spikes, J.D. (1975) Ann. New York Acad. Sci. 244, 496-508.
    A review of porphyrins as photosensitizers.
6. Spikes, J.D. and MacKnight, M.L. (1970) Ann. New York Acad.
    Sci. 171, 149-162. A review of the sensitized photooxidation
    of proteins.
7. Jori, G. (1975) Photochem. Photobiol. 21, 463-467. A brief
    recent review of photosensitized reactions of amino acids
    and proteins.
8. Musajo, L. and Rodighiero, G. (1972) Photophysiology 7,
    115-147. A review of the mechanism of furocoumarin photo-
    sensitization.
9. Fitzpatrick, T.B., Pathak, M.A., Harber, L.C., Seiji, M.
    and Kukita, A. (editors) Sunlight and Man, Univ. of Tokyo
    Press (1974). This volume contains a number of papers on
    photosensitization in man.

THE ROLE OF OXYGEN IN PHOTOSYNTHETIZED REACTIONS OF PROTEINS AND NUCLEIC ACIDS

Jehuda Feitelson

Department of Physical Chemistry

The Hebrew University of Jerusalem, Israel

SUMMARY

Based on a broad definition of photosensitization the follow-ing topics are discussed:

The excited states and primary reactions in photosensitiza-tion and the role that oxygen plays in these reactions.
Photosensitization of amino acids and nucleotides.
Photosensitizers which form part of biological macromolecules.
Problems related to reactivity within the three-dimensional network of a biological macromolecule.
Effects of oxygen in macromolecular systems and the informa-tion which can be obtained from photosensitization experiments in such systems.

INTRODUCTION

Photosensitized reactions have been thoroughly discussed in the VI International Congress on Photobiology and review articles on photosensitization in the absence and presence of oxygen, with special emphasis on singlet oxygen, appear in the Proceedings of that Congress (1-5) as well as in the more recent literature (6-8). We shall therefore touch only briefly upon the above well document-ed phenomena and concentrate upon some aspects of photosensitiza-tion which have not been extensively discussed previously.

PRIMARY AND SECONDARY REACTIONS IN PHOTOSENSITIZATION

We shall use a quite general definition of photosensitization,

235

proposed by Turro, according to which:

Photosensitization is a phenomenological effect in which
light absorption by a sensitizer induces the reaction of an acceptor molecule, which does not itself absorb the incident radiation[9]

This definition does not require the sensitizer to be restored to its previous form at the end of the reaction (i.e. play the
part of a catalyst only) but allows also for those cases where
the sensitizer itself undergoes a chemical change.   The following
scheme describes the initial steps in photosensitization grouped
according to the type of reaction.

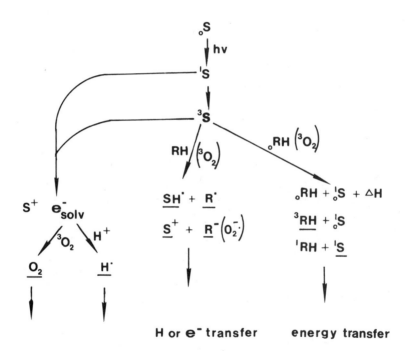

S indicates the sensitizer and RH is the primary substrate molecule.   Superscripts denote the multiplicity of a state and sub-
script o indicates the ground state of a molecule.   The potentially reactive species formed are underlined.

It is seen that oxygen can react directly with the sensitizer
triplet either by charge transfer or by energy transfer.   It can
of course also participate in the subsequent reactions following
the primary step.

The right hand column in the scheme describes energy transfer
reactions in which a comparatively long-lived sensitizer triplet

either a) transfers its energy to a ground state substrate molecule yielding ground state sensitizer and substrate triplet, or
b) reacts with another triplet to form one excited singlet and
one singlet ground state molecule.   Also collisional quenching of
triplets belongs to this class of reactions.   By far the most
important reaction in this group is the formation of singlet oxygen by

(1) $^3S + {}^3O_2 \rightarrow {}^1O_2(^1\Delta_g) + {}^1_oS$

Singlet oxygen reacts then further as an oxidant.

        The middle column describes hydrogen or charge transfer
between the triplet sensitizer and a ground state molecule.   The
radicals formed in the hydrogen abstraction by the sensitizer are
often oxidised by O_2 to yield peroxy radicals.   An example for
the charge transfer reaction is the formation of the superoxide
ion

(2) $^3S + {}^3O_2 \rightarrow S^{+\cdot} + O_2^{-\cdot}$

        The reactions in the left hand column have not hitherto been
generally described as photosensitization.   Usually solvated
electrons, e^-_{aq}, have been thought to originate either from high
energy irradiations or in systems illuminated by high intensity
light flashes.   In the latter case the molecule absorbs a photon,
crosses over into the triplet manifold and, subsequently, absorbs
a second photon.   It thus reaches a triplet state of sufficient
energy to release an electron into the solution.   Much of the
interest in solvated electrons in photobiology is due to the fact
that it can be formed in solution also at low light intensities.
It has been proposed (10) and since verified (11) that indole
derivatives at neutral pH release e^-_{aq} into aqueous solution
directly from their first excited singlet state.   Since most proteins contain one or more tryptophan residues, illumination into
the 280 nm absorption band can produce solvated electrons within
the protein or in solution.   Tryptophan therefore in our scheme
is a photosensitizer which produces upon illumination a species,
e^-_{aq}, capable of further reaction.

        The solvated electron itself is primarily a reducing species.
Two of its reactions which form secondary reactive species in
solution are

(3) $e^-_{aq} + H_3O^+ \rightarrow H^\cdot + H_2O$ $k_H \simeq 3\times10^{10}$ M^{-1} sec^{-1}

(4) $e^-_{aq} + O_2 \rightarrow O_2^{-\cdot}$ $k_{O_2} \simeq 2\times10^{10}$ M^{-1} sec^{-1}

Reaction (3) is of course pH dependent.   In the absence of oxygen below pH 5 it is the main decay process of e^-_{aq} since all
electrons are transformed into $H^\cdot$ radicals in t $\leq$ 3 μsec.   In

aerated aqueous solution where the concentration of O_2 is about 3×10^{-4} M and at neutral pH reaction (4) predominates and $O_2^-\cdot$, the superoxide radical ion, is formed.

$O_2^-\cdot$ can act both as a reducing and as an oxidising agent as can be seen from its redox potentials at pH 7

$$O_2 + e_{aq}^- = O_2^-\cdot \quad -0.33 \text{ V}; \quad O_2^-\cdot + e_{aq}^- = O_2^= \quad +0.87 \text{ V}$$

Thus it can oxidize Heme(Fe^{2+}) to Hemin(Fe^{3+}). Benzoquinone on the other hand is reduced by $O_2^-\cdot$. The superoxide radical ion is, however, less reactive than e_{aq}^- as can be seen from a comparison of their rate constants for the reduction of benzoquinone ($k_{e^-} \approx 10^{10}$ M^{-1} sec^{-1}; $k_{O_2^-} \approx 10^8$ M^{-1} sec^{-1}).

The solvated electron, e_{aq}^-, reacts destructively with a diffusion controlled rate constant with the various nucleosides. It also reacts readily with the aromatic amino acids tyrosine and tryptophan as well as with histidine and with cystine and other R-S-S-R compounds with whom it forms a $RSSR^-$ adduct. Other amino acids react very slowly ($k \approx 10^6$ M^{-1} sec^{-1}) with e_{aq}^-. Hydrogen atoms on the other hand react readily both with amino acids and with nucleosides ($k_H \approx 5 \times 10^8$ to 3×10^9 M^{-1} sec^{-1}) (12).

The reactions involving energy, charge and H atom transfer (middle and right hand columns in the Scheme) have conventionally been grouped in Type I or radical, and Type II or oxygen reactions as follows

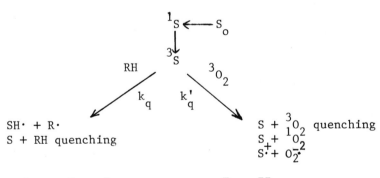

$$^1S \longleftarrow S_o$$
$$\downarrow ^3S$$

RH $\qquad$ $^3O_2$

k_q $\qquad$ k_q'

SH$\cdot$ + R$\cdot$ $\qquad\qquad\qquad$ S + $^3O_2$ quenching
S + RH quenching $\qquad\qquad$ S + $^1O_2$
$\qquad\qquad\qquad\qquad\qquad$ S$^+$ + O_2^-

Type I $\qquad\qquad\qquad\qquad$ Type II

The question arises whether a given sensitizer-substrate system will react via the Type I or Type II reaction mechanism. Quite generally the answer to this question depends on the nature of the sensitizer triplet, on the substrate as well as on the experimental conditions such as pH and concentration. It has been suggested (13) that sensitizers whose lowest triplet is of a $\pi\pi^*$ character (thiazines, for example) react preferably with oxygen while substances whose lowest triplet is a $n\pi^*$ state (aromatic ketons, for example) are good hydrogen abstractors.   The reactivities of two sensitizer triplets, eosin and thionine, towards a variety of substrates, have recently been studies by Kraljic and Lindqvist (14).

If one assumes that in the Type II reactions the formation of singlet oxygen predominates and that furthermore $^1O_2$ reacts in a diffusion controlled reaction with the substrate, then a comparison between the quantities $k_q[RH]$ and $k_q'[O_2]$ will indicate whether photosensitization will proceed by the radical or by the oxygen pathway.   For example, for 3×10^{-4} M tryptophan with eosin as sensitizer, $k_q[TRP] = 3 \times 10^5$ sec^{-1} equal to the value of $k_q'[O_2]$ in an aerated solution, meaning that under these conditions the two reaction probabilities are roughly equal.   For methionine at the same concentration, however, $k_q[Met] \simeq 10^3$ sec^{-1} which clearly indicates that methionine under these conditions will be oxydised via singlet oxygen.   Foote has lately estimated the probabilities for Type I and Type II reactions from kinetic data (6).   Unless all the rate constants in the reaction scheme are known it is important to note that k_q' yields only an upper limit for the formation of singlet oxygen and that the reactivity of the latter towards the substrate might also be important.   Thus it appears that nucleosides are rather non-reactive towards $^1O_2$. Even for guanosine, which reacts more readily than other nucleosides, the rate constant with singlet oxygen is only $k \simeq 10^7$ M^{-1} sec^{-1} (15).   In the absence of oxygen, photosensitized dimerization of purines is often observed.

With respect to amino acids: in tyrosine, tryptophan and histidine singlet oxygen causes an oxidative ring cleavage while cysteine is oxidised to cysteic acid and methionine to methionine sulfoxide.   The radical mechanism affects tyrosine and tryptophan under appropriate conditions while cystine and other RSSR compounds can act as electron acceptors with tyrosine or tryptophan triplets as electron donor (16).

PHOTOSENSITIZED REACTIONS IN PROTEINS AND NUCLEIC ACIDS

The general principles of photosensitization in low molecular weight systems in solution apply of course to reactions in biological macromolecules.   It can, however, be assumed that the three dimensional structure of the biopolymer influences greatly the reactivity of the system.   It has recently been shown (17) that the movement of small molecules, in particular oxygen, into a protein network is often much slower than in aqueous solution. Also, inherent potential sensitizer and substrate groups are located at fixed distances and orientations within the highly structured polymer network.   We must therefore ask whether and to what ex-

tent reactive groups are accessible to photosensitizing agents
when buried within the polymer network and whether any structural
information can be derived from photosensitized reactions within
the polymer? Whether a mechanism involving solvated electrons,
free radicals or singlet oxygen will operate is largely dependent
upon this question of accessability. We shall illustrate photosensitization in biopolymers with respect to the above structural
problems by only a few examples out of a vast number of experiments
described in the literature.

Solvated electrons formed by pulse radiolysis in the absence
of oxygen have been shown to reduce cytochrome C in a diffusion
controlled reaction ($k \simeq 10^{10}$ M^{-1} sec^{-1}) (18) although the heme
group is partly buried within the protein fold. Moreover, the
reaction proceeds with a quantum yield of $\phi = 0.8$ meaning that
only 20% of the e_{aq}^- present attack potentially susceptible amino
acid residues in the protein. In the presence of oxygen $O_2^- \cdot$ ions
are formed (equ.4). O_2^- reduces cytochrome C with a much lower
rate constant ($k \simeq 10^8$ M^{-1} sec^{-1}) than solvated electrons. The
reduction of the porphyrin-Fe^{3+} becomes, however, highly specific
in this case with a quantum yield close to unity. This means
that the amino acid residues of cytochrome C are not at all
attacked by the superoxide ion, the sole reaction being the reduction of the trivalent Fe^{3+}. Recently solvated electron experiments were conducted in a purely photochemical system in which
e_{aq}^- were formed by illuminating an electron releasing sensitizer
(naphthylamine) in aqueous solution containing cytochrome C (19).
Such experiments, it is hoped, will shed light on the above high
reactivity and specificity of the reaction.

That certain parts of a protein are not easily accessible
has been shown by Kenkare and Richards (20) in the photooxidation
of ribonuclease-S in the presence of methylene blue and oxygen.
It was found that two histidine residues (105 and 109) of the
RNase were rapidly destroyed while a third residue (#12) is photooxidized at a much slower rate, which shows that either the
methylene blue or the oxygen can reach it only with difficulty.

The sensitizers of particular interest to us are those forming part of the biopolymer itself. The tryptophan residues in a
protein play such a part, where they can act as photosensitizers
in a variety of ways. When illuminated in the 280 nm band tryptophan can release solvated electrons into the solution as shown
in our general scheme. It can also act as direct electron donor
to a suitable acceptor species and, when photolysed directly or
by photosensitization, it yields formylkynurenin which itself
again is a photosensitizer in the formation of singlet oxygen.

In papain Baugher and Grossweiner (21) were able to show that
the excited tryptophan triplet transfers directly an electron to

an (apparently near by) cystine S-S bond, as suggested by Bent and
Hayon (11). The formation of this -S-S- ion is not influenced by
the presence of oxygen and hence indicates that the reaction takes
place in a region of the protein structure not accessible to O_2.
However, in another protein, bovine carbonic anhydrase, where internal photosensitization by tryptophan residues takes place, it
was found that oxygen takes part in the destruction of a number of
histidine and tryptophan residues (22). The results suggest that
in this case oxygen is able to permeate large portions of the enzyme and that photooxidation takes place both by singlet oxygen
and by the O_2^- pathways.

From the few examples presented here it is already obvious
that the interior of a biological macromolecule is accessible to
oxygen in a selective manner and that the general "rules" applicable in free solution have to be reinvestigated in each case anew.

Another group of sensitizers which forms an inherent part of
a protein are the heme porphyrins. Their photosensitizing properties have been successfully exploited for structural studies.
Lately it has been shown that the sensitizing efficiency of porphyrins depends on the magnetic moment of their central metal ion
(23). Hemoporphyrins with large magnetic moment (high spin)
are poor photosensitizers while low spin (small magnetic moment)
hemoporphyrins do sensitize the photooxidation of amino acids
readily. The effect is explained by the rapid decay of the porphyrin triplet due to spin-orbit coupling, induced by the paramagnetic center of the high spin compound. Low spin porphyrins, on
the other hand, have long lived triplet states, which thus are
capable of acting as sensitizers. The photooxidation of myoglobin sensitized by its own heme group was studied by Folin, Gennari
and Jori (23). No sensitization takes place in deaerated solution. In presence of oxygen the photooxidation of methionine 131
and of the heme-linked histidine residues 93 and 64 takes place,
most likely by the singlet oxygen mechanism. Only low spin myoglobins with O_2, CO as an axial ligand were photoreactive while
the heme group in the high spin Met-myoglobin and in Met-myoglobin-
F did not act as sensitizers. The authors were able to obtain
from their data structural information about the environment of
the prostetic group in different myoglobins and to monitor conformational changes due to the binding of different ligands to the
heme group.

In the field of polynucleotides photosensitization often involves the formation and the splitting of pyrimidine dimers.
Meistrich and Lamola (24) have shown that, in the absence of oxygen, acetophenone sensitizes the dimerization of thymine in T4
phage heads. It was found that acetophenone had to be present in
the phage head prior to illumination for the reaction to take
place and it was concluded that an acetophenone triplet formed

outside the phage could not diffuse rapidly enough through the
phage membrane.   It would be of interest to know whether oxygen
might or might not be able to interfere with the reaction inside
the phage head.

Helene and Charlier (25) have found that positively charged
Lys-Trp-Lys peptides can, in the absence of oxygen, sensitize the
splitting of thymine dimers in DNA.   An internal charge transfer
complex between the tryptophan residue as electron donor and the
thymine dimer as acceptor is proposed as an intermediary in the
reaction.   Recently these authors were able to show that a protein from the T4 phage coded by gene 32 can replace the tryptophan
containing peptide in the reaction complex and sensitize the
splitting of thymic dimers in DNA.   It is most significant that
the reaction could be carried out in an aerated solution.   Experiments such as these might allow us to assess the accessibility of
oxygen to the site of reaction within a protein-DNA complex.

In summing up, it can be said that the role of oxygen in high
molecular weight biopolymers must be ascertained in each case
separately.   On the other hand data on photosensitization in the
presence and in the absence of oxygen can be used profitably to
study structural and functional aspects in biological macromolecules.

REFERENCES

1.   C.S.Foote. Proc.VI Intern.Congr on Photobiol.1972,Paper 005
      Deutsche Gesellschaft fur Lichtforschung,Frankfurt 1974
2.   H.Berg and F.A.Gollmick, ibid, Paper 006
3.   M.Koizumi and Y.Usui,   ibid, Paper 007
4.   H.Tsubomura,   ibid, Paper 008
5.   J.D.Spikes and F.Rizzuto,ibid, Paper 009
6.   C.S.Foote, in Free Radicals in Biology (W.A.Pryor,ed.)
      Academic Press, New York (in press).
7.   J.D.Spikes, in The Science of Photobiology (K.C.Smith,ed.)
      Plenum Press, New York (in press).
8.   L.I.Grossweiner, Current Topics in Radiation Res.Quarterly
      $\underline{11}$, 141 (1976).
9.   N.J.Turro, Proc.VI Intern.Congr. on Photobiology 1972, Paper
      010, Deutsche Gesellschaft fur Lichtforschung, Frankfurt 1974
10.   J.Feitelson, Photoch. and Photobiol. $\underline{13}$, 87 (1971).
11.   D.V.Bent and E.Hayon, J.Am.Chem.Soc. $\underline{97}$, 2612 (1975).
12.   N.Anbar and P.Neta, Internat.J.of Applied Radiation and
      Isotopes, $\underline{18}$, 493 (1967).
13.   J.Bourdon and B.Schnuriger, in Physics and Chemistry of the
      Organic Solid State, Vol.III, p.59 (D.Fox, M.M.Labes,
      A.Weissberger ed's), Interscience N.Y.
14.   I.Kraljic and L.Lindqvist, Photochem.and Photobiol.$\underline{20}$,351
      (1974).
15.   C.Foote, Proc.VII Internat.Congr. on Photobiology 1976,P-136.

16. J.Feitelson and E.Hayon, Photochem.and Photobiol.17, 265
     (1973).
17. M.L.Saviotti and W.C.Galley, Proc. Nat. Acad. Sci. U.S.A.,
     71, 4154 (1974).
18. H.Seki, Y.A.Ilan, Y.Ilan and G.Stein, Bioch.Bioph.Acta, in
     press.
19. G.Stein, In: Lasers in Physical Chemistry and Biophysics,
     p.511, Elsevier, Amsterdam, 1975.
20. U.W.Kenkare and F.Richards, J.Biol.Chem. 241, 3197 (1966).
21. J.F.Baugher and L.I.Grossweiner, Photoch. and Photobiol.
     22, 163 (1975).
22. P.Walrant, R.Santus and L.I.Grossweiner, Photochem. and
     Photobiol., 22, 63 (1975).
23. M.Folin, G.Gennari and G.Jori, Photochem. and Photobiol.
     20, 357 (1974).
24. M.L.Meistrich and A.A.Lamola, J.Mol.Biol. 66, 83 (1972).
25. M.Charlier and C.Hélène, Photochem. and Photobiol., 21,
     31 (1975).

NEW CHEMICAL ASPECTS OF THE PHOTOREACTION BETWEEN
PSORALEN AND DNA

F.DALL'ACQUA

INSTITUTE OF PHARMACEUTICAL CHEMISTRY OF THE

PADUA UNIVERSITY-PADOVA-ITALY

SUMMARY

Various aspects of the photochemical interactions
between furocoumarins and DNA are considered. Evidence
concerning the following topics are reported:formation
of the intercalated complexes between furocoumarins and
DNA, kinetic investigations on the photoreaction between
psoralen and DNA, photobiological significance of monoadducts and cross-linkages, receptor sites for the
preliminary intercalation and the successive photoaddition of psoralen.

INTRODUCTION

Furocoumarins or psoralen are a group of photo-
biologically active substances. Many furocoumarins are
widely distributed in nature while others have been
prepared by synthesis (1,2). The photoreactions (365 nm)
between DNA and psoralens have been widely studied
because they represent the molecular basis of the photosensitizing properties of these substances (3,4).
Fig.1 reports a scheme of the photoreactions between
psoralen and DNA. Psoralen in part is complexed with DNA
and in part is free in aqueous solution; photoaddition
to DNA involves only the part of psoralen which is
complexed with DNA. Therefore the preliminary step of
the photoreaction is the complex formation.

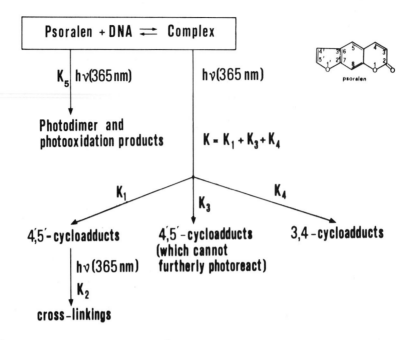

Fig.1 — Representation of the photoreaction between
         psoralen and DNA.

COMPLEX FORMATION

The complex formation has been shown by various
experimental evidence (5,6). Recently an experiment
concerning the possible intercalation has been obtained
by flow dichroism measurements (see Fig.2). In aqueous
solution 5-methoxy-psoralen does not show any dichroism,
while in the presence of DNA it shown an evident negative
dichroism in the range of wavelenghts (300-350 nm) in
which DNA does not show any dichroism. The negative
dichroism indicates that the furocoumarin has assumed
an ordered position parallel to that of base pairs of
DNA, strongly suggesting the intercalation between two
base pairs. The amount of furocoumarin complexed with
DNA at various different macromolecule concentrations,
has been evaluated both by spectrophotometrical
measurements and by equilibrium dialysis determinations.
The binding plot, $\underline{r}$ versus $\underline{c}$, is reported in Fig.3.This

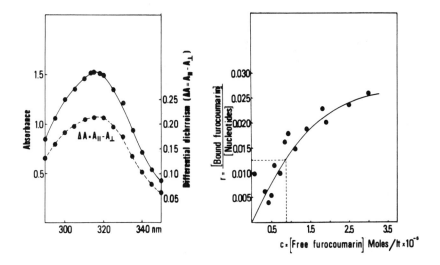

Fig.2 – Dichroic spectrum of 5-methoxypsoralen in the presence of DNA.

Fig.3 – Binding of 5-methoxypsoralen by calf thymus DNA.

figure shows that the number of sites useful for intercalation of 5-methoxy-psoralen are 2.5 per 100 nucleotides, while the association constant is of the order of 10^5.

KINETIC STUDIES ON THE PHOTOREACTION BETWEEN PSORALEN AND DNA.

For obtaining a detailed picture of the photoreaction this has been represented by means of a mathematical model consisting of a system of differential equations. Utilizing the experimental data which were possible to obtain directly (see Table I) the system has been resolved using a computer. The rate contant values of all the photoproducts as well as the amounts of the photocompounds formed as function of time of irradiation are reported in Table II and Fig.4. The cycloadducts which are formed with higher yield are 3,4-monofunctional adducts. 4',5'-adducts not able to give cross-linkages take place in a lesser extent while 4',5'-adducts forming cross-linkages are efficaciously converted into

bifunctional adducts, remaining present only in a very
small amount. The formation of cross-linkages occurs to
a good extent: they represent the second photocompound
after 3,4-adducts, from a quantitative point of view.
The rate constant of their formation is very high,
confirming that 4',5'-adducts are immediately converted
into bifunctional adducts (7,8). The photodimer and the
photooxidation products, which take origin from the
irradiation of free psoralen, are formed with relatively
low yield (see Fig.4 and Table II) and represent a
marginal aspect of the photoreaction(9).

TABLE I- Amount of psoralen transformed into various photoproducts

Time of irradiation (minutes)	Total amount of psoralen linked to DNA µMoles/l	Cycloadducts -4',5' (b+d) µMoles/l	Cross-linkages (c) µMoles/l	Photodimer and photooxidation products (f) µMoles/l
0.5	1.00	0.29	0.085	0.075
1	2.02	0.490	0.300	0.150
1.5	2.93	0.635	0.500	0.230
3	5.38	1.075	1.05	0.440
5	7.95	1.520		0.730
10	12.34			

TABLE II- Rate constant values

$k = k_1 + k_3 + k_4$	$12.08 \times 10^{-2} \times min^{-1}$
k_1 (rate constant for 4',5'-cycloadducts formation, type b)	2.55×10^{-2} "
k_2 (rate constant for the cross-linkages formation, c)	310.87×10^{-2} "
k_3 (rate constant for 4',5'-cycloadducts formation, type d)	2.17×10^{-2} "
k_4 (rate constant for 3,4-cycloadducts formation, e)	7.35×10^{-2} "
k_5 (rate constant for the formation of photodimer and photooxydation products, f)	0.28×10^{-2} "

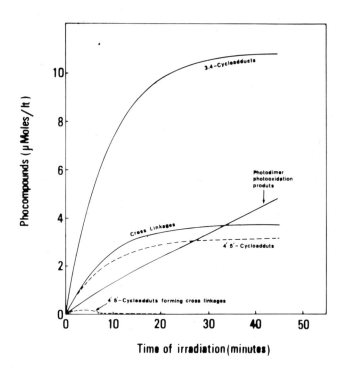

Fig.4 – Picture of the
various photocompounds
formed in the photo-
reaction between psoralen
and DNA as a function of
time of irradiation.

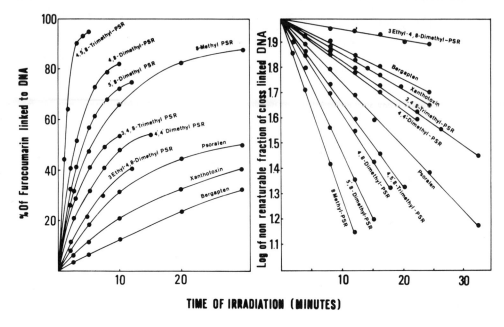

Figure 5

BIOLOGICAL SIGNIFICANCE OF MONO- AND BI-FUNCTIONAL ADDUCTS.

Fig.5 reports the photoreactivity (total binding) and the capacity of forming cross-linkages shown by various psoralens (8,10,11). The data show that a close corre‍lation between these two aspects of the photoreaction does not exist: e.g. 3-ethyl-4,8-dimethylpsoralen shows a relatively high capacity of photobinding to DNA but a much low ability of forming cross-linkages. Regarding this it would be interesting to know which is the biological role of monofunctional and of bifunctional adducts. Cole observed that inactivation of bacteria was correlated with the extent of cross-linkages formed during irradiation (12). A good correlation exists also between the cross-linkages formation and the skin-photo-sensitizing activity (13). Therefore in these cases bifunctional adducts have a more important role than monofunctional ones.

The biological role of monofunctional adducts has been
investigated studying angelicin, an angular furocoumarin
which, for steric reasons, can form only monofunctional
adducts with native DNA (8). The capacity of inhibition
of cellular DNA and RNA synthesis has been clearly
evidenced, even if it is lower than that of psoralen(14).
The inhibition of DNA and of RNA synthesis produced by
irradiation in the presence various furocoumarin
derivatives, having a different capacity of photobinding
to DNA and of forming cross-linkages, has been studied
in Ehrlich ascites tumor cells. The results obtained
indicate that a correlation exists between the total
photobinding to DNA and the inhibition of RNA and DNA
synthesis. This correlation demonstrates that the
template activity of DNA is affected both by monofunctional and by bifunctional adducts (15).In general
it appears that photobiological effects evaluated a
short time after irradiation, like DNA and RNA synthesis
inhibition (14,15) are produced more or less in the same
extent by monoadducts and by cross-linkages, while effects
evaluated by long lasting experiments, like skin-photo-
sensitization (2) killing of bacteria (12) and inhibition
of transmitting capacity of Ehrlich ascite tumor cells
(3) seem more correlated with the damage produced by
cross-linkages. May be that different repair processes
are involved in determining this difference.

RECEPTOR SITES IN DNA FOR THE PHOTOADDITION WITH PSORALEN

Investigations have been made for putting in evidences
if in DNA some receptor sites exist for the preliminary
intercalation and the successive photoaddition.

a) SITES for COMPLEX FORMATION. Some synthetic polynucleotides having known sequence have been used for
evaluating their capacity of interaction with 8-methyl-
psoralen. The results obtained using poly dA·poly dT
and poly d(A-T) clearly show that while the polymer
having linear sequence did not show any evident capacity
of the complex formation, poly d(A-T) showed a good
ability to forming the complex with 8-methylpsoralen.
Analogous results have been obtained using poly dC·poly
dG and poly d(C-G). Fig.6 reports the binding of 8-methyl
psoralen (evaluated spectrophotometrically) by the two
polynucleotides having alternate sequences and by calf
thymus DNA as reference sample. The binding curves show

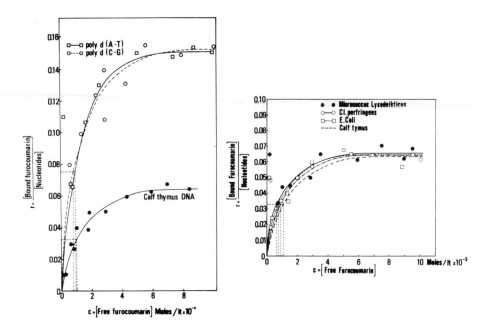

Fig.6 – Binding of 8-methylpsoralen by poly d(A-T),
        poly d(C-G) and calf thymus DNA.
Fig.7 – Binding of 8-methylpsoralen by various DNA
        samples having different base pair composition.

that the two synthetic polynucleotides have a number of
sites suitable for intercalation (15-16 per 100 nucleotides) higher than that of calf thymus DNA (7 per 100
nucleotides) while the association constant value is
of the same order of magnitude both for synthetic
polymers and for DNA ($\simeq 1\times10^5$). These data suggest that
adenine-thymine and cytosine-guanine play the same role
for the complex formation.
Further studies in this direction have been performed
using various DNA samples having different base pair
composition as shown in Fig.7. It is possible to observe
that DNA from Cl. perfringers having about 80% of A-T
has the same capacity of forming the complex as DNA from
Micrococcus lysodeicticus having about 28% of A-T. The
four samples examined demonstrate the same behaviour both
as regard the number of sites suitable for intercalation

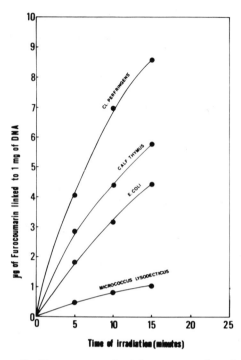

Fig.8 – Photobinding
capacity of psoralen twards
various DNA containing diffe
rent amounts of A–T.

and the association constant.
These data lead to the conclusion that receptor sites
exist in DNA for the complex formation and they are
represented by the segments having an alternate sequence
of bases without distinction of the base pair composition.

b) SITES for THE PHOTOADDITION. The receptor sites for
photoaddition have been also investigated using synthetic
polynucleotides. The highest photoreactivity has been
shown by poly d(A–T), that is by the polynucleotide
containing adenine and thymine in alternate sequence.
Much lower was the photoaddition capacity shown by poly
d(C–G), while an intermediate photoreactivity is shown
by calf thymus DNA, used as reference sample. The two
polynucleotides having linear sequence demonstrated a
much lower photoreactivity. These results seem to
indicate that a receptor site for photoaddition of
psoralen exists in DNA and should be represented by the
segments of macromolecule having an alternate sequence
like that of poly d(A–T) (16). This conclusion has been
confirmed using various DNA samples containing increasing
amounts of A–T. The photoreactivities towards psoralen

shown by the various samples are much different (see
Fig. 8). It is evident that DNA from Cl. perfringens,
having the highest extent of A-T shows the highest
photobinding capacity towards psoralen; when the extent
of A-T is decreased, like in the case of calf thymus
DNA, E.coli DNA and M. lysodeicticus DNA, the photoreactivity decreases in a parallel way.

c) SITES for CROSS-LINKAGES FORMATION. The possible
receptor site for the cross-linkages formation has been
investigated using synthetic polynucleotides. The
evaluation of cross-linkages has been performed on the
basis of the increased melting temperature and on the
decreased hypochromicity. Both poly dA·poly dT and poly
dC·poly dG do not form cross-linkages, and this fact is
in agreement with their molecular structure.
Poly d(A-T) by contrast showed a high formation of cross-
linkages; also poly dC-G) showed the capacity of forming
cross-linkages but in a lower extent. The alternate
segments of DNA highly rich in A-T seem to be the possible
receptor site for the cross-linkages formation (16).

REFERENCES
1. Musajo,L.,Rodighiero,G., and Caporale,G.,Bull.Soc. Chim.Biol.,36, 1213 (1954).
2. Musajo,L., and Rodighiero,G., Experientia, 18, 153 (1962).
3. Musajo,L., and Rodighiero,G., in Photophysiology, vol.VII, Edited by C.Giese, pag. 115-147, Academic Press, New York (1972).
4. Pathak,M.A., in Sunlight and Man (Edited by M.A. Pathak, L.C.Harber, M.Siji and A.Kukita) pag.495, University Tokyo Press (1974).
5. Rodighiero,G., Caporale,G., and Dolcher,T.,Rend.Atti Accad.Naz.Lincei, 30, 84 (1961).
6. Dall'Acqua,F.,and Rodighiero,G., Rend.Atti Accad.Naz. Lincei, 40, 411 (1966).
7. Dall'Acqua,F.,Marciani,S., and Rodighiero,G., FEBS Letters, 9, 121 (1970).
8. Dall'Acqua,F., Marciani,S., Ciavatta,L., and Rodighiero,G., Z.Naturforsch., 26b, 562 (1971).
9. Dall'Acqua,F.,Marciani,S., Zambon,F., and Rodighiero, G., unpublished results.
10. Rodighiero,G., and Musajo,L. et al., Biochim.Biophys. Acta, 217, 40 (1970).
11. Dall'Acqua,F.,Marciani,S.,Vedaldi,D., and Rodighiero, G., Biochim.Biophys.Acta, 353, 267 (1974).
12. Cole, R.S.,J.Bacteriol., 107, 846 (1971).
13. Dall'Acqua,F.,Marciani,S., Vedaldi,D., and Rodighiero, G., Z.Naturforsch., 29c, 635 (1974).
14. Bordin,F., Marciani,S., et al., Ital.J.Biochem., 24, 258 (1975).
15. Baccichetti,F.,Bordin,F., et al., Z.Naturforsch.,31c, 207 (1975).
16. Marciani,S.,Dall'Acqua,F.,Vedaldi,D.,and Rodighiero, G., Il Farmaco,Ed.Sci.,31, 140 (1976).

EXCITED STATES OF SKIN-SENSITIZING PSORALENS AND THEIR REACTIONS WITH NUCLEIC ACIDS[+]

Ching-Nan Ou, Chee-Hway Tsai and Pill-Soon Song*

Department of Chemistry, Texas Tech University

Lubbock, Texas 79409

[+]This paper is dedicated to the late Professor Luigi Musajo. This work was supported in part by the R. A. Welch Foundation (D-182), National Inst. of Health (CA 13598, GM 23089) and NSF (BMS75-05001). *To whom communication regarding this paper should be addressed.

ABSTRACT

Electronic excited states and structural specificity of skin-sensitizing psoralens and related derivatives are reviewed from various points of view in order to establish the photochemically meaningful structure-activity correlation. In addition, photodynamic effects of various psoralens and coumarins in <u>Bacillus</u> subtilis mutants are investigated as a model system in order to ascertain the biological consequence of photochemical modifications of nucleic acids.

In the present study, it is shown that 5-fluorouracil is much more reactive than thymine with respect to addition to the photo-excited state of psoralens.

Calf thymus DNA solution was mixed with 8-methoxypsoralen (MOP). From the fluorescence analysis, the binding ratio of MOP to DNA base (P) was determined to be 1:50. The mixture of DNA and ^{3}H-MOP was also irradiated with 365 nm light. The photoreacted DNA mixture was fractionated on an Ultrogel column.

The incorporation of $[^3$H]-MOP into tRNA structure under irradiation at 365 nm without O_2 was determined by gel filtration,

millipore filter and liquid scintillation counting techniques.
The maximum incorporation of [^{3}H]–MOP into 5–fluorouracil–containing
tRNA is about 3.5 moles of MOP per mole of t–RNA. The tRNA–MOP
adducts have been characterized by absorption, fluorescence and CD
spectra.

Chemical analyses of photoadducts have been carried out. The
t–RNA–[^{3}H]–MOP was digested with an enzyme mixture and the products
were separated by TLC and visualized on x–ray film by low–temperature fluorography. The results indicated that there were several
types of photoadducts formed during photolysis.

The photomodification of DNA and its photodynamic effects in
B. subtilis possibly involve the covalent photo–cross–linking of
duplex DNA by psoralens. However, we also found that 5,7–dimethoxy-
coumarin (DMC) with only one functional group is highly active as
an oxygen–independent photosensitizer in B. subtilis; it also
strongly intercalates into DNA, unlike other coumarin derivatives.
The ability to complex with DNA can be explained on the basis of a
steric compatibility of DMC relative to psoralens (e.g.,
5–methoxypsoralen).

INTRODUCTION

Furocoumarins (psoralens) are well known for their photosensitizing ability to induce skin erythema[1-3], carcinogenesis (see
review by Giese)[4] and mutagenesis[5-7]. A number of laboratories
reported on various experimental approaches to elucidating the
role of psoralens in the photo–induced effects on biological systems, but many of the relevant reports cannot be reviewed in this
paper due to the restricted space.

Our attempt in this field has been to establish a molecular
basis of the structure–activity relationship of these compounds,
and to elucidate photochemical mechanism(s) of their action in
vitro (with DNA and RNA) and in vivo (with Bacillus subtilis
mutants). Initially, photophysical characteristics of psoralen
and its derivatives are reviewed in this paper. Photochemical
reactivity and photobiological studies with nucleic acids and B.
subtilis, respectively, are presently being investigated. Some of
these studies are discussed in this report.

MATERIALS AND METHODS

All chemicals, solvents and spectroscopic measurements were
as described in our previous publications. Fluorescence lifetimes
were measured on an SLM Model 480 subnanosecond phase–modulation
spectrofluorometer.

RESULTS AND DISCUSSION
Photophysical Properties of Psoralens

(a) $^1\pi,\pi*$ and $^1n,\pi*$ States. Coumarin serves as a useful model for elucidating spectroscopic characteristics of psoralen and its derivatives[8,9,10]. The lowest singlet excited state of coumarin and psoralens is assigned to $^1\pi,\pi*$, with the $^1n,\pi*$ state lying slightly above the lowest singlet state.

The $^1n,\pi*$ state localized on the carbonyl group must be in the vicinity of the lowest $^1\pi,\pi*$ state in coumarin and psoralens, as suggested by the low fluorescence polarization and short fluorescence lifetime (ca. 0.4 nsec) at 77 K. The location of the $^1n,\pi*$ state determines not only the relative populations (and their lifetimes) of singlet and triplet excited states, but it also affects the photoreactivity of coumaryl compounds toward nucleic acid bases.

In order to resolve the $n\rightarrow\pi*$ band, we added monocations which complex with the carbonyl group, thus shifting the $n,\pi*$ state to a higher energy. It was observed that the intensity and fine structure of coumarin fluorescence were dramatically enhanced[11]; corresponding enhancement of the fine structure in absorption was also observed (unpublished data). These effects can be best interpreted as arising from the energy separation between the lowest $^1\pi,\pi*$ and the nearby $^1n,\pi*$ state due to monocation complexing. No similar effects were observed with psoralens.

The fluorescent state of psoralen is also of the $^1\pi,\pi*$ type. The fluorescence lifetime in ethanol at 298 K is 1.8 nsec, which agrees with the value reported recently[12].

In the case of coumarin, the lowest $\pi\rightarrow\pi*$ excitation is strongly localized in the pyrone moiety[13]. Thus, both $S_1(\pi,\pi*)$ states of coumarin and psoralen contain significant charge transfer character in the C=C bond of the pyrone moiety (32% CT to the C=C vacant MO in coumarin and 22% in psoralen), whereas CT character at the furyl C=C bond of psoralen is 14%, according to configuration analysis calculations[14].

(b) $^3\pi,\pi*$ State. Like coumarin, psoralens show stronger phosphorescence emission than fluorescence, indicating an efficient intersystem crossing from the excited $^1\pi,\pi*$ state. The phosphorescent state of psoralens is assigned to $^3\pi,\pi*$ type[8,9], and the coumarin triplet serves as a strikingly useful model for the psoralen triplet state, both in energy and vibrational fine structure and polarization characteristics of the phosphorescence. Both phosphorescence spectra of coumarin and 8-methoxypsoralen are nearly superimposable. In fact, the 0-0 phosphorescence energy of many coumarins and psoralens remains roughly constant, independent of substituents[8]. This is all the more striking, since the absorption and fluorescence spectra of these compounds are very sensitive to substituents.

The above results can be interpreted in terms of a strong localization of the triplet excitation in the region of the C=C bond of the pyrone moiety[8,9]. Recent theoretical[14] and EPR/ODMR[15,16] data are consistent with the localized π,π^* state. No significant localization is predicted at the furyl C=C bond[14].

Photochemical Reactivity of Psoralens with Bases, DNA and tRNA

Since FU (5-fluorouracil) is the most reactive among pyrimidine bases tested for photoreaction(s) with coumaryl compounds[17], we have compared the photoreactivity of various coumaryl compounds with this pyrimidine base. The order of photoreactivity found is: trans-benzodipyrone (BDP) > cis-BDP > 8-methoxypsoralen > coumarin[17].

Theoretically, the photoreactivity of the pyrone C=C bond is more susceptible to cyclo-addition to thymine than the furyl C=C bond. However, the reverse is predicted for photosubstitution reactions[9]. In this regard, trans-BDP is of particular interest, as it possesses two equally reactive pyrone C=C bonds for potential cross-linking with nucleic acids, as will be discussed later.

The relative importance of the singlet vs. triplet state reactivity of psoralens toward pyrimidine bases remains to be established. Bevilacqua and Bordin demonstrated a moderate quenching effect of the photo-C_4-cycloaddition of psoralen to pyrimidine bases by oxygen and paramagnetic ions[18]. Recently, McInturff[17] obtained more definite quenching data, using 0.04 M FU as the substrate. She showed a nearly complete quenching of the photoaddition between MOP and FU, as determined from the initial rate measurements based on the absorbance change of the former. It appears that the triplet psoralen plays an important role in the photoreactions with pyrimidine bases.

In this connection, we have proposed that the impotency of hydroxypsoralens (HP) as skin sensitizers is attributable to the lack of efficient triplet formation due to the competing ionization process in the S_1 state and the delocalized triplet nature of the anionic forms of HP[19]. In vitro photoreactivity studies based on the photoreactions between psoralens and FU (0.02–0.04 M) showed the following order (in order of decreasing relative reactivity in parentheses): 8-MOP in water, pH 6 (1.0), 8-HP in ether (0.91), 5-MOP in water (0.86), 8-HP in water (0.35) and 5-HP (0.22)[17]. Thus, it is apparent that HP in water is considerably less reactive than the corresponding methoxypsoralens toward FU.

5,7-Dimethoxycoumarin, which does not have the furyl C=C
bond for interstrand cross-linking of DNA, intercalates strongly

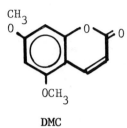

DMC

with DNA in the gound state and photobinds
covalently to DNA bases[20]. Unlike most other
coumarins and psoralens, this coumarin derivative
shows a high ϕ_F and a low ϕ_P. Since the self-
photodimerization of this coumpound may proceed
via the singlet excited state with or without
oxygen[20,21], it is quite probable that the photoaddition of DMC to DNA also occurs via the singlet state.

In order to ascertain the relative reactivity of singlet vs.
triplet states of psoralen toward DNA, we carried out photolysis
of a DNA-^{3}H-MOP mixture, followed by isolation of the photoadducts
on an Ultrogel column as well as on autoradiographic 2-dimensional
TLC of the enzyme digest from the DNA-MOP photoadducts. The photobinding ratio of 8-MOP to nucleotide was found to be 1:26[22].

The aerobic solution of DNA-MOP, photolyzed and analyzed as
described above, showed 37% quenching of the photobinding ratio,
possibly implicating the involvement of triplet 8-MOP. The effects
of KI on the photobinding were less conclusive; 50 mM KI inhibited
the photolysis by only 13%[22]. It is possible that both singlet and
triplet states are involved in the photobinding reaction, particularly in the interstrand cross-linking reaction which may proceed
in two steps[23], i.e., the first step via the triplet 3,4-photocy-
cloaddition of MOP to DNA followed by the singlet 4',5'-photocyclo-
addition of the former to the complementary strand of DNA. Furthermore, at least four different 8-MOP-DNA base photoadducts are formed,
as shown by 2-dimensional autoradiography-TLC and fluorescence
lifetime measurements of the photoadducts[22]. Thus, it is premature
to conclude that either triplet or singlet is the predominant photoreactive state, and that the photobinding to DNA exclusively involves
cycloaddition (see Concluding Remarks). However, there is no doubt
that psoralens photochemically cross-link DNA[23,24] probably via
cycloaddition[25].

The photobinding ratio of 8-MOP to tRNA and FU-enriched tRNA
was determined by gel filtration and tritium counting. Normal
tRNA showed sigmoidal photobinding kinetics, in contrast to FU-
enriched tRNA. The sigmoidal kinetics disappeared when preirradia-
tion induced the photoaddition of 4-S-Ur to cytidine in tRNA. No
more than about 3.5 molecules of 8-MOP were found to bind to tRNA
and FU-tRNA. Mizuno et al. reported photobinding up to 14 8-MOP/
tRNA, based on absorbance changes at 260 and 300 nm, which seems
unusually high. Chemical analyses of the photoirradiated 8-MOP/
tRNA's showed at least 8 major types of photoadducts. The sites
of photobinding are pyrimidine bases (including minor bases; random binding) in tRNA and exclusively FU base in FU-enriched tRNA[22].

Photobiological Effects in Bacillus subtilis

Photodynamic, photocarcinogenic and photomutagenic effects of psoralens[5-7], are likely to involve photochemical interactions with nucleic acids, particularly DNA. We have chosen numerous Bacillus subtilis strains in our attempt to draw a correlation between the molecular structure and the photobiological reactivity of coumarin derivatives described in the previous section.

Fig. 1 shows the survival curves of three B. subtilis strains treated with near UV and photosensitizers. First, it can be seen that 8-MOP is drastically more effective as a photo-killing agent than 8-hydroxypsoralen (also 5-hydroxypsoralen: HP). This observation is readily justified in terms of (a) the in vitro photoreactivity of (HP) discussed earlier and (b) the fact that HP are not potent skin sensitizers.

The photochemical reactivity of BDP in vivo is reflected by its greater ability to elicit a near UV-induced mutation[7] (and photolethality) than coumarin. Near UV irradiation of strains 168ind (Trp⁻, Hcr⁺, Rec⁺) and mc-1 (Trp⁻, Rcr⁺, Rec-2⁻) in the presence of BDP did not kill or mutate these strains. This result suggests that cyclobutane type photoadducts between BDP and DNA are

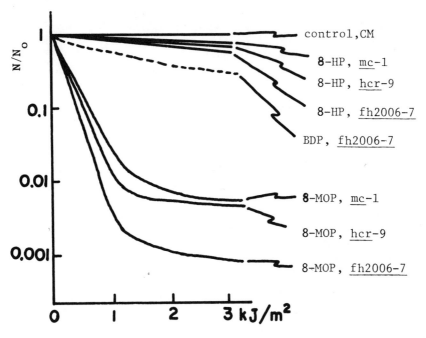

Fig. 1. Survival of B. subtilis at 365 nm in the presence
of photosensitizing compounds. CM;coumarin.

possibly involved, as these strains effectively remove such photoproducts to repair the near UV damage. On the other hand, other
excision-repair-deficient strains are sensitive to near UV–BDP
(and psoralens) treatments (Fig. 1), while cis-BDP is practically
ineffective. We therefore suggest that the former may involve
cross-linking of DNA strands as follows:

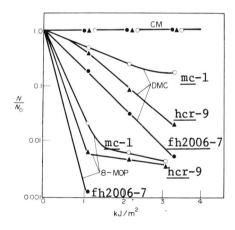

As mentioned earlier,
the photoaddition of DMC to
DNA in vitro occurs without
cross-linking of duplex
strands, since this compound
has only one functional C=C
group. A specific or limited
photoaddition of DMC to DNA
could adequately account for
the lesions in B. subtilis
strains (Fig. 2). Thus, DMC
provides a mechanism of photodynamic action alternative to
the cross-linking lesion[23].

The near UV effects
observed above may be due to
oxygen-dependent photodynamic
action as the result of singlet
oxygen via triplet energy
transfer from sensitizers to
oxygen. However, this is not
very likely for the observed
effects, particularly with
DMC, since other coumarins
with much higher efficiency
of triplet formation show no
photo-induced killing of
B. subtilis strains.

Fig. 2. Survival of B. subtilis
mutants at 365 nm.
Comparison between DMC
and 8-MOP is shown.

CONCLUDING REMARKS

In spite of continued efforts by several laboratories including our group, the chemical nature of the photobinding as to (a)
specific sites on DNA and RNA and (b) covalent linkage (cycloaddition vs. substitution) is yet to be definitively established.
Although a psoralen-Ur photoadduct is likely to be of a cycloaddition type[25], similar treatments (255 nm irradiation) did not result
in photodissociation of the DNA-8-MOP adduct.[26] Furthermore,
2-anthraquinone sulfonate (a triplet sensitizer for photodissociation of T̂T) did not sensitize the dissociation of the DNA- and tRNA-
MOP photoadducts. However, these negative results do not necessarily
implicate a non-cycloaddition type bonding for the photoadducts,
since the triplet state of the cyclobutane ring populated by direct
UV irradiation or by triplet energy transfer from the sensitizer
can be effectively quenched intramolecularly by the conjugated
residue (benzofurylfomate 1 if 3,4-cycloadded or 7-methoxycoumarin
2 if 4',5'-cycloadded) whose triplet level is likely to be lower
than that of the cyclobutane ring.[27] m-Methoxyphenylformate is a
possible quencher
in the case of
cross-linked
photoadducts.

$$E_T(\underset{\sim}{1}) < E_T(\underset{\sim}{2})$$

$$\underset{\sim}{1} \qquad\qquad\qquad\qquad\qquad \underset{\sim}{2}$$

Molecular mechanisms of psoralens' action in various biological systems (e.g., skin tissue and microorganisms) are still incom-
pletely understood. Generally, it is accepted that interactions
between psoralens and nucleic acids are prerequisite for the oxygen-
independent photodynamic action. As we have demonstrated in this
connection with DMC, cross-linking is not the exclusive DNA damage
for the photodynamic action of psoralens. Several preliminary
experiments on biological effects are noteworthy. 8-MOP-UVA
inhibited DNA synthesis in vivo[3,28]. No apparent early inhibition
of RNA and protein synthesis was observed. On the other hand, the
photobinding of 8-MOP to tRNA significantly affected the aminoacylation[3] and the amino acid accepting activity of tRNA[22], possibly due
to a conformational change of tRNA induced by the photomodification[3,22]. Clearly, more work is necessary before the mechanism(s)
of psoralen photobiology can be delineated at the molecular level.

REFERENCES

1. L. Musajo, Ann. Ist. Super. Sanita 376 (1969) and references
   therein.
2. M. A. Pathak, J. H. Fellman and K. D. Kaufman, J. Invest.
   Dermatol. 35, 165 (1960).
3. N. Mizumo, S. Tsuneishi, S. Matsuhashi, S. Kimura, Y. Fujimma

and T. Ushijima, in "Sunlight and Man" (edited by M. A. Pathak, et al.), p. 389 (1974), Univ. of Tokyo Press, Tokyo.

4. A. C. Giese, Photophysiology 6, 77 (1971) and references therein.

5. W. L. Fowlks, D. G. Griffith and E. L. Oginsky, Nature 181, 571 (1958).

6. M. M. Mathews, J. Bacteriol. 85, 322 (1963).

7. M. L. Harter, I. C. Felkner, W. W. Mantulin, D. L. McInturff, J. N. Marx and P. S. Song, Photochem. Photobiol. 20, 407 (1974).

8. W. W. Mantulin and P. S. Song, J. Am. Chem. Soc. 95, 5122 (1973).

9. P. S. Song, M. L. Harter, T. A. Moore and W. C. Herndon, Photochem. Photobiol. 14, 521 (1971).

10. T. A. Moore, M. L. Harter and P. S. Song, J. Mol. Spectrosc. 40, 144 (1971).

11. P. S. Song and Q. Chae, J. Luminesc. 12/13, 831 (1976).

12. W. Poppe and L. I. Grossweiner, Photochem. Photobiol. 22, 217 (1975).

13. P. S. Song and W. H. Gordon, J. Phys. Chem. 74, 4234 (1970).

14. P. S. Song, C. A. Chin, I. Yamazaki and H. Baba, Int. J. Quantum Chem. QBS No. 2, 1 (1975).

15. E. T. Harrigan, A. Chakrabarti and N. Hirota, J. Am. Chem. Soc. 98, 3460 (1976).

16. T. A. Moore, A. B. Montgomery and A. L. Kwiram, Photochem. Photobiol. 24, 83 (1976).

17. D. L. McInturff, M. S. Thesis, Texas Tech University, Lubbock (1975).

18. R. Bevilacqua and F. Bordin, Photochem. Photobiol. 17, 191 (1973).

19. P. S. Song, W. W. Mantulin, D. McInturff, I. C. Felkner and M. L. Harter, Photochem. Photobiol. 21, 317 (1975).

20. C. N. Ou, P. S. Song, M. L. Harter and I. C. Felkner, Photochem. Photobiol. 24, (1976), in press.

21. M. L. Harter, Ph.D. Dissertation, Texas Tech University (1974).

22. C. N. Ou, C. H. Tsai and P. S. Song, to be published.

23. R. S. Cole, Biochim. Biophys. Acta 254, 30 (1971).

24. F. Dall'Acqua, S. Marciani, L. Ciavatta and G. Rodighiero, Z. Naturf. 26b, 561 (1971).

25. C. H. Krauch, D. M. Krämer and A. Wacker, Photochem. Photobiol. 6, 341 (1967).

26. C. N. Ou and P. S. Song, Abstract, Am. Soc. Photobiol. 4th Ann. Mtg. (1976), WPAB6.

27. S. C. Shim, private communication (1976).

28. J. H. Epstein and K. Fukuyama, Photochem. Photobiol. 21, 325 (1975).

PHOTOCHEMOTHERAPEUTIC, PHOTOBIOLOGICAL, AND PHOTOCHEMICAL PROPERTIES OF PSORALENS

Pathak M.A., Fitzpatrick T.B., Parrish J.A., Biswas R.

Dept. of Derm., Harvard Med. Sch. & Mass. Gen. Hosp.

Massachusetts General Hosp., Boston, Mass., U.S.A. 02114

INTRODUCTION

A large number of drugs are known to interact strongly with nucleic acids (e.g., several antitumoral agents such as actinomycin-D, adriamycin, and daunomycin, and drugs such as chloroquin, quinacrine, miracil D, ethidium bromide, etc). Many of these drugs show specific, non-photochemical binding to double-stranded nucleic acids through intercalation between adjacent base pairs. Psoralens, which are particular isomers of furocoumarins, are an ubiquitous group of naturally occurring compounds (Fig. 1). Recently, certain photoactive agents such as 8-methoxypsoralen (8-MOP) and 4, 5', 8-trimethylpsoralen (TMeP), in combination with exposure to longwave ultraviolet light (UV-A, 320-400 nm), have been shown to be very effective in treating patients with disabling, generalized psoriasis and patients with vitiligo who have lost the normal melanin pigmentation of their skin (1-3). The beneficial therapeutic effects, particularly in clearing and controlling psoriasis, probably result from the in vivo photoconjugation of psoralens to DNA. Oral psoralen photochemotherapy involves the ingestion of a photoactive psoralen derivative such as 8-MOP (40-50 mg/70 kg) followed at a specified interval by exposure of the skin to ultraviolet light. Two hours after ingestion of 8-MOP, the patient is exposed to whole-body radiation using a high-intensity UV-A phototherapy unit. Initial doses are calculated to approach threshold erythema or MPD (minimum phototoxic dose) and range from 1.0 to 6.5 J/cm^2. The exposure dose is gradually increased during subsequent treatments, which are given two to four times per week until clearing of psoriasis is achieved. This photochemotherapy with psoralens is unique in the sense that the inhibition of DNA synthesis, the covalent conjugation of psoralen

267

The photosensitizing potency of psoralens and related compounds

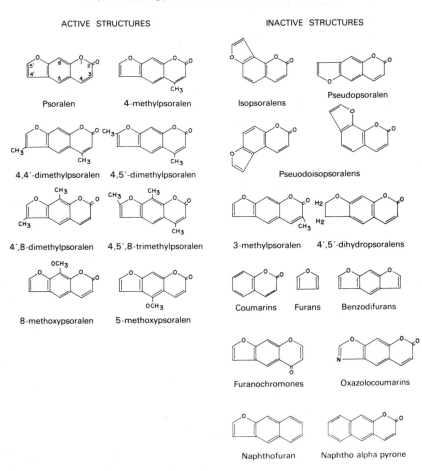

ACTIVE STRUCTURES   INACTIVE STRUCTURES

Psoralen   4-methylpsoralen   Isopsoralens   Pseudopsoralen

4,4'-dimethylpsoralen   4,5'-dimethylpsoralen   Pseuodoisopsoralens

4',8-dimethylpsoralen   4,5',8-trimethylpsoralen   3-methylpsoralen   4',5'-dihydropsoralens

8-methoxypsoralen   5-methoxypsoralen   Coumarins   Furans   Benzodifurans

Furanochromones   Oxazolocoumarins

Naphthofuran   Naphtho alpha pyrone

Fig. 1. Structural characteristics of active and inactive furocoumarins and other related compounds.

with DNA, and the clearing of psoriasis occur only in the presence
of a specific spectrum of ultraviolet light (320-380 nm, otherwise
known as the UV-A spectrum). In this presentation, we will briefly
discuss clinical, photobiological, and molecular aspects of psoralen
photosensitization and summarize some data concerning: a) the nature
of the photosensitization reaction in mammalian skin after oral administration of psoralens; b) the absorption, biotransformation, and
excretion of orally administered psoralens; and c) the molecular aspects of the psoralen photosensitization reaction in relation to its
therapeutic effectiveness.

RESULTS

Time Course of the Psoralen Photosensitization Reaction
in Laboratory Animals and Human Subjects

Oral administration of a single dose of 8-MOP (dose range, 5 to
10 mg/kg) to adult albino guinea pigs (700 to 1,000 g) and subsequent
exposure of the epilated back skin to UV-A radiation (exposure dose
range, 5 to 16 J/cm^2) results in photosensitization of the skin.
This photosensitivity reaction is visible between 24 and 48 hours
after exposure and is manifested by an augmented erythema and edema
reaction of the exposed skin. Serial exposures of the skin at 1, 2,
3, 4, 5, 6, 8, 10, 12, and 24 hrs after administration of 8-MOP show
the following characteristic reactions: a) the skin is sensitive to
light within 1 hr, and is maximally photosensitive between 2 and 4
hr; b) the photosensitivity reaction decreases significantly at 5
and 6 hr and is absent at 8 hr; c) the augmented erythema reaction
is visible within 3 to 4 hr after exposure and reaches its peak intensity between 36 to 48 hr; and d) the skin shows no erythemal response to a 16 J/cm^2 irradiance dose of UV-A in the absence of oral
8-MOP. Likewise, adult male or female human volunteers receiving
40 mg (or 0.6 mg/kg) of oral 8-MOP showed maximum photoreactivity at
between 1 1/2 to 3 hr. The photosensitivity reaction decreases after
4 hr, and by 6 hr after oral ingestion of 8-MOP the reaction can
barely be induced. By 7 to 8 hr after the initial oral dose of 8-MOP,
the skin fails to respond to UV-A exposure. With 4,5',8-trimethyl-
psoralen, similar observations can be reproduced, but the oral dose
should be in the range of 10 to 12 mg/kg for guinea pigs and 70 to
120 mg/70 kg for human subjects. A distinct characteristic of the
photosensitization reaction caused by UV-A + 8-MOP should be emphasized. In guinea pigs, the erythema (and inflammation) response
reaches its peak intensity approximately between 36 to 48 hr after
exposure of the skin to UV-A radiation. In human subjects, the erythema response is maximum between 48 to 72 hr. The color of the skin
after the photosensitization reaction is red to deep red, and some-
violaceous red, suggesting that the capillaries and larger vessels
of the papillary and reticular dermis are also involved in this

photosensitization reaction.

A study on the rate of excretion of orally administered 8-MOP, TMeP, or psoralen carried out in mice (dose 0.6 to 6 mg/kg) and in human volunteers (40-60 mg/70 kg), has revealed the following information that is consistent with the observations concerning the photosensitization reaction of skin presented above: a) The blood levels of 8-MOP in human volunteers, and of TMeP and 8-MOP in mice, are maximum between 2 and 3 hr after ingestion of the drug; b) Over 95% of the orally administered drug is absorbed by the gastrointestinal tract; c) The major pathway for the excretion of orally administered psoralens is the renal pathway through the kidneys. Over 50% of the drug is excreted in the urine between 2 to 3 hr and between 80 to 90% of the administered drug can be recovered in the urine as metabolites within 8 to 12 hr. The excretion of psoralens through the feces is small (4 to 14%); d) There is no accumulation or retention of the administered drug; mice receiving a daily dose of either 8-MOP, TMeP, or psoralen (6 mg/kg) for a 5 to 7 day period revealed no accumulation or retention of these drugs in various organs of the animals examined; and e) Furocoumarins are metabolized primarily in the liver. Metabolic studies with tritium-labelled 8-MOP, TMeP, or psoralen have revealed that the mixed function oxidase enzyme system of the liver is involved in the biotransformation of these furocoumarins (Mandula and Pathak, unpublished observations).

Biotransformation of Furocoumarins

Clinically useful furocoumarins such as trimethylpsoralen and 8-methoxypsoralen, which are generally hydrophobic and poorly soluble in water, are, upon oral administration, transformed in vitro to water soluble, hydrophilic moieties. In this biotransformation, glucuronide formation, carboxylation, and hydroxylation reactions of the administered furocoumarin appear to play a major role in the detoxification and excretion of psoralens. We have succeeded in isolating and characterizing one major metabolite which occurs in the urine of both mice and human subjects receiving oral TMeP. This metabolite has a molecular weight of 258 and is referred to as 4,8-dimethyl, 5'-carboxypsoralen, or DMeCP (4). This major metabolite was found to be inactive and non-photosensitizing when applied and tested on guinea pig skin. Another metabolite, presumably a hydroxylated moiety of DMeCP with a molecular weight of 274, was also detected. 8-MOP is excreted in the form of four fluorescent metabolites, one of which appears to be a hydroxylated moiety. Another metabolite of 8-MOP has a molecular weight of 232, but its structure remains to be established. 8-MOP also appears to be excreted as a glucuronide derivative and also as an open chain moiety which can be referred to as 8-methoxyfurocoumaric acid (Pathak et al., unpublished observations).

These in vivo biotransformation reactions are of clinical and photochemical importance to our understanding of the photoreactivity of psoralen molecules for their usefulness in systemic and topical photochemotherapy. For example, 4,5',8-trimethylpsoralen is a very potent photosensitizing agent when it is applied topically (5 to 10 mg/6 cm^2) to guinea pig skin or to human skin, but when administered orally (10 mg/kg to guinea pigs and 40 to 60 mg/70 kg to human subjects), it is undoubtedly a weak photosensitizer. These observations are consistent with our clinical experience, where patients with generalized psoriasis were treated either with 8-MOP and longwave ultraviolet light or with TMeP and longwave ultraviolet light. Photochemotherapy with oral 8-MOP was very effective in rapidly clearing psoriasis; TMeP with ultraviolet light or with sunlight was distinctly less effective and at times virtually inactive in clearing psoriasis. Recently, we have also observed that the drug metabolizing mixed function oxidase system present in mouse liver homogenate, guinea pig liver homogenate, and the microsomal fraction of these liver homogenates was able to convert TMeP to DMeCP. The epidermal homogenates of the same animals, however, showed no ability to transform TMeP to DMeCP (Mandula and Pathak, unpublished observations). This agrees well with the clinical observation that topically applied TMeP is more photoreactive than orally administered TMeP. Thus, in vivo biotransformation can play a major role in altering the photoreactivity of the drug.

Molecular Aspects of the Psoralen Photosensitization Reaction

The nature of psoralen + UV-A (320-400 nm) photochemical reactions, which are of fundamental importance to our understanding of the interaction of the drug and the light in producing certain biologic reactions (e.g., cutaneous photosensitization) and therapeutic changes in the skin (e.g., inhibition of DNA synthesis in normal skin and in proliferative diseases such as psoriasis) are undoubtedly related to the photochemical reactions of psoralens with cellular DNA. This interaction of the drug (e.g., psoralen, 8-MOP) and the light (320-380 nm) to produce these biologic changes is more specifically related to: a) the chemical structure of the photosensitizing molecule; b) the effects of group substitutions (e.g., $-OCH_3$, $-CH_3$, $-OH$, $-NH_2$, $-H_2$, etc.) in certain positions of the parent psoralen ring structure; c) the solubility and rate of diffusion of the topically applied photosensitizing furocoumarins (in topical therapy); d) in vivo absorption, and the resulting concentration of the orally administered furocoumarins in the blood and in the extracellular fluid that bathes the epidermal cells (in oral therapy); e) the photoreactivity of psoralens with DNA and with pyrimidine bases (thymine, cytosine, and uracil), including their nucleosides and nucleotides as established under in vitro and in vivo conditions (3); f) the non-reactivity of psoralens with purine bases (adenine, guanine and their nucleosides and nucleotides) and with proteins such

as albumin or keratin (3); and g) the metabolism and rate of biotransformation of the psoralens to either an active (capable of inducing photosensitization) or an inactive (non-photosensitizing)
moiety. These molecular aspects of the psoralen photosensitization
reaction are briefly discussed below.

The relationship between the chemical structure of furocoumarins
and their skin photosensitizing activity has been extensively examined by several investigators (3,5-8). From these studies (Fig. 1)
the following conclusions about the relationship between the structure and the activity of furocoumarins can be made.

The ability to sensitize cutaneous tissue to light (320-380 nm)
appears to be a unique characteristic of the psoralen-ring system
that exhibits linearly annulated, tricyclic structure (Fig. 1). Only
those compounds whose structure is of the psoralen type (e.g., psoralen, 8-MOP, 5-MOP) have been found to be significantly active. The
non-linear furocoumarins that exhibit angular structure (e.g., isopsoralen or angelicin) and the synthetic allopsoralens or pseudoiso-
psoralens, are inactive. Introduction of methyl groups in the 4,
the 4', the 5' or the 8 positions of the psoralen molecule results
in the enhancement of the activity of the parent molecule (e.g.,
4-methylpsoralen, 4,5',8-trimethylpsoralen). Introduction of a methyl group at the 3 position, however, does significantly reduce the
photosensitizing ability (e.g., 3,5',8-trimethylpsoralen, 3-methyl-
psoralen, 3,4,5',8-tetramethylpsoralen). Any other group substitution at the 3 position renders the parent psoralen molecule quite
weak (e.g., 3,4-benzo-5',8-dimethylpsoralen, 3-n-butyl-4,4',8-tri-
methylpsoralen). Introduction of a methoxy group at either the 8
or the 5 position renders the parent molecule slightly less active
than psoralen (e.g., 8-MOP and 5-MOP). Substitution with a hydroxy
group makes the psoralen molecule inactive (e.g., 5-hydroxypsoralen,
8-hydroxypsoralen). Lengthening of the alkyl chain at the 8 position
significantly reduces the activity to zero. Introduction of nitro,
amino, acetyl, cyano, chloro or $N(CH_3)_2$ groups at the 8 position
renders the parent molecule completely inactive. Hydrogenation of
the furan ring of psoralen results in complete loss of the photosensitizing potency of the psoralen molecule. This loss of photosensitizing activity can also be correlated with the inability of these
nonphotosensitizing molecules to conjugate and cross-link photochemically with DNA (9-11, Pathak and Biswas, unpublished observations).

The solubility and the rate of diffusion of the topically applied furocoumarins also influence the degree of photoreaction. Although the permeability properties (e.g., diffusion constants) of
these furocoumarins in mammalian skin are not known, it is important
to recognize that, in the induction of the skin photosensitization
reaction, the topically applied agent must penetrate to the viable
cells of the epidermis through the dead layer of stratum corneum
cells—the rate-limiting diffusion barrier for the applied molecules.

Some of the compounds that have been labelled as inactive or non-photosensitizing may not have penetrated under the test conditions involving the diffusion period of one hour and the exposure period of 45 to 60 minutes (7,8). Compounds such as psoralen, 8-MOP and TMeP are lipophilic and do penetrate the epidermis and render the skin photosensitive within 30 to 60 minutes after topical application.

Likewise, orally administered substances must be absorbed through gastrointestinal epithelium and be present in the blood and the extracellular fluid that bathes the epidermal cells. Dietary factors (e.g., lipids) do play a role in the rate of absorption of psoralens. A drug like 8-MOP is a weak acid, and its non-ionized form is more lipid soluble. It is absorbed quite readily. Absorption of the drug from the alimentary tract may be delayed if the drug is given after the ingestion of food, especially after a meal rich in fats and carbohydrates. Some of the differences related to the increased photoreactivity of a psoralen derivative after topical application, and to the decreased photoreactivity of the same drug after oral ingestion, are undoubtedly related to poor absorption of the orally administered drug.

Many of the photobiological effects of psoralens (e.g., photosensitization reaction manifested by increased redness, edema, epidermal cell disorganization, inhibition of DNA synthesis, cell death and subsequent desquamation) can be explained on the basis of their photoreactivity with DNA of the epidermal cells (keratinocytes) and dermal cells (fibroblasts and endothelial cells of blood vessels). Other events can also occur simultaneously in the skin and can be partially responsible for some of the effects that are seen in photosensitization reactions (e.g., photoexcitation of psoralens to singlet and triplet state and subsequent generation of free radicals in the skin that can damage the cell membrane; liberation of vasodilating substances; leakage of the capillaries and blood vessels; damage to lysosomes; etc.).

The photoreaction of psoralens involves first the intercalation or formation of molecular complexes between DNA and furocoumarins (10,12). Formation of these complexes involving weak bonding appears to be a very important dark reaction (occurring in the absence of light) for subsequent photoreaction with DNA. Irradiation of the biologic system containing pyrimidines or DNA with UV-A gives three types of photoadducts, that are referred to as C_4-cycloadducts of furocoumarins to the 5,6-double bonds or pyrimidine bases (3,10,13-15). Furocoumarins such as psoralen and 8-MOP can react either at their 3,4-double bond or with their 4',5'-double bond and form monofunctional photoadducts. The former type of monofunctional adduct (with 3,4-double bond) is non-fluorescent, whereas the photoadduct with 4',5'-double bond is fluorescent (Fig. 2). Among the monofunctional photoadducts, the 3,4-cycloadducts occur more frequently and with higher yields than 4',5' ones, especially when the reaction

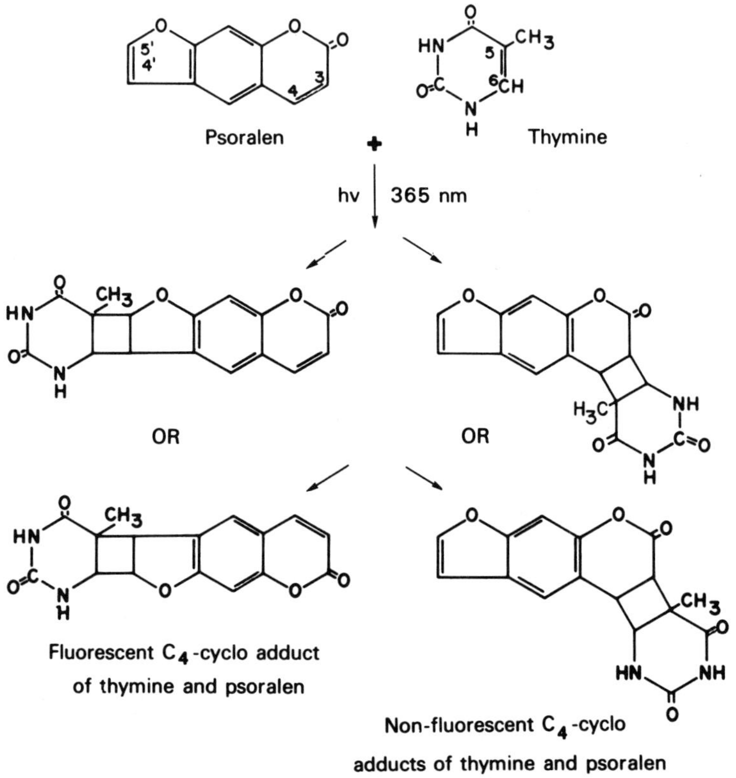

Fig. 2. Photoaddition products of psoralens with pyrimidine — thymine after UV irradiation (365 nm) under in vitro conditions.

occurs at room temperature or at 37° C in the presence of methyl-psoralens. Compounds like 4,5',8-trimethylpsoralen and 4,8-dimethyl-psoralen form more monofunctional adducts than 8-MOP, and they are generally non-fluorescent. In addition, the 3,4-double bond and the 4',5'-double bond of psoralens can photoreact with native DNA and form interstrand cross-links (3,10,16-18). Thus, psoralens can be-have as photoreactive bifunctional agents (one furocoumarin react-ing with two pyrimidines of the opposite strand of DNA). This in-terstrand cross-linking has been demonstrated both under in vitro conditions with native DNA isolated from calf thymus, and under in vivo conditions in guinea pig skin (3,10,16-20).

The interstrand cross-linking of one furocoumarin with two py-rimidine bases of each strand of DNA has been demonstrated in our laboratory with the following three techniques:

1) Determination of the sedimentation profiles of DNA in su-crose density gradient after irradiation with UV-A (12-16 J/cm^2) of solutions containing furocoumarins (e.g., psoralen, 8-MOP, TMeP). The photosensitized DNA is denatured with alkali (pH 11.0) and sub-sequently sedimented in sucrose density gradient (Fig. 3). The sedi-mentation pattern of psoralen cross-linked DNA is distinctly dif-ferent from that of non-cross-linked DNA.

2) By evaluation of the presence of furocoumarin cross-linked DNA on a hydroxylapatite column. The hydroxylapatite column chroma-tography technique is now selectively used for the separation of single and double-stranded DNA, and for demonstrating cross-links. This is achieved by first irradiating the native form of DNA in the presence of a furocoumarin like 8-MOP or TMeP. The irradiated mix-ture is then boiled and rapidly cooled. Subsequently, the separa-tion of the cross-linked DNA and of the single-stranded or denatured DNA is carried out on a hydroxylapatite column using a linear and a continuous gradient of 0.05M and 0.3M potassium phosphate buffer at pH 7.05 (Fig. 4).

3) Determination of the hyperchromicity of the photoreacted DNA after boiling and rapid cooling. Psoralen cross-linked with the double-stranded DNA shows minimal hyperchromicity after heat treat-ment. The denatured single-stranded DNA shows maximum hyperchromi-city.

Some of these examples are illustrated in Figures 3 and 4. It is quite apparent that potent photosensitizing molecules, like 8-MOP and psoralen, undergo cross-linking with DNA after irradiation with longwave ultraviolet. It is of interest to note that non-photosen-sitizing and therapeutically inactive molecules like isopsoralen (angelicin) and 8-hydroxypsoralen (xanthotoxol) will not show photo-conjugation and cross-linking with DNA.

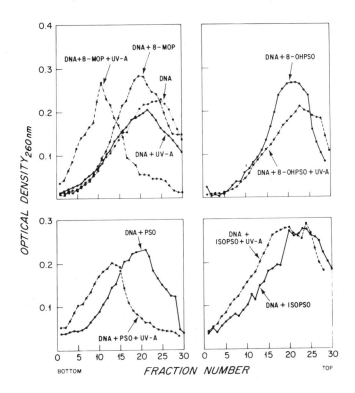

Fig. 3. Sedimentation profiles of calf thymus DNA in an alkaline
sucrose density gradient after photoreaction with: a) 8-methoxypso-
ralen (8-MOP, top left); b) 8-hydroxypsoralen (8-OHPSO, top right);
c) psoralen (PSO, lower left), and d) isopsoralen (ISOPSO, lower
right). Cross-linked pattern of DNA is seen with active 8-MOP and
psoralen. Cross-linking could not be detected with inactive 8-OHPSO
and isopsoralen (reaction volume 3 ml; DNA 1mg/ml, furocoumarin
10 ug/ml and UV-A 16 J/cm^2).

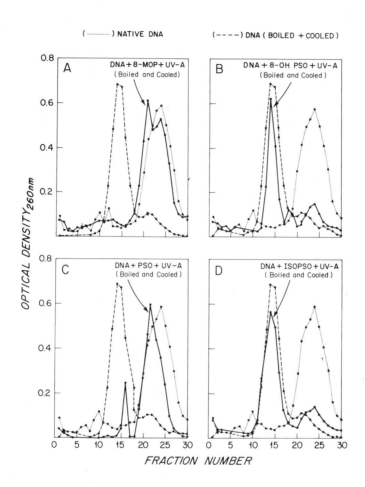

Fig. 4. Hydroxylapatite column chromatography of DNA after photo-reaction with certain furocoumarins. Using a linear gradient of 0.05 - 0.3M potassium phosphate buffer (pH 7.0) and collecting fractions of 3.5 ml, the elution profile of native DNA (double-stranded) is shown with a dotted line and that of heat denatured DNA (single-stranded) with a dashed line. The solid lines are the elution profiles of DNA after photoreaction with a) 8-methoxypsoralen (8-MOP); b) 8-hydroxypsoralen (8-OHPSO), c) psoralen (PSO); and d) isopsoralen or angelicin. 8-MOP and PSO are active; 8-OHPSO and ISOPSO are inactive furocoumarins.

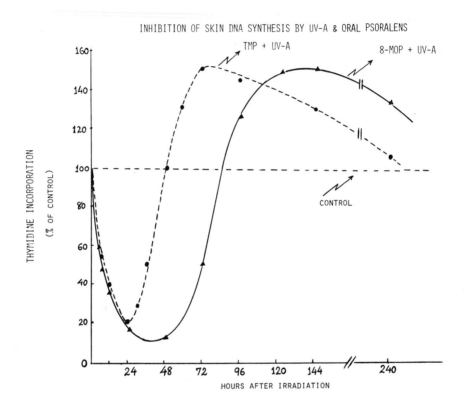

Fig. 5. Inhibition of DNA synthesis by UV-A plus oral psoralens
(8-MOP and TMeP) in guinea pig skin. Epidermal cells were made proliferative with epilation. Me-[^{3}H]-thymidine incorporation into epidermal DNA at various time intervals was examined before and after
oral administration of 8-MOP or TMeP and subsequent irradiation with
UV-A (16 J/cm^2). DNA was isolated from the control (non- irradiated)
and from the 8-MOP or TMeP irradiated skin sites. The radioactivity
of the isolated DNA from the irradiated sites is expressed as percentage of the control (J. Invest. Dermatol. 62:388, 1974).

This covalent linking of photoactive furocoumarins like 8-MOP, psoralen, and trimethylpsoralen to DNA results in the formation of both monofunctional and bifunctional photoadducts (cross-links). These events are presumably responsible for the inhibition of DNA synthesis in the proliferative epidermal cells (17-23). The therapeutic efficacy of photoactive psoralens like 8-MOP in the treatment of proliferative diseases such as psoriasis is at least partially, if not totally, due to these photochemical events (1-3,23).

The formation of monofunctional adducts of both types and the interstrand cross-linking of psoralens to DNA appear to be the major events responsible for the inhibition of DNA synthesis and death of cells. The inhibition of DNA synthesis in the cutaneous photosensitization reaction of guinea pig skin after oral administration of equimolar amounts of either 4,5',8-trimethylpsoralen or 8-methoxypsoralen and subsequent irradiation with UV-A is illustrated in Fig. 5. Both compounds show significant inhibition of DNA synthesis in normal epidermis made proliferative after epilation. It is of interest to note that 8-MOP showed greater inhibition of DNA synthesis than that shown by TMeP. The inhibition of DNA synthesis lasted nearly 42 to 48 hours in the case of 8-MOP, and only up to 24 hours in the case of TMeP. These observations have important bearings on the relative therapeutic effectiveness of 8-MOP and TMeP in the photochemotherapy of psoriasis. Although TMeP has undoubtedly a very low solubility in water ($\simeq$2 mg/100 ml) as comapred to 8-MOP (4.8 mg/100 ml), and this may have a certain limiting influence on the amount of TMeP that can be linked to DNA, there are some other factors (e.g., rapid biotransformation of orally administered TMeP to inactive, non-photosensitizing moieties) that are equally important in revealing a weaker therapeutic effectiveness of TMeP in the photochemotherapy of psoriasis (4).

REFERENCES

1.   Parrish, J.A., Fitzpatrick, T.B., Tanenbaum, L., Pathak, M.A.: New Eng. J. Med. 291:1207, 1974.

2.   Wolff, K., Fitzpatrick, T.B., Parrish, J.A., Gilchrest, B., Honigsmann, H., Pathak, M.A., Tanenbaum, L.:   Arch. Derm. 112:943, 1976.

3.   Pathak, M.A., Kramer, D.M., Fitzpatrick, T.B.:   Sunlight and Man: Normal and Abnormal Photobiologic Responses, M.A. Pathak, L.C. Harber, M. Seiji, A. Kukita, editors.   University of Tokyo Press, Tokyo, 1974, p. 335.

4.   Mandula, B.B., Pathak, M.A., Dudek, G.:   Science, in press, 1976.

5. Musajo, E.L., Rodighiero, G., Caporale, G.: Bull. Soc. Chim.
    Biol. 36:1213, 1954.

6. Musajo, E.L., Rodighiero, G.: Experientia 18:153, 1962.

7. Pathak, M.A., Fitzpatrick, T.B.: J. Invest. Derm. 32:256, 1959
    and 32:509, 1959.

8. Pathak, M.A., Fellman, J.H., Kaufman, K.D.: J. Invest. Derm.
    35:165, 1960, and 48:103, 1967.

9. Dall'Acqua, F., Marciani, S., Rodighiero, G.: Z. Naturforsch.
    [13] 24:307, 1969; FEBS Letters 9:121, 1970.

10. Musajo, L., Rodighiero, G., Caporale, G., Dall'Acqua, F.,
    Marciani, S., Bordin, F., Baccichetti, F., Bevilacqua, R.:
    Sunlight and Man: Normal and Abnormal Photobiologic Responses,
    M.A. Pathak et al, editors. University of Tokyo Press, Tokyo,
    1974, p. 369.

11. Rodighiero, G., Musajo, L., Dall-Acqua, F., Marciani, S.,
    Caporale, G., Ciavatta, L.: Biochim. Biophys. Acta 217:40,
    1970.

12. Dall'Acqua, F., Rodighiero, G.: Rend. Accad. Naz. Lincei.
    (Rome) 40:411, 1966, and 40:595, 1966.

13. Musajo, L., Bordin, F., Caporale, G., Marciani, S., Rigatti, G:
    Photochem. Photobiol. 6:711, 1967.

14. Musajo, L., Bordin, G., Bevilacqua, R.: Photochem. Photobiol.
    6:927, 1967.

15. Krauch, C.H., Kramer, D.M., Wacker, A.: Photochem. Photobiol.
    6:341, 1967.

16. Dall'Acqua, F., Marciani, F., Rodighiero, G.: FEBS Letters
    9:121, 1970.

17. Cole, R.S.: Biochem. Biophys. Acta 217:30, 1970, and 224:660,
    1970.

18. Dall'Acqua, F., Marciani, S., Ciavetta, L., Rodighiero, G.:
    Z. Naturforsch. [13] 26:561, 1971.

19. Dall'Acqua, F., Marciani, S., Vedaldi, D., Rodighiero, G.:
    FEBS Letters, 27:192, 1972.

20. Pathak, M.A., Kramer, D.M.: Biochim. Biophys. Acta 195:197,
    1969.

21. Baden, H.P., Parrington, J.M., Delhanty, J.D.A., Pathak, M.A.:
     Biochim. Biophys. Acta 262:247, 1972.

22. Ben-Hur, E., Elkind, M.M.: Biochim. Biophys. Acta 331:181,
     1973.

23. Walter, J.F., Voorhees, J.J., Kelsey, W.H., Duell, E.A.:
     Arch. Derm. 107:861, 1973.

    This work was supported by U.S. National Institutes of Health,
National Cancer Institute Grant # 2RO1CA 05003-17. We wish to
extend our thanks to Dr. Barbara Mandula for her collaboration.

SYMPOSIUM   VII

REPAIR OF RADIATION DAMAGE

DNA REPAIR SCHEMES IN BACTERIA AND IN HUMAN CELLS

Philip C. Hanawalt

Department of Biological Sciences

Stanford University, Stanford, California 94305

This symposium on Repair of Radiation Damage deals with our current understanding of the various mechanisms by which living systems respond to damage in their genetic material. Most of the important discoveries in the field of DNA repair have resulted from the study of the response of cells to radiation. Also we know more about the repair of one particular type of ultraviolet light induced damage than any other. Ultraviolet light produces a variety of photo-products in DNA, intrastrand pyrimidine dimers being the most abundant. Dimers serve as model lesions for the study of DNA repair processes. An unrepaired dimer poses a block to normal DNA replication *in vivo* although replication may resume at a new initiation site beyond the dimer leaving a gap. Unless the dimers are removed or the post-replication gaps are filled the system becomes inviable.

One paper in this symposium concerns recent evidence for the induction of an "error-prone" DNA repair system in bacteria. The other papers deal with three distinct DNA repair schemes in human cells. In the study of DNA repair mechanisms the results obtained with the bacterium *Escherichia coli* have generally led the way to the discovery of similar processes in humans. Notable in this regard were the studies of James Cleaver (1) in which, by analogy with *E. coli* repair defective mutants the molecular basis for xeroderma pigmentosum in man was shown to involve a deficiency in DNA repair. In *E. coli* a single unrepaired pyrimidine dimer causes lethality; in man an unrepaired pyrimidine dimer may lead to malignancy and death. Thus, patients with the hereditary defect, xeroderma pigmentosum, are unusually sensitive to sunlight and develop skin cancer upon brief exposure to the sun. In this introductory paper I should like to summarize our current view of

the pathways for repairing damaged DNA and to outline the methodology for studying DNA repair to set the stage for the papers that
follow.

The classic test lesion for the study of DNA repair processes is
the intrastrand pyrimidine dimer, formed with a relatively high
quantum yield in DNA upon the absorption of ultraviolet light.
Although this covalent linkage of adjacent pyrimidines is the photoproduct formed with the highest efficiency under most physiological
conditions it is important to appreciate that it is not the only
lesion. Its relative biological significance may depend upon where
it appears in the genome and upon its accessibility to the various
repair enzymes even though the presence of a single pyrimidine dimer
in the genome can be lethal to the cell. Dimers can also be mutagenic and carcinogenic, and there is some evidence that the operation of repair systems can either increase or decrease the likelihood of these events. This is documented in other papers in this
volume. Of the various classes of pyrimidine dimers, the thymine-
thymine product is formed most abundantly. This is fortuitous for
the biochemical analysis of DNA repair because of the ease of
radioactive labeling of DNA with thymine (or thymidine).

At least three molecular recovery mechanisms for dimer-containing
DNA have now been well-documented in *E. coli* and evidence for all
three has also been obtained in human cells (2). These schemes are
illustrated in Figure 1 along with several other more speculative
mechanisms. Enzymatic photoreactivation is the simplest, the most
specific, and probably the best understood DNA repair process. It
involves one enzyme which recognizes only pyrimidine dimers in DNA
and that enzyme binds to the dimer containing regions in the dark.
Upon illumination with visible light a photocatalytic cleaving of
the linked pyrimidines occurs *in situ,* restoring the DNA to normal.
The phosphodiester backbone linkages remain intact throughout this
process. Since photoreactivation is so specific for dimers and
because it can be regulated simply with a light switch it can be
used to determine whether a given biological effect is due to dimers.

Excision-repair is also initiated by an enzyme that recognizes
pyrimidine dimers although a variety of other structural distortions
in DNA can be recognized as well. In this case the phosphodiester
backbone of the DNA is broken in the damaged strand by the enzyme,
an endonuclease that produces an incision on the 5' side of the
damage. In *E. coli* an endonuclease activity that recognizes dimers
is specified by the *uvr*A and *uvr*B loci (3). The subsequent steps
in excision-repair involve deletion of the damage by a 5' → 3'
exonuclease and repair replication from the 3' OH terminus at the
incision. DNA polymerase I with its associated 5' - 3' exonuclease
can perform both of these steps in coordinated fashion as shown *in
vitro* (4). However, alternative enzymes can also carry out either

of these steps *in vivo* as discussed below. The excision-repair
process is completed by the action of polynucleotide ligase to join
the repair patch to the contiguous parental DNA strand.

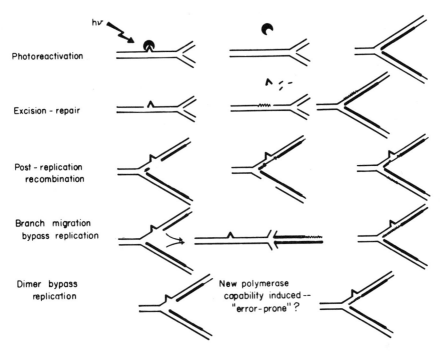

Figure 1   Schemes for repairing or bypassing dimers in DNA.
Configuration of growing point region shown by light lines representing DNA strands at the time of ultraviolet irradiation to
introduce pyrimidine dimer (inverted "V"). Normal replication
subsequent to irradiation shown by heavy lines. Repair or bypass
replication illustrated by wavy lines.

The last three repair schemes shown in Figure 1 all involve events
that may occur as the normal replication fork encounters the
damaged region. A dimer poses a block to the normal DNA polymerizing activity at the growing point but does not evidently interfere
with the unwinding of parental strands. Replication can resume
along the damaged parental strand when the next initiation event
takes place at a site beyond the dimer. Thus, a gap is left in
the daughter strand opposite the region containing the dimer.
Evidence for these post-replication gaps was first obtained in
E. coli by Rupp and Howard—Flanders (5) who proposed a model for
their repair by recombination. Physical evidence for the model
was provided by density-transfer experiments (6) and by the studies

of Ganesan (7) in which dimers were detected in the daughter strands synthesized after the irradiation. Meneghini in our laboratory has obtained evidence for the presence of gaps in the newly-synthesized DNA in UV irradiated human cells using a single strand specific endonuclease from *Neurospora crassa* as probe for such regions (8). The gaps were further shown to disappear upon continued incubation of the cells. We have also detected dimers in daughter strands as indicative of recombinational events in HeLa cells (9) and in WI38 diploid human cells (10). Buhl and Regan (11) reported evidence for such events in UV irradiated fibroblasts from xeroderma pigmentosum patients. Earlier studies had demonstrated *de novo* synthesis in the repair of gaps in daughter DNA strands in mammalian cells (12). Of course some repair synthesis may well accompany recombinational events, as shown in Figure 1, so the observations of *de novo* synthesis and dimer transfer are not mutually exclusive.

A model for the possible bypass of dimers in DNA by branch migration at the replication point has been recently suggested by Higgins *et al*. (13), and some experimental evidence consistent with this interpretation has been obtained in human cells.

The last model shown in Figure 1 is essentially that of "SOS repair" as proposed by Radman (14) for an inducible "error-prone" mode of repair in bacteria. The resultant effect is the replication around the damaged region in the parental DNA. This could involve a modification of DNA polymerase III or the replication "complex" to enable it to perform this unusual bypass mode. Alternatively but less likely it could require a new DNA polymerase. As yet there is no convincing evidence for an inducible "error-prone" mode of repair in human cells. Inducible repair is one of the most exciting areas of study in the DNA repair field just now. If the process were to be demonstrated in human cells there could be important practical aspects--men might receive preconditioning treatments to enable them to tolerate unusually high fluences of radiation or other deleterious environmental situations that result in repairable damage to DNA.

With the possible exception of photoreactivation all of the repair processes illustrated in Figure 1 are certainly more complex than indicated. For example, consider the excision-repair system in *E. coli*. The importance of a functional excision-repair system in *E. coli* is emphasized by the fact that other enzymes can substitute for either the polymerase or the 5' → 3' exonuclease activity of DNA polymerase I in dimer repair (15, 16, 17). The presumed alternative pathways for excision-repair of dimers are illustrated schematically in Figure 2. The studies of Cooper in our laboratory have shown that in the presence of functional DNA polymerase I most of the dimer containing regions can be replaced by short stretches

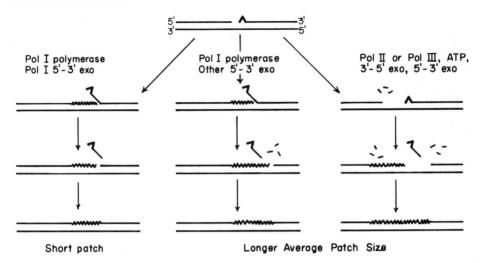

Figure 2   Alternate systems for excision-repair in *E. coli*.

of nucleotides, while in the absence of either activity of pol I
more DNA degradation and more extensive repair synthesis occurs
(15, 16, 17). Even in a conditionally lethal mutant deficient
in *both* pol I activities dimers are excised efficiently (17).
The details of the DNA degradation and repair resynthesis steps
differ in the different classes of pol I mutants however. It is
not clear which other 5' → 3' exonuclease removes dimers when the
pol I associated 5' → 3' exonuclease activity is deficient but it
has been shown that either DNA polymerase II or polymerase III can
substitute for the polymerizing activity of pol I (18). In fact,
an ATP dependent repair synthesis mode (in toluene-permeabilized
cells) is the only phenotype thus far associated with pol II (19).
Since both pol II and III require a gap (instead of a nick) for
binding to DNA it is likely that some degradation in the 3' → 5'
direction occurs in this pathway. Grossman *et al.* (20) have suggested that the *rec*BC exonuclease might perform this function
although other exonucleases, such as exonuclease III, could also
be responsible. The alternative DNA polymerase might then drive
the pol I 5' → 3' exonuclease to remove the dimer, while
resulting in more extensive degradation and resynthesis because of
inefficient coordination of the two activities (17).

In the characterization of excision-repair *in vivo* it is important
to examine a number of criteria. The different procedures outlined
in Figure 3 look at different aspects of the process and no one of
them adequately defines the completion of effective repair (2, 21).
For example, the demonstration of selective removal of dimers does
not ensure that excision gaps are being filled, and the observation
of repair replication does not ensure that dimers have been removed.

1. LOSS OF DIMERS FROM DNA
   A. CHROMATOGRAPHIC ANALYSIS OF ^{3}H-DIMERS/^{3}H-T RELEASED INTO ACID SOLUBLE FRACTION.
   B. CHROMATOGRAPHIC ANALYSIS OF ^{3}H DIMERS/^{3}H-T REMAINING IN DNA.
   C. TREATMENT OF EXTRACTED DNA WITH T4-ENDONUCLEASE V AND SEDIMENTATION IN ALKALINE SUCROSE GRADIENT.

2. NICKING AND EVENTUAL CLOSURE OF DAMAGED DNA STRANDS MONITORED BY SEDIMENTATION OF EXTRACTED DNA IN ALKALINE SUCROSE GRADIENTS AT DIFFERENT TIMES AFTER IRRADIATION.

3. REPAIR REPLICATION
   A. INCORPORATION OF ^{3}H-5BU DENSITY LABEL INTO REPAIR PATCHES IN ^{14}C PRELABELED PARENTAL DNA, DISTINGUISHED FROM SEMICONSERVATIVE REPLICATION BY BOUYANT DENSITY IN CsCl EQUILIBRIUM DENSITY GRADIENT. REBANDING OF PARENTAL DNA IN ALKALINE CsCl TO IMPROVE RESOLUTION.
   B. INCORPORATION OF 5BU; THEN 313 NM IRRADIATION OF EXTRACTED DNA TO CAUSE STRAND BREAKS AT 5BU SITES, ANALYZED BY SEDIMENTATION IN ALKALINE SUCROSE GRADIENT.

4. UNSCHEDULED DNA SYNTHESIS.
   SINGLE CELL AUTORADIOGRAPHY FOLLOWING INCORPORATION OF ^{3}H-T.
   ANALYSIS OF EXPOSED GRAIN DISTRIBUTION OVER NUCLEI NOT IN S-PHASE OF NORMAL CELL DIVISION CYCLE.

5. ^{3}H-T INCORPORATION IN THE PRESENCE OF HYDROXYUREA TO SELECTIVELY INHIBIT SEMICONSERVATIVE DNA REPLICATION WHILE PERMITTING REPAIR.

Figure 3   Procedures for determining excision-repair.

The detection of unscheduled DNA synthesis does not ensure that the synthesis is of the repair mode. Also, the quantitation of repair synthesis can be affected by the reagents (*e.g.*, hydroxyurea) used to selectively suppress normal replication (22). This problem can be particularly serious in procedures in which the two modes of replication are not qualitatively distingusihed by density labeling.

Careful quantitative comparisons of the extent of repair replication in various human diploid cell lines have been carried out by C. A. Smith in our laboratory using the combined radioisotopic and density labeling approach. Repair replication in response to UV irradiation has been studied in normal human diploid fibroblasts, WI38 cells, and in an SV40 transformant, VA13. We find no significant difference in the amount of repair replication performed, its dose response, or the time course of this synthesis when we compare growing vs. confluent WI38 cells, early passage vs. late passage (senescent) cells, or in the normal WI38 cells vs. the transformed VA13 cells (23).

Figure 4 lists procedures used for examining post-replication repair. Again it is important to appreciate that the different techniques measure different aspects of the process.

In bacterial systems the availability of a large catalogue of repair-deficient mutant strains has aided the elucidation of repair pathways and the formulation of models to guide the analysis of

PRELABEL DNA UNIFORMLY WITH ONE RADIOISOTOPE, ^{14}C-T

LABEL DNA AFTER IRRADIATION WITH SECOND RADIOISOTOPE, ^{3}H-T

A. SHORT PERIOD LABEL, THEN UNLABELED THYMIDINE FOR LONG PERIOD
SEDIMENTATION ANALYSIS IN ALKALINE SUCROSE GRADIENT AT DIFFERENT
TIMES TO FOLLOW CHANGES IN MW OF LABELED DNA SEGMENTS.

B. UNLABELED THYMIDINE FOR LONG PERIOD, THEN LABELED THYMIDINE
SEDIMENTATION ANALYSIS IN ALKALINE SUCROSE GRADIENT TO DETERMINE
MW OF DNA SYNTHESIZED AT LATE TIMES.

C. USE T4-ENDONUCLEASE v AS PROBE FOR DIMERS REMAINING IN PARENTAL STRANDS
(^{14}C) AND THOSE TRANSFERRED TO DAUGHTER STRANDS (^{3}H) BY SEDIMENTATION
ANALYSIS BEFORE AND AFTER TREATMENT OF EXTRACTED DNA WITH THE ENZYME.

D. AFTER SHORT PERIOD LABELING WITH ^{3}H-THYMIDINE, INCUBATION WITH 5BU AND
EXPOSURE OF EXTRACTED DNA TO 313 NM LIGHT. ALKALINE SUCROSE GRADIENT
SEDIMENTATION TO DETERMINE DE NOVO SYNTHESIS IN GAP FILLING.

Figure 4   Procedures for determining post-replication repair.

repair in eucaryotes. It is likely that the excision-repair system
in human cells is at least as complex as that in *E. coli*, discussed
above. Unfortunately, the repair deficient mutants in human cells
have been limited to only a few known genetic defects. It is
curious that in xeroderma pigmentosum a deficiency is sometimes
noted in *each* of the three distinct repair pathways (post replication repair, excision-repair, and photoreactivation) as documented
in the following three papers. This syndrome may thus involve a
defect in the regulation of a number of enzymatic processes
including those of DNA repair. More different classes of repair
defective human cell lines are needed to help clarify the repair
pathways in humans.

Acknowledgements

The research cited from my laboratory is supported by grants from
the American Cancer Society and the National Institute of General
Medical Sciences and by a contract with the U. S. Energy Research
and Development Administration.

References

1. Cleaver, J. E. *Nature 281*:652-656 (1968).

2. Hanawalt, P. C. In *Molecular Mechanisms for Repair of DNA*
     (P. Hanawalt and R. Setlow, Eds.) pp. 421-430 (1975)
     Plenum Press, N. Y.

3. Braun, A. and Grossman, L. *Proc. Nat. Acad. Sci. U. S. 71*:
     1838-1842 (1974).

4. Kelly, R. B.; Atkinson, M. R.; Huberman, J. A. and Kornberg, A.
     Nature 224:495-501 (1969).

5. Rupp, W. D. and Howard-Flanders, P. *J. Mol. Biol. 31*:291–304
    (1968).

6. Rupp, W. D.; Wilde, C. E.; Reno, D. L. and Howard-Flanders, P.
    J. Mol. Biol. 61:25–44 (1971).

7. Ganesan, A. K. *J. Mol. Biol. 87*:103–119 (1974).

8. Meneghini, R. *Biochim. Biophys. Acta 425*:419–427 (1976).

9. Meneghini, R. and Hanawalt, P. In *Molecular Mechanisms for
    Repair of DNA* (P. Hanawalt and R. Setlow, Eds.) pp. 639–642,
    (1975) Plenum Press, N. Y.

10. Meneghini, R. and Hanawalt, P. *Biochim. Biophys. Acta 425*:428–
    437 (1976).

11. Buhl, S. N. and Regan, J. D. *Nature 246*:484 (1973).

12. Lehman, A. R. *J. Mol. Biol. 66*:319–337 (1972).

13. Higgins, N. P.; Kato, K. and Strauss, B. *J. Mol. Biol. 101*:
    417–425 (1976).

14. Radman, M. In *Molecular Mechanisms for Repair of DNA*
    (P. Hanawalt and R. Setlow, Eds.) pp. 355–367, (1975)
    Plenum Press, N. Y.

15. Cooper, P. K. and Hanawalt, P. C. *Proc. Nat. Acad. Sci. U. S.
    69*:1156–1160 (1972).

16. Hanawalt, P. C.; Burrell, A.; Cooper, P. and Masker, W. In *DNA
    Synthesis and Its Regulation* (M. Goulian, P. Hanawalt and
    C. F. Fox, Eds.) pp. 774–790, (1975) W. A. Benjamin.

17. Cooper, P. K. *Mol. Gen. Genetics* (submitted) (1976).

18. Masker, W. E.; Simon, T. J. and Hanawalt, P. C. In *Molecular
    Mechanisms for Repair of DNA* (P. Hanawalt and R. Setlow, Eds.)
    pp. 245–254 (1975) Plenum Press, N. Y.

19. Masker, W. E.; Hanawalt, P. and Shizuya, H. *Nature New Biol.
    244*:242–243 (1973).

20. Grossman, L.; Braun, A.; Feldberg, R. and Mahler, I. *Ann. Rev.
    Biochem. 44*:19–43 (1975).

21. Setlow, R. B. In *Molecular Mechanisms for Repair of DNA* (P.
    Hanawalt and R. Setlow, Eds.) pp. 711–717, (1975) Plenum
    Press, N. Y.

22. Smith, C. A. and Hanawalt, P. C. *Biochim. Biophys. Acta 432*:
    336–347 (1976).

23. Smith, C. A. and Hanawalt, P. C. *Biochim. Biophys. Acta* (in
    press) (1976).

POSTREPLICATION REPAIR IN HUMAN FIBROBLASTS

A. R. LEHMANN, S. KIRK-BELL AND C. F. ARLETT

MRC Cell Mutation Unit

University of Sussex, Falmer, Brighton, England

SUMMARY

Postreplication repair is defective in fibroblasts from all
the xeroderma pigmentosum (XP) complementation groups with the
exception of group E, the defect being most pronounced in the
excision-proficient XP variants.   It is normal in cells from
patients with a variety of other disorders.   During postreplication
repair pyrimidine dimer sites become associated with daughter
strands in both normal and XP cells.

INTRODUCTION

UV-induced pyrimidine dimers are removed from the DNA of
normal human fibroblasts by excision repair.   In cells from most
patients with the light-sensitive genetic disorder, xeroderma pigmentosum (XP), the ability to carry out excision repair is either
reduced or completely absent (1).   These cells can nevertheless
survive doses of UV which produce many thousands of pyrimidine
dimers in their DNA.   Even in normal cells dimer excision is a
slow process, so that a substantial proportion of the dimers remain
in the DNA during replication.   Postreplication repair (PRR) is
the term used to describe the processes by which the replication
machinery copes with these unexcised lesions (2).   In most
instances gaps are left in the daughter DNA strands presumably
opposite pyrimidine dimers.   These gaps are subsequently sealed.
They are manifested as a reduction in the molecular weight of the
newly-synthesized DNA strands in UV-irradiated cells as compared
with unirradiated cells, and the sealing of the gaps as a conversion of this low molecular weight DNA into high molecular weight.

The present communication will in the first section summarize work on PRR in various defective human cells, and in the second section discuss some recent results on the mechanism of PRR.

PRR IN CELLS FROM PATIENTS WITH XP AND OTHER DISORDERS

We have studied PRR in cells from patients with XP and a variety of other disorders. The newly-synthesized DNA in UV-irradiated fibroblasts is pulse-labelled with ^{3}H thymidine, "chased" in unlabelled thymidine and the mol. wt. of the radio-actively labelled DNA strands is measured by sedimentation in alkaline sucrose gradients. Cells with defective PRR take a longer time to convert low mol. wt. DNA into high mol. wt. than do normal cells, so that at any given time after pulse-labelling the mol. wt. of the labelled DNA will be lower in the defective than in the normal cell.

Genetic studies have shown that XP patients fall into at least five complementation groups, designated A - E, all defective in excision repair to a greater or lesser extent (1). A sixth class, termed XP variants, whilst showing the same clinical symptoms, have normal levels of excision repair (1,3). XP cells are also defective in PRR, the defect being most extreme in XP variants (3). As shown in the experiment of Table 1, the mol. wts. of the newly-synthesized DNA in a pulse-chase experiment in UV-irradiated normal cells and in XP cells from group E were 120-140 x 10^6, in XP cells from groups A, B, C, D 80-115 x 10^6 and in cells from XP variants about 70 x 10^6.

Cell Strain	1BR	XP4LO	XP11BE	XP4RO	XP5BE	XP2RO	XP30RO
Complementation group	Normal	A	B	C	D	E	Variant
10^{-6} x mol. wt. - caffeine	126	113	81	105	91	121	71
10^{-6} x mol. wt. + caffeine	141	77	60	77	69	129	27

TABLE 1. Defective PRR in XP cells. The cells were UV-irradiated, pulse-labelled with ^{3}H thymidine and chased for 2·5 h with unlabelled thymidine, in the presence or absence of 1·6 mM caffeine. The weight average mol. wts. were determined by sedimentation of the DNA in alkaline sucrose gradients.

These defects become even more marked if caffeine is present
during the postirradiation period (3). Caffeine has little effect
on the newly-synthesized DNA in UV-irradiated cells from normal
donors or an XP from group E, whereas the mol. wt. of the DNA from
excision-defective XPs in groups A - D is reduced to 60-80 x 10^6 and
that of the XP variants to about 30 x 10^6 (Table 1). Thus caffeine
has very little inhibitory effect on PRR in normal cells, an intermediate effect in the excision-defective XPs and a drastic effect in
XP variants. This behaviour is highly reproducible.

The findings that XP cells have (in most cases) low levels of
excision repair (1), defective PRR (3) and also reduced levels of
photoreactivating enzyme (4) might imply that DNA repair functions
are subject to some kind of coordinate control.

The defect in PRR is highly specific for XP cells. We have
found an extreme defect in five XP variants, an intermediate defect
in seven excision-defective XPs and a normal response in fibroblasts
from donors with a variety of other disorders. These disorders
include several hereditary diseases showing some manifestations of
light or radiation sensitivity, or susceptibility to multiple malignancies, such as ataxia telangiectasia, Fanconi's anaemia, Cockayne's
syndrome, basal cell nevus syndrome, progeria and familial retino-
blastoma.

Cell survival studies have shown that excision-defective XPs
from groups A, B, C, D are very sensitive to the lethal effects of
UV-irradiation (1,5). XP variants on the other hand are only
slightly if at all sensitive (6,7). Caffeine, whilst having no
effect on the survival of normal cells, does reduce the survival of
the XP variants (6,7), these results correlating well with those
obtained at the biochemical level.

Both excision-defective XPs and XP variants have a higher
level of UV mutability than normal cells (8,9). Thus, although
the defective PRR in XP variants is of little importance for the
survival of the cells, it can give rise to an increased level of
induced mutations which might in turn be responsible for the
increased frequency of skin malignancies found in the patients.

THE INVOLVEMENT OF SISTER STRAND EXCHANGES IN PRR

Turning our attention to the molecular mechanism of PRR, there
is convincing evidence that the sealing of daughter strand gaps in
UV-irradiated Escherichia coli is associated with gross exchanges
of parental and daughter strand DNA. Presumably the piece of
parental DNA from the sister duplex, which contains the missing
nucleotide sequence is inserted in the gap opposite the pyrimidine
dimer. Physicochemical evidence has shown that there is end-to-end

association of parental and daughter DNA after UV-irradiation (10)
and Ganesan has shown that sites sensitive to an endonuclease specific
for pyrimidine dimers (see below) appear in the newly-synthesized
strands (11).   Her work indicates that during the strand exchange
process pyrimidine dimers become equally distributed between parent
and daughter strands (11).   Similar experiments have been carried
out with mammalian cells.   Attempts to detect association between
parental and daughter DNA strands have been unsuccessful (12,13,14).
It was concluded from these experiments that if any exchanges occur,
they must either involve very small amounts of DNA or they must
occur only at a very small proportion of lesions, and therefore
escaped detection by the tests used.

On the other hand, sites sensitive to the dimer-specific
nuclease have been found in the daughter DNA strands.   UV-irradiated
cells are pulse-labelled with radioactive thymidine, followed by a
chase with unlabelled thymidine.   After this treatment a cell
lysate or DNA extracted from the cells is treated with a preparation
containing an endonuclease which specifically nicks DNA close to
pyrimidine dimers.   These enzymically-induced nicks, are taken as
indications of the existence of pyrimidine dimers.   Such dimer
sites have been found associated with the daughter strands of UV-
irradiated human cells (15,16).   The endonuclease-induced nicks
are not merely the result of the enzyme acting on the strand opposite the pyrimidine dimer, since they have also been observed in the
daughter DNA strands after denaturation (15).

Dimer sites have been found in DNA synthesized immediately
after irradiation at a frequency of approximately 20% of those in
the parental strands.   We have been able to detect dimer sites
in the daughter strands of fibroblasts from normal, excision-
defective XP, and XP variant donors.   The sedimentation profiles
obtained from alkaline sucrose gradients in these experiments must
however be interpreted with extreme caution, since a large number
of parameters affect the mol. wt. distributions obtained.   From
our results and those of Meneghini and Hanawalt (15) we can draw
the following conclusions:

1.   Pyrimidine dimer sites can be detected associated with daughter
DNA strands synthesized immediately after UV-irradiation of normal,
XP and XP variant cells.

2.   In the excision-proficient strains these dimer sites, as well
as those in the parental strands, disappear on prolonged incubation.

Because of the complexity in interpretation of the experimental
results we have as yet been unable to answer the following questions:

1.   Are these sites which appear to be associated with the daughter

strands really contained in the DNA synthesized after UV-
irradiation (C in Fig. 1a), or are they in reality pyrimidine
dimers produced in the strands that were growing at the time of
irradiation.   These would then become attached to DNA synthesized
after irradiation, simply by the normal process of chain elongation
(D in Fig. 1) (the possibility of this trivial explanation of the
data was raised by Dr. R. Meneghini)?   In either case treatment
with endonuclease would reduce the mol. wt. of the newly-synthesized
DNA (by acting at sites C and D in Fig. 1).

2.   Are these sites detectable in DNA synthesized several hours
after irradiation or only in DNA synthesized immediately after
irradiation?

3.   Is there a difference in the kinetics of the production of
daughter strand dimer sites in normal cells and the different types
of XP cells?

    If it can be rigorously shown that pyrimidine dimers are
actually contained in the daughter strands (Fig. 1a) this implies
that at least some sister strand exchanges are associated with PRR
in human cells.

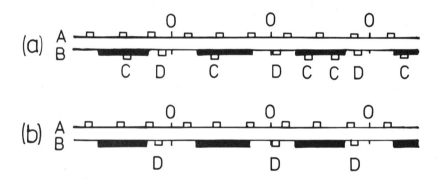

FIG. 1.   Dimer sites associated with daughter strands.   The
diagram shows a section of a DNA molecule after UV-irradiation,
radioactive labelling and completion of replication.
A: Parental strand; B: Daughter strand; O: origins of
replication; ———: DNA synthesized before UV; ▬▬: DNA synthesized
after UV and labelled; ⊓: Pyrimidine dimers.   Dimer sites in
the daughter strands synthesized after irradiation (C), and before
irradiation (D).
(a)   Sites in the DNA synthesized before and after irradiation.
(b)   Sites only in the DNA already in existence at the time of UV
      irradiation.

REFERENCES

1. Cleaver, J. E. & Bootsma, D. Ann. Rev. Genetics 9, 19-38 (1975).

2. Lehmann, A. R. Life Sci. 15, 2005-2016 (1974).

3. Lehmann, A. R. et al. Proc. Nat. Acad. Sci. U.S. 72, 219-223 (1975).

4. Sutherland, B. M., Rice, M. & Wagner, K. Proc. Nat. Acad. Sci. U. S. 72, 103-107 (1975).

5. Andrews, A. D., Barrett, S. F., & Robbins, J. H. Abs. 2nd International Workshop on Repair Mechanisms in Mammalian Cells (1976).

6. Arlett, C. F., Harcourt, S. A. & Broughton, B. C. Mutation Res. 33, 341-346 (1975).

7. Maher, V. M., Ouellette, L. M., Mittlestat, M. & McCormick, J. J. Nature 258, 760-763 (1975).

8. Maher, V. M. & McCormick, J. J. in Biology of Radiation Carcinogenesis (eds. Yuhas, J. M., Tennant, R. W. & Regan, J. D.) 129-145. Raven, New York, 1976.

9. Maher, V. M., Ouellette, L. M., Curren, R. D. & McCormick, J. J. Nature 261, 593-595 (1976).

10. Rupp, W. D., Wilde, III, C. E., Reno, D. L. & Howard-Flanders, P. J. Mol. Biol. 61, 25-44 (1971).

11. Ganesan, A. K. J. Mol. Biol. 87, 103-120 (1974).

12. Lehmann, A. R. J. Mol. Biol. 66, 319-337 (1972).

13. Fujiwara, Y. and Kondo, T. in Sunlight and Man (ed. Fitzpatrick, T. B.) 91-102, University, Tokyo (1974).

14. Rommelaere, J. & Miller-Faures, A. J. Mol. Biol. 98, 195-218 (1975).

15. Meneghini, R., & Hanawalt, P. C. Biochim. Biophys. Acta, 425, 428-437 (1976).

16. Buhl, S. N. and Regan, J. D. Nature 246, 484 (1973).

STUDIES ON THE ENZYMOLOGY OF EXCISION REPAIR IN EXTRACTS OF MAMMALIAN CELLS

E. C. Friedberg, K. H. Cook, K. Mortelmans and J. Rudé

Laboratory of Experimental Oncology, Department of Pathology, Stanford University, Stanford, California 94305 USA

SUMMARY

Damaged bases in DNA can be removed by at least two distinct mechanisms referred to as nucleotide excision repair and base excision repair. Our studies on the molecular mechanisms of excision repair in mammalian cells have used UV damage to bases as a model system. We have shown that crude extracts of human cells selectively remove thymine dimers from purified UV-irradiated DNA by a nucleotide excision mechanism which presumably involves the action of dimer specific endonuclease and exonuclease activities. Extracts of XP cells from complementation groups A, C, D and the XP variant are not qualitatively defective in this process. However, extracts of XP group A cells appear to be defective in the excision of thymine dimers from their own DNA present in the extracts, presumably as chromatin.

INTRODUCTION

The DNA repair mode most clearly established in studies on mammalian cells in vivo is that generally referred to as excision repair. At the outset, we would emphasize that recent studies (1, 2) have drawn attention to the fact that damaged bases in DNA can be removed by two distinct enzymatic mechanisms, which we refer to as nucleotide excision repair and base excision repair.

Figure 1 shows diagrammatically the familiar model of the excision of pyrimidine dimers from DNA based on studies in prokaryote systems (3). The essential feature of this model is that the

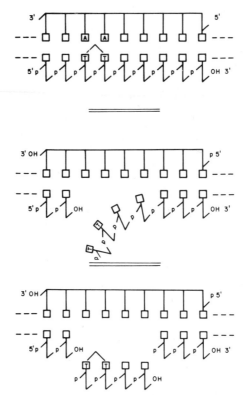

Fig. 1. Diagrammatic representation of nucleotide excision repair
of a cyclobutyl thymine dimer in DNA.

damaged bases are removed as nucleotides (probably as part of a
larger oligonucleotide) through the sequential action of a specific endonuclease and an exonuclease activity. The resulting gap
left in the DNA is repaired by DNA synthesis and ligation. In contrast to this mode of excision repair, Fig. 2 shows diagrammatically
a duplex DNA molecule in which a monoadduct type of base damage has
occurred. In the example chosen, a cytosine base has undergone
deamination to uracil; an event that can occur spontaneously (4) or
by treatment with chemicals (5) including the potent mutagen nitrous acid (6). It has been shown recently (5,7,8) that a variety
of prokaryote cells possess an enzyme activity (uracil-N-glycosi-
dase) that can remove uracil from DNA as the _free_ base, by catalyzing the hydrolysis of the N-glycosidic bond linking the base to
the sugar-phosphate backbone. Figure 3 shows the presence of this
enzyme activity in extracts of human cells. Preliminary observations indicate that the specific activity of this enzyme is significantly greater in cell lines than in cell strains.

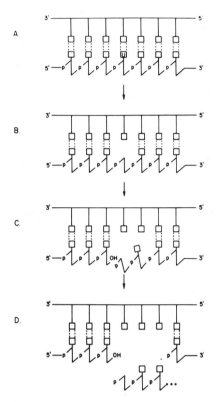

Fig. 2. Diagrammatic representation of base excision repair of
uracil from DNA.

     A second type of N-glycosidase with specificity for alkylated
purines has also been reported (2,9) and it is intriguing to speculate that a variety of forms of monoadduct base damage to DNA may
be excised by distinct N-glycosidase activities present in prokaryote and eukaryote cells.

     The removal of a free base (Fig. 2) would leave an apurinic or
apyrimidinic site in the DNA which could be attacked by an enzyme
specific for these sites. Such an enzyme activity (apurinic endonuclease) has been well characterized by a number of workers(10-12).
Following the endonucleolytic cleavage, exonucleolytic degradation
could remove the sugar-phosphate residue leaving a single-stranded
gap to be repaired by repair synthesis and ligation as in the later
steps of nucleotide excision repair.

     Using UV radiation damage as a model system of repair of base
damage, our interest is focused on the understanding of the molecular mechanisms of excision repair in mammalian, particularly human
cells.

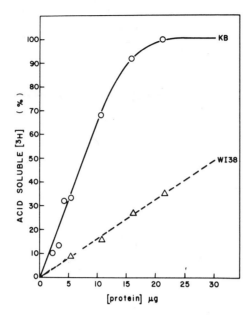

Fig. 3. Presence of uracil N-glycosidase activity in extracts of
human KB and WI-38 cells. Extracts were incubated with [³H] uri-
dine-labeled PBS2 DNA in the presence of EDTA. All acid-soluble
radioactivity is in the form of free uracil.

METHODS

In order to avoid the potential problem of selecting for enzyme activities that may have specificity for minor photoproducts
in DNA or for nonspecific conformational distortions induced by UV
radiation, we have measured the selective loss of thymine dimers
from UV-irradiated DNA substrates by extracts of mammalian cells as
an assay of enzyme activity. Two types of UV-irradiated DNA substrate were used.

a) highly purified DNA derived from bacteria or human cells;

b) DNA present in crude extracts of mammalian cells prepared
by sonication. Such DNA presumably exists as chromatin, except to
the degree that sonication and possible proteolysis disrupts normal
DNA-protein interactions. Details of the experimental procedures,
including the measurement of thymine dimers in DNA, are described
elsewhere (13,14).

RESULTS

Previously published data have demonstrated that crude extracts of HeLa or WIL II cells (a human lymphoblast line) effect the loss of 30-60% of the thymine dimers from their own chromatin with the release of less than 5% of the total DNA as acid soluble material (15).   Similar results have been obtained with a variety of human fibroblast strains, including WI-38 (13) and strain 629, derived from human foreskin. Figure 4 shows a chromatographic analysis of the acid-soluble fraction of a typical experiment using UV-irradiated chromatin as the substrate. All the radioactivity was present either at the origin of the chromatogram (indicating its existence as nucleotide or oligonucleotide) or as thymidine and thymine. No radioactivity was detected as free thymine dimers. We conclude, therefore, that as in prokaryotes, thymine dimers are excised by a nucleotide excision mechanism. Similar results have been obtained using purified DNA as the substrate instead of chromatin.

The assay we have employed measures both a dimer specific

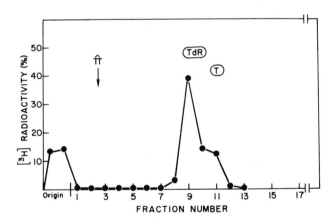

Fig. 4.   Thin-layer chromatographic separation of [³H]-labeled acid soluble components after incubation of crude extract of WI-38 cells with labeled UV-irradiated chromatin. The arrow indicates the position of purified thymine dimers in this chromatographic system.

endonuclease activity and an exonuclease activity. The former enzyme, as previously demonstrated (15), is extremely labile to freezing and thawing. The enzyme is currently being purified using the dimer excision assay with purified DNA as the substrate.

Exonuclease activity in extracts of mammalian cells that excises thymine dimers can be studied independently by using a UV-irradiated DNA substrate previously incised at dimer sites with purified T4 UV endonuclease (15). The use of the T4 endonuclease specifically defines excision activity that removes thymine dimers in the 5' → 3' direction since the endonuclease catalyzes incision on the 5' side of pyrimidine dimers (16). Such 5' → 3' exonuclease activity has been partially purified and characterized in our laboratory. As indicated in Table I, based on a number of physico-chemical criteria there appear to be at least two distinct 5' → 3' dimer excision exonuclease activities in extracts of human KB cells. In addition enzyme I contains a fraction with distinct chromatographic properties on phosphocellulose which may represent an isozyme or a distinct third exonuclease activity.

In addition to our efforts in the isolation and characterization of the components of the nucleotide excision repair system in normal mammalian cells, we are interested in understanding the biochemical basis of the excision repair defect in xeroderma pigmentosum (XP) cells. We have made direct comparisons between the capacity of

TABLE I

5' → 3' Dimer Excising Exonucleases
in Human KB Cells

Physico-Chemical Characteristic	Enzyme I	Enzyme II
Divalent cation optimum	1 mM Mg^{++}	5 mM Mg^{++}
Inhibition by 0.1 mM PCMPSA	0%	40%
S value	3.3	3.5
Isoelectric point	9.05	6.0
[NaCl] for 50% inhibition	50 mM	80 mM

extracts of normal diploid fibroblasts and of XP fibroblasts to effect thymine dimer excision from both purified DNA and from their own prelabeled DNA.

With respect to the use of purified DNA as a substrate, extracts of XP cells from complementation groups A, C, and D and from the XP variant excise thymine dimers as efficiently as do extracts of normal cells (13). However, when UV-irradiated chromatin is used as the substrate, extracts of XP group A cells fail to lose thymine dimers from the acid-insoluble fraction (13). More recent experiments using the identical conditions of sonication to those previously described (13) have shown that extracts of XP group D cells effect dimer excision from chromatin at normal levels; however, extracts of XP variant cells are defective in this process. Using preincised UV-irradiated DNA as a substrate for 5' → 3' exonuclease activity, measurements on crude extracts of a number of XP cells have shown no defect in total activity (18). However, since multiple such enzymes are now known to be present in crude extracts, this question needs to be reexamined in XP cells.

                              DISCUSSION

Our studies with extracts of human cells in tissue culture have demonstrated that cyclobutyl thymine dimers are excised from DNA or chromatin as nucleotides, as has been previously shown in prokaryotes. It is reasonable to assume that the enzymes involved in the process of excision include a dimer specific endonuclease and one or more exonuclease activities. Both of these enzyme activities are currently being purified and characterized in our laboratory. Our studies with XP cell extracts indicate that the molecular mechanism of dimer excision from chromatin may be more complex than from UV-irradiated purified DNA. At present, there are at least two possible interpretations that merit consideration. If the endonuclease activity that is involved in thymine dimer excision from DNA is the same as that from chromatin, then apparently other (possibly non-enzymatic) factors are involved in the molecular mechanism of dimer excision in mammalian cells. Alternatively, it is possible that the endonuclease active on DNA is distinct from that active on chromatin, in which case the defect in XP group A cells may be an enzyme function required for dimer excision from chromatin.

Our results with crude extracts of XP variant cells are difficult to interpret at present, since studies on living cells suggest no defect in dimer excision (17). However, it is possible that the variant may represent a form of XP which carries out quantitatively normal levels of excision but is qualitatively defective in a specific molecular component. This defect may be readily manifested

<u>in vitro</u> as a lability to certain modes of cell breakage such as
sonication. Further studies using defined chromatin substrates
and defined methods of cell breakage will be required to address
this problem more definitively.

ACKNOWLEDGMENTS

These studies were supported by grants from the U.S.P.H.S. and
American Cancer Society and by a contract with the U.S.E.R.D.A.

REFERENCES

1. Friedberg, E. C., Cook, K. H., Duncan, J. and Mortelmans, K.
    Photochemistry and Photobiology Reviews. Ed. K. C. Smith.
    Plenum Press, New York (In press).
2. Lindahl, T. Nature 259:64-66 (1976).
3. Grossman, L., Braun, A., Feldberg, R., and Mahler, I. Ann.
    Rev. Biochem. 44:19-43 (1975).
4. Lindahl, T., and Karlström, O., Biochemistry 12:5151-5154 (1973).
5. Lindahl, T. Proc. Natl. Acad. Sci. (USA) 71:3649-3653 (1974).
6. Schuster, T. Z. Naturforsch. 15b:298-304 (1960).
7. Friedberg, E. C., Ganesan, A. K. and Minton, K. J. Virol. 16:
    315-321 (1975).
8. Duncan, J., Hamilton, L. and Friedberg, E. C. J. Virol. 19:
    338-345 (1976).
9. Kirtikar, D. M. and Goldthwait, D. A. Proc. Natl. Acad. Sci.
    (USA) 71:2022-2026 (1974).
10. Verly, W. G. and Rassart, E. J. Biol. Chem. 250:8214-8219
    (1975).
11. Ljundquist, S. and Lindahl, T. J. Biol. Chem. 249:1530-1535
    (1974).
12. Ljungquist, S., Andersson, A. and Lindahl, T. J. Biol. Chem.
    249:1536-1540 (1974).
13. Mortelmans, K., Friedberg, E. C., Slor, H., Thomas, G. and
    Cleaver, J. E. Proc. Natl. Acad. Sci. (USA) 73:2757-2761
    (1976).
14. Cook, K. H. and Friedberg, E. C. Anal. Biochem. 73:411-418
    (1976).
15. Duncan, J., Slor, A., Cook, K. and Friedberg, E. C. In Mole-
    cular Mechanisms for Repair of DNA (eds. P. C. Hanawalt and
    R. B. Setlow). Plenum Press, New York, pp. 643-649 (1975).
16. Minton, K., Durphy, M., Taylor, R. and Friedberg, E. C. J.
    Biol. Chem. 250:2823-2829 (1975).
17. Lehmann, A. R., Kirk-Bell, S., Arlett, C. F., Paterson, M. C.,
    Lohman, P. H. M., de Weerd-Kastelein, E. A. and Bootsma, D.
    Proc. Natl. Acad. Sci. (USA) 72:219-223 (1975).
18. Cook, K., Friedberg, E. C., Slor, H. and Cleaver, J. E.
    Nature 256:235-236 (1975).

HUMAN PHOTOREACTIVATING ENZYMES

Betsy M. Sutherland

Department of Molecular Biology and Biochemistry

University of California   Irvine, California   92717

INTRODUCTION

The photoreactivating enzyme (PRE) repairs UV-damaged DNA in
a two-step reaction: the enzyme binds to a damaged region of the
DNA containing a pyrimidine dimer, the enzyme-DNA complex absorbs
a photon in the wavelength range 300 - 600 nm, resulting in photolysis of the dimer to constituent monomers and restoration of biological activity of the DNA (1,2,3). Although the photoreactivating
enzyme was known to be widely distributed through both plant and
animal kingdoms, most studies indicated that the enzyme was absent
in placental mammals (4). Recent evidence indicates, however,
that placental mammals do possess photoreactivating enzyme activity,
that this activity can monomerize pyrimidine dimers in cellular
DNA, and thus prevent UV-induced biological damage (5,6,7). I
shall discuss these data with emphasis on the human enzyme, its
deficiency in some human cells, and the possible role of the
enzyme in DNA repair in man.

BIOCHEMISTRY OF THE HUMAN PHOTOREACTIVATING ENZYME

The first indication that human cells might contain photoreactivating enzyme came in 1974, with the isolation of a photoreactivating enzyme from human leukocytes (5). The human enzyme
is remarkably similar to that purified from E. coli; their behavior
in purification procedures, molecular weight, isoelectric pH, pH
optimum for activity, and tendency to aggregate in solution are
highly similar. There are two striking differences between the
two enzymes, however: first, the ionic strength optimum of the
human enzyme ($\mu=0.05$) is much lower than that of the E. coli

enzyme (μ=0.2), and second, the action spectrum of the human
enzyme extends to at least 600 nm and thus includes light emitted
by yellow bulbs commonly used as safelights for work with other
photoreactivating enzymes (8). These factors may well have
impeded the detection of photoreactivating enzyme in cells from
placental mammals. The enzyme is found in human blood monocytes
and polymorphonuclear cells (but not in erythrocytes, serum or
lymphocytes) in bovine bone marrow, leukocytes of rabbit and cow,
feline cornea, as well as in cultured human and murine fibroblasts (9,10).

It is clear that choice of enzymatic assay methods is a
crucial step in study of photoreactivating enzymes in mammalian
cells (11). Three assay methods for photoreactivation are in
widespread use: H. influenzae transformation (12), nuclease
digestion (13), and chromatographic analysis (14). Although the
chromatographic method has the great advantage of chemical identification of substrate (dimer) and product (monomer), its insensitivity (requirement for large amounts of dimer photolysis
for detection) and slowness are serious drawbacks. The transformation assay is exquisite in sensitivity but is subject not only
to interference by non-specific nucleases in cell extracts, but
also to an unknown substance produced in some mammalian cell
extracts when exposed to photoreactivating light. This interference in some cases, is of sufficient magnitude to completely
obscure measurement of photoreactivation of transforming DNA. The
nuclease digestion method, used in my laboratory, combines the
virtues of speed, specificity, sensitivity, and insensitivity to
interference from other components in cell extracts. Its requirement for highly purified, ^{32}P-labeled DNA of high specific activity,
however makes inconvenient its use as as occasional experimental
procedure.

PHOTOREACTIVATING ENZYME ACTIVITY IN HUMAN CELLS

Since the enzyme might be expected to function in repair of
sunlight-induced damage in skin cells, we examined PRE content in
cells of normal individuals and of xeroderma pigmentosum patients
(6). [XP patients have extreme sun sensitivity, pigmentation
abnormalities, and high susceptibility to sunlight-induced malignant
growths (15)]. We found that all XP cells examined had less than
the normal levels of photoreactivating enzymes. The PRE content
in the cells ranged from 0 (in CRL 1199, PoCo,XP11BE) to about
50% (CRL 1259, XP2RO) of normal. Since the specific activity of
the enzyme might well be influenced by culture conditions, we
tested the following: growth rate of the cells, levels of enzymes
not involved in DNA repair, age of cell donor at biopsy, passage
number of the cells, or possible mycoplasma contamination of the

normal cells. None of these factors were responsible for the
depressed enzyme levels in the XP cells.

We also examined photoreactivating enzyme levels in cells from
xeroderma "variants", individuals with all clinical signs of XP
but normal levels of unscheduled DNA synthesis (16). We found
that three of the four cell types tested contained very low ($<$ 20%)
levels of photoreactivating enzyme, while one (with mild to moderate
XP) had a much higher PRE content. The apparent correlation between
severity of XP and degree of PRE depression led us to an examination of the total repair capacity of normal and XP individuals.
As a first approximation, we can sum the individual repair pathways, excision repair, post-replication repair and photoreactivation - to give a measure of the total repair capacity. Although
the three paths are doubtless not of equal importance in DNA repair,
assignment of a ten-fold weighting factor to any one of the three
led to predictions of phenotypes unlike those actually observed;
thus they will be given equal weight for purposes of discussion.
Comparison of the clinical signs of XP, XP variant and normal
individuals with their "total repair capacity" index indicates a
continuous variation in both, with a rather good correlation of the
two. We hypothesized that the balance of the individuals' repair
capacity on the plus side and his UV burden on the minus determines
his susceptibility to sunlight-induced skin malignancy. This hypothesis awaits testing by the evaluation of repair capacity and
clinical appearance of large numbers of individuals.

If the depressed PRE level we detect in XP cells is a true
characteristic of the individual (and not merely of his cells in
culture), we might be able to find a simple pattern of inheritance
of the enzyme in families with XP members (17). Cells from members
of one such family are available from the American Type Culture
Collection; we tested the father [CRL 1167, BeAr), mother (CRL
1165, CeAr), three XP sons (TaAr (XP2BE), PeAr (XP8BE), GeAr
(XP9BE)] and one normal son (CRL 1168, EmAr). Both the father
and normal son contained high levels of PRE, while the mother
and XP sons contained only about one-half as much. We thus suggested that the individuals with normal enzyme levels could be
designated as PP, and those with half that amount of enzyme as Pp.
Although cells of no other such families are available, we devised
another method of testing our hypothesis; we determined the PRE
activity in the parents of children with very low enzyme levels
(pp). All the parents should be either Pp or pp; no PP parents
should appear. In each of the normal parent - XP child pairs
tested, only intermediate level (Pp) parents were found. The
absence of PP parents is in agreement with our hypothesis of a
simple pattern of enzyme inheritance; the absence of pp individuals from the population of normal parents might indicate that
those individuals total repair capacity would be sufficiently

reduced for them to have developed XP and thus not appear in our
"normal parent" class. The availability of cells from more entire
family groups containing XP members would allow further testing
of this inheritance scheme.

We have also examined photoreactivating enzyme levels in cells
of individuals afflicted with other diseases; the cells of a
progeroid (syndrome of premature ageing and early death) boy contained 88% of the normal PRE content, not distinct from the range
of normals. Samples from a normal, 73 year old donor contained
75% of other normal controls; whether this individual is significantly lower than normal, and whether the apparent decrease is
a result of his age at biopsy or would have been present throughout life has yet to be determined.

BIOLOGICAL ACTION OF THE PHOTOREACTIVATING ENZYME

Although we can easily measure photoreactivating enzyme
activity in cell extracts, these results do not give any information on action of the enzyme in the cell. The biochemical
measurement of enzyme activity in a given cell type under a particular set of experimental conditions is, however, an absolute
prerequisite to examination of the biological effects of the
enzyme: if the enzyme is not present or is not active in the cells,
examination of biological action of the enzyme will be for naught.
Recent data indicate that cells from the same individual grown in
different media but otherwise apparently identical conditions
differed greatly in PRE content (enzyme present in one case, not
measureable in the other (18). In addition, different cell types
from the same individual vary in enzyme content (9). With these,
and possibly other, unknown, factors affecting PRE content in
mammalian cells, it is clear that biochemical measurements of
enzyme activity must preceed biological PR experiments.

We have examined the cellular action of the human PR enzyme
in three types of experiments: photoreactivation of dimers in
[3]H-thymidine-labeled DNA in the cells (6,7), photoreactivation of
survival of UV-irradiated Herpes simplex virus (19), and photo-
photoreactivation of UV-induced inhibition of DNA synthesis. In
all cases the biological results corresponded to the biochemical
measurements; if enzyme activity was detectable in cell extracts,
biological photoreactivation was found by any of the three assay
methods. If enzyme activity was not detectable in vitro no
cellular PR was found.

A representative experiment showing dimer photoreactivation
in normal human cells is shown in Figure 1. Cells were exposed to
UV, then to photoreactivating light or kept in the dark. Although
dimers were removed very slowly from the cellular DNA in the dark,

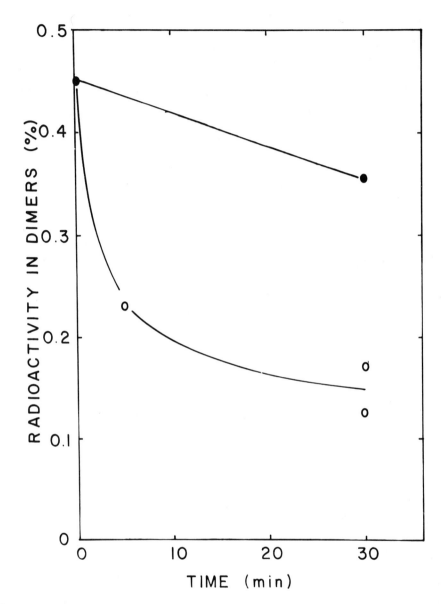

Fig. 1. Photoreactivation of pyrimidine dimers in normal human cells. Cells were labeled with ^{3}H-thymidine, exposed to 254 nm radiation and kept in the dark (●) or exposed to photoreactivating light (○).

samples exposed to photoreactivating light removed over 66% of
their cellular dimers within thirty minutes. No such dimer photoreactivation could be detected in the same cell type if grown in
minimal medium (PRE activity is not measureable in extracts of
these cells), nor could it be found in cells of the XP line Sally
G, also found to be low in PRE content (even if grown under
optimal conditions)(7).

Although cellular dimer photoreactivation is an index of
successful repair events, its effect must be sought by biological
assays. One such assay is viral survival, and Figure 2 shows
data on photoreactivation of survival of UV-irradiated Herpes
simplex virus in Jay Tim cells. The survival of the virus was
markedly increased by exposure to yellow photoreactivating light;
in the PRE⁻ cell type no such increase could be detected. A
serious technical problem encountered in these experiments was
the adverse effect of white photoreactivating light on cellular
capacity for virus production (even of HSV not exposed to UV).
The problem was alleviated in these experiments by the use of
yellow photoreactivating light, which are almost as effective in
photoreactivation but greatly reduce cell damage.

A second test of biological photoreactivation is the examination of the effect of photoreactivating light on UV-induced
DNA synthesis inhibition. (The adverse effects of photoreactivating
light on the cells were minimized by use of yellow bulbs and by
using only 10 min. of PR exposure, sufficient in dimer experiments
to photoreactivate over 65% of the cellular dimers.) Figure 3
shows that the DNA synthesis inhibition produced at very low UV
exposures can be almost completely prevented by 10 min. of photoreactivating light exposure. We also tested the possibility that
photoprotection might be responsible for the increased DNA synthesis rate; however, samples exposed to PR light before the UV
showed no difference from thos exposed to UV along.

CONCLUSION

The human photoreactivating enzyme shows great similarity to
those from other organisms: molecular weight, isoelectric pH,
aggregation tendency purification behaviour and pH optimum. Its
differences of ionic strength optimum and action spectrum present
slight but easily conquered inconveniences to the photobiologist.
Since measurement of enzyme activity can be made by a variety of
methods, it is important that the enzyme content of cells of
interest be tested under one's own experimental conditions. The
cellular PRE content depends not only on the experimental conditions, but (for cells from different individuals grown in identical conditions) on the cell donor - whether, normal, XP, XP variant
or progeroid. Still further deficiencies of this enzyme may

appear upon examination of other syndromes. The enzyme seems to function much as do PREs from other organisms: it monomerizes dimers in DNA, thus restoring the biological integrity of the DNA.

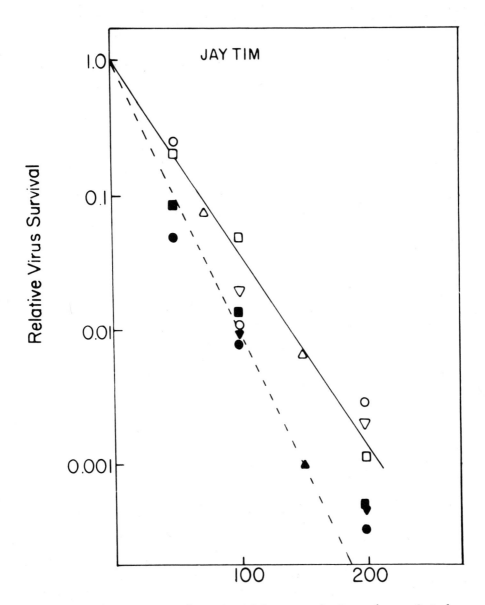

Fig. 2. Survival of uv-irradiated herpes simplex virus plated on Jay Tim fibroblasts and (closed symbols) kept in the dark, or (open symbols) exposed to photoreactivating light.

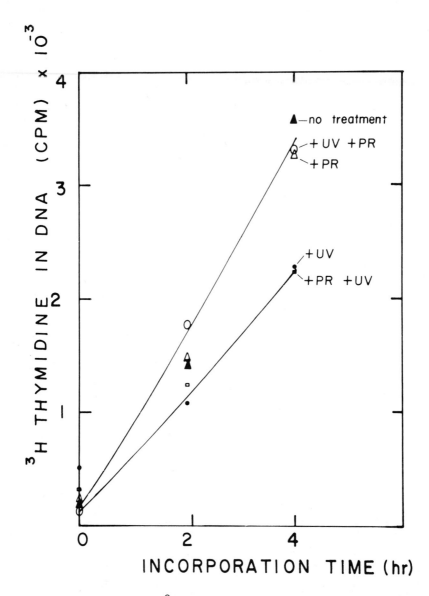

Fig. 3. Incorporation of ^{3}H-thymidine into the DNA of Jay Tim fibroblasts exposed to uv and kept in the dark (●), photoreactivating light followed by uv (■), photoreactivating light only (△), uv followed by photoreactivating light (◐) or no treatment (▲). Each point is the average of 6 independent determinations.

Acknowledgments

Figures 1 and 2 were reprinted with permission from B. M. Sutherland et al. Biochemistry 15, 402-406, 1976 (copyright by the American Chemical Society) and Wagner et al., Nature, 254, 627-628, 1975 (copyright Nature, Macmillan Journals,Ltd), respectively. This research was supported by grants from the American Cancer Society (NP-154) and the National Institutes of Health (CA14005-04) and a Research Career Development Award 5K04 CA00009-2.

References

1. Rupert, C. S., J. Gen. Physiol. 43, 573 (1960).
2. Setlow, J. K. and R. B. Setlow, Nature 197, 560 (1963).
3. Setlow, J. K., M. E. Boling and F. J. Bollum, Proc. Nat. Acad. Sci. U.S.A. 53, 1430 (1965).
4. Cook, J. S. and J. R. McGrath, Proc. Nat. Acad. Aci. U.S.A. 58, 1359 (1967).
5. Sutherland, B. M., Nature 248, 109 (1974).
6. Sutherland, B. M., M. Rice and E. K. Wagner, Proc. Nat. Acad. Sci. U.S.A. 72,103 (1975).
7. Sutherland, B. M., R. Oliver, C.O. Fuselier and J. C. Sutherland, Biochem. 15, 402 (1976).
8. Sutherland, J. C. and B. M. Sutherland, Biophys. J. 15,435 (1975).
9. Sutherland, B. M., P. Runge and J. C. Sutherland, Biochem. 13, 4710 (1974).
10. Harm, H., Abstracts, Amer. Soc. for Photobiol. Meeting, Vancouver (1974).
11. Sutherland, B. M., Life Sci. 16, 1 (1975).
12. Rupert, C. S., S. H. Goodgal and R. M. Herriott, J. Gen. Physiol. 41, 451 (1958).
13. Sutherland, B. M. and M. J. Chamberlin, Anal. Biochem. 53, 168 (1973).
14. Carrier, W. L. and R. B. Setlow, Methods in Enzymol. 210, 230 (1971).
15. Robbins, J. H., K. H. Kraemer, M. A. Lutzner, B. W. Festoff and H. G. Coon, Ann. Int. Med. 80, 221 (1974).
16. Sutherland, B. M. and R. O. Oliver, Nature 257, 132 (1975).
17. Sutherland, B. M. and R. Oliver, Photochem. Photobiol., in press (1976).
18. Sutherland B. M. and R. Oliver, Biochim. Biophys. Acta, in press, (1976).
19. Wagner, E. K., M. Rice and B. M. Sutherland, Nature 254, 627 (1975).

EFFECTS OF ULTRAVIOLET-LIGHT AND OF SOME MAJOR CHEMICAL CARCINO-

GENS OF E. COLI K12 (λ)

Raymond DEVORET

Laboratoire d'Enzymologie, C.N.R.S.

91190 GIF-SUR-YVETTE, FRANCE

SUMMARY

UV-irradiation of E. coli bacteria produces DNA lesions, which are repaired by excision and recombinational repair. These two processes, however, do not remove all the DNA lesions. The remaining lesions induce a complex metabolic pathway (SOS functions). SOS functions lead to cellular effects such as prophage induction, induced phage reactivation and mutagenesis.

Chemical carcinogens induce mutations in bacteria. They do so if i) the cells are permeable to them ii) the chemicals are metabolized by mammalian microsomal enzymes. Major chemical carcinogens induce prophage λ to develop vegetatively. For instance, aflatoxin B1, one of the most potent carcinogens, is fully efficient at doses as low as 1 nanogram. Aflatoxin B1 metabolites also produce induced phage reactivation.

In conclusion, aflatoxin B1, as many other chemical carcinogens, mimics the action of UV-light on bacteria in inducing SOS functions. It is postulated that, in eucaryotes, chemical carcinogens induce a metabolic pathway analogous to bacterial SOS functions and that such a process leads to carcinogenesis. The characterization at the molecular level of bacterial SOS functions should provide a theoretical model useful for the understanding of the mechanism of cell transformation.

INTRODUCTION

The use of radiations to produce or to cure cancer dates
back from the discovery of radium and of its biological effetcs.
Cellular radiation biology has made rapid progress when it was
discovered that pyrimidine dimers formed in the DNA are the main
lesions produced by UV-light. This finding led to the discovery and
characterization at the molecular level of two repair processes :
excision repair (1,2) and post-replication recombinational repair
(3). A third repair process acting on UV-damaged phage DNA has
been recently defined at the genetical (4-6) and at the molecular
level (7) (Radman, this book). It is an inducible error-prone type
of repair called "induced phage reactivation" (the use of this
term will be justified further). It is very likely that this repair process is not restricted to phage DNA and acts on chromosomal DNA to restore cell survival and produce mutagenesis.

It must be emphasized that repair processes are but one
facet of the many physiological events that occur in a UV-damaged
cell. UV-irradiation of E. coli bacteria also triggers other cellular phenomena, the most spectacular of which being the induced
development of a dormant virus (lysogenic induction) (8). There
is a close physiological relationship between prophage induction
and induced phage reactivation ; these two phenomena are under the
control of the same bacterial genes. This led Radman to propose
his theory of "SOS repair" (9, 10). This theory gained immediate
experimental support (4, 5, 11, 12) and has been developed further
(13, 14).

The aim of this paper is to draw the attention of the reader
to some as yet mysterious physiological cellular processes I call
"SOS functions", the manifestations of which can clearly be seen
at the biological level : filament formation, mutagenesis, lysogenic induction, cessation of respiration, etc... Not only are "SOS
functions" triggered by UV-irradiation of E. coli cells, they are
also induced by cell treatment with chemical carcinogens.

Bacteria have been used for long as biological tools to uncover the cellular action of chemical carcinogens. This type of
study has recently gained momentum when it was realized that to
be active : 1) most carcinogens must be activated with liver enzymes 2) the cells must be permeable to the carcinogen or to the
carcinogenic metabolites. Using permeable bacteria along with a
metabolizing system Ames et al. were able to demonstrate the mutagenic action of major carcinogens (15). In our lab we have shown
that major carcinogens induce the vegetative multiplication of
bacteriophage λ (16, 17). Aflatoxin B1, the most potent hepato-
carcinogen, also induce phage reactivation (18). In short, major
carcinogens mimic all the effects of UV light irradiation on E.
coli K12 (19).

In E. coli the mechanism of prophage induction and of induced phage reactivation by UV-light or chemical carcinogens is
being studied at the molecular level ; such research should ultimately provide a model useful to the understanding of the mechanism of cell transformation in mammalian cells.

UV INDUCTION OF PROPHAGE λ

Most bacterial species, if not all, carry the DNA of a virus in a dormant state (prophage). Bacteria carrying a prophage
are said to be lysogenic. Prophage λ in E. coli K12 has been extensively studied (16). The dormant state of prophage λ is maintained by a protein called the repressor. The repressor binds to
two operator sites, thus preventing transcription of the prophage
genes except for the one (cI) which codes for the repressor itself.

UV-irradiation of lysogenic bacteria triggers the development of the prophage into vegetative phage : this is lysogenic
induction (8). Lysogenic induction is caused by the irreversible
inactivation of the repressor, which can no longer block DNA
transcription : all prophage genes are then expressed, a progeny
of about 100 phage particles per cell is then produced.

Prophage λ induction is not due to a direct action of UV-
light on the prophage DNA or on the repressor itself (20, 21).
Lesions of the bacterial chromosome trigger a complex metabolic
pathway whose final step is the inactivation of the repressor
(22). Repressor inactivation may result from proteolytic cleavage
(23).

UV-REACTIVATION

When host-cells are exposed to a low UV dose before phage
infection, the survival of UV-irradiated phage λ is increased,
the reactivated phage being heavily mutated (24). This is UV-
reactivation (25).

It has taken about twenty years to begin to grasp some knowledge on the mechanism of UV-reactivation. Until recently, research on this subject has mostly consisted in determining not
what UV-reactivation really was but rather what it was not.

We now know that UV-reactivation does not result from an enhancement of excision repair (26). It is not the product of an
illegitimate recombination between phage and host chromosome DNAs
as Weigle believed it was (27). It does not result from a post-
replicative recombinational repair process operating on UV-damaged
phage λ DNA since single-stranded phage can be UV-reactivated (27).

In fact, UV-reactivation is the expression of an inducible error-prone type of repair. This has been established by two lines of evidence. At the genetical level, it has been shown that UV-reactivation can be produced indirectly by conjugal transfer of UV-damaged DNA to the host cell before infection (29). Chloramphenicol treatment of the host cell before infection prevents the occurrence of UV-reactivation (30).

At the molecular level, it has been shown that upon UV-reactivation of phage λ, DNA replication does not stop at pyrimidine dimers for a notable fraction of UV-damaged single-stranded DNA molecules (Radman, this book). The fraction of DNA molecules which can be replicated corresponds to the fraction of repaired phage. Therefore, replication can occur on UV-damaged templates. However, the surviving phage is mutated : mutagenesis is the price to be paid for survival.

A COMMON MECHANISM FOR UV-INDUCTION AND UV-REACTIVATION OF PHAGE λ : SOS FUNCTIONS

UV-reactivation of phage λ and development of prophage λ are both inducible. The two phenomena can be produced by conjugal transfer of UV-damaged DNA (5, 29) ; they are controlled by the same bacterial genes. A few bacterial mutations such as recA, lexA, lexB or zab are known to prevent the occurrence or prophage induction as well as UV-reactivation (31-35). Interestingly enough a bacterial mutation tif-1 located in the recA region (33) can induce the two phenomena when tif bacteria are shifted from 32° to 42° (33).

Such a striking correlation led Radman to put forward the hypothesis that UV-reactivation was an inducible error-prone type of repair, SOS repair, that was also involved in the mechanism of lysogenic induction and bacterial mutagenesis. Witkin extended Radman's idea in postulating that a number of cellular and host mediated phage events were all under the control of repressors and that they are expressed when those repressors are inactivated (13). In short, UV-irradiation in producing DNA lesions provides a signal for the coordinated expression of cellular events through the destruction of repressors.

In fig. 1 is represented as a function of time the sequence of events starting at the production of DNA lesions and ending with such cellular phenomena as lysogenic induction, filament formation, mutagenesis, etc...

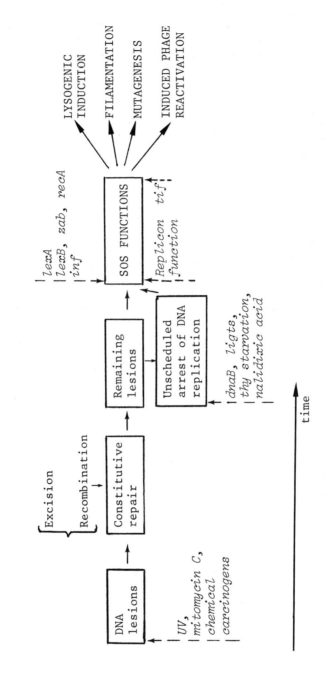

Figure 1.

I define as "SOS functions" a still unknown enzymatic pathway controlled by genes recA, lexA, lexB, tif, inf (36),... This
pathway has been previously called by Witkin rec lex pathway (11)
(12). The new terminology implies that SOS repair (10) is only
one of the many consequences of SOS functions.

The signal for SOS functions is a trouble at the replication fork (unscheduled inhibition of DNA replication). This can
result from the presence of persisting DNA lesions that have not
been or cannot be removed by constitutive repair mechanisms. This
can also be caused by a modification of some of the many proteins
that participate with the polymerases in the replication process
(dnaB, nalidixic acid treatment) or by lack of a DNA building
block (thymine deprivation).

SOS functions may result in the inactivation of repressors
by proteases. The only repressor we know for which it actually
occurs is the λ repressor. However, according to Witkin's hypothesis (15), septation and error prone repair might also be controlled negatively by repressor proteins.

The concept of SOS functions is useful because it provides
a unitary mechanism for various cellular events resulting from
radiation damage. From it one can derive a definite set of working
hypotheses that can be tested at the biochemical level.

MAJOR CARCINOGENS INDUCE PROPHAGE λ

Aflatoxin B1, a mycotoxin synthesized by some strains of
Aspergillus flavus is one of the most potent carcinogen (37).
Aflatoxin B1 metabolites induce prophage λ in the totality of the
cell population (17) (45). Aflatoxin B1 alone is not active.
Prophage λind⁻, not inducible by UV-light, was not induced by the
drug metabolites but showed a high level of mutagenesis. Likewise,
the phage λ progeny produced by induction contained also a high
level of clear plaque mutants (45).

These results prompted us (17) to set up a simple, inexpensive and sensitive test for potential carcinogens based upon the
property of carcinogens to induce prophage λ. By using chemicals
activated with microsomal enzymes, and E. coli K12 permeable (envA)
bacteria also deficient in DNA repair (uvrB), the range of carcinogens detected in a lysogenic induction test (inductest) has been
extended.

The evidence has been provided that, after activation, carcinogenic polycyclic hydrocarbons such as benzo(a)pyrene and 7,12-
dimethylbenz(a)anthracene induce prophage λ. Three variants of
the test have been developed (inductests I, II and III) which are

as sensitive as the mutagenicity test of Ames et al. (15). Induction of envA uvrB lysogens has been detected with input concentrations of aflatoxin B1 as low as 1 ng per assay (17).

Inductests II and III provide a quantitative estimation of
the inducing activity of a carcinogen. With the latter test, one
can determine : i) the cellular toxic effect of a carcinogen,
ii) the kinetics of appearance and disappearance of active metabolites.

For two series of chemicals, aflatoxins and benz(a)anthra-
cenes, there is a good correlation between their carcinogenic activity in rodents and their prophage inducing activity in bacteria.

The fact that the majority of the cell population is induced gives the possibility to test the inducing activity of carcinogens at the biochemical level, e.g. by measuring λ repressor inactivation.

The results of mutagenicity and lysogenic induction tests
show that there is a corroboration of carcinogenic activity in
eucaryotes with the ability of the very same chemicals either to
induce a prophage or to produce mutations in procaryotes. Such a
correlation leads to postulate that : i) the in vivo active forms
of carcinogens are alike those generated in vitro by mammalian
liver enzymes, ii) DNA lesions induced in eucaryotes by carcinogens are similar to those induced in procaryotes.

One is tempted to surmise that if a chemical responds positively to the test of mutagenesis (mutatest) and to the test of
lysogenic induction (inductest) it is likely to cause cell transformation.

INDUCED PHAGE REACTIVATION

Pretreatment of wild type host cells with activated aflatoxin B1 increases the survival ability of UV-damaged phage λ (18).
The reactivation effect is roughly proportional to the input concentration of the drug in the range used. Twenty five percent of
the lesions are repaired in this process, and mutagenesis, scored
as the formation of clear plaques, is increased 10-fold. Repair
and mutagenesis of UV-damaged phage λ can be obtained at a time
when aflatoxin B1 active metabolites are no longer present in
the metabolizing mixture. One can conclude that the repair process is induced by the presence of DNA lesions in the host before
phage infection ; phage λ reactivation is obtained in uvr⁻ but
not in recA bacteria and it is very similar to direct (27) or
indirect UV-reactivation (5).

I propose to designate from now on UV-reactivation and si-

milar phenomena as the one reported above, i.e. <u>induced phage</u>
<u>reactivation</u>.

        The term UV-reactivation restricts the phenomenon to the
particular case or UV-irradiation of the host-cell. A slight damage to the host-cell produced by X-ray irradiation (Errera, personal communication), thymine starvation (38), exposure to mitomycin
C (39) or thermal shift of tif-1 bacteria (40) also enhance the
survival and concomitant mutagenesis of UV-damaged phage λ. The
rather obscure terms W- or Weigle reactivation should be rejected
on the ground that they might imply a mechanism of repair (favored
by Weigle) which has been proven wrong (4).

MAJOR CARCINOGENS ARE MUTAGENS

        More than a hundred carcinogens have recently been shown to
be mutagens in <u>Salmonella</u> <u>typhimurium</u> (15). The correlation between
carcinogenicity and mutagenesis is very striking. However, it must
be emphasized that there are two types of mutagenesis : 1) <u>direct</u>
<u>mutagenesis</u> resulting from e.g. incorporation of a base analog ; this
type of mutagenesis is not a consequence of SOS functions ; 2) <u>SOS</u>
<u>mutagenesis</u> resulting from the induction of SOS functions ; in
this case, the expression of <u>recA</u> and <u>lex</u> functions is required
to produce mutagenesis. Mutagenesis by ultraviolet light involves
such a mechanism.

        It is expected that all chemical compounds that, in bacteria,
induce mutagenesis <u>and</u> lysogenic induction (phenomena which both
result from cellular SOS functions) will be found to be carcinogenic.

CONCLUSIONS

        Some major carcinogens induce prophage λ ; reactivation of
UV-damaged phage λ is also produced by aflatoxin B1 metabolites.
Most of the major carcinogens seem to mimic the cellular effects
of UV-light on <u>E. coli</u>. DNA lesions produced by carcinogenic metabolites interfere with DNA synthesis as UV lesions do. Persisting
lesions in the DNA (whether produced by UV-light or by carcinogenic metabolites) induce a complex metabolic pathway, SOS functions, which result, at least, in the inactivation of the phage λ
repressor and in error-prone polymerisation of a damaged DNA
template and, therefore, in mutagenesis. These SOS functions are
governed by a rather large set of bacterial genes.

        Cell transformation has been claimed to be the result of
either a somatic mutation (41) or the induction of a dormant virus
(42, 43, 44). It is postulated that a metabolic pathway, similar
to the bacterial SOS functions, might be induced in mammalian cells.
Eucaryotic SOS functions might also lead to, at least, mutagenesis

and induction of a provirus through the coordinated inactivation of a few cellular repressors. The inactivation of cellular repressors through eucaryotic SOS functions may be the basic mechanism of cell transformation.

ACKNOWLEDGEMENTS

I thank M. RADMAN and E. WITKIN for many delightful discussions. I am grateful to my collaborators A. GOZE, P. MOREAU, A. BAILONE and P. MORAND for their contribution and to J. GEORGE, A. SARASIN and Y. MOULE for a fruitful collaboration.

REFERENCES

1. Setlow, R.B. and Carrier, W.L. (1964) Proc. Natl. Acad. Sci. US 51, 226–231.
2. Boyce, R.P. and Howard-Flanders, P. (1964) Proc. Natl. Acad. Sci. US 51, 293–300.
3. Rupp, W.D. and Howard-Flanders, P. (1968) J. Mol. Biol. 31, 291–304.
4. Blanco, M. and Devoret, R. (1973) Mutation Res. 17, 293–305.
5. George, J., Devoret, R. and Radman, M. (1974) Proc. Natl. Acad. Sci. US 71, 144–147.
6. Devoret, R., Blanco, M., George, J. and Radman, M. (1975) in Molecular Mechanisms for repair of DNA. Plenum Publishing Co., New York, 155–171.
7. Radman, M., Caillet-Fauquet, P., Defais, M. and Villani, G. (1976) in Screening Tests in Chemical Carcinogenesis. IARC Scientific Publications, Lyon, 12, 537–548.
8. Lwoff, A. (1953) Bacteriol. Reviews 17, 269–337.
9. Radman, M. (1974) Molecular and Environmental Aspects of Mutagenesis. C.C. Thomas Publishing, Springfield (III.) USA.
10. Radman, M. (1975) in Molecular mechanisms for repair of DNA Plenum Publishing Co., New York, 355–367.
11. Witkin, E. and George, D.L. (1973) Genetics Supplement 73, 91–108
12. Witkin, E. (1975) Genetics 79, 191–213.
13. Witkin, E. (1974) Proc. Natl. Acad. Sci. US 71, 1930–1934.
14. Witkin, E. (1976) Bacteriol. Rev. (in press).
15. Ames, B.N., Mac Cann, J. and Yamasaki, E. (1975) Mutation Res. 31, 347–364.
16. The Bacteriophage Lambda (1971) Cold Spring Harbor Laboratory, New York.
17. Moreau, P., Bailone. A. and Devoret, R. (1976) Proc. Natl. Acad. Sci. US 73, 3700–3704.
18. Sarasin, A., Goze, A., Devoret, R. and Moulé, Y. (1977) Mutation Res. in press.
19. Devoret, R., Goze, A., Moulé, Y. and Sarasin, A. Repair and carcinogenesis, C.N.R.S., Paris (1976) in press.

20. Borek, E. and Ryan, A. (1957) Proc. Natl. Acad. Sci. US 44, 374-377.
21. Devoret, R. and George, J. (1967) Mutation Res. 4, 713-734.
22. Shimagawa, H. and Itoh, T. (1973) Mol. gen. Genetics 126, 103-110.
23. Roberts, J.W. and Roberts, C.W. (1975) Proc. Natl. Acad. Sci. US 72, 147-151.
24. Kellenberger, G. and Weigle, J. (1958) Biochim. Biophys. Acta 30, 112-124.
25. Weigle, J.J. (1953) Proc. Natl. Acad. Sci. US 39, 628-636.
26. Radman, M. and Devoret, R. (1971) Virology 43, 504-506.
27. Weigle, J.J. (1966) Phage and the Origins of Molecular Biology. Cold Spring Harbor Laboratory, New York.
28. Ono, J. and Shimazu, Y. (1966) Virology 29, 295-302.
29. George, J. and Devoret, R. (1971) Mol. gen. Genetics 111, 103-119.
30. Defais, M., Caillet-Fauquet, P., Fox, M. and Radman, M. (1976) Virology (in press).
31. Hertman, I. and Luria, S. (1967) J. Mol. Biol. 23, 117-133.
32. Brooks, K. and Clark, A.J. (1967) J. Virol. 1, 283-293.
33. Castellazzi, M., George, J. and Buttin, G. (1972) Mol. gen. Genetics 119, 139-152.
34. Castellazzi, M., George, J. and Buttin, G. (1972) Mol. gen. Genetics 119, 153-174.
35. Blanco, M., Levine, A. and Devoret, R. (1975) in Molecular Mechanisms for repair of DNA. Plenum Publishing Co., New York, 379-382.
36. Bailone, A., Blanco, M. and Devoret, R. (1975) Molec. gen. Genetics 136, 291-307.
37. Goldblatt, L.A. (1969) Aflatoxin. Academic Press, New York.
38. Hart, M.G.R. and Ellison, I. (1970) J. Gen. Virology 8, 197-208.
39. Otsuji, N. and Okubo, S. (1960) Virology 12, 607-609.
40. George, J., Castellazzi, M. and Buttin, G. (1975) Molec. gen. Genetics 140, 309-332.
41. Boveri, T. (1929) The origin of malignant tumors. Williams and Wilkins, Baltimore.
42. Gross, L. (1974) Proc. Natl. Acad. Sci. US 71, 2013-2017.
43. Huebner, R.J. and Todaro, G.J. (1969) Proc. Natl. Acad. Sci. US 64, 1086-1094.
44. Todaro, G.J. and Huebner, R.J. (1972) Proc. Natl. Acad. Sci. US 69, 1009-1015.
45. Goze, A., Sarasin, A., Moulé, Y. and Devoret, R. (1975) Mutation Res. 28, 1-7.

SYMPOSIUM VIII

SOLAR ENERGY CONVERSION SYSTEMS

HYDROGEN METABOLISM IN PHOTOSYNTHETIC ORGANISMS, THE MECHANISM OF HYDROGEN PHOTOEVOLUTION[1]

Dan King, David L. Erbes, Ami Ben-Amotz[2], and
Martin Gibbs

Institute for Photobiology, Brandeis University
Waltham, Massachusetts   02154   USA

Studies utilizing cell-free systems and dealing with H2 photo-evolution have been few. Abeles (1) reported H_2 formation from reduced pyridine nucleotides in the presence of DCMU with a preparation of <u>Chlamydomonas</u> <u>eugametos</u>. Benneman <u>et</u> <u>al</u>.(4) demonstrated light-dependent H2 evolution from water and from reduced dichlorophenolindolphenol with a reconstituted spinach chloroplast-soluble ferredoxin-clostridial hydrogenase system. We have adapted a recently published method of brief sonication of algal cells in a medium of low osmolarity (6) to obtain stable and sustained rates of H2 photoevolution in order to evaluate and characterize photosystem participation in this process. We also present results concerning the effects of soluble ferredoxin, electron donors, and inhibitors on the photoevolution of hydrogen.

MATERIALS AND METHODS

<u>Chlamydomonas</u> <u>reinhardi</u> (Wt) and <u>C</u>. <u>reinhardi</u> strain F60 (9) (F60 is missing phosphoribulokinase and is thus incapable of CO_2 fixation) were grown as previously described (3). Cells were harvested by centrifugation and resuspended in 10mM KH_2PO_4, 20mM KCl, 2.5mM $MgCl_2$, pH 7.4, at a final chlorophyll concentration of 0.5 mg/ml. After adaptation under N_2 the cells were broken with a Branson Sonifier 200, 75 W, 15-20 sec in the dark at 0 C. The suspensions were then centrifuged at 500g for 10min under N_2. The supernatant was drawn into syringes and transferred into serum bottles

[1]Supported by National Science Foundation Grant GMS71-00978 and Energy Research and Development Administration Grant ET(11-1)3231.
[2]Present Address: Israel Oceanographic & Limnological Research, Ltd., Haifa Laboratories, Haifa, 21 Hativat Golani-Road,Israel

anaerobically.   Three milliliters were routinely used for mano-
metric assays.

For preparations of hydrogenase, C. reinhardi was grown under
fluorescent lighting on TAP medium (8) and harvested by continuous
flow centrifugation.   Cell yield averaged 4g wet cell paste/liter
of culture.   About 200g of wet cell paste were suspended in 50mM
Tris-Cl, pH 8.0, containing 10μM DCMU[3] to give a final volume of
1.2 liters.   After evacuation for 5min followed with H2 flushing
(repeated three times) the suspension was frozen in liquid N2. The
frozen suspension was thawed in the presence of 240g of (NH4)2SO4
under H2 at room temperature.   After dissolving the pellet and anaerobic centrifugation at 15,000g for 1 hr, the supernatant was de-
canted and saturated with 600g of (NH4)2SO4 under H2.   Anaerobic
centrifugation yielded a precipitate containing both ferredoxin
and hydrogenase.   The pellet was redissolved in the appropriate
buffer for use.

To check for activity during purification and handling, hydrogenase was assayed by methyl viologen reduction followed at 605nm.
An extinction coefficient of $9.3mM^{-1}cm^{-1}$ was used in calculating
the rate of methyl viologen reduction.

Hydrogen evolution was measured manometrically under N2 with
alkaline pyrogallol in each Warburg flask center well.   All manipulations were carried out under conditions of strict anaerobicity.
Nitrogen was used only after being assayed at less than 5μl/1 O2.
Hydrogen was purified by passage through Deoxo cartridges.

Spinach chloroplasts were prepared according to Avron and Gibbs
(2) and were broken osmotically in the assay reaction mixture.
Sonically shocked spinach particles were prepared anaerobically by
a 10sec sonication at 0 C at 75W.   C. reinhardi ferredoxin was prepared as a byproduct of the hydrogenase preparation. Chlorella pyrenoidosa ferredoxin was a gift from A. Schmidt and J. Schiff.

RESULTS

Brief sonication of anaerobically adapted algae followed by a
500g centrifugation to remove whole cells resulted in a cell-free
preparation of chloroplast particles capable of light-dependent
production of H2 upon the addition of hydrogenase and suitable electron donors.   Problems incurred with whole cells producing O2 during
adaptation were eliminated by utilizing the F60 strain of C.rein-
hardi.   Longer sonication periods resulted in a loss of stable
rates of H2 photoevolution.   Broken particles of chloroplasts could
be prepared, using only 8-10 sec sonication, that demonstrated H2
photoevolution sensitive to DCMU inhibition. The reduced pyridine

---
[3] Abbreviations: Asc, Ascorbic acid; DBMIB, Dibromothymoquinone;
DCMU, Dichlorophenylmethylurea; DPC, Diphenylcarbazide; DPIP, Dichlorophenolindolphenol; DTT, Dithiothreitol; MeV, Methyl viologen;
PMS, Phenazine methosulfate; PSI, PSII, Photosystem I, II.

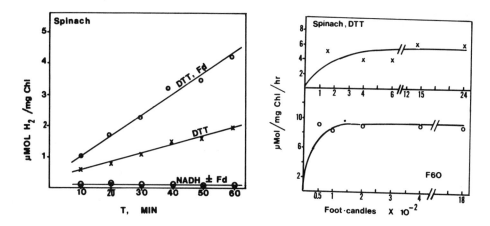

Fig.1. Effects of Ferredoxin on Hydrogen Photoevolution Utilizing Sonicated Particles and Chlamydomonas Hydrogenase
Fig.2. Effects of Light Intensity on Hydrogen Photoevolution in Intact Chlamydomonas and Reconstituted Spinach Particles.

nucleotide, NADH, and DTT demonstrated the highest activity as donors for H_2 evolution, 6.1 and 7.4 µMoles/mg Chl/hr, respectively. Photoevolution of H_2 from NADH and DTT was eliminated by a low concentration of PMS (7µM). The inhibition of ferredoxin by DSPD (10) abolished photoevolution from both DTT and NADH. The presence of 10µM DBMIB, an inhibitor of plastoquinone (5), eliminated evolution from NADH, but only partially inhibited (45%) the rate of evolution from DTT. The addition of NADP had no effect on H_2 evolution, whereas the presence of NADPH resulted in a 30% decrease in the rate of H_2 evolution. As a donor, NADPH would not support H_2 photoevolution. Other donors tested were relatively poor electron sources.

Preparations of sonicated spinach particles yielded low rates of hydrogen evolution with DTT as electron donor(Fig.1). With NADH as donor, no evolution could be observed. The presence of additional ferredoxin (30µM) stimulated evolution from DTT by 3-4 fold, but no stimulation was observed with NADH. The preparations used in these experiments were 3-4µM with respect to ferredoxin. In later experiments utilizing ferredoxin-free hydrogenase, no photoevolutionary activity could be observed. The addition of ferredoxin to these reconstituted systems restored activity. Preliminary data indicates that the algal system also requires ferredoxin.

Although aerobic photosystems generally do not saturate with respect to light below about 2000 footcandles, photosystems involved in H_2 evolution appear to saturate at low light intensities. Both whole cells and reconstituted spinach particles-hydrogenase systems appear to saturate at 100-200 footcandles (Fig.2). In Table 1, data are represented for a particle demonstrating electron flux through both photosystems. The water-splitting act appeared to

TABLE I. The Effect of Light Intensity on Electron Flow in
Chlamydomonas reinhardi

Donor	Acceptor	μMol/mg Chl/hr	
		100 footcandles	1500 footcandles
DPC	Protons (H_2)	4.0	5.5
DPC	O_2 (MeV)	4.1	52
DPC	NADP	3.4	30
Asc/DPIP	O_2 (MeV)	---	810
Asc/DPIP	NADP	---	80

have been destroyed by the sonication. All activities measured at
low light intensity showed about the same rate of reaction. When
the light was increased to 1500 footcandles, H_2 evolution did not
increase in rate. The rates of O_2 comsumption via MeV and the reduction of NADP increased and are comparable with those of Curtis
et al.(7).

Light intensities were held at low levels (50-100 footcandles)
since in the absence of DCMU and despite the presence of alkaline
pyrogallol, higher intensities inhibited H_2 evolution in these algal systems. Gas evolution was inhibited 50% by the presence of
10μM DCMU.

The role of ferredoxin in the mechanism of H_2 photoevolution
was also studied with the use of purified ferredoxins from C.rein-
hardi and C. pyrenoidosa. A comparison of apparent Km values of
these ferredoxins for dark H_2 evolution from dithionite using a
hydrogenase from C. reinhardi indicated compatibility among algal
ferredoxins and hydrogenase from Chlamydomonas (data not shown).

CONCLUSIONS

Characterizations on the pathways of hydrogen evolution have
been done mainly on studies with intact cells. We have now utilized cell-free preparations of sonicated algae capable of light-
dependent H_2 evolution to evaluate photosystem dependency and the
requirement for soluble ferredoxin. Briefly sonicated, anaerobically adapted algae demonstrated high, reproducible, and stable
rates of H_2 photoevolution with exogenous electron donors. Our results indicated that longer periods of sonication caused a progressive destruction of the particles. It would appear that the best
donors would be those compounds which are poor electron acceptors.
Mediators of flow around PS I, eg PMS, apparently inhibited H_2
evolution by an electron drainage to a cyclic flow. An inhibitor
of ferredoxin, DSPD, inhibited H_2 evolution by preventing the flow
of electrons from PS I through ferredoxin to hydrogenase. The presence of the plastoquinone antagonist, DBMIB, resulted in an indication of the difference between sites of possible electron donation. The donation by NADH was abolished, indicating that the donation from NADH occurs at the site of, or before, plastoquinone.
Donation of electrons from DTT was only partially inhibited,

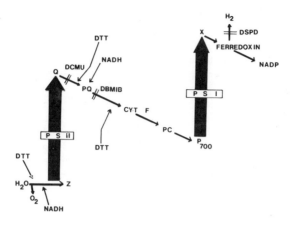

Figure 3. Suggested Mechanism for Hydrogen Photoevolution in Algae

indicating that donation from DTT can occur on either side of plastoquinone. The rate of H_2 evolution from NADH was not affected by
the presence of NADP, suggesting that flow of electrons through
flavoprotein from ferredoxin to NADP is not occuring to any extent
under conditions favoring H_2 evolution. The inhibition of gas evolution from NADH in the presence of NADPH might be due to a competition for the site of donation in the electron chain. The absence
of donation by NADH in spinach systems may reflect a lack of a catalyst necessary to transfer electrons from the nucleotide to a site-
specific point in the electron transfer chain. Dithiothreitol
probably donates non-specifically to the photosynthetic pathway in
both the algal and spinach systems.

It would appear that both photosystems are involved in the
flow of electrons for H_2 photoevolution. The inhibition by DCMU is
observed in both in vivo and in vitro systems. Whether electrons
were derived from water for donation to PS II remains open. In our
preparations, it is clear that electrons are donated to the oxidizing sides of PS II. If, in addition, water is split under our
conditions, the resulting O_2 evolution could inactivate the hydrogenase. At very low light intensities, the rate of O_2 evolution
may be so low that the rate of respiratory consumption prevents
inactivation by O_2. In this respect, the in vitro preparations
would mimic intact cells.

The requirement for soluble ferredoxin now appears to be established for both algal and spinach particles(5). Both the inhibition by DSPD and the requirement for ferredoxin when ferredoxin-
free hydrogenase is used support this contention.

Total anaerobiosis during particle preparation resulted in
higher rates and may reflect oxygen lability of components of the

electron transport chain. Sonication under air could lead to extensive oxidation of these components.

The effects of light intensity on H_2 evolution in broken particles reflect the observations in intact algae. The rate of electron flow to PS I appears to be the same under anaerobic or aerobic conditions. Therefore we assume that the rate-limiting event involves the transfer of electrons from ferredoxin to an acceptor. If the acceptor is NADP-reductase, the flow is relatively high; if the acceptor is hydrogenase, then the flow is restricted and rates of hydrogen evolution are correspondingly low.

The mechanism shown in Fig. 3 is an interpretation of currently available information. Photosystems are seen as photoevolving H_2 from reduced pyridine nucleotide by the introduction of electrons either into plastoquinone or into the oxidizing side of PSII. DCMU would block any PS II-dependent activity, and DBMIB would interfere with any plastoquinone-dependent activity. Hydrogen evolution is thought to involve the transfer of electrons through ferredoxin and hydrogenase to protons within PS I.

REFERENCES

1. Abeles, F.B. (1964) Plant Physiol. 39: 169-176.
2. Avron, M., and M. Gibbs. (1974) Plant Physiol. 53:136-139.
3. Ben-Amotz, A., D.L. Erbes, M.A. Riederer-Henderson, D.G.Peavey, and M. Gibbs. (1975) Plant Physiol. 56:72-77.
4. Benemann, J.R., J.A. Berenson, N.O. Kaplan, and M.D. Kamen. (1973) Proc. Nat. Acad. Sci. 70:3217-2320.
5. Bohme, H., S. Reimer, and A. Trebst. (1971) Z.Naturforsch. 26b: 341.
6. Brand, J.J., V.A. Curtis, R.K. Togasaki, and A. San Pietro. (1975) Plant Physiol. 55:187-191.
7. Curtis, V.A., J.N Siedow, and A. San Pietro. (1972) Arch. Biochem. Biophys. 94:838-902.
8. Gorman, D.S., and R.P. Levine. (1965) Proc.Nat.Acad.Sci.54:1665.
9. Moll, B. and R.P. Levine. (1970) Plant Physiol. 46:576
10. Trebst, A., and M. Burba. (1967) Z. Pflanzenphysiol. 57:419-433.

CHEMICAL REACTIONS FOR THE UTILIZATION

OF SOLAR ENERGY

GABRIEL STEIN

Casali Institute of Applied
Chemistry and Department of Physical Chemistry
Hebrew University, Jerusalem, Israel

In the present survey we shall discuss particularly
the developments in the last two or three years which led
to the recognition of chemical reactions for the utilization of solar energy. Many of these systems include
the possibility of fuel formation and storage. We shall
compare such chemical systems with the efficiency of
photovoltaic devices. These devices are already commercially available and are used profitably on a small scale
under special conditions to convert solar energy directly
to electric voltage and current. They do not provide
storage so that they provide electricity only when the
sun shines, unless an additional capital investment is
used to attach to them storage devices, e.g. lead acid
batteries. Both physical, photovoltaic devices in which
electrons are excited to conduction bands and photochemical systems are quantum devices in which the high
free energy of the solar photons are not degraded to
heat. Such quantum devices have to compete with some
major systems proposed for solar energy utilization -
heating and cooling, sea-based large power stations, etc.
Quantum devices which provide directly electricity
including fuel formation and energy storage have in principle a promising future. Hence the growing interest in
the photochemical utilization of solar energy.

PHOTOVOLTAIC DEVICES

Fig. 1 shows the principle of such a device and a
commercial device (Solar Power Corporation) which is now
being sold in increasing quantities. At present melt

335

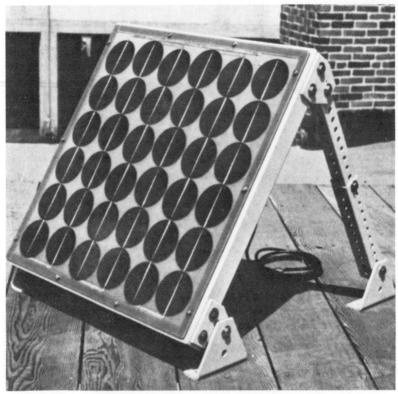

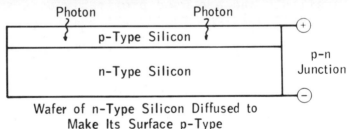

Wafer of n-Type Silicon Diffused to
Make Its Surface p-Type

Figure 1.

grown single crystals of lower than space quality are
cut to provide the thin wafers, which develop up to 0.5
volts in sunlight. Fully packaged long-lived installations cost at present $15-20 per peak watt, i.e. some
$80 per average watt, not including storage. Forecast
is that the cost of such devices may be halved within a
reasonably short time but several disappointments with
alternative techniques make it at present uncertain when,
if at all, decrease by a factor of 10 may be expected.

The cost of some \$80,000 per kilowatt installed has to
be compared with \$300 per kilowatt installed in fossil
fueled power stations and \$700 in nuclear power stations
at present. Of course no fuel requirements arise in
solar installations. Remarkably with this vast differential photovoltaic devices are commercially competitive in
small decentralized installations where fuel supply and
servicing would be the major cost factors.

HETEROGENEOUS PHOTOCHEMICAL DEVICES

In the following we shall divide photochemical
devices into two main groups according to their manufacturing technology. In one group heterogeneous light-
sensitive, semi-conductor surfaces, e.g. TiO_2 immersed
into solutions, usually aqueous. At the illuminated TiO_2
electrode oxygen is evolved, and a voltage is developed
vis-a-vis another, e.g., Pt electrode, at which hydrogen
evolves. Thus, the process leads to heterogeneously
sensitized photoelectrochemical water-splitting, and
includes as an inherent feature fuel formation and energy
storage. This is, of course, a very great step forward
towards the development of technologically competitive
solar energy devices. Some limiting factors are that
light absorption is proportional to the manufactured semi-
conductor surface, the relative expense of the manufacture
of the electrode systems and the fact that even using an
electrode such as titanium dioxide (where light absorption
occurs only for quanta of 3eV or greater) complete water
splitting does not occur except in systems in which additional chemical reactions, for example the reaction between an acid and a base, are also provided. Such devices
have been described among others by Honda (Fig. 2) (1)
and also the group at the Lincoln Laboratories. One of
the significant departures that has occurred very recently
is the realization that it might be desirable to switch
from such sensitized water-splitting mechanisms to other
photoelectrochemical processes in which substances other
than water and its components undergo the reversible
chemical changes with energy storage. Gerischer (2) has
discussed the possibility of using a cycle based on the
reversible redox reactions of sulphur on a cadmium sulfide
semiconductor electrode. Gerischer pointed out that the
chemical stability of the electrode is of crucial importance in such devices.

$$CdS + 2h^+ + solv. \rightarrow S + Cd^{2+} \cdot solv.$$
$$Cd^{2+} - S^{2-}_{surf} + h^+ \rightarrow Cd^{2+} - S^-_{surf}$$
$$Cd^{2+} - S^-_{surf} + Red \cdot solv \rightarrow Cd^{2+} - S^{2-}_{surf} + solv$$
$$Cd^{2+} - S^-_{surf} + h^+ \rightarrow Cd^{2+}_{surf} + S_{surf}$$

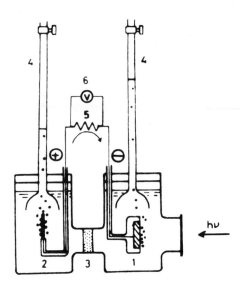

Figure 2. Schematic diagram of electrochemical photo-
cell. 1. TiO_2 electrode; 2. Pt electrode; 3. Diaphragm;
4. Gas burette; 5. Load resistance; 6. Voltmeter.

This problem has been attacked and considerable stability
achieved by Wrighton (3) and by Manassen and his group
(4) using the provision of polysulfides in the solution.
The polysulfides participate in fast electron transfer
processes or precipitate on the electrode surface thus
providing stability of this light-sensitive surface.
Manassen and coworkers also indicate how storage can be
obtained. Doubtlessly such electrochemical devices providing relatively long-range stability and storage of
energy provide a very considerable step forward in con-
ception. Their inherent major limiting factor is that
light absorption remains proportional to the absorbing
semiconductor electrode surface, thus putting a limit to
manufacturing costs. Doubtlessly such devices will be
among those which may reach a technologically worthwhile
device stage within the next few years, including the
possibility of light concentration to decrease the relative cost of the light-sensitive electrode.

SEMICONDUCTOR DEVICES BASED ON PHOTOSYNTHETIC
APPARATUS MODELS

The considerable efficiency of the photosynthetic
apparatus as evolved by nature created interest in the
possibility of constructing man-made devices which would

employ similar principles and lead to economically viable
solar energy utilization devices, incorporating if possible
storage. In these systems it is to be remarked that
nature usually does not split water into oxygen and hydrogen but evolves oxygen and stores the reducing equivalents
in the form of chemical fuel, rather than gaseous hydrogen.
This approach of partial water splitting in the sense of
not attempting to evolve two gases but only one gas and
producing one other chemical species which stores energy,
leads to the separation of the products and has some
thermodynamic advantages. The remarkable properties of
chlorophyll as the light harvesting agent have led several
groups, for example Albrecht at Cornell (5), Tomkiewicz
at IBM (6), and Katz at Argonne (Fig. 3) to develop manufactured devices in which the light accepting surface is
some heterogeneous system incorporating chlorophyll.

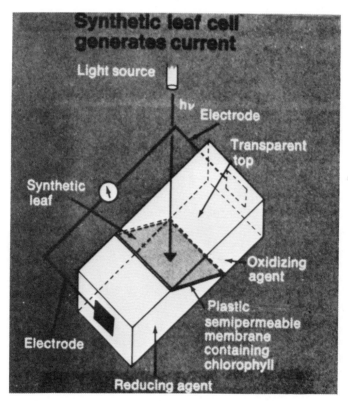

Figure 3.

Nature operating under the constraints of organic
evolution has evolved very complicated systems for locomotion, vision, flight, and solar energy utilization which
were constrained by the inability of organic nature to
achieve certain mechanical or chemical components and
used evolutionary time as the money to pay for the development of devices. By contrast when human technology
reached the appropriate stage it could develop devices,
such as the wheel, photomultipliers, jet flight and now
the challenge of technologically useful solar energy
conversion and storage devices which in the first three
cases are clearly based on principles different from and
not achievable by organic evolution. It is not at all
certain that following the path prescribed by organic
evolution will prove the technologically fruitful approach
for the best device for human technological utilization
of solar energy in chemical devices, although one can
probably learn very much from a study of such systems.
The manufacturing difficulties of complex photosynthetic
apparatus-like devices, in which light absorption is once
again proportional to a manufactured surface, does indicate
that this approach might be very suggestive and of very
great interest but not necessarily the best way to follow
with confidence. On the other hand improvements in
natural photosynthetic systems may be the key to most
useful biomass techniques of solar energy utilization.

HOMOGENEOUS CHEMICAL SYSTEMS FOR SOLAR ENERGY UTILIZATION

Compared to the previous surveyed devices in which
light absorption is proportional to a manufactured surface
it is tempting to consider whether homogeneous chemical
systems, in particular aqueous solutions might not be
relatively cheap and efficient devices for the chemical
utilization of solar energy and fuel formation for its
storage.

PHOTOGALVANIC DEVICES

In this wet device, compared with the dry device of
the photovoltaic silicon cell an aqueous solution of
solutes is illuminated so that the absorption of quanta
produces through electronic excitation labile chemical
intermediates which are energy-rich. The chemical
reaction of such intermediates on electrodes provides an
electrical potential, but no storage. Of the hitherto
used devices the ferric-thionine chemical cell has been
most fully investigated by Lichtin, Hoffman and coworkers
at Boston and by the research group at Exxon (7). Rapid
chemical back reaction in the bulk of the solution

decreases the available chemical energy very considerably
and necessitates the use of relatively large electrode
area to solute volume ratios. One development was the
introduction of semi-transparent NESA glass (SnO_2) electrodes which are selective with regard to one intermediate
and thus enable the cell to deliver an appreciable
voltage even in the absence of selective membrane components. Gomer (8) in a recent paper has analyzed,
following the original work of Rabinowitch some 30 years
ago, some of the inherent limitations of such systems
and its potential promise. One of the advantages of the
system is the absence of light absorbing manufactured
electrode surfaces. However without energy storage built
into the system it may be difficult to make it competitive
with photovoltaic devices as they are being developed.
With storage the photogalvanic type of solar quantum
utilization could by the heart of practically important
devices.

PHOTOSENSITIZED WATER-SPLITTING IN HOMOGENEOUS SOLUTIONS

    The classical review article of Marcus (9) surveyed
the aqueous systems in which photochemical reactions may
lead to sensitized decomposition of water into hydrogen
and oxygen. From this point of view water is a near
ideal solvent, its reactions being reversible to the
original starting material without loss. If inorganic
components e.g. transition metal ions are used as photosensitizing solutes, they are not used up and are not
attacked irreversibly in the course of photochemical and
chemical reactions involved. More recently such systems
have been reconsidered from the thermodynamic point of
view by Balzani and coworkers (10) and led to the conclusion that the quantum region within which such processes
can be carried out is limited to relatively short wavelengths thus making such cycles not too attractive from
the point of view of solar energy utilization. We have
recently (11) undertaken a full analysis of this problem
from the basic thermodynamic point of view introducing
correcting factors including consideration of entropy
change and the existence of a Franck-Condon strain factor,
which in the vertical transition required on the absorption of a quantum leads to a photochemical activation
energy. This may add an energy requirement of up to 30
kilocalories, i.e. up to about 1.3eV to the quantum
required to carry out the process. Therefore all one
quantum processes are relatively inefficient and expensive in energy input. This appears to be a thermodyna-
mically inherent property of all one-quantum processes,

in which necessarily one electron equivalent changes
produce free radical, H atom or OH radical intermediates,
by the breaking of covalent bonds in water. A departure,
which could improve the situation would be the utilization
of two electron equivalent processes, so that intermediate
free radicals are not liberated as such into the solution.
A system in which such steps may in fact occur may be
the ruthenium tris-bipyridyl system in which Sutin's
work (12) indicates a possibility of such a reaction.
It may be pointed out that the relatively lower energy
requirements in photoelectrolysis in photoelectrochemical
reactions to decompose water are connected with the fact
that intermediate free radicals are not necessarily
released into the bulk and that radical recombination
and utilization of bond energy may occur on the heterogeneous surface. Such two-electron equivalent processes
should be actively searched for, in homogeneous solutions.
In the case of homogeneous systems aiming at fuel formation through products other than the components of water,
H_2 and O_2, has led to the recent development of alternative systems. One practical example which is already
practically developed is the ferric-bromide system (13)
in which the complex between ferric iron and bromide is
the light absorbing species in aqueous solution and the
products which store electrochemical energy are free
bromine, which is driven out of the solution while ferrous
iron accumulates and can be separated.

$$Fe^{3+} + Br^- \rightleftharpoons FeBr^{2+}$$

$$FeBr^{2+} \xrightarrow[\rightarrow 600nm]{h\nu} (Fe^{2+} + Br^-) \rightarrow Fe^{2+} + Br$$

$$Br + Br^- \rightarrow Br_2^-$$

$$2Br_2^- \rightarrow Br_2 + 2Br^-$$

$$Br_2^- + FeBr^{2+} \rightarrow Br_3^- + Fe^{2+}$$

It is possible that such a concept, which just as the
sulfur cycle in the heterogeneous cycles, introduces into
homogeneous aqueous solution devices the possibility of
producing fuels other than hydrogen, may prove profitable
in the future.

THERMALLY ASSISTED PHOTOCHEMICAL REACTIONS AND PHOTOCHEMICALLY ASSISTED THERMAL REACTIONS

Many of the chemical devices discussed previously
utilize only part of the solar spectrum and leave the
longer wavelengths, containing a considerable part of

the solar energy spectrum unutilized. In photovoltaic
devices the heating effect of this part of the spectrum
is undesirable and the efficiency of for example silicon
cells decreases with increasing temperatures. Not so
for some of the chemical devices. It is of interest to
introduce the concept of integrated systems in which the
photochemical reaction is promoted and assisted by temperature rise obtained by the utilization of the remaining
part of the solar spectrum. This is for example the case
for the ferric bromide reaction where the enthalpy of the
complex formation, as well as the separation of dissolved
bromine both benefit from an increase in temperature. In
this way the overall engineering efficiency of the utilization of the total solar spectrum is somewhat improved.
Experiments are now in progress to elucidate the exact
contribution of the thermal promotion of such essentially
photochemical processes. In connection with this we may
also consider photochemically assisted thermal processes.
The interesting problem of the thermal catalytic splitting
of water, utilizing the waste heat of nuclear reactors,
reached for the time being a limit owing in part to the
fact that in the entire thermal cycle there was often one
step, and one only, which required relatively high temperatures (for example of the order of 700 to 800°C) while
the rest of the steps did proceed at much lower temperatures, say 300 to 400°C. The necessity for higher temperatures in one step made the entire cycle economically
not yet viable. Recently (14) an analysis of a thermal
cycle in which one step was photochemically assisted was
carried out. In this analysis the authors reach unfavor-
able economic judgment concerning this particular process.
Nevertheless it appears worthwhile to think in the future
of the possible utilization of thermally-assisted photochemical conversion devices and the photochemical assisted
thermal devices according to the relative importance in
role of the two components in the useful chemical cycle.

SOME GUIDELINES

The developments of the last few years indicate some
fairly obvious new criteria. Homogeneous devices, if
achievable, including electricity generation and fuel
formation could inherently be technologically superior
to heterogeneous manufactured devices. Full water-
splitting into oxygen and hydrogen is probably less
attractive than the evolution of one gas, O_2 or H_2, and
storage of another accumulated chemical species. Cycles
based on water as a solvent (which is very stable) but
providing electricity and fuel formation through cycles

not involving the elements of water (exactly because of
their great stability) may be attractive as indicated by
the recent developments of the sulfur cycle in heterogeneous systems and bromo cycle in homogeneous. These
new considerations may lead in the near future to better
systems.

Silicon photovoltaic devices have reached 15 per
cent overall engineering efficiency in providing electrical power (without storage). Both the sulfur cycle and
the bromine cycle give preliminary results giving overall
efficiencies of the order of one per cent or possibly
better. Should these results be substantiated and improved
photochemical devices may approach competitiveness with
photovoltaic devices.

Photochemical electricity generation may be most
advantageous for relatively small (1-5 MW) decentralized
stations which may be economically and socially systems
of choice in some parts of the world. Were the cost per
installed kilowatt of the order of $3000 a large market
would open and considerable demand would arise at twice
this figure in situations where the specific advantages
of such installations would make them genuinely competitive with very large centralized, fuel demanding generating stations.

References

1. A. Fujishima, K. Kohayakawa and K. Honda, J. Electro-
    chem. Soc. 122, 1487 (1975); J.G. Mavroides, D.I.
    Tchernev, J.A. Kafalas and D.F. Kolesar, Maler. Res.
    Bull. 10, 1023 (1975); D.I. Tchernev in ref. 7.
2. H. Gerischer, Electroanal. Chem. 58, 263 (1975).
3. A.B. Ellis, S.W. Kaiser and M.S. Wrighton, J. Am.
    Chem. Soc. 98, 1635 (1976).
4. G. Hodes, J. Manassen and D. Cahen, Nature, 261, 403
    (1976).
5. C.W. Tang and A.C. Albrecht, Nature, 254, 507 (1975).
6. M. Tomkiewicz and G. Corker, Proc. Third Mt. Congress
    Photosynth. Ed. M. Avron, Elsevier, Amsterdam, 1974,
    p. 265 ff.
7. N.N. Lichtin, ed. The Current State of Knowledge of
    Photochemical Formation of Fuel, NSF-Washington, 1974.
8. R. Gomer, Electrochim. Acta, 20, 13 (1975).
9. R.J. Marcus, Science, 123, 399 (1956).

10. V. Balzani, L. Maggi, M.P. Manfrin, F. Bolletta and M. Gleria, Science, 189, 852 (1975).
11. G. Stein and A. Zeichner, Casali Inst. Report, 1976, and to be published; G. Stein in ref. 7.
12. C. Creutz and N. Sutin, Proc. Nat. Acad. Sci. USA, 72, 2858 (1975).
13. S.N. Chen, N.N. Lichtin and G. Stein, Science, 190, 879 (1975), G. Stein, Israel J. Chem. 14, 213 (1975).
14. A. Schlatter, E. Plattner and Ph. Javet. Energy Conv. 14, 43 (1975). cf. also T. Ohta et al. IECEC Record, 1975 p. 772.

PHOTOSYNTHESIS - A PRACTICAL ENERGY SOURCE?

D.O. HALL

Department of Plant Sciences, University of London King's

College, 68 Half Moon Lane, LONDON SE24 9JF

SUMMARY

The process of photosynthesis supplies us with practically all our food, fuel and fibre - derived directly from present day photosynthesis or indirectly from fossil fuels. A better understanding of its mechanism should enable us to realise its maximum potential in the future. Each year about 3×10^{21} J of energy is fixed as carbon (2×10^{11} tonnes) by photosynthesis representing <u>stored</u> solar energy. This is ten times the world's 1970 energy consumption. The scope for increasing the total utilization and for using photosynthesis in other natural and artificial ways is enormous - if we can increase energy output:input ratios by improved productivity, changed crop types, efficient harvesting, decreased post-harvest deterioration, and so on. Photosynthesis <u>in</u> <u>vivo</u> and <u>in</u> <u>vitro</u> will be discussed from the following points of view: efficiency, food vs. fuel, leaf protein, energy plantation, cellulose and waste disposal, greenhouse production, plant breeding, nitrogen fixation, regulation of metabolism and products, hydrogen production, carbon reduction, chlorophyll membranes, and bacteriorhodopsin.

INTRODUCTION

Photosynthesis is the conversion of solar energy into fixed energy: $CO_2 + H_2O \longrightarrow$ organic material $+ O_2$. The products of photosynthesis represent stored energy. Photosynthetic conversion efficiencies of 0.5% to 3% thus represent the efficiency of the total process; sunlight $\longrightarrow$ fixed chemical energy. By contrast, for example, photovoltaic conversion efficiencies of 12-15% represent the process: sunlight $\longrightarrow$ electric power, without

347

including any energy storage.

Only fifty years or so ago CO_2 fixed in photosynthesis would
have been used as food, fuel and fibre. However, now with
abundant oil the products of present day photosynthesis are mainly
used as food. We should re-examine and, if possible, re-employ
the previous systems; but, with today's increased population and
standard of living we cannot revert to old technology, but must
develop new means of utilising present day photosynthesis more
efficiently.

ENERGY AVAILABLE

The scope for increasing the total utilisation and for using
photosynthesis in other ways is enormous (Figure 1) – if we can
improve yields, change crop types, harvest efficiently, decrease
post-harvest deterioration, and so on.

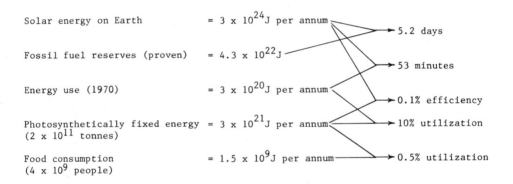

Solar energy on Earth	$= 3 \times 10^{24}$ J per annum
Fossil fuel reserves (proven)	$= 4.3 \times 10^{22}$ J
Energy use (1970)	$= 3 \times 10^{20}$ J per annum
Photosynthetically fixed energy (2×10^{11} tonnes)	$= 3 \times 10^{21}$ J per annum
Food consumption (4×10^9 people)	$= 1.5 \times 10^9$ J per annum

5.2 days

53 minutes

0.1% efficiency

10% utilization

0.5% utilization

Figure 1.

EFFICIENCY OF PHOTOSYNTHESIS

Plants use radiation between 400 and 700 nm, the so-called
photosynthetically-active radiation (P.A.R.). This P.A.R.
comprises about 50% of the total sunlight which on the earth's
surface has an intensity of about 800–1000 W/m^2 (5–6 J/cm^2/min;
also equivalent to 10^{-2} cal/cm^2/sec or 42×10^4 ergs/cm^2/sec for
P.A.R.). The overall practical maximum efficiency of photosynthetic energy conversion is 5–6%.

TABLE 1

	Available light energy
At sea level	100%
50% loss as a result of 400–700 nm light being the photosynthetically useable wavelengths	50%
20% loss due to reflection, absorption and transmission by leaves	40%
77% loss representing quantum efficiency requirements for CO_2 fixation in 680 nm light (assuming 10 quanta/CO_2) and that the energy content of 575 nm red light is the radiation peak of visible light	9.2%
40% loss due to respiration	5.5%
	Overall PS efficiency

If the minimum quantum requirement is 8 quanta/CO_2, then this loss factor becomes 72% (instead of 77%) giving a final photosynthetic efficiency of 6.7% (instead of 5.5%).

Under optimum field conditions values between 3% and 5% conversion are achieved by plants: however, often these values are for short-term growth periods and when averaged over the whole year fall to between 1% and 3%. In practice, photosynthetic conversion efficiencies in temperate areas are typically between 0.5% and 1.3% of the total radiation when averaged over the whole year, while values for sub-tropical crops are between 0.5% and 2.5%. Fig. 2 shows the yields which can be expected under various sunlight intensities at different photosynthetic efficiencies

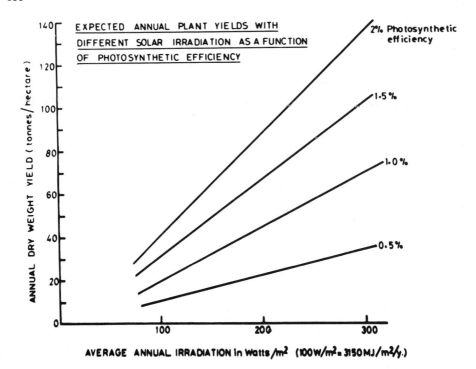

Figure 2.

AREAS REQUIRED FOR SOLAR ENERGY

There are problems in collecting solar energy, the most obvious of which is its diffuse nature and the fact that it is intermittent; therefore any solar energy system has to have a storable component. If a 10% solar conversion efficiency was achieved (solar cells vary between 12% and 15% efficiency already) the land areas required in various countries to provide total energy requirements can be calculated (Table 2). It is not implied that any country will ever achieve a complete solar energy economy, however, but it shows the magnitude of the land areas involved. Net energy output of any system is essential; so-called solar breeder systems might accomplish this tantalizing target.

TABLE 2

Land area (approx. % of total) required to provide total
energy requirements (1970) from solar energy at a 10%
conversion efficiency

Australia	0.03%
South Africa	$\frac{1}{4}$%
Norway	$\frac{1}{2}$%
Sweden	$\frac{3}{4}$%
Eire	1%
Spain	1%
U.S.A.	$1\frac{1}{2}$%
Israel	$2\frac{1}{2}$%
France	$3\frac{1}{2}$%
Italy	4%
Denmark	$4\frac{1}{2}$%
U.K.	8%
W. Germany	8%
Netherlands	15%

FOOD VERSUS FUEL

The good agricultural efficiency achieved over the last thirty
or more years has primarily been through the greater use of fossil
fuel, e.g. the use of fertilisers such as nitrate, and mechanical
operation systems. The following values show the comparative
figures for the food energy output per unit of energy input: low
intensity agriculture 20; intensive field crops 2; livestock
production 0.2; greenhouse production 0.02. It has been calculated
that for every calorie of food that we eat at our table it has
taken 5 calories of energy to get in onto our plates – this is with
a western standard of living.

Calculations in the United States on energy output:input ratios
in the production of maize grain have shown that this ratio has
fallen from 3.7 in 1945 to 2.8 in 1970; that is a doubling of
of yield has been achieved by a trebling of energy input.

The aim is to maximise energy output:input ratios. At the farm
level we must be sure that we are not just converting oil into food
without any net gain in energy, since it is the process of solar
radiation via photosynthesis which increases energy output. Many
of the agricultural systems which have been considered unprofitable
in the past may now become more profitable due to the combined
increased costs of food and fuel.

LEAF PROTEIN

Leaves are potentially a large source of protein.
Traditionally they are composted, discarded as waste, or fed to
animals for conversion to meat, which is a very inefficient
process. Techniques have been developed for the extraction of
leaf protein which yield at the same time other useful products,
namely fibre and soluble components such as carbohydrates,
nitrogen and inorganic nutrient compounds. The composition of
leaf protein is about 60-70% true protein, 20-30% lipid, 5-10%
starch. Yields of two tonnes of dry extracted leaf protein per
hectare have been obtained without irrigation and three tonnes can
be expected. In the south-west US it has been proposed that if
the yields of alfalfa (grown in an enclosed environment) reached
that of sugar cane at 100 tonnes dry weight/hectare/year, about
25 tonnes of protein could be extracted per hectare from alfalfa.

ENERGY PLANTATIONS

This implies the growing of plant materials for their fuel
value, and is the only known operation that offers a renewable
source of liquid fuel and organic chemicals. Energy plantations
may be considered as a long-term alternative to fossil and nuclear
energy and fossil-derived chemicals providing us with the energy
options we may require in the next century. They have been subject
to feasibility studies in the US and Australia and the following
advantages have been identified: (a) capable of storing energy for
use at will; (b) renewable; (c) dependent on technology already
available, with minimal capital input; (d) can be developed with
our present manpower and material resources; (e) reasonably priced;
(f) ecologically inoffensive and free of hazards other than fire
risk.

Traditionally we think of energy plantations as forests, but
increasingly we should consider alternatives, such as shrubs,
weeds, agricultural crops, grasses, and algae (fresh-water and
marine); for example, in Australia five species have been selected,
namely Eucalyptus, Cassava, Hibiscus, Napier Grass and Sugar Cane
as being potentially the most desirable high-yielding crops which
can be harvested over the whole year. Recent calculations show
that alcohol produced from cassava (starch-rich) is an economically
viable system but that if processing to destroy cell walls is
required, the costs become too high. In the US one group has
opted for fast-growing deciduous trees which re-sprout from

stumps when cut (hybrid poplars).

CELLULOSE

This is probably the most abundant single organic compound on Earth (about 10^{11} tonnes are produced annually). It could be exploited as a source of energy, or food, or as a source of chemicals in the chemical industry. Technology for converting cellulose to glucose is now well advanced. This may be done with acid or alkaline treatment in order to break down the cellulose, but significant advances have been made in the utilisation of enzymes, or enzyme extracts. Costly milling processes need to be avoided.

WASTE DISPOSAL

Many of the liquid and semi-solid wastes from our houses, industries and farms are ideal for the growth of photosynthetic algae. Under good conditions rapid growth with about 3-5% solar conversion efficiency can be obtained. The harvested algae may be fed directly to animals, burnt to produce electricity, or fermented to produce methane. Simultaneously, waste can be disposed of and water purified; it is estimated that such algal systems are half to three-quarters as expensive as conventional waste disposal systems in California.

GREENHOUSE PRODUCTION

This is often considered a very uneconomic process, except for specialist crops. However, utilising cheap and efficient greenhouse structures (the majority of greenhouses now used are very inefficient) may become feasible for a much larger and widespread production of ordinary crops. In the south-west US where photosynthesis efficiencies of 3% have been obtained with sugar cane outdoors, it is calculated that these efficiencies could easily be doubled under greenhouse conditions, utilising cheap plastic structures and hydroponic-type fertilisers and water facilities. Such systems could be increasingly used in countries who wish to become self-sufficient agriculturally. There is also a large potential in temperate countries for green-house-type production which fulfils the requirements of maximum energy output for minimum energy input.

PLANT SELECTION AND BREEDING

In order to obtain the maximum energy output from plants in a given area, photosynthesis needs to be optimized. Considering all or individually those factors which limit production, plants could be developed or selected that will give integrated maximum yields of food, fuel and fibre over the whole year. Utilising our knowledge of C_4 characteristics of plants and of photo-respiration <u>may</u> allow the breeding and selection of efficient

photosynthetic plants. Chollet and Ogren put it strongly: "The
control of this process (photorespiration) and the associated
oxygen inhibition of photosynthesis has emerged as representing
one of the most promising avenues for dramatically increasing the
world supply of food and fibre". Genetic engineering using plant
cell tissue cultures is a recently developed technique which has
great promise for improving plants. Species and genus crossings
and creation of new hybrids by mutations induced in the cultures
are possible.

REGULATION OF PLANT REACTIONS AND PRODUCTS

Usually crops are grown for one final product, such as grain
or root, containing constant proportions of carbohydrate, protein
and fats. The possibility that we could alter biochemical
reactions at defined times during the growing season in order to
obtain more or less of a given constituent. It is also possible
that we could regulate detrimental processes in the plant, such as
photorespiration and water losses, giving greater net plant yields.
There is some work on algae but whole plant systems should be
investigated.

NITROGEN FIXATION

It is thought that one of the major limitations of N_2 fixing
capability in both symbiotic and associative symbiotic systems is
an inadequate supply of carbohydrate to the N_2 fixing bacteria:
more efficient photosynthesis could overcome some of these
problems. The very interesting discovery of associative symbiotic
N_2 fixation in grasses such as maize (and possibly wheat and rice)
has led to the realisation that improved carbohydrate production
may be the prerequisite for useful extension of biological N_2
fixation to other crops.

When soya beans are grown under greenhouse conditions, a 3-
fold increase in CO_2 concentration in the atmosphere resulted in
a 6-fold increase in the amount of N_2 fixed/ha: yields increased
from 75 to 425 kg of fixed N_2/ha. CO_2-enriched soya beans fixed
85% of their N_2 requirement, whereas the unenriched plants fixed
only 25% obtaining the rest from the soil in the form of nitrate
fertilisers. It is possible that this increased net production
of photosynthesis,and hence N_2 fixation, is made possible by a
decrease in photorespiration due to the increased CO_2 concentration
in the atmosphere.

BIOCATALYTIC HYDROGEN PRODUCTION SYSTEMS

There is a direct interest in both living and non-living
systems which emulate the biological production of H_2 gas via the
breakdown of water - analogous to the electrolysis of water.
Certain algae produce H_2 gas under specific conditions and contain
the enzyme hydrogenase. Thus with a hydrogenase any plant type

system could have the ability to produce H_2 gas. This has been demonstrated in the laboratory using components extracted from leaves and bacteria. The ultimate object should probably be to use a completely synthetic system mimicking the algal or plant-bacterial systems. In this case an Fe-S catalyst would be used instead of a hydrogenase, a chlorophyll layer membrane or vesicle instead of the chloroplast, and a manganese catalyst to evolve the O_2 from H_2O.

CARBON REDUCTION

In vitro systems which emulate the plant's ability to reduce CO_2 to the level of carbohydrate are a very attractive proposition and are being actively investigated by biochemists and synthetic chemists. A recent report claims the formation from CO_2 of keto-acids (and then amino acids) using an alkyl-mercaptan, an Fe-S protein analogue, and an inorganic reducant.

ARTIFICIAL CHLOROPHYLL MEMBRANES

The chlorophyll-containing membranes of all photosynthetic organisms are able to separate positive and negative charges on either side of the membrane under the influence of light. This basic photogalvanic system is a key to photosynthesis which we might be able to use directly for the production of electricity or the storage of energy. Artificial chlorophyll-containing membrane bilayers and vesicles have been used and shown to produce currents and charge separation. The possibility of utilising such artificial membranes or vesicles for direct photo-chemical systems has scope, even though the efficiencies so far achieved are low.

BACTERIORHODOPSIN MEMBRANES

Very stable "purple membranes" have been isolated from the bacterium Halobium which grows naturally under very high salt concentrations and in hot, sunny areas. The isolated membrane can withstand 6N HCl, high temperatures and prolonged exposure to the atmosphere. These purple membranes "function as proton pumps in the bacterium"; this capability has been proposed as a potentially useful means of converting solar energy. Laboratory systems have been constructed which can: (a) produce photopotentials of 200 mV or more across a membrane (b) produce pH gradients which may well result in the production of H_2 and O_2 in separate compartments (c) act as desalting devices with Na^+ and K^+ exchanging H^+ (d) produce ATP is an ATPase enzyme is incorporated into the membrane.

CONCLUDING REMARKS

Photosynthesis is a key process in the living world and will continue to be so for the continuation of life as we know it. The development of photobiological energy conversion systems has

long term implications from both energy and food points of view.
Their applicability might be immediate in some tropical areas and
countries with large amounts of sunshine. However, in more
temperate climates there is still a large potential for the
utilization of the ever-abundant solar energy - even recognising
land use constraints resulting from high population densities and
intensive agriculture. For example, Europe should not consider
that it does not have sufficient solar energy - the difference in
total annual solar radiation between the UK (105 W/m^2 continuous)
and Australia (200 W/m^2) or the US (185 W/m^2) is only a factor of
2. The difference between the UK and the Red Sea area (the area
with the most amount of solar energy in the world - 300 W/m^2) is
only a factor of 3. Whatever solar energy systems are developed,
these could provide viable alternatives to other types of energy
production in the next century. Whatever systems are devised in
the temperate zones could be applicable to those countries that
have more sunshine,and these are predominantly the developing
countries of the world. Thus the temperate countries could help
themselves by becoming self-sufficient and help the other countries
of the world by not competing for their food and raw material.
Lastly, we might have an alternative way of providing ourselves
with food and fuel in the next century and we should consider all
our energy options and not put all our money and effort into only
one or two energy systems as we have in the past.

REFERENCES

1.   "Agricultural and biological systems" (1976) Chapter 9 of
       "Solar Energy: a U.K. assessment". UKISES, 21 Albemarle
       Street, London W1.
2.   J.A. Alich and R.E. Inman (1976) "Energy from agriculture",
       Energy 1, 53-61.
3.   J.A. Bassham (1971) "The control of photosynthetic carbon
       metabolism", Science 172, 526-534.
4.   J.A. Bassham (1976) "Mechanism and efficiency of photosynthesis
       in green plants" Symposium papers. Inst. Gas Technology,
       3424 South State St., Chicago, Ill. 60616,U.S.A.
5.   J. Berry (1975) "Adaptation of photosynthetic processes to
       stress". Science 188, 644-650.
6.   "Biological methods of conversion". Chapter 6 of the "Solar
       Energy research in Australia". Australian Academy of
       Sciences, Canberra. Report No. 17 (1973).
7.   N.K. Boardman and A.W.D. Larkum (1975) "Biological conversion
       of solar energy" Chapter 3 of "Solar Energy" H. Messel
       and S.T. Butler eds. Shakespeare Head Press, Sydney.
8.   A.W.A. Brown et al (1976) "Crop productivity - research
       imperatives" Michigan-Kettering Conference, Mich. Agric.
       Expt. Stn. East Lansing, Mich. 48844 U.S.A.

9. J.F.W. von Bulow and J. Dobereiner (1975) "Potential for nitrogen fixation in maize genotypes in Brazil". Proc. Natl. Acad. Sci. U.S.A., 72, 2389-2393.

10. M. Calvin (1974) "Solar energy by photosynthesis" Science 184, 375-381: (1976) "Photosynthesis as a resource for energy and materials". Photochem. Photobiol. 23, 425-444.

11. P.S. Carlson and J.C. Polacco (1975) "Plant cell cultures: genetic aspects of plant improvement". Science 188, 622-625.

12. W.J. Chancellor and J.R. Gross (1976) "Balancing energy and food production, 1975-2000". Science 192, 213-218.

13. R. Chollet and W.L. Ogren (1975) "Regulation of photo-respiration in C_3 and C_4 species". Bot. Rev. 41, 137-179.

14. R.K. Clayton (1976) "Photosynthetic reaction centers: photo-chemical mechanism and potential utilisation". Am. Chem. Soc. Centennial Meeting, New York, April 1976. Abstr. INOR 102.

15. "Clean fuel from biomass, sewage, urban refuse and agric. wastes" (1976) Symposium papers. Inst. Gas Technology, 3424 South State St., Chicago, Ill. 60616, U.S.A.

16. J.P. Cooper, ed. (1975) "Photosynthesis and productivity in different environments" Cambridge University Press.

17. J. Coombs (1975) "Total utilization of the sugarcane crop" Proc. Conf. C-7, UKISES, 21 Albemarle St. London W1 pp 25-27.

18. R.O.D. Dixon (1976) "Hydrogenase and efficiency of nitrogen fixation in aerobes" Nature 262, 173: see also: K.R. Schubert and H.J. Evans (1976) "Hydrogen evolution, a major factor affecting the efficiency of nitrogen fixation in nodulated symbionts" Proc. Natl. Acad. Sci. U.S.A. 73, 12071211.

19. K-E Eriksson (1976) "Enzyme mechanisms involved in fungal degradation of wood components" Am. Chem. Soc. Centennial Meeting, New York, April 1976. Abst. CELL 074.

20. I.S. Goldstein (1975) "Potential for converting wood into plastics" Science 189, 847-852.

21. J. Gorman (1975) "A source of self-sufficiency" The Sciences, October 1975, 25-29.

22. C. Gudin (1976) "Method of growing plant cells" U.S. Patent No. 3,955, 317: May 11, 1976.

23. F. Hakahashi and R. Kikuchi (1976) "Photoelectrolysis using chlorophyll electrodes" Biochim. Biophys. Acta 430, 490-500.

24. D.O. Hall (1976) "Photobiological energy conversion" FEBS Letters, 64, 6-16.

25. R.W.F. Hardy and U.D. Havelka (1975) "Nitrogen fixation research: a key to world's food?" Science 188, 633-643.

26. G.H. Heichel (1976) "Agricultural production and energy

resources" Am. Scientist <u>64</u>, 64-72.

27. R.H. Holm (1975) "Iron-sulphur clusters in natural and
       synthetic systems" Endeavour <u>34</u>(121) 38-43.

28. M. Kitajima and W.L. Butler (1976) "Microencapsulation of
       chloroplast particles" Pl. Physiol. <u>57</u>, 746-750 see also:
       M. Mangel (1976) "Properties of Liposomes that contain
       chloroplast pigments" Biochem. Biophys. Acta <u>430</u>, 459-466.

29. G. Leach (1975) "Energy and food production" Intl. Inst.
       Environ. Devel.,27 Mortimer St. London W1.

30. S.P. Long and H.W. Woolhouse (1975) "C_4 photosynthesis in
       plants from cool temperate regions with particular
       reference to <u>Spartina Townsendii</u>" Nature, <u>257</u>, 622-624.

31. D.J. McCann and H.D.W. Saddler (1976) "Photobiological energy
       conversion in Australia" Search <u>7</u>, 17-23.

32. T. Nakajima, Y. Yobushita and I. Tobushi (1975) "Amino-acid
       synthesis through biogenetic-type CO_2 fixation. Nature
       <u>256</u>, 60-61.

33. D. Oesterhelt (1976) "Bacteriorhodopsin as an example of a
       light-driven proton pump" Angew. Chemie Intl. Edn.
       <u>15</u>, 17-24; see also: R. Lewin (1976) "Exotic bacterium
       unravels energy problems" New Scientist, April 1, 1976
       pp. 28-29; W. Stoeckenius (1976) "The purple membrane of
       salt-loving bacteria" Sci. Amer. <u>234</u>(6), 38-47.

34. W.J. Oswald (1976) "Gas production from micro-algae" Symposium
       papers. Inst. Gas Technology. 3424 South State St. Chicago,
       Ill 60616, U.S.A. pp 311-324: see also: (1974) "Productivity
       of algae in sewage disposal" Solar Energy <u>15</u>, 107-117.

35. D. Pimentel et al (1973) "Food production and the energy crisis".
       Science <u>182</u>, 443-449; (1975) Science <u>190</u>, 754-761.

36. N.W. Pirie (1975) "Leaf protein: a beneficiary of tribulation"
       Nature <u>253</u>, 239-241.

37. G. Porter and M.D. Archer (1976) "In vitro photosynthesis"
       Interdisc. Sci. Rev. <u>1</u>, 119-143.

38. R. Revelle (1976) "Energy use in rural India" Science, <u>192</u>,
       969-975.

39. K.Y. Sarkanen (1976) "Renewable resources for the production of
       fuels and chemicals" Science <u>191</u>, 773-776.

40. M. Slesser (1973) "Energy subsidy as a criterion in food policy
       planning" J. Sci. Food Agric. <u>24</u>, 1193-1207. M. Slesser
       and I. Houman (1976) Nature <u>262</u>, 244-245.

41. L.A. Spano (1976) "Enzymatic hydrolysis of cellulosic wastes
       to fermentable sugars for alcohol production". see Bassham
       ref. pp 325-348.

42. H.A. Wilcox (1975) "The ocean food and energy farm project".
       Intl. Conf. Marine Technology Assessment - Man and the
       Oceans, Monaco. Conference Proc. (in press).

43. S.W. Wittwer (1974) "Maximum production capacity of food crops"
       Bioscience <u>24</u>, 216-224: (1975) "Food production:

technology and the resource base" Science <u>188</u>, 579-584.
44. I. Zelitch (1975) "Improving the efficiency of photosynthesis
Science <u>188</u>, 626-633.

PHOTOPRODUCTION OF HYDROGEN IN PHOTOSYNTHETIC SYSTEMS

A.A. Krasnovsky

A.N. Bakh Institute of Biochemistry of the

USSR Academy of Sciences, Moscow, U.S.S.R.

A review is presented on the experiments performed in our laboratory on the photochemical hydrogen
evolution by algal cells and in model systems.

PHOTOEVOLUTION OF HYDROGEN BY GREEN ALGAE

In 1942 Gaffron and Rubin discovered that illumination of unicellular algae in anaerobic conditions
may lead to the formation of gaseous hydrogen (I, see
review 2). In 1949 Gest and Kamen (3) found that photosynthesizing bacteria were capable of photochemical
evolution of hydrogen. In this case, however, hydrogen
is donated not by water but by various organic and
inorganic substances used in bacterial metabolism. The
extensive literature dealing with the production of
hydrogen by bacteria and algae has been surveyed in
recent reviews by Kondratieva and Gogotov (4) and
Oshchepkov and Krasnovsky (5).
The evolution of hydrogen by unicellular algae
has been studied in our laboratory (6). For this purpose, a device was constructed on the basis of a gas
chromatograph and a monochromator (7): a cuvette with
a magnetic stirrer containing I ml of Chlorella suspension was connected to the gas chromatograph. Illumination of the algal suspension in the air resulted
in oxygen evolution; after bubbling through a flow of
argon, hydrogen evolution occured without any adaptation period.
In accord to Gaffron, addition of glucose or

other exogenous hydrogen donors sharply increased the
production of hydrogen; carbon dioxide was released
simultaneously. Oxygen and hydrogen were evolved alternatively: under steady-state conditions no simul-
taneous stoichiometric release of oxygen and hydrogen
was usually observed. However, in the recent study by
Efimtzev, Boichenko and Litvin (8) who used a sensitive amperometric method of hydrogen recording, simul-
taneous evolution of hydrogen and oxygen was recorded
in many photosynthesizing organisms during the induction period. Evolution of hydrogen by Chlorella was
measured as a function of wavelength of incident light
(6). The action spectrum of hydrogen evolved proved
to be close to that of oxygen evolution; some differences were noticed in the far-red region where hydrogen was evolved more effectively. These measurements
have shown that no more than 5 quanta of red light are
required to release one mole of hydrogen upon illumination of a Chlorella suspension. To evolve one mole
of oxygen during photosynthesis, 8-IO quanta are required, so the data given above present an impression
that the photoevolution of hydrogen is more effective
from the energetic point of view. However, the results
of the above measurements should not be overestimated,
since hydrogen is evolved as a result of photometabolism of the organic substances stored in the cell during ordinary photosynthesis. It is well known that
this process may occur in general by the way of enzymatic reactions without any use of light energy. So,
a "dark" evolution of hydrogen is observable in Chlorella and, especially, in blue-green algae (9). In the
case of latter organisms heterocysts can probably
function by releasing hydrogen while normal cells can
effect photosynthesis by giving off oxygen (IO).

It is significant that oxygen inhibits the photoevolution of hydrogen either by interacting with reduced products or by inhibiting the enzyme hydrogenase.
The mechanism of hydrogen photoevolution could be visualized in the simplest way according to a scheme in
which oxygen is given off at one "end" of the photosynthetic electron transport chain while hydrogen is
released at the other "end". Such a simple scheme,
however, is not consistent with experimental evidence.

That the hydrogen evolution is invariably accompanied with a release of carbon dioxide was already
demonstrated in Gaffron's experiments. In the case of
mutants deficient in photosystem II (II) hydrogen was
effectively evolved. Thus, no rigid connection exists
between photolysis of water (in photosystem II) and
evolution of hydrogen. This is consistent with the

action of diuron which in a concentration of 10^{-6} M
suppresses the release of oxygen without affecting
that of hydrogen. It follows that the photoproduction
of hydrogen is closely linked with the carbon metabolism of the cell (I,2), that was confirmed recently in
M. Gibbs laboratory. Hydrogen is evolved by way of a
number of enzymatic and photochemical intermediate
reactions with participation of reduced compounds formed during the operation of carbon cycles of photosynthesis and respiration. However, in the overall process, the evolved molecular hydrogen is derived from
water molecules since no other hydrogen source is present in the system. It is most likely that in the
course of carbon cycles of photosynthesis and respiration active hydrogen donors are formed which enter
photosystem I of the electron transport chain. Reduced
pyridine nucleotides are formed in the Krebs cycle.
These compounds may enter a locus of photosystem I
where chlorophyll sensitizes electron transport to
ferredoxin. That such a mechanism is possible is indicated by model experiments described below.

The question arises as to whether photolysis of
water can occur without carbon cycles participation.
To achieve this, it would be necessary to carry out
the reactions in isolated structures such as chloroplast lamella where the photosynthetic electron transport chain is localized.

CHLOROPLASTS: REDUCTION OF VIOLOGENS AND HYDROGEN EVOLUTION

The redox potential of methylviologen, $E'_0 = -0.455v$,
is close to that of ferredoxin which is the final electron acceptor of photosystem I, and is higher than the
E'_0 of the hydrogen electrode (-0.42v). The reduced viologen is easily oxidized by oxygen to form hydrogen
peroxide; to observe accumulation of reduced viologens
by chloroplasts it is necessary either to inactivate
the production of oxygen (i.e., photosystem II), for
example by heating, or to introduce an oxygen-consu-
ming system in the chloroplast suspension.

Arnon (I2) reported photoreduction of methylviologen by chloroplasts in the presence of cysteine and
dichlorphenolindophenol which acted as electron donor.
Kok and his co-workers (I3) observed photoreduction of
viologens by chloroplasts in the presence of glucose
and glucose oxidase. Zweig and Avron (I4) and Black
(I5) described reduction of various viologens by chloroplasts in the presence of ethanol and catalase.

We have studied photoreduction of methylviologen
by chloroplasts using hydrazine as hydrogen donor in
the course of progressive disruption of chloroplast
structure by organic solvents (16). This reaction proceeded most actively at pH close to 8.5. With 10% of
the solvent, the process was activated, and with 50%
completely inhibited; at 70-80% of the solvent chlorophyll was extracted and active photoreduction of
viologen was resumed. Illumination of chloroplasts in
the presence of oxygen led to oxygen reduction with
formation of hydrogen peroxide; addition of methylviologen resulted in a manyfold increase of the amount of
peroxide formed; viologen competes with oxygen for
electron of photosystem I and the formed cation radical of methylviologen is reversibly oxidized by oxygen
to form hydrogen peroxide (17). Photoreduction of viologens by chloroplasts indicates that the system has
reached the hydrogen electrode potential which is a
precondition for a release of molecular hydrogen if
the system includes a catalyst (hydrogenase) promoting
the reaction $2H + 2e \longrightarrow H_2$.

Boichenko reported (18) that isolated chloroplasts
of higher plants were capable of releasing hydrogen in
the presence of glucose. In Arnon's experiments (12)
photorelease of hydrogen by chloroplasts was observed
in the presence of bacterial hydrogenase and of cysteine as electron donor. The production of hydrogen
was accompanied with formation of ATP.

Benemann et al. (19) have described photoproduction of hydrogen in a similar system (but without cysteine) consisting of isolated chloroplasts, ferredoxin
and hydrogenase isolated from a Clostridium pasteuri-
anum culture. The authors believe that the reaction is
accompanied by the production of oxygen which, however,
they were unable to measure. It is presumed that oxygen may be used to oxidize exogenous electron donors
such as glucose. It is possible, however, that in this
study, as in Arnon's experiments, endogenous electron
donors other than water were used. Hall and co-workers
(20) described recently a prolonged hydrogen evolution
in similar systems and concluded that H_2O was the source of electrons for H_2 production. Ben-Amotz and
Gibbs (21) revealed H_2 evolution in cell-free preparations from algae mixed with hydrogenase and dithioerith-
rol as electron donor.

In our laboratory photoproduction of hydrogen by
chloroplasts was observed in the presence of hydrogenase isolated by Gogotov et al. from photosynthesizing
bacteria (22), with $NADH_2$ acting as electron donor (23).

Table I   Hydrogen evolution in red (600-750 nm) and white (400-700 nm) light by bean leaves chloroplasts containing 0.05 mg/ml chlorophyll in the presence of $NADH_2$ ($10^{-3}M$) and methylviologen ($10^{-3}M$). Light intensities: $7 \cdot 10^5$ (red light) and 10^6 (white light) erg/$cm^2 \cdot sec$

System	Hydrogen, $\mu l/min$	
	Red light	White light
Without $NADH_2$	0.000	0.000
$NADH_2$	0.007	0.015
$NADH_2$ + MV^{2+}	0.072	0.150

CHLOROPHYLL SOLUTIONS: PHOTOREDUCTION OF VIOLOGEN AND PHOTOEVOLUTION OF HYDROGEN

Studies carried out in our laboratory many years ago revealed that during the photoreduction of chlorophyll intermediate is formed with an E'_o close to that of the hydrogen electrode. In 1949 we revealed the possibility of chlorophyll sensitized reduction of NAD (24). In a reaction of this type chlorophyll acts as a light-excited electron carrier from electron-donating molecules to electron acceptors (see review (25). More recently, our laboratory has investigated the photosensitized reduction of methylviologen under the action of red light absorbed by chlorophyll in the presence of a number of electron donors; the reactions were done in organic solvents and in aqueous solutions of detergents where chlorophyll and other reaction components were solubilized (25,26,27). The most efficient photoreduction of methylviologen under anaerobic conditions was observed when phenylhydrazine, cysteine and $NADH_2$ were used as electron donors; thiourea was inactive under these conditions. Efficient photosensitized reduction of methylviologen in the presence of thiourea did, however, occur in experiments without preliminary evacuation of air (28). The mechanism of this reaction was revealed: as a result of photosensitized oxidation of thiourea by oxygen active long-lived reductants were formed capable of reducing me-

thylviologen; the anaerobiosis which had developed due
to photosensitized reduction of oxygen, prevented re-
oxidation of reduced viologen. In pyridine solution,
the "red" photoreduced form of chlorophyll is capable
of a dark reaction with viologen; in this medium, a
possible mechanism of the reaction consists in the
photoreduction of chlorophyll by electron donor followed by a reaction between the reduced chlorophyll and
viologen. On the other hand, observations of fluorescence quenching by chlorophyll and its analogs by methylviologen point to a possibility of primary photooxidation of the sensitizing pigment (29). Thus, it
is possible to achieve photoreduction of methylviologen in a chlorophyll solution at the expense of the
light absorbed by chlorophyll. In order that molecular
hydrogen might be released in the reactions described
above, a catalyst of the reaction $2H + 2e \longrightarrow H_2$
should be introduced into the system. Indeed, addition
of bacterial hydrogenase to an aqueous solution of
Triton X-I00 containing chlorophyll and cysteine (or
$NADH_2$) did result in a release of molecular hydrogen
upon illumination (23). Addition of methylviologen
considerably activated the reaction, as in the case
of chloroplasts.

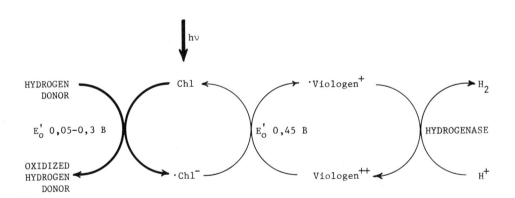

Fig. I   Chlorophyll photosensitized hydrogen
         evolution

Table 2   Chlorophyll-sensitized photoevolution of
          hydrogen in aqueous solution of Triton X-I00
          in the presence of hydrogenase and various
          electron donors (I.4·I0^{-3}M) and methylviolo-
          gen (I.4·I0^{-3}M) upon illumination with white
          light (400-700 nm) I0^{6} erg/cm^{2}·sec and red
          light (600-750 nm) 5.I0^{5} erg/cm^{2}·sec

System	Hydrogen, μl/min	
	Red light	White light
NADH$_2$	0.700	0.I00
NADH$_2$ + MV^{2+}	0.I50	0.260
Cysteine	0.006	0.0I0
Cysteine + MV^{2+}	0.I25	0.200

PHOTOACTIVATION OF REDUCED PYRIDINE NUCLEOTIDES: REDUCTION OF VIOLOGEN AND HYDROGEN EVOLUTION

In the reactions described above, light-excited
chlorophyll reacts with non-excited NADH$_2$. However,
absorbing a light quantum in its own absorption region
(at 340 nm), NADH$_2$ and NADPH$_2$ are activated and their
redox potential becomes more positive than the E$_0^'$ of
the hydrogen electrode. Thus excited NADH$_2$ is capable
to reduce methylviologen and ferredoxin (30). The
activation of NADH$_2$ by light has been investigated in
a series of studies undertaken in our laboratory. Just
as in the above-mentioned reactions, introduction of
hydrogenase led to a release of hydrogen (23) upon
illumination of aqueous solutions of NADH$_2$ with ultraviolet light at 365 nm.

THE USE OF INORGANIC PHOTOCATALYSTS FOR VIOLOGEN REDUCTION AND HYDROGEN EVOLUTION

Inorganic photocatalysts (electronic semiconduc-
tors) such as titanium dioxide and zinc oxide, can act
as photosensitizers upon their excitation in the ultraviolet region of the spectrum. Using these photocatalysts we have been able to simulate the Hill reaction

Table 3   Hydrogen evolution upon illumination of aque-
          ous solution of $NADH_2$ $(I.4 \cdot IO^{-3}M)$ + hydroge-
          nase + methylviologen $(I.4 \cdot IO^{-3}M)$. Light
          intensity: $2 \cdot IO^5$ erg/$cm^2 \cdot$sec at 365 nm

System	Hydrogen, μl/min
$NADH_2$	0.I2
$NADH_2$ + MV^{2+}	0.40

occurring in chloroplasts: when UV light (365 nm) acted upon an aqueous suspension of titanium or zinc
oxides containing oxidants (ferric compounds or quinones) an evolution of oxygen was observed (3I). The
use of water labeled with a heavy oxygen O^{18} showed
that the oxygen evolved was derived from water molecules indeed (32). Experiments performed in anaerobic
conditions revealed the photoreduction of viologens
photosensitized by zinc and titanium oxides (33). The
next step was to introduce hydrogenase in the system
in order to set molecular hydrogen evolution. Indeed,
in the presence of methylviologen and preparations of
bacterial hydrogenase, molecular hydrogen was released
in aqueous suspensions of titanium and zinc oxides
upon illumination with ultraviolet light (34). Hydrogen was probably originated from water molecules but
experimental verification of this is necessary.

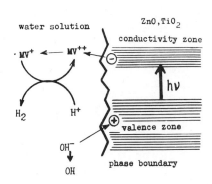

Fig. 2   Photosensitized reduction of methylviologen

Table 4   Evolution of hydrogen upon illumination of aqueous suspension of photocatalysts in the presence of hydrogenase and methylviologen (10^{-3}M). Illumination time I min, 365 nm, $2 \cdot 10^5$ erg/$cm^2 \cdot$sec

System	Hydrogen, μl/min
Titanium dioxide	0.009
Titanium dioxide + MV^{2+}	0.094
Zinc oxide	0.002
Zinc oxide + MV^{2+}	0.005

These model experiments can help in understanding the mechanism of hydrogen and oxygen photoevolution and may serve as a prototype of systems using solar energy to effect photolysis of water.

REFERENCES

I. H. Gaffron and J. Rubin, J.Gen.Physiol.,26,2I9,I942
2. H. Gaffron, in "Horizons of Bioenergetics", Academic Press, N.Y., I972
3. G. Gest and M. Kamen, Science,I09,558,I949
4. E.N. Kondratieva and I.N. Gogotov, Izvestiya AN SSSR, Ser.Biol.,No.I,69,I976
5. V.P. Oshchepkov and A.A. Krasnovsky, Izvestiya AN SSSR, Ser.Biol.,No.I,87,I976
6. V.P. Oshchepkov and A.A. Krasnovsky, Fiziol.Rastenii,I9,I090,I972; 2I,462,I974
7. V.P. Oshchepkov and A.A. Krasnovsky, Prikl.Biochim. Mikrobiol.,I0,760,I974
8. E.I. Efimtsev, E.A. Boichenko, and F.F. Litvin, Dokl.AN SSSR,220,986,I975
9. V.P. Oshchepkov, K.A. Nikitina, M.V. Gusev and A.A. Krasnovsky, Dokl.AN SSSR,2I3,739,I973
I0. J.R. Benemann and N.M. Weare, Science,I84,I74,I974
II. N.I. Bishop and H. Gaffron, in "Photosynthetic Mechanisms in Green Plants", Nat.Acad.Sci., Nat. Res.Counc.,44I,I963
I2. D.I. Arnon, A. Mitsui and A. Paneque, Science, I34,I425,I96I
I3. B. Kok, H. Rurainski and O. Owens, Biochim.Biophys.Acta,I09,347,I965

14. G. Zweig and M. Avron, Biochem.Biophys.Res.Comm.,
    19,379,1965
15. C.C. Black, Biochim.Biophys.Acta,120,332,1966
16. G.P. Brin and A.A. Krasnovsky, Dokl.AN SSSR,
    204,1253,1972
17. V.A. Shuvalov and A.A. Krasnovsky, Biokhimiya,
    40,358,1975
18. E.A. Boichenko, Dokl.AN SSSR,64,545,1949
19. J.R. Benemann, J.A. Berenson, N.O. Kaplan and
    M.D. Kamen, Proc.Nat.Acad.Sci. USA,70,2317,1973
20. K.K. Rao, L. Rosa and D.O. Hall, Biochem.Biophys.
    Res.Comm.,68,21,1976
21. A. Ben-Amotz and M. Gibbs, Biochem.Biophys.Res.
    Comm.,61,335,1975
22. I.N. Gogotov, N.A. Zorin and L.V. Bogorov,
    Microbiologia,43,5,1974
23. A.A. Krasnovsky, V.V. Nikandrov, G.P. Brin, I.N.
    Gogotov and V.P. Oshchepkov, Dokl.AN SSSR,225,
    711,1975
24. A.A. Krasnovsky and G.P. Brin, Dokl.AN SSSR,67
    325,1949
25. A.A. Krasnovsky, "Transformation of Light Energy
    in Photosynthesis", "Nauka" Publishing House,
    Moscow, 1974
26. A.A. Krasnovsky and G.P. Brin, Dokl.AN SSSR,163,
    761,1965
27. G.P. Brin, A.N. Luganskaya and A.A. Krasnovsky,
    Dokl.AN SSSR,174,221,1967
28. A.N. Luganskaya and A.A. Krasnovsky, Molekul.
    Biologiya,4,848,1970
29. A.A. Krasnovsky and N.N. Drozdova, Dokl.AN SSSR,
    167,928,1966
30. A.A. Krasnovsky, G.P. Brin and V.V. Nikandrov,
    Dokl.AN SSSR,220,1214,1975
31. A.A. Krasnovsky and G.P. Brin, Dokl.AN SSSR,147,
    656,1962; in "Molecular Photonics", "Nauka" Pub-
    lishing House, Leningrad, p. 161, 1970
32. G.V. Fomin, G.P. Brin, M.V. Genkin, A.K. Liubimova,
    L.A. Blumenfeld and A.A. Krasnovsky, Dokl.AN SSSR,
    212,424,1973
33. A.A. Krasnovsky and G.P. Brin, Dokl.AN SSSR,213,
    1431,1973
34. A.A. Krasnovsky, G.P. Brin and V.V. Nikandrov,
    Dokl.AN SSSR,229,990,1976

DETERMINANTS OF FEASIBILITY IN BIOCONVERSION OF SOLAR ENERGY

William J. Oswald

Professor of Sanitary Engineering and Public Health

University of California, Berkeley, California 94720

ABSTRACT

The fundamental determinants limiting the feasibility in bioconversion of solar energy for food, feed, fertilizer, and fermentable substrates may be categorized as technical, economical and
sociological.   Technical determinants are related to the amount
of solar energy available and the efficiency of bioconversion.   The
amount of solar energy is mainly determined by location on the surface of the earth whereas photosynthetic efficiency is closely related to species specific growth rates.   Light availability, nutrient base, temperature, and organism size and metabolic characteristics are among the factors influencing growth rate and efficiency.
Because of low efficiency and other adverse characteristics of
higher plants, it appears likely that the microalgae are the only
organisms which can be produced at sufficiently high and sustained
photosynthetic efficiencies to make biological transformation of
solar energy, for energy alone, economically feasible.   There is
little question that algae produced on a large-scale would now be
economically feasible for human food, animal feed, and for fertilizer if competitive markets were available.   However, economical,
political and sociological restrictions appear to be major barriers
to current widespread applications.   It seems likely that as the
world food and protein crisis and the inadequacies of conventional
agriculture to meet this crisis become more widely recognized
acceptance and growing application of algal technology will occur
in the developed and developing nations of the world.   This, in turn,
should permit new levels of environmentally sound abundances of food
in the world.

DETERMINANTS OF FEASIBILITY IN BIOCONVERSION OF SOLAR ENERGY

Introduction

In the bioconversion of solar energy, the efficiency of primary fixation is of major importance. With a few exceptions most higher plants now used in agriculture attain primary fixation efficiencies of less than 1 percent of the total annual incident photosynthetically active radiation (PAR). These cannot, in the long run, be competitive with microalgae which can attain sustained efficiencies of conversion of PAR in the range of 3 to 4 percent.

The microalgae may be defined as those members of the plant kingdom which carry out photosynthesis with the release of oxygen and are usually distinguishable as to species only with the aid of a microscope. The cultivation and use of microalgae appears now to be on the threshold of worldwide application and it is worthwhile to explore limitations as well as potentials in such an important field.

Before discussing the determinants of feasibility of microalgae production in detail, it is worthwhile to review the current status of algal applications and production technology. The microalgae are, of course, recognized as the base of the aquatic food web[1] and are well established as indicators of water quality.[2] The application of wild strains of microalgae growing on wastes of various sorts to produce oxygen for microbial oxidation of the waste is an established technology currently applied in waste management.[3] Microalgae can also be used to remove nutrients from wastes by assimilative uptake.[4] The removal of heavy metals and toxicants from wastes through adsorption by algae and removal of the algae also bears promise but has not been extensively investigated. Also technically well established and of great promise is the use of algae to provide protein, vitamins, and minerals in animal feeds.[5] Certain microalgae have been used since prehistoric times for human food in various parts of the world and their growth under controlled hygienic conditions as a protein staple or supplement in the nutrition of mankind has increasing promise.[6] Because algae undergo methane fermentation and certain species under certain conditions may be induced to produce hydrogen current studies are underway to explore their full potential in biological transformation of solar energy as an energy resource.[7,8] In addition to those algae which assimilate nitrogen from wastes, there are those which fix atmosphere nitrogen.[9] Both are of great potential importance in conservation of fertility and in de novo fertilizer production.[10]

Because of these broad and diverse existing and potential applications of microalgae it is worthwhile now to examine from the engineers viewpoint the feasibility of microlage in large scale bioconversion of solar energy. The determinants of feasibility to be

explored may be divided into technological, economical, and sociological categories.

Technical Determinants of Feasibility

The technological feasibility of photosynthetic bioconversion
is dependent on location on the surface of the earth which, in turn,
determines the amount of solar energy, the mean temperature, water
quality, and the probable need for solar energy which, of course,
is related to population.

The relationship that position on the surface of the earth has
to bioconversion is evident from the data presented in Table 1 in
which the yields of biomass are related to latitude and conversion
efficiency of PAR.   Table 1 is independent of species, but, as
noted above, the microalgae seem to be the only plants that can
maintain sustained production over all of the surface exposed to
the sun all of the time and thereby attain a high overall efficiency.
Most higher plants go through cycles of growth that diminish overall efficiency when considered on an annual basis.

The species of microalgae influences many factors including
harvestability, specific growth rate, response to environmental
factors, and fermentability. Although all microalgae are harvestable by centrifugation and coagulation which are the most expensive
techniques, the smaller unicellular microalgae such as Chlorella
and Selenastrum are harvestable only by coagulation or centrifugation. Many of the multicellular microalgae which form clumps such
as Microcystis and which form filaments such as Spirulina and
Oscillatoria are separatable by 75 to 100 mesh per cm screens.
Screening is the least expensive separation technique other than
natural sedimentation. Algae of intermediate size such as Scenedes-
mus quadricauda (10μ) and Euglena gracilis (30μ) can be separated by
very fine filters and pass all but the finest screens (150 meshes
per cm). If separation alone were the major determinant of feasibility in algae production one might conclude that large organisms
would be most desirable for propagation. Unfortunately, there is
evidence that the larger algae grow more slowly than smaller algae.

In Figure 1 evidence is presented which indicates that smaller
algae have higher maximum specific growth rates than larger forms.
As is shown, other factors equal

$$\hat{\mu} = 2.7 - 0.64 \ln D \tag{1}$$

in which D is the cell diameter in microns and $\hat{\mu}$ is the maximum SGR.
Chlorella pyrenoidosa has a SGR greater than 2 per day whereas the
SGR of Euglena gracilis is much less than 1 per day. It has also
been pointed out previously that for several of the microalgae

TABLE 1

Potential Algal Production at Various Latitudes and Sunlight[1]
Energy Conversion Efficiencies[2]

North Latitude Degrees	Condition[4]	Quantities of Dry Algae in Metric Tons Per Hectare for Assumed Visible Light Energy Conversion Efficency in %[3]									
		1	2	3	4	5	6	7	8	9	10
0	Max	17.1	34.2	51.3	68.4	85.5	103	119	137	154	171
	Min	12.3	24.5	36.8	49.1	61.3	74.0	85.1	98.4	110	123
10	Max	16.7	33.5	50.3	67.0	83.8	100	117	134	151	167
	Min	11.6	23.3	35.0	46.6	58.3	70.0	81.2	92.8	105	116
20	Max	15.9	31.8	47.8	63.7	79.7	95.4	112	127	143	159
	Min	10.4	20.8	31.3	41.7	52.1	62.6	72.8	83.4	93.6	104
30	Max	14.6	29.1	43.7	58.2	72.8	87.4	102	116	131	146
	Min	8.71	17.3	26.0	34.7	43.3	52.0	60.7	69.4	78.0	86.7
40	Max	12.3	24.5	36.8	49.0	61.3	73.6	85.8	98.1	110	123
	Min	6.71	13.3	20.0	26.7	33.3	40.0	46.7	53.4	60.0	66.7
50	Max	10.3	20.6	30.9	41.2	51.5	61.8	72.2	82.3	92.8	103
	Min	5.11	10.2	15.2	20.3	25.4	30.5	35.6	40.6	45.7	50.8
60	Max	6.0	12.0	18.0	24.0	30.0	36.0	42.0	48.0	54.0	60.0
	Min	3.0	6.0	9.0	12.0	15.0	18.0	21.0	24.0	27.0	30.0
70	Max	2.0	4.0	6.0	8.0	10.0	12.0	14.0	16.0	18.0	20.0
	Min	1.0	2.0	3.0	4.0	5.0	6.0	7.0	8.0	9.0	10.0

[1]Sunlight refers to the visible portion of the solar spectrum which penetrates a horizontal clear water surface at sea level.

[2]Practical conversion efficiencies up to 5% are attainable with existing technologies; efficiencies between 5 and 10% can only be attained through continuing research and development.

[3]Algae energy is assumed at 5.5 Kcal per gram of volatile material.

[4]Maximums and minimums are determined by variations in available sunlight due to climatological factors.

To convert hectare to acres, multiply by 2.471. To convert units per hectare to units per acre, divide by 2.471. To convert metric tons to tons, multiply by 1.1025. Example: 100 metric tons per hectare per year = 100 x 1.1025 = 110.25 tons per hectare per year = 110.25/2.471 = 44.6 tons per acre per year.

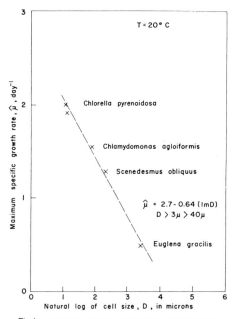

Fig. I — Observed variation in maximum growth rates in continuous cultures of the indicated species cultured in domestic sewage

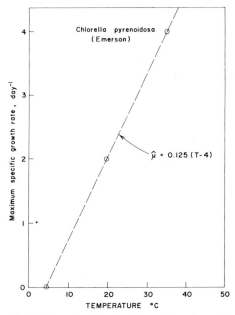

Fig. 2 — Observed relationships between temperature and maximum growth rate for continuous cultures of Chlorella pyrenoidosa

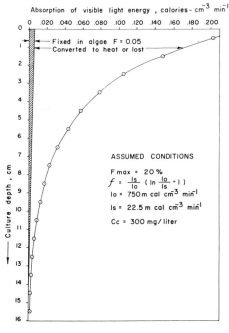

Fig. 3 Integration of the Beer–Lambert Law and the Bush limitations for a culture of algae under outdoor conditions.

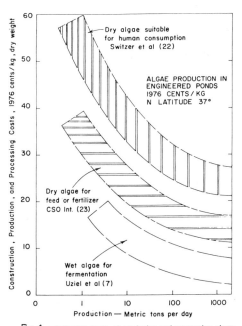

Fig. 4 — Estimated costs of producing and processing algae of various grades.

maximum photosynthetic efficiency is attained at about ½ the maximum SGR.[12] Because of the relationship between efficiency and productivity shown in Table 1 it appears that the smaller more efficient algae would be most productive. There is then, at least theoretically, an algae of optimum size that is both harvestable and has
a high SGR. Unfortunately SGR data is lacking for those filamentous
algae which are easily harvestable by screens. No conclusion can
be drawn, therefore, as to an optimum species of algae at this time
but we at Berkeley are both actively seeking it and ways of maintaining it once it is found. Both Oscillatoria and Spirulina are
prime candidates.

Environmental factors -- light and temperature -- also importantly influence SGR. As pointed out by Myers[13] Shelef[14] and others,
the specific growth rate begins at a threshold light intensity and
increases as a linear function of light intensity to a point where
it tends to become constant. This light intensity is termed the
saturation intensity, I_s. Bush[15] has shown mathematically that the
fraction of the maximum photosynthetic efficiency that can be attained is dependent on I_s and I_o, the incident intensity, thus:

$$f = \frac{I_s}{I_o}\left(\ln \frac{I_o}{I_s} + 1\right) \tag{2}$$

As the magnitude of I_o and I_s converge, the value of f tends to a
maximum of 1.0. However, out of doors in real system I_o is an
independent variable and hence I_s must be varied to effect f. The
maximum photosynthetic efficiency Fmax itself is determined by the
quantum requirement which is the number of mole Einsteins (Nhν) required per mole of carbon fixed, generally regarded to be 8 to 10.
The overall photosynthetic efficiency F is then defined as:

$$F = f \cdot Fmax \tag{3}$$
in which all terms are as previously defined.

Because of the quantum requirement, and the spectral distribution of sunlight, F_{max} outdoors is in the range of 20 to 25% of the
incident PAR or about 8.5 to 11% of the total solar energy. In currently studied outdoor systems f averages about 0.2 whence F = 0.2 x
.25 = .05 of PAR or .025 of the total solar radiation.

Because of these low levels improvements are sought. Myers[13]
has shown that temperature is one of the major factors influencing
I_s and therefore F. As is shown in Figure 2, other factors equal,
the approximate relationship between growth rate and temperature
for C. pyrenoidosa grown on wastes is:

$$\mu = 0.125 \ (T - 4) \tag{4}$$

in which T is the temperature in °C and μ is the maximum observed
SGR in continuous cultures. Presumably algae with higher growth

rates have higher I_s values and may attain higher values of f and
hence higher photosynthetic efficiencies in sunlight. Also, at
least theoretically, higher efficiencies and higher productivities
result when the thermal environment is such that I_s is at its highest value for a given species. We have, therefore, suggested that
the temperature of large scale cultures should be increased by using
waste heat from power plants or industries to warm large scale algal
cultures.

Returning to considerations of light alone, its penetration
into cultures containing algae closely follows the Beer-Lambert Law:

$$\frac{I_d}{I_o} = e^{-\alpha\, C_c d} \qquad\qquad (5a)$$

or in a form useful for design

$$\alpha \simeq \frac{\ln I_o - \ln I_d}{C_c \alpha} \qquad\qquad (5b)$$

in which I_d is the intensity at depth d, I_o the surface light intensity, α is an absorption coefficient, C_c is the concentration of
algae, and d is the depth. I_d and I_o are conveniently expressed in
gram cal cm^{-2} $min.^{-1}$. If C_c is in mg $liter^{-1}$ and d is cm, the
value of α is about 1×10^{-3} liter mg^{-1} cm^{-1}.

Natural factors described by Equation 5 and Equation 2 function
together to limit the overall photosynthetic efficiency, F, and
hence the productivity of a culture out of doors. By integrating
these functions for sunlight the energy fixed is predicted to be
a linear function of depth approaching zero at about two-thirds the
maximum depth of light penetration. This relationship is shown in
Figure 3. Thus, in engineering design of algae production ponds,
depths must be selected which permit efficiency and productivity to
be at a maximum. The optimum depth for this condition appears to
be in the range of 20 to 30 cm for real systems in sunlight[16].

Another factor of importance in engineering design is selection of residence time or detention period. Detention period θ is
defined as follows:

$$\theta = \frac{V}{Q} \qquad\qquad (6)$$

in which V is the culture volume and Q is the quantity of feed or
withdrawal per day. The unit of θ is day and at steady state the
SGR of a continuous culture, μ, is:

$$\mu = \frac{1}{\theta} = \frac{Q}{V} \qquad\qquad (7)$$

According to Equation 7 selection of θ limits the SGR of algae that can be maintained in a system and in accordance with relationship between SGR and efficiency also influences efficiency.

Another requirement is that θ must be selected to permit integration of sufficient solar energy to satisfy the needs of the algae for energy; thus,

$$hC_c = \theta ASF \qquad (8)$$

in which h is the heat of combustion of the algae in cal mg^{-1}, C_c is the algal concentration in $mg\text{-}liter^{-1}$, θ is the detention time in days, A is the area in cm^2 occupied by 1 liter, S is the solar energy flux, Cal cm^{-2} day^{-1}, and F is overall the photosynthetic efficiency.[17]

To introduce the factor d or depth to Equation 7, we may substitute 1,000/d for A, whence Equation 7 rearranged is:

$$\frac{d}{\theta} = \frac{1,000 \ SF}{hC_c} \qquad (9)$$

Selection of a concentration for C_c is based on the required concentration of cells, F is determined as previously discussed (currently about 4%), d is predicted from Equation 5 and S is the visible solar energy in cal cm^{-2} day^{-1}. By substituting typical values for these parameters a numerical value may be found for θ. Values of θ usually range from 2 to 6 days, depending chiefly on S and F.

Mixing of large scale algal cultures is essential for two reasons -- to prevent thermal stratification and to prevent bottom anaerobiosis. Thermal stratification deprives the organism of nutrients in the presence of excess light causing inactivation whereas anaerobiosis causes odors and may deprive the cells of both light and nutrients. The former can be prevented by using a flow velocity of 5 cm per second during daylight hours, the latter by using a velocity of 30 cm per second, preferably during nocturnal hours. However, a 2 hr period of mixing at 30 cm per second each 24 hours is adequate to prevent anaerobiosis.[18]

In large multi-pond systems recirculation is valuable to transfer nutrients or cells from one pond to another pond. For example, when an organic waste stream is used as media, it is beneficial to utilize a primary settling pond for nutrient hydrolysis by bacterial action and to draw the soluble nutrients into the algal growth pond by recirculation.[18]

Economic Determinants of Feasibility

The economic determinants of large scale bioconversion of solar energy are cost of land, water power and nutrients, the cost of pond construction, interest rates, the cost of operations and maintenance including algal separation, the size of the system, the productivity of the system and the value of the algae. The most economical sources of water and nutrients are organic wastes.[19] However, in areas where water is plentiful, the addition of nutrients which are lacking to natural waters is quite economical. Many natural waters are rich in bicarbonate alkalinity nitrates and trace minerals. Under these circumstances, only phosphorus and perhaps iron would needs be added. An example of this is the remarkable work of Durand-Chastel of Soso Texcoco, Mexico[20] in which the water in salt evaporation beds enriched with phosphorus and iron produces large crops of _Spirulina_. Domestic sewage and animal wastes are another source of water and nutrients. The quality of wastes or other nutrient solutions for algal growth can be determined with bioassays in which various dilutions of a waste are inoculated with algae and incubated in the light. An ideal dilution is that for which photosynthetically produced oxygen will satisfy or exceed the biochemical oxygen demand of the waste. If algal growth is insufficient in the assays, the waste may be supplemented with critical nutrients to determine which is lacking. The lacking nutrient may then be added to the culture.

Pond costs are made up of excavation, pond lining, channelization and pumping equipment. Production costs include labor, nutrients, power, harvesting and drying if needed. These factors have recently been summarized in an NSF MIT Protein Resources Study[21] and are integrated into a continuum of costs shown in Figure 4. As is evident from the figure, product quality and system size are the major determinants of cost.

The market value of algae is, at present, somewhat more uncertain than the production cost data. However, assuming the cost data in Figure 4 to be reliable, certain other facts are sufficiently known to permit an estimate of market values of algae for various purposes. If algae are to be fermented one may reliably obtain 12,000 BTU's per kilogram of algae in the form of methane.[7] Such gas may have a value of about 5 cents, which according to Figure 4 may be the median cost of a kilogram of fermentable algae in a 100-metric ton per day facility.

In the category of animal feed according to Hintz, et al.[5] algal meal is equivalent to soy bean oil meal for livestock feed. The current value of soy bean oil meal is about 22 cents per kilogram which, according to Figure 4, is the median estimated cost of algae in a 10-metric ton per day facility.

In the category of human food the microalgae <u>Spirulina</u> grown under
hygienic conditions, is useful as a food protein for humans;[22]
judging from fish protein concentrate such proteins generally sell
for about 50 cents per kilogram. According to Figure 4, this is
the median estimated cost of <u>Spirulina</u> suitable for human consumption in a one ton per day facility. It is thus evident that feed
or food algae production facilities in the 100 ton per day category
could significantly decrease the cost of human and animal protein.[23]

Sociological Determinants of Feasibility

The sociological determinants of feasibility in microalgae
technology are complex and comprised of negative public attitudes
toward algae and the need for governmental support to overcome
these attitudes. In many developed countries of the world, microalgae are scorned as a nuisance, relegated to fish food, or regarded
as science fiction "food for the future". On the other hand, it is
well established that microalgae, harvested from natural populations,
are a basic food in several societies. To produce algae as an agricultural crop on a large scale will require new and large uses of
capital, land, and other resources, and, while the expected capital
return is positive, it will not be a high percentage of the original
investment.[23] There is, therefore, little incentive for industries
or individuals to engage in algal production and the only remaining
entity is government. But the incentive for government is clear. As
is evident from Table 1 enormous quantities of algal protein can be
produced quickly with relatively small amounts of the essential resources compared with other protein production methods. Urgency
also exists. World grain stocks, according to Brown[24], are at an
all time low and concurrent consecutive droughts or other weather
anomalies in the grain producing countries could bring on a relatively sudden and potentially disasterous famine condition. This
would touch all nations, rich and poor, developed and developing
alike and could lead to new confrontations, political upheavals,
and indeed even to the ultimate nuclear holocaust. The capability
of producing large amounts of protein quickly, as is possible with
microalgae, is, therefore, a goal which the developed nations of
the world should undertake now before greater crises arise. To
accomplish this we need to have in reserve a well developed algal
production capability that need not be economically feasible now
but which could be applied quickly to forestall or ameliorate local
or world wide protein famines. We feel certain that such a capability would inevitably lead to a world in which protein is no
longer a limiting factor in human development.

Except for the few societies that already consume microalgae,
the concept of humans being forced by necessity to directly eat
microalgae would, in many societies, be regarded as a disaster in
itself. However, livestock production or chemical extraction can

eliminate this problem. The use of algae as a feedstuff for livestock is established as highly acceptable and awaits only the first
large-scale application to demonstrate its many benefits in intensive meat, milk and egg production. If needs be, livestock could
be eliminated as an intermediary for human consumption through
chemical processing of algae to produce colorless and tasteless
amino acids, vitamins, and minerals which would be acceptable to
everyone. However, this in turn would require additional capital
and technical development would be more costly than livestock production and probably would be much less acceptable to those who enjoy
meat, milk and eggs for their protein.

Conclusions

Those who work daily with the microalgae know that this
technology like all others has distinct limitations in geography,
climate and economics; limitations in the technology itself and
limitations in the sociological mores of our societies. We also
know, however, that this microalgal technology through more
efficient solar energy utilization can increase the available food
feed, fertilizer and fermentable substrate in the world by an order
of magnitude, at costs that on a large scale even now would be
approximately equal to those of comparable products currently in
use. The major determinants for application of large scale algal
culture are no longer technical nor even economical, they are mainly
social, educational and political. The term "algal technology"
means little or nothing to the average politician who eventually
must advocate the technology if essential government subsidies for
initial large scale development.

Large scale federally supported algae production centers have
been recommended in the United States as a method of integrating
technical demonstrations with public information, education, research,
and political support. These centers remain to be funded. In the
face of environmental deterioration, the energy crisis, and an
urgent crisis in world feed and protein resources the current
dearth of support and lack of recognition of the potential for algal
technology cannot long endure.

Acknowledgments

This research was supported in part by the Lawrence Berkeley
Laboratory and the Sanitary Engineering Research of the University
of California Laboratories, by the U.S. National Science Foundation,
Research Applied to National needs and by the Energy Resources and
Development Administration of the United States. I am indebted to

IUBS for providing funds for my travel and for my stay in Rome.

References

1. Ryther, John H. "Potential Productivity of the Sea," Science,
   130(3376)6, 65-69 (1959).

2. Sladecek, Vladamir. "Zur biologischen Gliederung der höheren
   Saprobitatsstufen." Arch. f. Hydrobiol. Bd. 58, 103-121
   (1961).

3. Oswald, W.J. "Complete Waste Treatment in Ponds," Progress in
   Water Technology, Water Quality Management and Pollution
   Control, 3:153, Pergamon Press, London (1973).

4. Brown, R.L. "The Occurrence and Removal of Nitrogen in Sub-
   surface Agricultural Drainage from San Joa uin Valley,
   California," Water Research, 9, 524 (1975).

5. Hintz, H.F., Heitman, H., Weir, W.C., Torell, D.T., and Meyer,
   J.H. "Nutritive Value of Algae Grown on Sewage." J. Animal
   Sci., 25, 675 (1966).

6. Waslein, C.I. "Unusual Sources of Proteins for Man," Critical
   Reviews in Food Science and Nutrition, 6:1, 77-151, C.R.C.
   Press (1975).

7. Uziel, M., W.J. Oswald, and C.G. Golueke. "Integrated Algal
   Bacterial Systems for Fixation and Conversion of Solar
   Energy." Presented before the Annual Meeting A.A.A.S.
   Symposium, "Energy New Sources," In press (1975).

8. Benemann, J.R. and Weare, N.M. "Hydrogen Evolution by Nitrogen-
   Fixing Anabaena cylindrica Cultures," Science, 184, 174-
   175, (1974).

9. Stewart, W.D.P. "Nitrogen Input into Aquatic Eco-systems," in
   Algae Man and the Environment, Daniel F. Jackson, Ed.
   Syracuse University Press (1968).

10. Fogg, G.E. "Studies in Nitrogen Fixation by Blue-Green Algae
    I Nitrogen Fixation by Anabaena cylindrica," J. Expl.
    Biology, 19, 78 (1942).

11. Golueke, C.G., W.J. Oswald, and H.K. Gee. "Harvesting and
    Processing Sewage-Grown Planktonic Algae," Journal of the
    Water Pollution Control Federation, 37:4 (1965).

12. Oswald, W.J. "Growth Characteristics of Microalgae Cultured
    in Domestic Sewage: Environmental Effects in Productivity."
    In Prediction and Measurement of Photosynthetic Productivity.
    Proceedings IBP/PP Technical Meeting Trebon, Centre for
    Ag. Publ. and Documentation, Wageningen, Netherlands (1970).

13. Myers, J. "Algal Cultures," Encyclopedia of Chemical Technology,
    2nd ed. 1, Intersience Encyclopedia, Inc., John Wiley (1963).

14. Shelef, G. "Kinetics of Algae biomass Production Systems with
    Respect to Light Intensity and Nitrogen Concentration,"
    Ph.D. Dissertation, Univ. of Calif., Berkeley (1968).

15. Burlew, J.S. Algal Culture from Laboratory to Pilot Plant,
    Carnegie Inst. of Washington, Pub. 600, Wash., D.C. (1953).

16. Oswald, W.J. "Ecological Management of Thermal Discharges,"
    J. of Env. Quality, 2, 203-207 (1973).

17. Oswald, W.J. and Gotaas, H.B. "Photosynthesis in Sewage Treat-
    ment," Trans Amer. Soc. of Civil Eng., 122, 73-105 (1957).

18. Oswald, W.J. "Experience with New Pond Designs in California,"
    Water Res Symp, No. 9, 257-273, Center for Water Res.,
    Univ. of Texas at Austin (1976).

19. Oswald, W.J. and C.G. Golueke. "Large-Scale Production of
    Algae," Single Cell Protein, 271-305 MIT Press (1968).

20. Durand-Chastel and Clement, G. "Spirulina Algae Food for
    Tomorrow," Proc. 9th Congress Nutrition Mexico, 3, 85-90
    Publisher Karger, Basil (1975).

21. Waslein, C., B. Kok, J. Myers and W.J. Oswald. "Photosynthetic
    Single Cell Protein," in Skrimshaw, N. and M. Milner.
    NSF/MIT Protein Resources Study. In press (1976).

22. Switzer, L. and N. Grisanti. "Spirulina Production in California
    Prosada report, Prosada, Inc. 1350 Summit Road, Berkeley,
    Calif., 94708, 29 pp. (1975).

23. CSO Int. Conversion of Animal Feed Grade Algae. An Engineering
    Report, prepared for Steve Marks Feedlot Zomora, Calif.,
    Suite 100, 2450 Stanwell, Concord (1975).

24. Brown, L.R. "The World Food Prospect." Science, 190, 1053
    (1975).

SYMPOSIUM IX

PHOTOBIOLOGY IN MEDICINE

PHOTOBIOLOGY IN MEDICINE

James D. Regan

Biology Division, Oak Ridge National Laboratory

P. O. Box Y, Oak Ridge, Tennessee   37830

I would like to spend a few minutes at the beginning of this symposium first, to put the involvement of photobiology in medicine in its historical context and secondly, to mention briefly some of the problems and advances in photomedicine that are active areas of research but which, because of exigencies of time could not be included in this symposium.

As other authors have pointed out (1) the use of light in the treatment of disease goes back to some of the earliest recorded history of man. And too, early man must have realized the harmful as well as the beneficial effects of sunlight. Sunburn, old farmer's (or sailor's) skin must certainly be as old as sunbathing, farming and sailing.

A fascinating aspect of photomedicine is its constant historical resurgence, particularly the therapeutic use of light. We see evidence of phototherapy in ancient writing as mentioned above. In the late nineteenth century there was a strong resurgence of phototherapy. Phototherapy was the treatment of choice in many diseases and crucial in certain others. As Daniels (2) has stated "Erysipelas was life-threatening disease and the ultraviolet phototherapist would be called in the middle of the night to treat it."

The achievements of photomedicine during its turn of the century rebirth were duly recognized by the award of the Nobel prize in 1903 to Dr. Niels Finsen the father of modern photomedicine.

Today we stand in the midst of a new resurgence of photomedicine. Table 1 presents examples of the involvement of light

TABLE 1
PHOTOBIOLOGY IN MEDICINE

IN DISEASE	IN THERAPY
UV Induced Skin Cancer	Phototherapy
Sunlight Sensitive Diseases	Dye-Light Therapy
Drug Induced Sunlight Sensitivity	Lasers

IN DIAGNOSIS	IN PREVENTIVE MEDICINE
Fluorescence Diagnosis	Full Spectrum Illumination
	Melatonin Secretion
	Calcium Metabolism

in medicine, in disease, in diagnosis, in therapy and possibly
even in preventive medicine.

Light in Disease
    Ultraviolet-induced skin cancer is a major world cancer problem.
It is extensively discussed in the scientific and lay press
especially nowadays because of the controversy over possible
depletion of the ozone layer by certain technological advances.

    In addition to skin cancer, sunlight promotes other disease
states, several of which will be discussed by forthcoming speakers.

Light in Diagnosis
    Fluorescence diagnosis is one of younger children of photo-
medicine and among the most promising, with capabilities for rapid
and accurate diagnosis of heritable storage diseases, heavy metal
poisoning and certain tumors.

Light in Therapy
    Light is employed extensively for phototherapy of hyperbili-
rubinemia, and now for certain tumors, two subjects which are
included in this symposium. Light in combination with photodynamic
chemicals finds a wide variety of application in photomedicine. Dye-
light therapy for herpes simplex lesions is a widely used modality and
is also widely critized as possibly being tumorogenic (3). Psoralens
and light are used to treat several skin conditions. Light in the
form of laser beams is a well-accepted clinical tool. The laser is
currently used routinely to treat retinopathies, and certain skin
conditions such as hemangiomas. The use of lasers in medicine is
a subject that could well occupy an entire symposium by itself (4).

Light in Preventive Medicine

As Wurtman (5) has pointed out artifically-lighted environments are apparently designed with one objective in mind and that is to provide significant reflected light to enable man to see. Until a few years ago it was not a consideration in the design of artifically-lighted environments that there may be other effects than simply vision engendered by the illumination that man perceives. Recently there have been a number of studies that suggest that artifically-lighted environments are not sufficient to produce some of the non-vision-related physiological responses in man generated by the full spectrum illumination of solar radiation. These responses include Vitamin D metabolism and its effect on calcium; melatonin secretion and its effects on several horomone systems of the body including sex hormones. Thus, it may be one of the objectives of the "new" science of photomedicine to consider, measure and prescribe the benefits and the limits of full spectrum illumination mimicking natural solar radiation in artifically-lighted environments.

Obviously we could not cover all the subjects I have mentioned in one symposium. We have therefore chosen topics which represent a cross-section of the active pursuits of modern photomedicine.

References:

1. M. A. Pathak, D. M. Kramer, and T. B. Fitzpatrick, Photobiology

   and Photochemistry of Furocoumarins (Psoralens), in "Sunlight

   and Man" (T. Fitzpatrick, M. Pathak, L. Harber, M. Seiji and

   A. Kukita, eds.) p. 336, University of Tokyo Press, Tokyo (1974).

2. F. Daniels, Physiological and Pathological Extracutaneous Effects

   of Light on Man and Mammals, Not Mediated by Pineal or Other

   Neuroendocrine Mechanisms, in "Sunlight and Man" (T. Fitzpatrick,

   M. Pathak, L. Harber, M. Seiji and A. Kukita, eds.) p. 252,

   University of Tokyo Press, Tokyo (1974).

3. Anon., Dye-light therapy for herpes virus, Med. World News 15:

   39-55 (1974).

4. D. Rounds, Laser Applications to Biology and Medicine, in "Laser

   Handbook" (F. Arecchi and E. Schulz-Dubois, eds.) Vol. 2, pp. 1864-

   1890 (1972).

5. R. J. Wurtman, The effects of light on man and mammals, Annual

   Rev. Physiol. 37: 467-508 (1975).

PHOTOPHYSIOLOGICAL RESEARCH - PREVENTIVE MEDICINE

H. Ippen

Direktor der Dermatologischen Klinik

v.-Siebold-Straße 3, D 3400 Göttingen

Numerous environmental factors have an effect upon man. One of the most important of these is sunlight. For centuries, probably even throughout millenia, the quantity of solar radiation affecting the inhabitants of a given region remained practically constant. It is most likely that a minimum of exposure to solar radiation was reached in the past few centuries in civilized countries because of expanding urbanisation, dress habits, etc. Around the beginning of the twentieth century, however, a drastic increase in exposure to sunlight occured. This phenomenon had the following reasons.

1. A rapid increase of life expectancy. This resulted in a greater total exposure of the individual to light.

2. Changes in dressing habits. Initially only in regard to the face, neck, hands, and the length of skirts; later by increasing exposure of larger and larger skin areas; and finally by increasing usage of synthetic materials for clothing which lack the light absorbent or reflecting qualities of traditional materials.

3. A radical change of attitude towards tanned skin. This was earlier carefully avoided as a stigma of physical labor. Now it became the sign of health, closeness to nature, and even affluence, and was intensified by voluntary increased exposure to light.

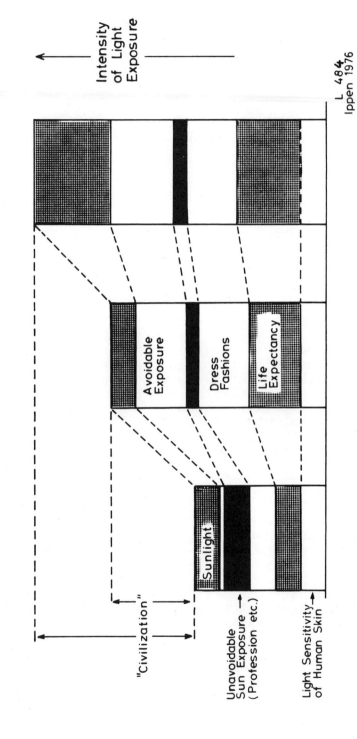

Figure 1. Factors influencing sun exposure of modern man.

4. The growing prosperity of a wider spectrum of
   society. Thus voluntary exposure is increased by
   travelling in sunny regions.

In a majority of the people, these factors cause a
rise in the quantity of light, which in the course of
their lives, pervades their skins. It is certainly,
nowadays many times more than that which affected
former gererations. An additional danger may arise
in consequence of civilization-caused diminuation of
the atmospheric ozone layer which may increase the
effective radiation to the surface of the earth and
is bound to cause a shifting of the radiation towards
the short wave range (ultraviolet C).

Medically, the consequence of this lifelong, immensely
increased exposure to light is without doubt, a rapid
increase of malignant tumors- basalioma, carcinoma,
and to a certain extent also malignant melanoma - in
the light damaged skin. This assumption is confirmed
by the frequency of skin cancer among those parts of
the white population which also, in the course of
civilization, settled in regions near the equator
(for example, Australia).

From this growing danger, there arise for the
physician, a number of questions which photobiological
basic research up until now can answer only in part
or not at all. At the time being, the physician is
able to fight the growing danger merely by means of
informing himself and his patients. In the process,
however, he has to combat such potent adversaries as
the fashion of tanned skin, the fondness for travel,
and sun addiction (whatever "Sonnen-Hunger" may be).

One of the few opportunities where the physician is
able to intervene, is the early diagnosis and
elimination of phototoxic drugs and other environmental hazards, which may add to the sum of light
damages in the individual. But in doing this, he can
only rely on the results of basic research which by
means of pre-clinical tests may recognize phototoxic
side-effects of new products beforehand. Furthermore,
it ought to be made clear, what kinds of effects the
residues of such phototoxic substances have, which
remain in the body for a longer period of time, for
example the furocoumarins, therapeutically utilized.
It is highly probable that they are not completely
eliminated in a short period of time.

One important area of photobiological research, the
light screening agents, has already met with greater
attention in recent years. Twenty years ago, when I
began researching in this area, there were already a
large number of preparations on the market. Their
effectiveness, however, was so slight, that after
application of most of these products, the time in
which the threshold of erythema was reached, was
seldom doubled and in only a few cases tripled.
Accordingly, the advertisements ignored light
screening, but lured the customer into the sun by
promises of a deep tan. Today, most of the sun tan
preparations on the European market lengthen the
threshold of erythema time from three to ten-fold and
advertisements strongly stress the necessity of skin
protection against sunlight. This development is
mainly due to the important contributions of the
unforgotten Rudolf S c h u l z e and Franz
G r e i t e r . Schulze, with the concept and method
of the "average light screening factor" developed a
standard for testing the effectiveness of light
screening agents, which by virtue of its simplicity,
has found general recognition. Greiter's merit
consists of the adoption of the term "sun screen
factor" for the description of his products. On the
basis of his own experiments, he expanded these
descriptions into an intensive and objective
information for the customers. As a consequence of
this pioneer's work, the majority of European light
screening products, indispensable from the physicians
point of view, are offered carrying an indication of
their light screening factor and frequently, other
relevant information as well.

At this point, a brief survey of the working
principle of light screening substances and the
technique of the screening factor determination is
appropriate. Light screening substances are
preparations, which on the surface of the skin and,
as my own experiments (4) show, also in the stratum
corneum form a layer, which reduces the amount of
light (predominantly ultraviolet B) that enters into
the living cells of the epidermis. With the exception
of some purines and pyrimidines (3), whose mode of
action is not yet decisively known, the reduction of
UV B radiation is based predominantly on the
application of substances with a marked absorption
maximum between 300 and 320 nm, in other words, the
maximum of the erythema effectiveness of the sunlight
today. Furthermore, it is desirable that such prepara-

tions have the highest possible water resistance, a
goal which has not yet been achieved satisfactorily.

The determination of the effectiveness of such
products in the form of the "medium screening factor"
theoretically, ought to be carried out with natural
sun light. But Schulze could already show the
agreement in comparison of values between sunlight
and certain artificial light sources (9). The use of
such light sources is necessary in the rather
scantily sunlit areas of middle Europe. Furthermore,
it is important that artificial light sources be
available, which permit the application of fifty or
even one hundred-fold the erythema dose in a
reasonable time. Further progress in the development
of light screening products cannot, at the present
time, dispense with the biological method of the
screening factor determination, since all the
experiments concerning the determination of the
effectiveness in vitro have been, up to now,
unsatisfactory. For one, the effectiveness of such
preparations depends not so much on their optical
properties, meaning the quantity of light filtering
substance they contain- it depends rather more on the
base of the preparation (2). And secondly, all the
experimental models concerning skin surface and
erythema reaction carried out on animals, are not
entirely applicable to human skin. It is, therefore,
still necessary to determine the "medium screening
factor" as the average of individual factors
(quotient of the time in which threshold of erythema
is reached in protected and unprotected skin.) of at
least twenty test subjects. In Europe this is
generally done using four "Ultravitalux" lamps
(Osram) at a distance of 40 cm. The time threshold
of the unprotected skin here is between one to four
minutes. It is, therefore, easily calculated that
the determination of a medium screening factor of
ten already requires a period of several hours of
exposure.

To meet these demands, the light bulb industrie is
required to produce sunlight-simulating lamps of much
greater capacity.

But there arise also in photobiological research, a
series of important questions concerning sun
screening products:

1. Is it at all sensible to continue development of
the current type of sun screening products which
are almost wholly restricted to ultraviolet B ?
First, it must be certain, whether ultraviolet B
and the acute light damage of the skin are
responsible for chronic light damages. It has not
been determined yet, whether the effect of UV B
and the adaptation to this range of light do not
play a positive role through interference with the
effects of other ranges of radiation (UV A etc.).

2. Are there more satisfactory ways to obtain a more
thorough protection of the skin against light
damage? One would, for instance, think of a lasting alteration in the optical properties of the
stratum corneum by chemical or physical means.
Also, we ought to search for simpler and more
reliable forms of application of light screening
agents. This might be done with baths containing
ingredients that go on by substantivity to the
stratum corneum of the skin (1).

3. Finally, further research of the light screening
effect of carotine against the effects of phototoxic porphyrins should lead to an internally
applicable screen in which a substance as a
radical scavenger, quencher or the like, interupts
the basic process of the light erythema.

But beyond that, other important areas must not be
neglected. Among these, the questions concerning the
pathogenesis of chronic light damages of the skin
will probably be treated intensively during the
Symposia X and XII.

One burning problem, however, does not yet receive
the necessary attention: The effect of ultraviolet C
on the skin, in connection with the possible change
in terrestric solar radiation, due to the
decomposition of the ozone layer.

The credit for basic research of the acute effects of
this short wave length, high energy radiation goes
mainly to R o t t i e r (8). But now it is
necessary to find out the chronic changes UV C causes
in the human skin. In connection with this, the
following questions suggest the impending problems:

1. Are there any changes in the stratum corneum under
   the influence of ultraviolet C ? Do its optical
   properties change and, as a result, its
   permeability for other ranges of light? Do the
   physico-chemical properties of the stratum corneum
   change, perhaps the permeability or water-binding
   capacity which are of fundamental importance for
   maintenance and quality of the skin?

2. What are the effects of the relatively small doses
   of UV C which permeate the stratum corneum and
   thus reach the cells of the epidermis? Especially,
   the fact that their quantity is seldom sufficient
   for the destruction of the cell, but still enough
   for manifold photochemical reactions with parts of
   the cell, especially the nuclei, ought to be
   reason enough to dedicate special attention to this
   question.

3. What are the effects of photo-products which are
   formed in the stratum corneum under the influence
   of UV C and later invade the organism?

Another area of research in photobiology must be
considered as a concession to fashion and widespread
ignorance. That is the further research on the
relationship between the influence of light and the
pigmentation of skin.

It is highly improbable that the fashion of tanned
skin will, in the near future, give way to
"aristocratic pallor" again. Therefore, one must also
consider, for the sake of preventive medicine, the
search for such substances which either artificially
or by means of the melanin, bring forth the desired
tan. This effect, however, ought no longer to require
the enormous quantities of light which people at the
time being have to try to find in the sun and
additionally in the so-called "Solaria".

Technology as well as ignorance and fashion increase
the exposure of fair skinned men to dangerous amounts
of radiation. This hazard presents a challenge for
the photo-biologist and the physician. It is of great
interest to a large section of the population that
research in the above mentioned field should soon
achieve widely hoped for successes.

References

1. Abbe, N.J.van, "The Substantivity of Cosmetic In-
    gredients to the Skin, Hair and Teeth" J. Soc.
    Cosmet. Chem. 25, 23-31, 1974.
2. Charlet, E. and Finkel, P., "Lichtschutzfaktor -
    Emulsionstype und Lichtschutzsubstanz als Einfluss-
    grösse" Angew. Kosmetik No. 5., 1976.
3. Ippen, H., "Erythemschutz durch externe Anwendung von
    Pyrimidin- und Purin-Derivaten" Arch. klin. exp.
    Derm. 235, 25-31, 1969.
4. Ippen, H., and Perschmann, U., "Zum Verhalten fluor-
    eszierender Lichtschutzmittel auf der Haut" Arch.
    Derm. Forschg. 236, 207-216, 1970.
5. Krinsky, N.I., "The Protective Function of Carotenoid
    Pigments" Photophysiology 3, 123-195, 1968.
6. Mathews-Roth, M.M., "Phytoene As A Protective Against
    Sunburn (>280 nm) Radiation in Guinea Pigs" Photochem.
    Photobiol. 21, 261-263, 1975.
7. Mathews-Roth, M.M., "Therapy of Human Photosensitivity"
    Photochem. Photobiol. 22, 302-303, 1975.
8. Rottier, P.B., "Was ist die Rottier'sche Theorie über
    das UV-Erythem bei 250 nm?" Arch. Derm. Forschg. 239,
    148-149, 1970.
9. Schulze, R., "Einige Versuche und Bemrkungen zum Problem
    der handelsüblichen Lichtschutzmittel" Parf. u. Kosmet.
    37, 310, 365, 1956.

ERYTHROPOIETIC PROTOPORPHYRIA - THE DISEASE, AND ITS TREATMENT WITH BETA-CAROTENE

Micheline M. Mathews-Roth

Channing Laboratory, Harvard Medical School

774 Albany Street, Boston, Mass., 02118

SUMMARY

The clinical, biochemical and photobiological characteristics of the photosensitivity disease, erythropoietic protoporphyria, are presented. Treatment of this disease with beta-carotene is also discussed.

INTRODUCTION

The porphyrias are a group of diseases in which abnormal amounts of various porphyrin compounds accumulate in blood and body tissues. The characteristics of the various kinds of porphyrias have recently been reviewed (1,2). Some porphyrias are characterized by sensitivity to light, some are not. In this paper, we will discuss one form of light-sensitive porphyria, erythropoietic protoporphyria (EPP).

CHARACTERISTICS OF THE DISEASE

EPP is characterized by abnormally elevated levels of protoporphyrin lX in erythrocytes, feces and plasma, and by sensitivity to visible light (380-560 nm)(1,2,3). This sensitivity, in which the protoporphyrin was shown to be the photosensitizer, manifests itself by a burning sensation in the skin, followed by varying degrees of erythema and edema. The disease is diagnosed by detecting the presence of abnormally high levels of protoporphyrin in blood and stool by chemical analysis. Contrary to what is found in the other porphyrias, urinary porphyrins remain within normal limits in EPP. When a smear of blood from a patient with EPP is examined under the fluorescence microscope, large numbers of red-

fluorescent erythrocytes are seen; these are not seen in normal
individuals. In addition, if the skin of the light-exposed areas
of the body is examined under the light microscope, an amorphous,
homogenous substance in and around the walls of small blood-vessels
of the upper papillary dermis will be seen (1,2,3).

The majority of patients with EPP have other members of their
families who also have the disease. There have been several genetic studies of EPP patients and their families. It was found that
some relatives of the patients have somewhat elevated erythrocyte
protoporphyrin levels, but yet were asymptomatic; this suggests
the existance of a carrier state for EPP. It was concluded that
EPP is genetically transmitted as an autosomal dominant trait with
variable penetrance and expressivity.

The majority of patients with EPP report that the onset of
photosensitivity began in early childhood - usually before age six,
some as early as 18 months. The predominent manifestations of sensitivity reported by the patients were, in decreasing order of frequency: burning, swelling, itching and redness of the skin. Some
patients develop shallow depressed scars over the nose and cheeks,
and on the backs of the hands, which developed after severe epi-
sodes of photosensitivity. Some patients report only subjective
symptoms of itching and burning, and have none of the objective
changes of redness, swelling and scarring; these patients are usually dismissed by their physicians as hypochondriacs, when in reality they have EPP. Thus, it is important for the physician to investigate for the presence of EPP all patients who report itching
and burning of the skin on light exposure, even in the absence of
objective findings.

The amount of sun exposure a patient with EPP can tolerate varies. Some report they can tolerate only a few minutes, others say
they can tolerate several hours. Characteristically, they report
being sensitive to light through window glass. About half of the
patients report decreases in photosensitivity during winter, but
those engaging in skiing report that the light reflected by the snow
can cause severe photosensitivity reactions.

In the majority of cases, EPP appears to be a benign disease.
Many patients have a mild degree of anemia characterized by somewhat decreased levels of hemoglobin and hematocrit. This usually
requires no treatment. There also seems to be an increase of incidence of cholelithiasis, with several patients requiring cholecys-
tectomies. Chemical analyses of the gallstones detected high levels
of protoporphyrin. An occasional patient, however, has been found
to develop fatal liver disease, probably due to massive deposition
of protoporphyrin in the liver. There are nine such cases recorded
to date (3). These patients all had extremely high levels of protoporphyrin, and had abnormal liver function tests, jaundice, and

microscopic evidence of cirrhosis.

On light-microscope examination of liver biopsies from EPP patients, both with and without cholelithiasis, some workers have noted porphyrin deposits in the liver cells, and in some cases, slight fibrotic changes; these were not associated with abnormal liver chemistries (3). Electron microscope examination of liver biopsies from EPP patients revealed that the liver cells of some EPP patients contain either cytoplasmic or mitochondrial inclusions, again in the absence of abnormal liver function tests (4). The significance of these various structural changes, and their relation if any to the development of serious liver disease are unknown at this time.

The source of the abnormal amounts of protoporphyrin which accumulate in EPP has been under study for several years. Some workers have postulated that a significant amount, if not all, of the protoporphyrin is synthesized in the liver, others have suggested that all the abnormal porphyrin in synthesized by the bone marrow. Recently Piomelli and his collaborators have calculated that all protoporphyrin excreted in the stool can be accounted for by synthesis by the reticulocytes in the bone marrow (5). They showed that there is rapid leakage into the plasma of this protoporphyrin from the reticulocytes during the process of their maturation into erythrocytes, and that the leaked protoporphyrin is rapidly cleared from the plasma by the liver, thus accounting for the protoporphyrin in the stool. They found that circulating erythrocytes of up to about 20 days of age contained protoporphyrin - older ones usually did not. They also suggested that those pathological liver changes seen in occasional EPP patients which we have just mentioned were secondary to the accumulation in the liver of this protoporphyrin leaked from the erythrocytes and cleared by the liver. They calculated that their theory would hold true even in the presence of a stool porphyrin level equivalent to the erythroid mass. Thus they suggested that there is no need to postulate synthesis of porphyrins by the liver as an integral finding in EPP. Their work also showed that there is only one line of erythrocytes present in EPP, as all reticulocytes were found to fluoresce.

The genetic lesion in EPP has also been studied. Several groups of workers have now found that there are markedly decreased levels of the enzyme ferrochelatase (heme synthetase) in the cells of patients with EPP (6,7). Enzyme levels have been studied in bone marrow, liver, and skin fibroblasts; decreased levels of enzymes have been found in all these tissues. Decreased levels in the bone marrow would lead to overproduction of protoporphyrin, but apparently not to the point of getting any significant degree of feedback inhibition of ALA synthetase, which would lead to severe anemia. Although many patients with EPP have some degree of anemia, it is very slight, and usually does not need treatment. There has

been one reported case of a serious hemolytic anemia in a patient
with EPP (8). Some workers have suggested that there might be some
episodic overproduction of protoporphyrin by the liver if there were
instances of sporadic increases of ALA-synthetase production (perhaps triggered by various chemicals, drugs, alcohol, etc.). Calculations suggest that even though there are decreased levels of ferrochelatase in liver in EPP, there would be enough heme made to keep
ALA-synthetase inhibited to avoid constant overproduction of this
enzyme, with concomitant increase in protoporphyrin production.

Other workers have suggested that perhaps the high levels of
protoporphyrin occur because of the presence in EPP patients of a
defective ferrochelatase, or m-RNA for the enzyme, which is destroyed rapidly, and thus protoporphyrin builds up (9).

Thus it would seem that in EPP there is either a decreased a-
mount of, or the presence of a defective, easily destroyed ferrochelatase or m-RNA, which leads to the accumulation of protoporphyrin in reticulocytes. This excess protoporphyrin leaks rapidly into
the plasma from the maturing reticulocytes and young erythrocytes.
The protoporphyrin is then cleared from the plasma by the liver,
and excreted into the bile (with or without some recirculation via
the enterohepatic circulation). Accumulation of this protoporphyrin
in the liver may, in some patients, lead to serious liver disease.
The levels of ferrochelatase in the marrow are usually sufficient
to allow enough hemoglobin synthesis to take place so as not to lead
to serious anemia, and the levels in the liver are sufficient to
keep ALA-synthetase inhibited. Further work still needs to be done
to determine if in actuality there may be episodic overproduction
of protoporphyrin by the liver in EPP under certain circumstances.

MECHANISM OF PHOTOSENSITIZATION

Studies to date suggest that the high levels of protoporphyrin
present in the erythrocytes, plasma, or extracellular fluid of the
skin can be stimulated to an excited (triplet) state by the visible
light, which penetrates the skin to the level of the dermal capillaries. The excited protoporphyrin may either destroy cellular
components directly, or may react with molecular oxygen to form
singlet oxygen, a moiety which can be highly damaging to cellular
components. Direct cell damage and/or damage to cells or cellular
organelles such as lysosomes, with release of chemical mediators
which in turn damage other cells, would then cause the manifestations of itching burning, swelling and erythema experienced by the
patient with EPP, and in time lead to the morphological changes
seen in light-exposed skin under the microscope. Work is in progress at this time in several laboratories using animal modesls to
try and determine the actual molecular and cellular events in photosensitization in EPP. Gschnait and his collaborators have been able

to reproduce the dermal blood vessel changes seen in EPP patients
in light-exposed animals made porphyric with griseofulvin (10,11).

TREATMENT OF EPP

Patients with EPP have found that the topical sunscreens effective in protecting against hypersensitivity to the sunburn range
are ineffective as protective agents in EPP. Various systemic a-
gents, such as antimalarials, inosine, and Vitamin E have been tried,
but with little success.

Because of the finding that carotenoid pigments could prevent
lethal photosensitization in bacteria (12), and that the administration of beta-carotene could prevent the lethal photosensitization
of mice made photosensitive with hematoporphyrin (13), high doses
of beta-carotene were administered to patients suffering from EPP,
in an attempt to ameliorate their photosensitivity (14,15). In a
collaborative study started in 1970 and completed in 1975, we have
now treated with beta-carotene in collaboration with 46 additional
physicians, a total of 133 patients with EPP, whose disease was
documented by the presence of elevated porphyrin levels in blood
and stool. Of these, 84% have at least tripled their ability to
tolerate exposure to sunlight, whereas the remaining 16% have reported little or no benefit from beta-carotene therapy (Table 1).
Other workers have also used beta-carotene in the treatment of EPP,
and have reported results similar to ours (16,17,18).

We have used the following starting dosage schedule of beta-
carotene; 1 to 4 yrs. – 60 to 90 mg/day; 5 to 8 yrs. – 90 to 120
mg.day; 9 to 12 yrs. – 120 to 150 mg/day; 13 to 15 yrs. – 150 to
180 mg/day; and 16 yrs. and older – 180 mg/day. The average dose
for the patient's age should be administered for 4 to 6 weeks, and
the patient should be instructed not to increase sun exposure for
either 4 weeks, or until some yellow discoloration of the skin,
especially of the palms of the hands, is noted. Then, exposure can
be increased cautiously and gradually until the patient determines
the limits of exposure to light that can be tolerated. If the degree of protection is not sufficient, the daily dose of carotene
should be increased by 30 to 60 mg for children under 16, and up
to a total of 300 mg/day for those over 16. If after two months
of therapy at these higher doses no significant increase in tolerance to sunlight exposure has occurred, it can be assumed that beta-
carotene therapy will not be effective in the patient, and the medication should be discontinued.

TABLE 1

EFFECT OF BETA-CAROTENE ON TOLERANCE TO SUNLIGHT
OF PATIENTS WITH ERYTHROPOIETIC PROTOPORPHYRIA

Number of Patients	%	Protection Index*
10		1
12	16	2
27		3 - 5
39		6 - 10
15	84	11 - 15
6		16 - 20
24		> 21
Totals: 133	100	

*Protection index: the number of minutes of summer sunlight toler-
ated after beta-carotene therapy divided by the number of minutes
of sunlight tolerated before therapy. A protection index of 1
indicates those patients who stated that beta-carotene proved of
no benefit at all to them. We have arbitrarily considered that
protection factors of less than three would not be held to represent significant improvement.

CLINICO-PHARMACEUTICAL CONSIDERATIONS

Beta-carotene was well-tolerated by the patients. There were
no reports of untoward side effects, other than a rare complaint of
occasional loose stools, which cleared up spontaneously, and was
never severe enough to warrant stopping the administration of beta-
carotene. The majority of the patients were not bothered by the
carotenodermia associated with beta-carotene intake. The patients'
blood carotene levels reached a maximal value, and carotenodermia
developed about 4 to 6 weeks after the start of therapy. No
patients developed abnormally high levels of vitamin A, and there
were no significant changes during therapy from pre-treatment values
of blood glucose, urea nitrogen, bilirubin or glutamate-oxaloacetate
transaminase. The complete blood count of the patients was also
not affected by beta-carotene intake. The mild anemia, which many

patients had before the start of therapy, was neither worsened nor improved by beta-carotene. No menstrual irregularities were reported by the women patients.

We do not recommend high intake of carotenoid-containing foods as a method of obtaining high levels of blood and skin carotenoids. Toxic reactions, such as leucopenia and methemoglobinemia have been found to occur in those who ingest large quantities of vegetables in the amounts which would be necessary to obtain the carotenoids' protective effect. When purified beta-carotene was given, neither of these toxic reactions were found to occur. This confirms the suggestions of earlier workers that these untoward effects were due to constituents of the vegetables other than carotenoids.

The form in which carotene is given is also important. It has been shown that the absorption of beta-carotene is more effective as the "beadlet" (Roche) preparation, rather than as crystalline carotene dissolved in oil.

We recently studied the livers of two patients who had been taking beta-carotene for photosensitivity. There was no evidence of the accumulation of large amounts of either carotene or vitamin A in the liver by chemical analysis, or evidence of morphological alterations attributable to carotenoid or vitamin A accumulation by light-microscopic examination (19).

MECHANISMS OF CAROTENOID PROTECTION AGAINST PHOTOSENSITIZATION

As we have mentioned previously, protoporphyrin, the photosensitizer in EPP can be raised to an excited state in the presence of light, and can react with oxygen to form singlet oxygen, and that these reactive moities can cause cellular damage.

Beta-carotene has been shown capable of quenching both free radicals and singlet oxygen in vitro (20,21), and the ability of bacterial carotenoids to confer protection to the organism containing them is correlated to their ability to quench singlet oxygen in vitro (22,23). However, some degree of free radical quenching may also be involved, but the evidence seems to indicate that it may play a rather minor role. No work on the mechanism of carotenoid function has yet been done in an animal model, as there are many technical difficulties. Hopefully, future studies will overcome these difficulties, and will elucidate the definitive mode of action of the carotenoids' protective effect in EPP.

REFERENCES

1. D. P. Tschudy, I. A. Magnus, and J. Kalivas, in "Dermatology in General Medicine" (T. B. Fitzpatrick, ed.) pp. 1143-1166, McGraw-Hill Book Co., Inc., New York (1971).

2. H. S. Marver and R. Schmid, in "Metabolic Basis of Inherited Disease" (J. B. Stanbury, et al., eds.) pp. 1087-1140, McGraw-Hill Book Co., Inc., New York (1972).

3. V. A. DeLeo, M. B. Poh-Fitzpatrick, M. M. Mathews-Roth, and L. C. Harber, Erythropoietic protoporphyria - 10 years experience, Amer. Jour. Med. 60:8-22 (1976).

4. K. Wolff, E. Wolff-Schreiner, and F. Gschnait, Liver inclusions in erythropoietic protoporphyria, Eur. Jour. Clin. Invest. 5:21-26 (1975).

5. S. Piomelli, A. A. Lamola, M. B. Poh-Fitzpatrick, C. Seaman, and L. C. Harber, Erythropoietic protoporphyria and lead intoxication: The molecular basis for difference in cutaneous photosensitivity: I. Different rates of disapppearance of protoporphyrin from the erythrocytes, both in vivo and in vitro, Jour. Clin. Invest. 56:1519-1527 (1975).

6. S. S. Bottomly, M. Tanaka, and M. A. Everett, Diminished erythroid ferrochelatase activity in protoporphyria, Jour. Lab. Clin. Med. 86:126-131 (1975).

7. H. L. Bonkowsky, J. R. Bloomer, P. S. Ebert, and M. J. Mahoney, Heme synthetase deficiency in human porphyria - demonstration of the defect in liver and cultured skin fibroblasts, Clin. Invest. 56:1139-1148 (1975).

8. F. S. Porter and B. A. Lowe, Congenital erythropoietic protoporphyria 1. Case reports, clinical studies and porphyria in two brothers, Blood 22:521-531 (1963).

9. K. G. A. Clark and D. C. Nicholson, Erythrocyte protoporphyrin and iron uptake in erythropoietic protoporphyria, Clin. Sci. 41:363-379 (1971).

10. K. Konrad, H. Honigsmann, F. Gschnait, and K. Wolff, Mouse model for protoporphyria II. Cellular and subcellular events in the photosensitivity flare of the skin, Jour. Invest. Derm. 65:300-310 (1975).

11. H. Honigsmann, F. Gschnait, K. Konrad, G. Stingl, and K. Wolff, Mouse model for portoporphyria III. Experimental production of chronic erythropoietic protoporphyria-like skin lesions, Jour. Invest. Derm. 66:188-195 (1976).

12. M. M. Mathews and W. R. Sistrom, Function of carotenoid pigments in non-photosynthetic bacteria, Nature 184:1892-1893 (1959).

13. M. M. Mathews, Protective effect of β-carotene against lethal photosensitization by hematoporphyrin, Nature 203:1092 (1964).

14. M. M. Mathews-Roth, M. A. Pathak, T. B. Fitzpatrick, L. H. Harber, and E. H. Kass, Beta-carotene as a protective agent in erythropoietic protoporphyria, New England Jour. Med. 282:

1231-1234 (1970).

15. M. M. Mathews-Roth, M. A. Pathak, T. B. Fitzpatrick, L. H. Harber, and E. H. Kass, Beta-carotene as an oral photoprotective agent in erythropoietic protoporphyria, Jour. Amer. Med. Assoc. 228:1004-1008 (1974).

16. H. Baart de la Faille, D. Suurmond, L. N. Went, J. van Steveninck, and A. A. Schothorst, β-carotene as a treatment for photo-hypersensitivity due to erythropoietic protoporphyria, Dermatologica 145:389-394 (1972).

17. G. Krook and B. Haeger-Aronson, Erythropoietic protoporphyria and its treatment with β-carotene, Acta Dermatovener 54:39-44 (1974).

18. F. Gschnait and K. Wolff, Die erythropoetische protoporphyrie, Der Hautarzt 25:72-80 (1974).

19. M. M. Mathews-Roth, A. A. Abraham, and T. G. Gabuzda, β-carotene content of certain organs from two patients receiving β-carotene, Clin. Chem. 22:922-924 (1976).

20. E. Fugimori and M. Tavla, Light-induced electron transfer between chlorophyll and hydroquinone, and the effect of oxygen and β-carotene, Photochem. Photobiol. 5:877-887 (1966).

21. C. S. Foote and R. W. Denny, Chemistry of singlet oxygen VII. Quenching by β-carotene, Jour. Amer. Chem. Soc. 90:6233-6235 (1968).

22. M. M. Mathews-Roth and N. I. Krinsky, Failure of conjugated octaene carotenoids to protect a mutant of Sarcina lutea against lethal photosensitization, Photochem. Photobiol. 11:555-557 (1970).

23. M. M. Mathews-Roth, T. Wilson, B. Fujimori, and N. I. Krinsky, Carotenoid chromatophore length and protection against photosensitization, Photochem. Photobiol. 19:217-222 (1974).

PHOTOCHEMOTHERAPY OF

PSORIASIS (PUVA)

Klaus   Wolff

Department of Dermatology, University

of Innsbruck, Anichstrasse 35, A-6020 Innsbruck

SUMMARY

This paper reviews photochemotherapy (PUVA) in over 300 patients with severe, generalized psoriasis, covering a follow-up period of up to two years. Results of treatment, side effects and laboratory studies are described and the principles, theoretical background, and optimum conditions of this treatment are discussed.

INTRODUCTION

Photochemotherapy is an approach to the treatment of psoriasis which is based on the interaction of long wave ultraviolet light (UV-A) and a systemically administered photoactive compound, 8-methoxypsoralen (8-MOP), within the skin (1); hence the designation PUVA (PUVA = 8-methoxy-Psoralen plus UVA) (2,3). As the photoactivation of 8-MOP depends on UVA (particularly wavelengths of 330-365 nm) and as the penetration of UVA is confined to the superficial layers of the skin, PUVA represents a successful attempt to localize systemic chemotherapy to the skin without risking cytotoxicity in other organs (1-3).

Photosensitizing agents have been employed, for some time, in the topical treatment of psoriasis; these include coal tar preparations followed by conventional ultraviolet light, as used in the Goeckerman regimen (4), and more recently, and as reported at the 6th International Congress of Photobiology (5), the topical application of psoralens followed by black light irradiation (6-10). However, it was not until high intensity UVA-irradiation

systems had been developed (1) that systemic photochemotherapy
became feasible, as a practical approach to treatment of disease,
and thus opened what appears to be a new dimension of photomedicine. The dramatic effectiveness of PUVA in inducing and maintaining clinical remissions of psoriasis has been documented in large
series of patients (2,3,11) and has been confirmed in several
centers (12-15).

METHODS

The rationale of PUVA therapy is to bring psoriasis into remission by repeated, controlled photosensitization reactions which
are monitored to remain within a therapeutically desired range
(3). Since phototoxic erythema is a limiting factor, careful
attention to dosimetry is essential. PUVA doses are administered
in 0.6 mg/kg body weight of 8-MOP and in J/cm^2 of UV-A and are
monitored according to the patients sensitivity to phototoxic reactions, as determined by phototesting (16). The light system employed has to deliver UV-A uniformely over the entire body surface
without significant intensities of UV-B or infrared; it should
provide means to accurately measure and reliable deliver predetermined UV-A doses to the patient and thus to guarantee a high degree of safety; most importantly, the intensity of irradiation
should be high enough to deliver adequate doses within reasonable
periods of time (16). Irradiation is performed two hours after
oral ingestion of the drug and a maximum of four such treatments
are given per week until clearing is achieved. During treatment
doses are increased or adjusted to the patients tolerance and
response (16).

RESULTS

Clearing: The clinical response to PUVA is unequivocal (1-3);
it usually becomes noticeable after the third or fourth treatment
and manifests as flattening of lesions and a decrease in scaling
and erythema. A mild erythema of the uninvolved skin is usually
the result of the first PUVA exposures but is soon followed by a
deep uniform tan (3). In a study of 305 patients with severe, generalized psoriasis (11) complete clearing was obtained in 289
(94%) requiring 13 treatments (mean) within a time period of 23
days (mean) (Figure 1). Fifteen patients were considerably improved and one patient failed to respond. Similar results have been
obtained in patients with psoriatic erythroderma (complete clearing in 13/14) and in patients with severe pustular psoriasis of
Zumbusch (complete clearing in 7/7) (11). Dosimetry may represent
a problem in these patients, when no uninvolved skin is available
for phototesting, and thus requires considerable experience. These
patients tend to flare when treated too aggressively and this ne-

cessitates a cautious approach both determining and adjusting PUVA doses. Treatment times are thus longer than in patients with chronic indurated psoriasis but the results are equally impressive.

One of the most gratifying clinical aspects of PUVA treatment is the fact that PUVA is effective in patients who have previously not responded to any type of conventional treatment and, more importantly, in those patients who had previously required therapy with systemic cytotoxic agents and / or systemic corticosteroids. Under PUVA these patients can be cleared and, at the same time, taken off their previous systemic medication which practically eliminates the risk of side effects attributable to these agents (17).

Maintenance therapy: After clearing patients are maintained in remission by PUVA treatments which are given once or twice a week and are gradually reduced to once a month or less. 83% of the patients have been kept free of psoriasis for periods of more than

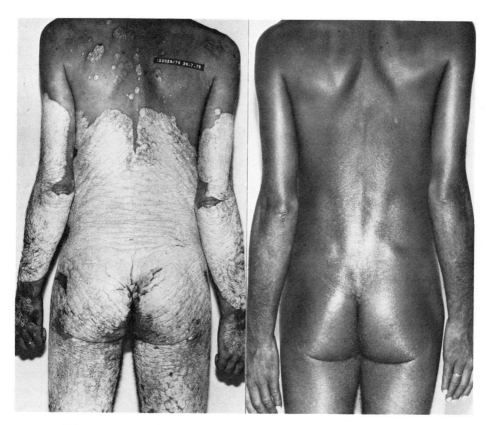

Figure 1: Patient before and after PUVA treatment.

two years; to achieve this result 27% had to be treated once/week, 27% once/2 weeks, 14% once/3 weeks, 12% requiring no treatment, and the rest at shorter or longer intervals. A mean cumulative irradiation load of 37 J/cm^2/month has been calculated for patients undergoing this type of maintenance treatment (11).

In an attempt to keep the total cumulative irradiation dose low patients have been arbitrarily considered maintenance treatment failures when more than two treatments per week were required to maintain remission over longer periods of time. In our experience, 8% of the patients fall into this category (11).

Short-term side effects: Short term side effects are a result of overdosage or manifest as nausea after ingestion of the drug. Careful observation of the guidelines for dosimetry have permitted to limit these side effects to a small percentage of the patients: +++ erythemas have been observed in 6%, localized blistering in 2% and light-induced Koebner reactions in 2%. Pruritus occurs in 21% of the patients and thus may represent a problem whereas nausea, observed in 7,5% of the patients (3,11), can usually be overcome by splitting the dose of 8-MOP or administering it with some food.

Potential long-term side effects: The fact that 8-MOP binds to DNA under the influence of UV-A light (18-22); that multiple cumulative treatments are required by the maintenance treatment schedule (2,3); and animal experiments on tumorgenesis and cataract formation* have given rise to some concern regarding long term side effects of PUVA treatment. Studies performed to date have failed to show any adverse effect of PUVA on routine laboratory tests performed during a follow-up period of 2 years (11);nor have they revealed clinical or histologic evidence for degenerative changes in the skin, the development of actinic keratoses or other changes that would suggest the development of skin tumors (11); there have been no ocular changes; and immunofluorescence studies of skin, studies on antinuclear antibodies and delayed type hypersensitivity reactions of patients undergoing long term PUVA-therapy have not revealed anything abnormal (11).

Since an increased number of chromosomal aberrations, sister chromatid exchanges, and micronuclei have been found in lymphocytes treated with 8-MOP and UV-A in vitro (12,23,24), there have also been some concerns as to what may happen to the blood cells of PUVA patients while circulating through superficial skin capillaries. Preliminary studies have failed to reveal an increased number of sister chromatid exchanges in lymphocytes of PUVA patients (25).

---

* for reference see ref. 3

COMMENT

PUVA puts into practice a principle long known to photobiologists - the interaction of light and a drug - as a systemic treatment to suppress disease. Its mechanism of action in psoriasis is believed to be based on the inhibition of the increased DNA synthesis within the psoriatic lesions by the interaction of 8-MOP and UV-A (330-336 nm) (26). Photoexcited 8-MOP (triplet state) can transfer the absorbed UV-A energy to DNA, forming monofunctional single-strand photoadducts with thymine bases and, on further irradiation, interstrand crosslinks (18-20). It has been postulated that this may interfere with DNA synthesis and thus with cell division in the rapidly dividing psoriatic epidermis (26). However, other mechanisms may be equally or even more important and it seems likely that PUVA treatment may also interfere with the inflammatory cell infiltrate in the superficial dermis of psoriatic patients. This infiltrate constitutes the first noticeable change in early psoriatic lesions (27) and may well represent the primary target of PUVA treatment. Support for this speculation may be derived from the fact that, according to our experience, PUVA is also effective in atopic dermatitis, lichen planus, and mycosis fungoides (28) where an increased epidermal turnover plays no major pathogenic role.

The fact that, under the influence of UV-A, psoralens can react with DNA poses the theoretical risk of mutagenicity and oncogenicity of the cell populations involved. Present concerns about possible long term sequelae have been discussed repeatedly and so have the notions that these concerns may be overrated (2,3). There is no doubt that risk benefit ratio decisions have to be made before a patient is admitted to PUVA treatment and that, for patients with severe psoriasis, PUVA appears to be a better choice than corticosteroids, methotrexate or other cytotoxic agents (3, 11). However, only long term studies will provide the final answer to the long-term safety of this treatment.

REFERENCES

1. J . A. Parrish, T. B. Fitzpatrick, L. Tanenbaum, and M.A.
    Pathak, Photochemotherapy of psoriasis with oral methoxsalen
    and longwave ultraviolet light, New Engl.J.Med. 291, 1207-
    1211 (1974).

2. K. Wolff, H. Hönigsmann, F. Gschnait, and K. Konrad, Photo-
    chemotherapie bei Psoriasis. Klinische Erfahrungen bei 152
    Patienten, Dt.Med.Wschr. 48, 2471-2477 (1975).

3. K. Wolff, T.B. Fitzpatrick, J.A. Parrish, F. Gschnait, B.
    Gilchrest, H. Hönigsmann, M.A. Pathak , and L. Tanenbaum,Pho-
    tochemotherapy of psoriasis with oral 8-methoxypsoralen,
    Arch.Derm.(Chic.) 112, 943-950 (1976).

4. H.O. Perry, C.W. Soderstrom, and R.W. Schulze, The Goeckerman
    treatment of psoriasis, Arch.Derm (Chic), 98, 178-182 (1968).

5. H. Tronnier and D. Schüle, First results of therapy with
    longwave UV after photosensitization of skin, Book of Ab-
    stracts, Symposia, and Contributed Papers, 6th International
    Congress of Photobiology, Bochum, Germany, GO Schenck,ed.1972.

6. S.M.A. Mortazawi and H. Oberste-Lehn, Lichtsensibilisatoren
    und ihre therapeutischen Fähigkeiten, 1. vorläufige Mittei-
    lung. Z.Hautkr, 48, 1-9 (1973).

7. H. Tronnier and D. Schüle, Zur Therapie von Dermatosen mit
    langwelligem UV nach Photosensibilisierung der Haut mit
    Methoxsalen. Z.Hautkr, 48, 385-393 (1973).

8.   G. Weber, Combined 8-methoxypsoralen and blacklight therapy
     of psoriasis, Brit.J.Derm., 90, 317-323 (1974).

9.   H. Oberste-Lehn and S.M.A. Mortazawi, Therapeutische Ergeb-
     nisse bei der Anwendung von 8-Methoxypsoralen (MOP) und UV-A,
     Z. Hautkr, 50, 559-751 (1975).

10. H. Tronnier and R. Löhning, About the current status of
     methoxsalen-UV-A-therapy in Dermatology. Castellania, 2, 267-
     271 (1974).

11. K. Wolff, F. Gschnait, H. Hönigsmann, K. Konrad, G. Strügl,
     E. Wolff-Schreiner and P. Fritsch, Oral photochemotherapy-
     results, follow-up, and pathology, Proceedings of the Second
     International Symposium on Psoriasis, Stanford, July 12-15,
     (1976), Plenum Press, in press.

12. G. Swanbeck, M. Thyresson-Hök, A. Bredberg, and B. Lambert,
     Treatment of psoriasis with oral psoralens and longwave
     ultraviolet light. Acta Dermatovener (Stockholm) 55, 367-376,
     (1975).

13. C. Hofmann, G. Plewig, and O. Braun-Falco, Technische Erfah-
     rungen mit der 8-Methoxypsoralen-Photochemotherapie bei Pso-
     riasis vulgaris, Hautarzt, 27, 277-285,(1976).

14. K. Weisman, J. Howitz, and A. Bro-Jorgensen, Treatment of pso-
     riasis with 8-methoxypsoralen and longwave ultraviolet light
     (PUVA). Clinical Medicine 4, 28, (1976).

15. H. Tronnier and H. Heidbüchel, Zur Therapie der Psoriasis
     vulgaris mit ultravioletten Strahlen. Z. Hautkr. 51, 405-424,

(1976).

16. K. Wolff, F. Gschnait, H. Hönigsmann, K. Konrad, J.A. Parrish and T.B. Fitzpatrick, Phototesting and dosimetry for photochemotherapy. Brit. J Dermatol, in press.

17. F. Gschnait, K. Konrad, H. Hönigsmann, and K. Wolff, Photochemotherapy in cortocosteroid- and methotrexate-treated psoriatics. Hautarzt, in press.

18. R.S. Cole, Light-induced cross-linking of DNA in the presence of a furocumarin (psoralen), Biochim Biophys Acta 217, 30-39, (1970).

19. F. Dall'Acqua, S. Marciani, L. Ciavatta et al, Formation of interstrand crosslinks in the photoreactions between furocomarins and DNA, Z Naturforsch 26, 561-569, (1971).

20. M.A. Pathak, D.M. Kramer, and T.B. Fitzpatrick, Photobiology and photochemistry of furocumarins (psoralens), in Pathak MA, Harber IG, Seiji M, et al, (eds.): Sunlight and Man: Normal and Abnormal Photobiologic Responses. Tokyo, University of Tokyo Press, pp 335-368, (1974).

21. J.H. Epstein, and K. Fukuyama, A study of 8-methoxypsoralen-induced phototoxic effects on mammalian epidermal macromolecule synthesis in vivo. Photochem Photobiol 21, 325-330, (1975).

22. H.P. Baden, J.M. Parrington, J.D.A. Delhanty, et al, DNA synthesis in normal and xeroderma pigmentosum fibroblasts following treatment with 8-methoxypsoralen and longwave ultravio-

let light. <u>Biochim Biophys Acta</u> 262: 247-255, (1972).

23. M.J. Ashwood-Smith and E. Grant: Chromosome damage produced
     by psoralen and ultraviolet light. <u>Brit Med.J, 272</u>, (1976).

24. D.M. Carter, K. Wolff, and W. Schnedl, 8-methoxypsoralen and
     UV-A promote sister chromatid exchanges. <u>J Invest Dermatol</u>,
     in press.

25. E. Wolff-Schreiner, D.M. Carter, W. Schnedl, and K. Wolff,
     Sister chromatid exchanges in psoriasis patients treated with
     photochemotherapy, <u>J Invest Derm.</u>, in press.

26. J.F. Walter, J.J. Voorhees, W.H. Kelsey, et al, Psoralen plus
     black light inhibits epidermal DNA synthesis, <u>Arch Dermatol</u>,
     <u>107</u>, 861-865, (1973).

27. O. Braun-Falco and E. Christophers, Structural aspects of
     initial psoratic lesions, <u>Arch.Derm.Forsch. 251</u>, 95-99,
     (1974).

28. H. Hönigsmann, K. Konrad, F. Gschnait, and K. Wolff, Photo-
     chemotherapy of mycosis fungoides, VII International Congress
     on Photobiology, August 29-September 3, (1976), Rome, Book of
     Abstracts, P 222.

ADVANTAGES AND DISADVANTAGES OF PHOTOTHERAPY (PT) IN
NEONATAL HYPERBILIRUBINEMIA

MARCELLO ORZALESI M.D.

Chair of Neonatology, Univ. of Naples, 2nd Med.

Sch., Via Pansini 5, Naples, Italy

SUMMARY

Phototherapy (PT) with white (day-light) or monochromatic (blue light) lamps is widely used for the prevention and treatment of neonatal hyperbilirubinemia.
PT has the following advantages: it is inexpensive and
easy to use; the breakdown products of bilirubin are
not toxic for the CNS and are rapidly eliminated through
the kidneys and liver; it is the most effective of all
modes of prevention and treatment of neonatal hyperbilirubinemia, with the exception of exchange-transfusion.
Despite the above mentioned advantages, there is still
some reluctance in the use of PT for fear of unknown
immediate and/or long term side effects. Immediate side
effects described in newborn infants include: decreased
intestinal transit time, with loose, greenish stools;
increased insensible water loss with slow weight gain;
skin rashes and, rarely, a brownish discoloration of
skin, urines and plasma. Other possible side effects,
described in vitro and/or in experimental animals, include: retinal damage; liver cell damage; alterations
of the pineal gland with modifications of serotonin metabolism; suppression of cyrcadian rithms; modifications
of hormonal secretion and plasma levels; acceleration
or delay in sexual maturation; photoxidation of structures or substances in the body, other than bilirubin

(i.e. vitamines, aminoacids, albumin, red blood cells,
etc.). Further studies are needed in order to exclude,
some of the above mentioned side effects. Meanwhile PT,
when indicated, should be used with caution and under
well controlled circumstances.

    Severe hyperbilirubinemia (i.e. levels higher than
18-20 mg/dl) has been known for a long time to produce
severe neurological lesions in newborn infants. In recent years, however, more and more emphasis has been
placed on the dangers associated with what were thought
to be low levels of bilirubin (1).
It has been shown now by many authors, that severe brain
damage can occur even with serum bilirubin levels as
low as 8-12 mg/dl, in some high risk infants with associated abnormalities such as prematurity, hypoxia, hy-
percapnia, acidosis, hypothermia, etc. (2). Furthermore
Boggs and Coworkers (3) have suggested that, most likely,
the neurological damage from hyperbilirubinemia is not
an "all or nothing" phenomenon, but that minor neurological and developmental abnormalities may be shown on
follow up in babies with neonatal serum bilirubin levels in the 15 to 20 mg/dl range.
Neonatal jaundice, therefore, is of great concern to
the pediatrician for at least 2 good reasons: 1) it is
a very frequent abnormality, probably the most frequent
one to be encounterd in the newborn period; 2) it may
have deleterious effects on the Central Nervous System
(CNS) and future development of the child.
The most effective and rapid way for reducing serum bilirubin levels in newborn infants is to perform an exchange-transfusion with fresh blood. This procedure,
which is widely used throughout the world, has saved
many lives but it is not entirely free from severe and
sometimes lethal complications. Therefore, various types of treatments have been proposed and used, either
prophylactically or therapeutically, as an alternative
to an exchange-transfusion (1, 4). These include: the
early introduction of nutrients and fluids, orally or
intravenously, to reduce hemolysis and bilirubin production; the administration of microsomal enzyme indu-
cers (such as phenobarbital) to enhance bilirubin conju-

gation and excretion by the liver; the oral administration of substances (such as agar or activated charcoal) to inhibit intestinal reabsorption of bilirubin; and phototherapy (PT).

Among all these treatments, and with the exception of exchange-transfusion, phototherapy is certainly the most effective one in reducing serum bilirubin levels, and also the one that has gained wider acceptance in many Countries (4).

While the action of the other treatments takes place through the usual pathways of bilirubin metabolism, the effect of PT is based on a completely different concept and consists of a direct photo-oxidation of the bilirubin molecule, with the formation of dipyrrols or similar compounds, which, being more polar, are water soluble and readily eliminated through the liver and kidneys (4, 5). This action of light takes place in the skin, although it is not known if it is intravascular or extravascular or both, and its effect is proportional to the intensity of incident light, particularly in the blue spectrum, to the length of exposure, to the area of skin exposed and does not seem to be affected by the color of the skin (4, 6).

Recent evidence from animal experiments (7) and from the work of Lund and Jacobsen (8) in newborn infants seems to indicate another site of action for PT. These Authors have shown a significant rise in the concentration of indirect bilirubin in the bile of newborn infants exposed to PT and have postulated that light might produce some slight modifications in the bilirubin molecule, which could facilitate excretion and/or leakage of bilirubin into the bile, without glucuronization.

There are a number of reasons for the wide acceptance of phototherapy (4, 9). First of all, there is no question that, among the various proposed therapies, PT is the most effective of all in reducing serum bilirubin (10). When used prophylactically PT is extremely efficient in preventing severe hyperbilirubinemia of prematurity. According to the experience so far accumulated, it has been estimated that PT can reduce by 20-60% the average serum bilirubin levels, depending on the duration and intensity of treatment. What is even more important PT can reduce by 90% the number of infants that

will develop a serum bilirubin level higher than 15 mg/
dl (10).
Secondly the split products which are produced are not
toxic for the CNS, do not bind to albumin and are rapidly eliminated (1, 4). However, most of the evidence for
these two statements has been obtained from in vitro experiments and has to be verified in vivo. Finally, until now, no serious immediate complications have been
reported. Minor reported side effects include skin rashes, decreased gut transit time with loose greenish
stools, increased insensible water loss with slower weight gain. An interesting, but rare, complication has
been reported by Kopelman (11) and by others, namely a
bronze discoloration of the skin, plasma and urines during PT (the so called "bronze baby"). The pathogenesis
of this syndrome is not fully understood, but is probably related to a poor excretion of split products and/
or other derivatives of oxidized bilirubin in some infants. In any case this condition is rapidly reversible
upon discontinuation of treatment. Another advantage of
phototherapy is that it is inexpensive and easy to use.

Despite all the advantages listed above, many pe-
diatricians and many centers in various Countries are
still reluctant to use PT for the fear of unknown unt-
ward effects (9, 12). It is important to realize, however, that most of the supposed damaging effects of light are either hypothetical or are based on experiments
performed in animals or in vitro under extreme conditions.

Light of the same intensity as that used for PT
has been shown to damage the retina in newborn piglets
(13), and for this reason we cover the eyes of our infants undergoing PT. It remains to be shown what could
be the effect of this sensory deprivation, but it is
difficult to conceive that anything terrible would happen. After all, this only mimics a situation which has
been physiologic for these babies throughout gestation
and it is highly unlikely that 2 or 3 days in the dark
would have any deleterious effect.

Alterations in the morphology and function of the
pineal gland, with consequent modifications in serotonin

metabolism, have been shown to occur in animals during intense light exposure (14, 15). However, since the effect of light on the pineal gland is mediated through the optic nerve, by covering the eyes we should take care of this problem too. Indeed, when the urinary excretion of 5-hydroxyindolacetic acid, a metabolite of serotonin which closely reflects the turnover of this substance in the body, was measured in a group of 15 low birth weight infants after 3 days of phototherapy, it was found to be similar to that of 16 control infants with the same gestational and post-natal age (16).

Prolonged exposure to light has also been shown to interfere with sexual behaviour and sexual maturation in certain animals (12). This effect is also probably mediated through the optic nerve. The possible relevance of this finding to the human situation must of course await careful, long term follow-up studies.

Experiments in rats have suggested that light may cause liver damage (17,18,19). Newborn rats, born to mother irradiated with fluorescent light during the second half of gestation and further exposed to light for 10 days after birth showed morphological evidence of liver cell damage, both in the nucleus and in the cytoplasm, with disruption and disorganization of the mitochondria, enzyme dispersion and increased glycogen content. We have therefore performed an investigation concerning the possible toxic effects of light on liver cells and on glycogen utilization, by studying the plasma enzyme patterns and the glucagon response test in newborn infants undergoing PT (20,21). The following plasma enzymes were measured in each subject at the beginning and at the end of phototherapy: Glutamate-oxalacetate transaminase (GOT), Alkaline phosphatase (AP), Glutamate-pyruvate transaminase (GPT), Lactate dehydrogenase (LDH), Leucynaminopeptidase (LAP) and Sorbitol dehydrogenase (SDH). The average plasma enzyme levels before PT in the 25 jaundiced infants studied were slightly higher than those obtained in a control group of 16 normal infants of the same age, but did not change significantly during and after PT (20). The response to a standard glucagon test was also inve-

stigated in a different group of 10 full term jaundiced
infants after 2-3 days of PT and in 10 controls of similar gestational age, birth weight and post-natal age.
Here again we failed to show any significant difference
in the average levels of blood glucose at different times after glucagon infusion, or in the maximal levels,
between the two groups of infants, suggesting that glycogen utilization is unaffected by PT (21).
Therefore, the results of our study failed to provide
any evidence that PT has a direct toxic effect on the
liver cells of human newborns, since light exposure did
not seem to influence appreciably the plasma enzyme pattern and glycogen utilization.

An alarming report of Ballowitz and Coworkers (22)
had suggested that continuous exposure to intense light
may produce somatic and brain growth retardation in newborn rats. These changes, however, were not due to the
direct effect of light on the newborn animals, but to
the fact that their mothers were also continuously exposed to light, which altered their circadian rhythm and
impaired lactation. When the experimental design was appropriately modified, no untoward effects were noted in
the newborn animals exposed to light (23,24). This is
in good agreement with follow-up studies performed so
far in infants, which failed to show any significant abnormality of growth or development in infants treated
with PT (10,25).

The most important and realistic side effect of PT
could be the photooxidation of substances or structures
in the body, other than bilirubin.
There are a number of substances that could undergo photo-oxidation, including drugs, hormones, vitamins, aminoacids, etc. (14). Of course, any oxidation of blood
constituents can take place only if light penetrates
deep under the skin and reaches the intravascular space,
and it is not well known where in the skin the photo-oxidation of bilirubin takes place. Blanching of the skin
suggests degradation of bilirubin in the interstitial
space. However, since bilirubin is bound to albumin and
since there is little albumin in the extravascular compartment, it is unlikely that this mechanism could account entirely for the very effective action of light.

It is also possible that blanching of the skin is due to
the degradation of the bilirubin contained in the subcutaneous fat. This in turn would imply a penetration of
light even deeper than what is needed to reach the vessels contained in the dermis. Therefore at the moment,
there is not clear cut evidence for either extravascular
or intravascular photo-oxidation, but it is likely that
both do in fact occur (4).
Concerning the photo-oxidation of the other substances
in the blood, Odell has suggested that the imidazole
ring of the albumin molecule could be broken by light,
and that the capacity of albumin to bind bilirubin could
be therefore affected (26). He has indeed shown that this
can occur when plasma or albumin solutions are exposed
to intense light in vitro. However, in vivo studies have
not shown any significant drop of the albumin binding
capacity during PT (1,10). He has also postulated (27)
that babies on PTcould become anemic as a result of an
increased hemolysis secondary to photo-oxidative damage
of the red blood cells (RBC). Indeed, we have found in
vitro that intense light produces swelling and increased
hemolysis of fetal RBC, both by increasing the K^+ leak
as well as by oxidizing the GSH, and that the damage is
greater in the presence of bilirubin (28). In a different set of experiments (29) we have also investigated
the in vitro effect of intense light on the lipid composition if the RBC membrane, since it is known that oxidative hemolysis is generally preceded by alterations
in membrane phospholipids. When compared to the basal
sample and to the control sample incubated in the dark,
the red cells exposed to intense light showed a significant reduction of the lecithin fraction with a concomitant rise in lysolecithin, It is therefore likely that
lecithin was oxidized and transformed in its lysoderiva-
tive.Phosphotydilserine and phosphotidil etanolamine were also reduced, while the sphyngomielin fraction was
increased. Here again the diminution of these 2 fractions
is probably due to their oxidation and transformation
into their lysoderivatives, which are known to migrate,
on thin layer chromatography, at the level of sphyngomielin, producing an apparent rise of this latter fraction. Therefore, there is some evidence that light can
produce oxidative damage of albumin and red blood cells

in vitro.

It should be noted, however, that both Odell's and our
observations were made in vitro, under extreme circumstances not to be found in the clinical use of light
and than when we performed an in vivo controlled study
we could not find any evidence of increased hemolysis
in infants treated with PT (30).

Sisson and Coworkers have recently shown that riboflavin
and sometimes G6PD in the RBC can be oxidized by light,
both in vitro and in vivo (31). However, the drop in riboflavin is transient and it is rapidly compensated when
the baby gets sufficient riboflavin through his feedings.
Therefore it is likely that in the usual clinical situation the disadvantages of a slight oxidative damage of
albumin or RBC is outweighted by the very definite advantages of an efficient reduction of serum bilirubin
(4). This view is also supported by the good results obtained by Meloni and Coworkers, who used PT for treatment
of neonatal jaundice due to G6PD deficiency (32).

In conclusion, the evidence so far accumulated indicates that PT has some effects on newborn infants, o-
ther than the simple reduction of serum bilirubin levels.
Further research is needed in order to define better the
ultimate significance of these and other hypothetical
side effects of PT. Meanwhile, since the clinical experience with PT has indicated that the benefit: risk ratio of this mode of treatment is great, PT can and should be used for the prevention and treatment of neonatal
hyperbilirubinemia; its use, however, should comply strictly to the guidelines and reccomendations published so
far (1,4,9,10), which are based on the data reported in
this brief review.

REFERENCES

1)Maisels M.J.: Bilirubin. On understanding and influencing its metabolism in the newborn infant. Ped.Clin.N.
Amer.,19:447,1972.
2)Stern L.,Doray B.,Chan G. and Schiff D.: Bilirubin metabolism and the induction of kernicterus.Natl.Found.
Birth Defects.Orig.Art.Ser. vol. 12,N°2,pg 255,1976.
3)Boggs T.R.Jr.,Hardy J.R. and Frazier T.M.: Correlation
of neonatal serum total bilirubin concentration and developmental status at age 6 months. J.Pediat. 71:553,

1967.
4)Sisson T.R.C.: Visible light therapy of neonatal hyperbilirubinemia. Photochem.Photobiol.Rev. 1:241,1976.
5)McDonagh A.F.: Photochemistry and photometabolism of
bilirubin IX . Natl.Found.Birth.Defects.Orig.Art.Ser.
vol.12,N°2,pg. 30,1976.
6)Indyk L.: Physical aspects of phototherapy. Natl.Found.
Birth.Def.Orig.Art.Ser. vol.12,N° 2,pg 23,1976.
7) Ostrow J.D.,Berry C.S.,Knodell R.G. and Zarembo J.F.:
Effect of phototherapy on bilirubin excretion in man and
the rat. Natl.Found.Birth.Def.Orig.Art.Ser. vol.12,N°2,
pg 81,1976.
8)Lund H.T. and Jacobsen J.: Influence of phototherapy
on the biliary bilirubin excretion pattern in newborn
infants with hyperbilirubinemia. J.Pediat.,85,262,1974.
9)Orzalesi M.,Panero A. e Tambucci S.: Vantaggi e svan-
taggi della fototerapia nell'ittero neonatale, in "Pro-
blemi Attuali di Nutrizione in Pediatria".Plasmon 1975.
10) Lucey J.F.: Neonatal Jaundice and Phototherapy. Ped.
Clin.N.Amer.,19,827,1972.
11)Kopelman A.E., Brown R.S. and Odell G.B.: The "bronze" baby syndrome: A complication of phototherapy. J.
Pediat., 81,466,1972.
12)Wurtman R.J. and Cardinali D.P.: The effects of light
on man. Natl.Found.Birth.Def.Orig.Art.Ser. vol. 12,N° 2,
pg 100,1976.
13)Sisson T.R.C., Glauser S.C.,Glauser E.M.,Tasman W.
and Kuwabara T.: Retinal changes produced by phototherapy. J.Pediat.,77,221,1970.
14)Halaris A. and Matussek N.: Effect of continuous illumination on mitochondria of the pineal body. Experientia, 25,486,1969.
15)Nir I.,Hirschmann and Sulman F.G.: The effect of light and darkness on lactic acid content of the pineal
gland. Proc.Soc.Exp.Biol.Med.,133,452,1970.
16)Spennati G.F.,Girotti F. and Orzalesi M.: Urinary excretion of 5-hydroxyindolacetic acid in low birth weight
infants with and without phototherapy. J.Pediat.,82,286,
1973.
17)Iester A.,Quazza G.F.,Bertolotti E.,Moscatelli P.,
Fregonese B.,Lanzavecchia G. e Cordone G.: Studio speri-
mentale sulle modificazioni indotte dalla fototerapia
sulle cellule epatiche. Min.Ped.,20,870,1969.

18)Fregonese B.,Moscatelli P.,Lanzavecchia L. e Cordone
G.: Variazioni indotte dalla fototerapia sul contenuto
in glicogeno del fegato di ratto neonato. Min.Ped.,21,
984,1969.
19)Quazza G.F.,Cordone G.,Iester A.,Fregonese B.,Mosca-
telli P. e Lanzavecchia C.: Enzimo-istochimica del fegato e fototerapia. Min.Ped.,21,1233,1960.
20)Orzalesi M., Natoli G.,Panero A. and Ciocca M.: Plasma hepatic enzymes in jaundiced newborn infants treated
with phototherapy. Natl.Found.Birth.Def.Orig.Art.Ser.
vol.12,N° 2, pg 93,1976.
21)Casadei A.M.,Panero A.,Tambucci S. and Orzalesi M.:
Blood glucose response to the administration of glucagon in jaundiced newborn infants treated with phototherapy. To be published.
22)Ballowitz L.,Heller R.,Natzschka J. and Oh H.: The
effect of blue light on infant Gunn rats. Natl.Found.
Birth.Def.Orig.Art.Ser. 6,106,1971.
23)Kendall S.,Golberg S. and Sisson T.: Influence of visible light on growth rates of Gunn rats. Clin.Res.,28,
377,1970.
24)Sisson T.R.C.,Goldberg S. and Slaven B.: The effect
of visible light on the Gunn rat: convulsive threshold,
bilirubin concentration and brain color. Pediat.Res.,
8,647,1974.
25)Hodgman J.E.:Clinical application of phototherapy in
neonatal jaundice. Natl.Found.Birth.Def.Orig.Art.Ser.,
vol. 12, N° 2, pg 3,1976.
26)Odell G.,Brown R. and Holtzman N.: Dye-sensitized
photo-oxidation of albumin associated with a decreased
capacity for protein-binding of bilirubin. Natl.Found.
Orig.Art.Ser.,6,31,1970.
27)Odell G.B.,Brown R.S.,Kopelman A.E.: The photodynamic
action of bilirubin on erythrocytes. P.Pediat.,81,473,
1972.
28)Blackburn M.G.,Orzalesi M. and Pigram P.: Effect of
light and bilirubin on fetal red blood cells in vitro.
Biol.Neonate,21,35,1972.
29)Castro M.,Tambucci S.,Panero A.,Giardini O. e Orzalesi M.: Studio in vitro degli effetti della luce sui li-
pidi del globulo rosso. Min.Ped.,28,391,1976.
30)Blackburn M.G.,Orzalesi M. and Pigram P.:Effect of
light on fetal red blood cells in vivo. J.Pediat.,80,

640,1972.
31)Sisson T.R.C.,Slaven B. and Hamilton P.B.: Effect of broad and narrow spectrum fluorescent light on blood constituents. Natl.Found.Birth.Def.Orig.Art.Ser., vol. 12,N° 2, pg 122,1976.
32)Meloni T.,Costa S.,Dore A. and Cutillo S.: Phototherapy for neonatal hyperbilirubinemia in mature newborn infants with erythrocyte G-6-PD deficiency. J.Pediat., 65,560,1974.

This work has been supported in part by CNR contract N° 740023604.

PHOTOTHERAPY OF NEONATAL JAUNDICE: EFFECT ON BLOOD BIORHYTHMS

Thomas R. C. Sisson, M.D.

Professor of Pediatrics and Director of Neonatal Research

Temple University School of Medicine, Phila., Penna.

The biologic effects of visible light upon the mamalian organism have proven to be as marked as those from any other agency in our environment. Because light pervades more than half our days and is a benign, not to say beneficial, force, and is even less tangible than the air, we have seldom had reason to consider its less benevolent activities.

The world-wide use of phototherapy for the treatment of neonatal hyperbilirubinemia following the classic studies of Cremer in the late 1950's (1), brought forward a field of study that had been neglected, if not ignored, by the medical profession for almost a century. The use of phototherapy has also raised many questions about the biologic effects of light on the human, quite aside from the more limited consideration of bilirubin photo-degradation. The wide range of investigation of bilirubin chemistry has been discouragingly difficult to solidify since its biochemistry is so entangled with its photochemistry. We now recognize that the light employed to reduce elevated levels of bilirubin in blood acts also on a series of photo-products themselves sensitive to light. They are not completely identified, nor are their biologic effects on tissue.

Diverse sources of light are used in the whole-body irradiation of the jaundiced newborn, with intensities and spectral characteristics that are also diverse. This has led to a realization that there is a dose/response relationship between light source and bilirubin photo-decomposition, and a further realization that phototherapy lights act with less or more effect upon other plasma and tissue compounds.

We have reported from our laboratory, for instance, that whole blood riboflavine is reduced in quantity during phototherapy regardless of light sources (2). In some infants, unpredictably, phototherapy produces hemolysis of red cells and a consequent increase of serum bilirubin concentrations, rather than the decrease expected. This is associated with a loss of glucose-6-phosphate dehydrogenase activity in the erythrocyte (2). Since glutathione reductase and G-6-PD activity in these cells is dependent upon the presence of riboflavine as a co-enzyme, we have postulated that light-induced rise of serum bilirubin is caused in fact by the loss of sufficient red cell riboflavine to permit the enzyme activity; cell respiration is thus compromised and hemolysis results.

Although visible light is absorbed to greater degree by the skin, giving rise to the erroneous notion that it will not penetrate significantly beyond the dermis, our findings that plasma and erythrocyte constituents are affected by phototherapy would seem to contradict this common belief. In order to determine if the light of phototherapy units does indeed penetrate beyond the skin, we studied the transmission of daylight and blue fluorescent light through the skin, subcutaneous tissue and musculature of the abdominal wall of living adult Gunn rats and 5 to 6 week old piglets. Approximately 25% of the incident light was transmitted through the dermis and epidermis (a thickness of about 0.2 cm.), and between 18 and 23% through the entire thickness of the abdominal wall, about 0.5 - 0.6 cm.

It has been of interest in our establishment that the use of phototherapy in the newborn has caused a distinctly unusual pattern of management of these patients. They are customarily maintained under constant irradiation except for short periods of time during feeding and bathing. The eyes are covered by opaque shields, and the usual tactile stimulation of the infants, swaddling, rocking, and so on, is not carried out. In a sense these infants are deprived of much of the sensory impact of normal nursery care during the several days of phototherapy. In this regard, one should note that the use of broad-spectrum fluorescent light requires a longer period of exposure for effective phototherapy than does the use of a monochromatic blue fluorescent light source.

We were concerned that phototherapy might affect biorhtyms of the neonate by constant light irradiation over a period of days. Previous studies in this laboratory (3) have shown that there is a circadian rhythm, as well as ultradian rhythmicity, of Human Growth Hormone in plasma of infants at least by the second day of life. Under a light:dark cycle of 14L : 10D hours an ultradian rhythm is readily apparent throughout the total 24 hours.

Appropriate analysis of the data (% difference from group means),
however, has revealed a circadian rhythm with peak plasma HGH
levels between midnight and 0200 hours, as is known in the adult.
We believe that this rhythm is endogenous in the neonate, not
entrained by the mother's own rhythm since HGH does not cross the
placenta.

In a subsequent study (4) we altered the lighting environment
of the nursery by maintaining constant illumination. Under this
constant radiation the ultradian rhythm of HGH persisted, but the
circadian rhythm was obliterated.

Having thus determined that a biorhythm does exist in the
neonate if the lighting environment is cycled between light and
dark, we undertook a study of the effects of phototherapy upon
plasma HGH, and blood calcium and blood glucose concentrations.
Three groups of 12 newborns each were investigated on the second
day of life. GROUP I consisted of 12 infants, born at term, who
were kept in our standard nursery light (100-120 ftcd.) from 0800
to 2000 hours, and in the dark from 2000 to 0800 hours; GROUP II,
12 infants, were in the same nursery light from 0700 to 2200
hours and in dim light (5 ftcd.) from 2200 to 0700 hours;
GROUP III consisted of 12 infants with physiologic jaundice who
were maintained under constant phototherapy for 24 hours. The
eyes were shielded throughout the period of study, and the infants
were fed while under the phototherapy light.

Blood samples were drawn every 8 hours for 24 hours on a
rotating basis, before feedings, so that samples were secured in
each group at each 4 hour interval. It was necessary to draw
samples in a fasting state since the blood calcium and glucose
would otherwise have been influenced by the intake of milk.

It was apparent from the data collected that no significant
ultradian or circadian rhythm was present in any of the three
groups of infants in respect to blood calcium and glucose,
although slightly lower mean values were observed in the phototherapy (Gp. III). We concluded that any possible rhythmicity of
these blood constituents is dependent more upon dietary intake
and normal metabolism than upon the influence of light.

An analysis of the data in respect to plasma Human Growth
Hormone, however, indicates that the lighting environment has a
profound effect. It was clear from the data that absolute
darkness is not a requirement in the cycling of light to produce
a circadian rhythm, for the same peak levels between 2400 and
0200-0400 hours were present when the nursery light was dimmed
to very low though discernible levels.

The infants in Group III, exposed to constant phototherapy

for the 24 hour period of study, under monochromatic blue fluorescent light with an irradiance of 3.9 uW/cm^2/nm. (420-470 nm. range), had neither ultradian nor ciradian rhythm.

Since the eyes of these infants were shielded with opaque masks, they had no effective photoreception. We cannot say whether the lack of rhythm of HGH under phototherapy is due to the constant light exposure or to the lack of sensory input to the retinas of the infants. However, since no circadian rhythm was observed in the previous study of infants under constant nursery light environment, and whose eyes were not shielded, it is a possibility that the lack of rhythm is due to light irradiation per se.

We conclude that variation of light intensity in the nursery, or in the management of phototherapy will not obliterate plasma HGH circadian rhythm in the newborn, and may be a more physiologic routine for the infant than constant light exposure. This is true at least for plasma HGH, whose concentrations depend on intrinsic (endogenous) production, not upon outside sources of supply.

REFERENCES

1. Cremer, R. J., Perryman, P. W., and Richards, D. H.,
    Influence of light on the hyperbilirubinemia of infants,
    Lancet 1:1994, 1958.

2. Sisson, T. R. C., Slaven, B., and Hamilton, P. B., Effect
    of broad and narrow spectrum fluorescent light on blood
    constituents, Birth Defects: Original Article Series
    12:122, 1976.

3. Sisson, T. R. C., Root, A. W., Kechavarz-Oliai, L., and
    Shaw, E., Biologic rhythm of plasma human growth hormone in
    infant os low birth weight, in "Chronobiology", Sheving, L.
    E., Halberg, F., & Pauly, J. eds., Igaku Shoin Ltd., Tokyo,
    1974, pp. 348-352.

4. Sisson, T. R. C., Katzman, G., Shahrivar, F., and Root, A. W.,
    Effect of uncycled light on plasma human growth hormone in
    neonates, Pediat. Res. 9:280, 1975.

PHOTOTHERAPY OF HUMAN TUMORS

T. Dougherty, D. Boyle, K. Weishaupt, C. Gomer
D. Borcicky, J. Kaufman, A. Goldfarb, and G. Grindey

New York State Department of Health
Roswell Park Division
666 Elm Street, Buffalo, N.Y. 14263

This paper presents a brief review and the current status of the use of phototherapy in malignancy.

Although photodynamic processes have been known since 1900[1]. and the ability of certain photodynamic agents (especially porphyrins) to accumulate in malignant tissue has been recognized for more than 30 years [2] there has been little published data relating to the use of in vivo photodynamic methods to treat cancer. Apparently the only attempt prior to 1972 was by Jesionek and Tappenier who reported in 1903 [3] that eosin and light could be used to treat skin tumors. Sixty-nine years later, Diamond and workers [4] reported that glioma tumors transplanted into rats responded destructively to the combined effect of hematoporphyrin and visible light. About the same time, working independently, Dougherty and workers [5] reported that combined effects of fluorescein and light of 488 nm could markedly retard growth of subcutaneous mammary tumors transplanted into mice. Neither group reported complete tumor control, however, until 1975 when Dougherty reported that hematoporphyrin derivative first described by Lipson [6] (an apparently modified or purified form of hematoporphyrin) and red light could be used to cure spontaneous or transplanted mammary tumors in mice and rats [7]. In 1974, Tomson and workers showed that acridine orange fed orally to mice and accumulated in transplanted epithelial tumors, could be activated by means of an argon laser to cause tumor destruction[8].
Later in 1974, Berenbaum, Kelly and Snell [9] demonstrated a highly specific destruction of human bladder carcinoma, (grown in immunosuppressed mice) following administration of hematoporphyrin derivative and exposure to visible light. Significantly, these workers reported that human bladder tissue grown and

435

treated in a similar way, was not destroyed by the combination of
dye and light. In 1976, Kelly and Snell reported the first clinical use of the photodynamic process to treat tumors. In a single
case, a patient with extensive recurrent superficial carcinoma of
the bladder was administered 2 mg/kg body weight of the hematoporphyrin derivative and 24 hr later a portion of the tumor was
exposed per urethram to visible light from a mercury vapor lamp
linked to a 5 mm diameter quartz rod. Forty-eight hr later necrosis of the exposed portion of the tumor area was observed
with no apparent change in the unexposed portions.

It appears that the hematoporphyrin derivative is the preferred photosensitizing dye for clinical use because of its very
low toxicity and well-known ability to accumulate quite specifically in a large number of human tumors [2,11]. So far only minor
photosensitization resulting from exposure to sunlight has been
seen as a complication.

Mechanisms Pertinent to Phototherapy of Tumors

Although it has long been known that certain porphyrins,
notably hematoporphyrin (especially the so-called hematoporphyrin
derivative) and tetraphenylporphinesulfonate [12] tend to be
found in higher concentration in malignant tumors than in other
tissue following systemic administration, the reasons for this
apparent specificity are unknown. In 1957, Kosaki and workers
[13] reported the presence of a lipid found only in malignant
cells which had a high affinity for porphyrins. There have been
no subsequent reports of this finding by Kosaki or others. Thus
the mechanism of this important property of these porphyrins
remains obscure.

More is known about the mechanisms of photodynamic processes.
It appears that in most cases, the effective oxidizing species is
singlet oxygen formed via energy transfer from the photosensitizer
[14].

$$Dye + hv \longrightarrow Dye^* (s)$$

$$Dye^* (s) \longrightarrow Dye^* (T)$$

$$Dye^* (T) + O_2 \longrightarrow {}^1O_2^* + Dye$$

$$^1O_2^* + substrate \longrightarrow Oxidation$$

$$^1O_2^* \longrightarrow O_2$$

where Dye* (s) is the excited singlet state of the dye, Dye* (T)
is the excited triplet state of the dye, O_2 is the normal triplet
ground state of oxygen and $^1O_2^*$ is singlet oxygen. In aqueous

media singlet oxygen has a life-time of about 1 μsec during which time it can interact with most any molecule in its environment. Matheson has shown that within cells the most likely targets are the protein molecules especially histidine and tryptophane residues which react at about 10^8 1 mole^{-1} sec^{-1} (15).

We have studied the efficiency and energy requirements for formation of singlet oxygen within cells in vitro by utilizing 1,3 - diphenylisobenzofuran as an intracellular singlet oxygen trapping agent (16). Cells were allowed to incorporate hematoporphyrin derivative to known concentrations and subsequently exposed for various times to a known intensity of red light. Survival was determined with and without the singlet oxygen trapping agent. It was found that the normally toxic effect of the combination of porphyrin and red light could be quenched until the trapping agent was consumed at which point normal cell kill commenced. By determining the rate of consumption of the furan, as well as rate of formation of its singlet oxygen product, we were able to show that approximately 10^9 singlet oxygen molecules formed within a cell resulted in a 99.9% probability of its destruction. Further, we were also able to show that the quantum yield for formation of singlet oxygen within the cells was 0.16 compared to a value of 0.76 for hematoporphyrin derivative in alcoholic solution.

The survival curves show an exponential response to the photodynamic effect of hematoporphyrin and light following a shoulder at the lower doses (Figure 1). A possible explanation for the shoulder is cellular repair of photodynamic damage, a phenomenon well established for ionizing radiation. We have examined this possibility in vivo using normal skin of the mouse. Twenty-four hr after receiving various doses of hematoporphyrin derivative, the foot of the animal was exposed to red light (15 mw/cm^2, 620-640 nm) either for a single 60 min period or for two 30 min periods separated by 3 hr. At a drug dose of 7.5 mg/kg a significant decrease in photodynamic damage to the foot was seen for the split dose exposure. Although not definitive, this effect is consistent with repair of damage during the 3 hr interval. Possible repair of damage in tumors is currently under study.

An important question with regard to clinical use of the photodynamic effect is the degree of penetration of visible light through tissue. This is a difficult parameter to measure directly because of the very high scattering properties of tissue. Thus direct measurements, e.g. using fiber optic light detectors in tissue, neglect the very important contribution due to backscatter. We have attempted to determine the importance of scattering in excised rat tumors using red light from a 2 mw He Ne laser. The light was fed into the tissue by means of a fiber optic embedded within a 22 gauge needle. A similar fiber with the distal end

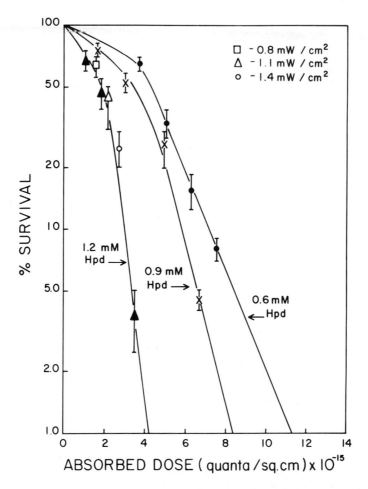

Figure 1. Survival of TA-3 mouse mammary carcinoma ascities cells
           as function of absorbed dose of red light, 620–640 nm.
           Bars indicate standard deviations of 3 or more experi-
           ments.
           Hpd = intracellular concentrations or hematoporphyrin
           derivative.

attached to a radiometer head served as detector. The distance
between the inlet fiber and detector fiber was varied to measure
transmission as a function of distance within the tissue (Figure
2). In addition, we varied the angle of the detector fiber to the
inlet fiber and found that the amount of light at a given distance
from the tip of the inlet fiber was essentially independent of
the angle, i.e. the incoming light was completely scattered.
Thus the actual light intensity at a given point within the
tissue was actually much higher (perhaps 10-15 times) than was detectable by simple transmission as represented in Figure 2.

Clinical Results with Hematoporphyrin Derivative and Red Light

Aside from the single case reported by Kelly and Snell[10]
there are no reports in the literature concerning the use of the
photodynamic effect to treat cancer. Our experience to date is
promising but very preliminary. To date we have treated only 9
patients. Two had chest wall metastasis following breast surgery,
radiation and chemotherapy, two had large abdominal masses recurrent following colo-rectal surgery, two had metastatic melanoma, one had local superficial squamous cell carcinoma recurrence,
one basal cell recurrence and one an undefined mass later shown
to be benign. Our purpose at this stage is to define conditions
for maximum tumor response without undue normal tissue injury
(skin in these cases) (i.e. drug dose, time interval, light
intensity, exposure time, etc). Hematoporphyrin derivative doses
have been 2.5 or 5.0 mg/kg, the interval between injection and
light exposure has ranged from one to six days, light intensities
have ranged from 5 to 15 mw/cm^2 (620-640 nm) and the treatment
time has ranged from 20 min to 2 hr. So far only single treatments of a given tumor or tumor area have been used. Our light
source is a 5,000 watt xenon arc lamp with appropriate filters to
eliminate the infrared and ultraviolet radiation and to pass only
visible wavelengths over 620 nm. Thermal effects of the light
are minimal and patients experience no sensation of heat during
treatment. (Animal tumors exposed to this source show temperature
rise of less than 3°). The hematoporphyrin derivative is essentially as described by Lipson [6].

The following generalizations can be made. If the tumors
are superficial, hematoporphyrin fluorescence can be detected for
several days post injection. Fluorescence intensity depends on
characteristics and amount of overlying tissue. If the skin over
the tumors is involved or shows erythema due to the tumor, fluo-
escence is generally intense.

All of the tumors treated to date have shown a marked and
rapid response. The smaller, superficial tumors (2-3 cm diameter)
appeared to regress completely within 6 or 7 days. In one
patient with squamous cell carcinoma of 5 x 2 cm diameter apparent

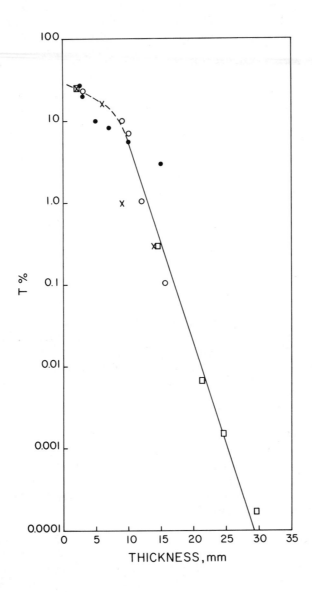

Figure 2.   Transmission of red light (633 nm, He Ne laser)
through a mouse mammary carcinoma tumor as function of
distance from light source. Data does not take scattering into account, see text.

on the surface, given a single treatment (60 min., 24 hr post 5.0 mg/kg hematoporphyrin derivative) at 10 mw/cm^2 (620-640 nm), residual or possibly recurrent tumor appeared about 7 weeks later. Retreatment of the patient is planned.

One patient with extensive chest wall metastasis, having failed on all conventional therapy (surgery, radiotherapy and chemotherapy) was treated daily in several different areas following a single injection of 2.5 mg/kg hematoporphyrin derivative. Treatment time was held constant at 20 min and light intensity at 10 mw/cm^2 (620-640 nm). We found marked tumor response each day for six days following the single injection. Several isolated nodules were treated to include the surrounding non-involved skin. These nodules became completely necrotic within 2-5 days following treatment (see slides 1-3). At times of 48 hr or more following the injection, the surrounding skin exhibited only moderate erythema and mild edema. At shorter times (e.g. 24 hr post injection) more marked reactions on the surrounding skin was seen. However it was clear that the degree of reaction on the skin surrounding the tumor was not typical of normal skin reaction, since in patients exposed to the light in areas remote from the tumor exhibit only very mild erythema under these same conditions. It thus appears that a very high therapeutic ratio can be obtained by proper choice of dose and time interval.

Another patient with chest wall metastasis, treated under the same conditions as described above, showed no tumor response at time periods exceeding 72 hr post injection. At 48 and 72 hr nodule necrosis occurred. This response is in contrast to that seen in the patient described above, where tumor response was observed up to 144 hr post injection.

A patient with small metastatic melanoma nodules was treated for 20 min, 48 hr post 2.5 mg/kg hematoporphyrin derivative at 15 mw/cm^2 (620-640 nm). Within one day no nodule was palpable and the surrounding skin exhibited only very mild erythema. Depth of penetration was studied in a patient with a very large abdominal mass (adenocarcinoma) protruding to the surface. Forty-eight hr following 5.0 mg/kg hematoporphyrin derivative, a small area of the tumor was exposed for 50 min to the light at 10 mw/cm^2 (620-640 nm). Seven days later the tumor was removed and histological sections made. It appeared that necrosis occurred to a depth of approximately 1.5 cm from the surface. We can assume about another 0.5 cm of tumor response. Below this depth there was viable tumor

It should be emphasized that no long term follow up has been possible with these patients to date. However it appears that this method may be useful in the treatment of chest wall metastasis which occurs after surgery, radiotherapy and chemotherapy.

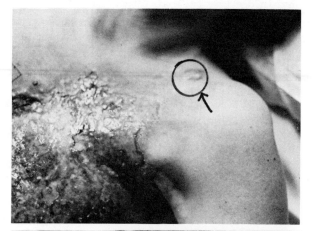

Slide 1.

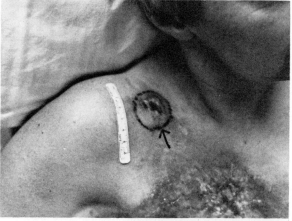

Slide 2.

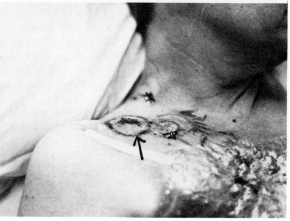

Slide 3.

We are currently expanding our number of tumor types and continuing to study the numerous variables. We are especially interested in increasing depth of penetration by using higher light
intensity, higher porphyrin doses and fiber optics methods.

←———

Slide 1. Metastatic breast carcinoma prior to treatment with
hematoporphyrin derivative. Note isolated module on right shoulder
(picture reversed right to left). Circle indicates light field.

Slide 2. Shoulder nodule, 3 days post 30 min of light (10 mw/cm^2,
620-640 nm) delivered 6 days post 2.5 mg/kg hematoporphyrin derivative. Note absence of normal tissue reaction.

Slide 3. Shoulder nodule, 6 days post treatment. No evidence of
tumor up to 8 weeks. Normal tissue in field not affected.

REFERENCES

1. O. Raab, Über die Wirkung Fluorescirender Stoffe auf Infuso-

    riera, Z. Biol., 39, 524 (1900).

2. H. Auler and G. Banger, Untersuchungen über die Rolle der

    Porphyrine bei geschwulstkranken Menschen und Tieren, Z.

    Krebsforsch, 53, 65-68 (1942).

3. A. Jesionek and V. Tappenier, Zur Behandlung der Hautcarcinome

    mit Fluorescirender Stoffen, Muench Med. Wochschr, 47,

    2042-2044 (1903).

4. I. Diamond, S. G. Granelli, A. F. McDonogh, S. Nielsen, C. B.

    Wilson and R. Jaenicke, Photodynamic Therapy of Malignant

    Tumors, Lancet 2, 1175-1177 (1972).

5. T. J. Dougherty, Activated Dyes as Antitumor Agents, J. Natl.

    Cancer Inst. 52, 1333-1336 (1974).

6. R. L. Lipson, E. J. Baldes and A. M. Olsen, The use of a

    derivative of hematoporphyrin in tumor detection, J. Natl.

    Cancer Inst. 26, 1-9 (1961).

7. T. J. Dougherty, G. B. Grindey, R. Fiel, K. R. Weishaupt and D. G. Boyle, Photoradiation therapy II. Cure of animal tumors with hematoporphyrin and light., J. Natl. Cancer Inst. 55, 115-129 (1975).

8. S. H. Tomson, E. A. Emmett and S. H. Fox, Photodestruction of mouse epithelial tumors after oral acridine orange and argon laser, Cancer Res. 34, 3124-3127 (1974).

9. J. F. Kelly, M. E. Snell and M. C. Berenbaum, Photodynamic destruction of human bladder carcinoma, Br. J. Cancer 31, 237-244 (1975).

10. J. F. Kelly and M. E. Snell, Hematoporphyrin derivative: a possible aid in the diagnosis and therapy of carcinoma of the bladder, J. Urology 115, 150-151 (1976).

11. H. B. Gregorie, E. O. Horger, J. L. Ward, J. F. Green, T. Richards, H. C. Robertson Jr., and T. B. Stevenson, Hematoporphyrin derivative fluorescence in malignant neoplasms, Ann. Surg. 167, 820-827 (1968).

12. J. Winkleman, G. Slater and J. Grossman, The concentration in

    tumor and other tissues of parenterally administered tritium

    - and ^{14}C-labeled tetraphenylporphinesulfonate, Cancer Res.

    27, 2060-2064 (1967).

13. T. Kosaki, T. Ikoda, J. Kotani, S. Nakagawa and T. Saka, A

    New Phospholipid, malignolipin, in human malignant tumors,

    Science 127, 1176-1177 (1958).

14. C. Foote, R. Denny, L. Weaver, M. Chang and J. Peters,

    Quenching of singlet oxygen, Annals N.Y. Acad. Sci. 171,

    139-148 (1970).

15. I. Matheson, R. Etheridge, N. Kratovich and J. Lee, The

    Quenching of singlet oxygen by amino acids and proteins.

    Photochem. Photobiol. 21, 165-171 (1975).

16. K. R. Weishaupt, C. J. Gomer and T. J. Dougherty, Identifi-

    cation of singlet oxygen as the cytotoxic agent in photo-

    destruction of a murine tumor, Cancer Res. 36, 2326-2329

    (1976).

SYMPOSIUM X

CANCEROGENIC EFFECTS OF RADIATION

CARCINOGENIC EFFECTS OF RADIATION-INTRODUCTION

R. B. Setlow

Biology Department
Brookhaven National Laboratory
Upton, New York 11973

The world has always been full of hostile physical and chemical agents. Living creatures have evolved and adopted life styles that permit them to survive this environment. However, as a result of man's activity the hostility is increasing rapidly-probably too rapidly for life to adapt to them. Some of the present hazards are shown in Fig. 1. It is important to evaluate quantitatively these potential dangers to life, so as to make reasonable regulatory decisions concerning them. The evaluation for man is a difficult one. One cannot legitimately do experiments on people and moreover people are not good experimental material. The solution to this difficulty is to have basic insight into the modes of action of physical and chemical agents on cells and organisms and to use our basic biological theories to extrapolate to man.

The recent results of photobiological research have altered our ways of looking at such extrapolation procedures and, for a number of reasons, light-induced skin cancer is the best example of

Figure 1. A schematic diagram illustrating some of the environmental hazards to which man is exposed.

how to approach the general problem of environmental carcinogenesis. The reasons are simple to enumerate.

1. The dosimetry of light radiation for both cells and for
populations of people is far superior to that for chemicals. The
superiority is illustrated schematically in Fig. 2. The average
number of photons incident and absorbed by cells is relatively
easy to measure. However, the concentrations of active carcinogens in the region of target organs are very difficult to ascertain because many carcinogens are not reactive by themselves but
must be activated by cellular enzyme systems to the ultimate
carcinogenic form. The tricks used for chemical dosimetry--the
number of products in DNA and the stability of these products--
are derived from those of photobiology.

2. The quantitative photochemistry of DNA is well worked
out and one of the photoproducts--cyclobutane pyrimidine dimers--
is associated with many of the lethal and mutagenic effects of UV
radiation (1). Such photoproducts are subject to a number of repair systems whose pathways in microorganisms are partially known
at the molecular level and which have their analogs in mammalian
cells (2). An important repair pathway is enzymic photoreactivation because this pathway does only one thing to UV photoproducts. It monomerizes pyrimidine dimers. Hence, if there is a
biological property that is affected by UV radiation and if the
effect is reversed by subsequent exposure to visible light, one
concludes safely that the initial change made by UV was the production of pyrimidine dimers in DNA. Photoreactivation is a trick
if you like, for assessing the biological importance of one of
a large number of photoproducts. Such a simple trick is not
available for chemical damage.

3. The fish, _Poecilia formosa_, may be grown as clones.
Cells from one animal may be treated _in vitro_ and injected into
isogenic recipients. If UV irradiated cells are injected, they

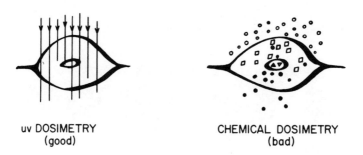

uv DOSIMETRY   CHEMICAL DOSIMETRY
(good)   (bad)

Figure 2. UV radiation and a chemical carcinogen (o) incident
on cells. The chemical carcinogen goes through a series of
changes to give the ultimate carcinogen (Δ).

give rise to thyroid carcinomas. However, no tumors appear if
the irradiated cells are exposed to photoreactivating ilumination
before injection (3). These data are clear evidence that UV
irradiation gives rise to transformed cells as the result of the
production of pyrimidine dimers in DNA. The importance of such a
finding is that many chemicals mimic the action of UV radiation
insofar as the type of dark repair system that operates on them
is similar to that that operates on dimers and the fact that
cells defective in the excision repair of UV damage are also defective in repair of a large class of chemical damages (4,5).

   4. Individuals with the genetic disease xeroderma pigmentosum (XP) have a very high risk of getting skin cancer--both
nonmelanoma and melanoma (Table 1). Such individuals are defective in one or more pathways involved in the repair of
pyrimidine dimers in their DNA. The defects in repair are discussed in this and in other symposia (Papers by Bootsma, Lehmann,

Table 1. Some characteristics of xeroderma pigmentosum (from
data on thirteen light exposed individuals described in ref. 6).

% with skin tumors	100
average age of first tumor	10 ± 5 years
tumors per individual	40 ± 30
% with malignant melanoma	54

and Sutherland). The existence of such individuals is the best
direct experimental evidence for the close association of DNA
damage with carcinogenesis.

   5. In bacteria UV radiation stimulates an error-prone
repair pathway (7). The pathway--associated with post replication repair (8)-- involves replication of DNA containing lesions.
A simple view is that the chance of an error (an initiating
carcinogenic event) during replication is the product of the
probability of a replication fork passing a lesion and the
probability of making a mistake if it passes one. Cells from
most XP individuals do not excise dimers from their DNA. In a
simple model the dimers remain for appreciable periods of time
and increase the probability of an error during subsequent replication. In XP variant individuals excision is normal but post-
replication repair is slow (9). It is an inference that in such
cells the probability of a carcinogenic event is high not because
many dimers remain but because replication past a dimer is more
error prone than in normal cells. In normal human cells
there are no defects in excision or postreplication repair and
the probability of a carcinogenic event taking place as a
result of an initial pyrimidine dimer is very small but not zero.

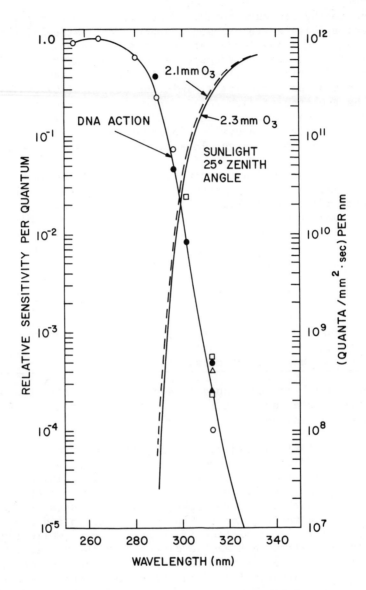

Figure 3. The action spectrum for affecting DNA compared to sunlight through two stratospheric thicknesses of ozone. As a result of absorption by the epidermis the presumptive carcinogenic spectrum decreases appreciably but has a similar shape to the DNA one above 300 nm. The different symbols represent different microbial, viral or molecular end points. See ref. 11 for details.

In cells from XP variants one can also observe a stimulation of postreplication repair as a result of small doses of UV radiation (10). It is tempting to suppose that this stimulated pathway is an error-prone one.

The weight of experimental evidence reviewed above indicates that UV damage to DNA--probably pyrimidine dimers--is the best molecular candidate for the initiating damage that leads to skin cancer. As a result, we can take a superior guess that the carcinogenic action spectrum should be similar to the DNA action spectrum filtered through the upper layer of skin (11). The use of such an action spectrum (Fig. 3,next page) is necessary to evaluate the effects of the change in UV striking the earth as a result of changes in stratospheric ozone--a subject considered in much more detail in Symposium XIII.

(This work was supported by the U.S. Energy Research and Development Administration.)

REFERENCES

1. Setlow, R. B. and Setlow, J. K. (1972) Ann. Rev. Biophys. Bioengineer. 1, 293-346.
2. Hanawalt, P. C. and Setlow, R. B. (1975) Molecular Mechanisms for Repair of DNA, Plenum, N. Y.
3. Hart, R. W. and Setlow, R. B. (1975) In ref. 2, pp. 719-724
4. Regan, J. D. and Setlow, R. B. (1974) Cancer Res. 34, 3318-3325.
5. Ikenaga, M., et al. (1975) In ref. 2 pp. 763-771.
6. Robbins, J. H., et al. (1974) Ann. Internal Med. 80, 221-248.
7. Witkin, E. M. (1975) In ref. 2 pp. 369-378.
8. Sedgwick, S. G. (1975) Proc. Nat. Acad. Sci. USA 72, 2753-2757.
9. Lehmann, A. R.,et al. (1975) Proc. Nat. Acad. Sci. USA 72, 219-223.
10. Setlow, R. B. and Grist, E. (1976) Biophys. J. 16, 183a.
11. Setlow, R. B. (1974) Proc. Nat. Acad. Sci. USA 71, 3363-3366

DEFECTIVE DNA REPAIR AND CANCER

D. BOOTSMA

Erasmus University, Dept. of Cell Biology and Genetics
P.O.Box 1738, Rotterdam, The Netherlands

SUMMARY

A relationship between defective DNA repair and actinic carcinogenesis is supported by the following observations:
1. As seen by complementation analysis, at least 5 different
mutations affecting excision repair in human cells result in the
genetic skin disease xeroderma pigmentosum, with a high incidence
of malignancies.
2. The XP patients with normal excision repair have shown to be
defective in post replication repair.
3. We have found an inverse correlation between the residual levels
of DNA repair in cells of different XP patients and the severity
of their clinical symptoms. This observation is not supported by
the findings of Robbins and coworkers (38) in their comprehensive
study of patients from the USA, whereas it is supported by the
study of Takebe (44) dealing with Japanese XP patients. This discrepancy might be the result of environmental factors which may
influence the phenotypic expression of the XP genotype, e.g. the
amount of sun exposure.
4. At least one other genetic disease in man, the Louis-Bar syndrome (Ataxia telangiectasia), in which an increased incidence of
malignancy is observed, exhibits a DNA repair defect. In this
disease evidence is presented for a decreased excision of gamma-
ray-induced DNA base lesions. A DNA repair defect, concerning the
repair of DNA interstrand cross links, might be present in Fanconi's anemia.
The mechanism by which defective DNA repair causes malignancy is
still subject of speculation. At present this question is approached by several groups of investigators studying a.o. the induction
of mutations. By comparing the different DNA repair mutants of

human origin, for their susceptibility for mutation,information is obtained concerning the error prone and error free components of DNA repair in human cells.

INTRODUCTION

It has been shown for many carcinogenic agents, that they react with the cellular DNA causing lesions of different type. These lesions are subject to repair mechanisms operating in euka-ryotic cells. There is increasing evidence that this damage to the DNA ultimately might result in the neoplastic transformation of the cell. Support for this hypothesis is obtained by the detection of defective DNA repair in some human diseases showing predisposition for malignancy. It is the purpose of this paper to summarize the evidence obtained from the study of these diseases that indicate a relationship between defective DNA repair and carcinogenesis.

DEFECTIVE EXCISION REPAIR IN XERODERMA PIGMENTOSUM

In his search for mutants in DNA repair in mammals and man Cleaver (8) studied cultivated cells from xeroderma pigmentosum (XP) patients. This disease was chosen because patients show a high susceptibility to sunlight. The intolerance of the skin and eyes to sunlight is manifested by abnormal pigmentation and freckling, actinic hyperkeratosis and skin malignancies. Cleaver found that the DNA synthesis which is observed following exposure of mammalian cells to ultraviolet light (UVL), the so-called Unscheduled DNA synthesis (UDS), was inhibited in XP cells. These decreased levels of UDS indicated that the repair of UV induced lesions in the DNA, which is present in normal cells, is impaired in XP cells. In addit-ion Cleaver (9) and others have shown that these XP cells are high-ly UVL sensitive in terms of cell survival. Moreover viruses ex-posed to UVL are inefficiently repaired following infection of XP cells (1, 2, 13, 28, 36).

GENETIC HETEROGENEITY IN XERODERMA PIGMENTOSUM

Although direct evidence was lacking it was attractive to associate this defective repair of DNA lesions with the high inci-dence of malignancies in these patients. Further evidence for this relationship was obtained by the detection of genetic heterogeneity in this disease. At the clinical level large differences in the severity of the disease were observed by comparing different patients. This heterogeneity was expressed in differences in the age at on-set of the disease, the frequency of tumors, the involvement of the nervous system and other clinical features.

In 1972 a complementation analysis was performed on XP by
means of fusion of cells from different patients (47). Following
the fusion of cells from two different patients, in some combinations normal UDS was observed in the hybrid binuclear cell after
exposure   of these cells to UVL. These experiments indicated the
presence of different complementary mutationsin the xeroderma syndrome. Recently it has been shown that 5 different mutations, most
probably affecting different genes in the human genome, might be
responsible for this disease (25). These 5 complementation groups
comprise a large number of families from the USA, England, The
Netherlands, Lebanon, Israel, Iran and Japan. Skin tumors and other
skin lesions are observed in all the 5 groups whereas neurological
abnormalities, including mental retardation, microcephaly etc.,
are only seen in the A, B and D group (Table 1). The decreased
levels of UDS found in these 5 groups indicate that mutations in
5 different genes all result in defective excision repair and is
associated with skin cancer. These observations support an
association between defective DNA repair and carcinogenesis. Unrepaired lesions remain in the DNA of XP cells for much longer
time intervals than in the DNA of normal human cells. These lesions
might ultimately result in the neoplastic transformation of the
cell.

TABLE 1

COMPLEMENTATION GROUPS IN EXCISION-DEFICIENT XERODERMA PIGMENTOSUM

COMPLEMENTATION GROUP	NUMBER OF FAMILIES	SKIN LESIONS	NEUROLOGICAL ABNORMALITIES
A	>10	+	+
B	1	+	+
C	>10	+	-
D	3	+	+
E	2	+	-

August 1976

THE UNREPAIRED LESION(S) IN XP

Biochemical analysis of XP cells should resolve the nature
of the lesions being responsible for actinic carcinogenesis in
the skin of the patients. By using different techniques it has
been shown that the excision of UV-induced pyrimidine dimers
occurs at a lower rate in XP compared with normal cells (11, 24,
34). A defect in the repair mechanism involved in the removal of
pyrimidine dimers seems to be well documented. Recently Fornace
and coworkers (16) presented evidence that XP cells (complementation group A) are also defective in the repair mechanism of UV
induced DNA-protein cross links. This observation suggests that the
pyrimidine dimer might not be the only UV-induced DNA lesion
which is responsible for the abnormalities in XP. Moreover, treatment of XP cells with carcinogenic agents like 4-nitroquinoline-1-
oxide (4NQO), N-acetoxy-2-acetylaminifluorene (4-acetoxy-AAF) and
many others (37) has demonstrated that XP cells repair the DNA
lesions produced by these substances less efficient than normal
cells. Apparently the repair mechanism which is defective in these
cells is required not only for the repair of UV induced DNA lesions,
but also for the repair of damage caused by these chemicals.
Robbins et al. (38) have suggested that the neurological involvement sometimes seen in XP (complementation group A, B and D) might
be the result of defective repair of chemically induced DNA damage in the neurons. The same explanation might hold for the induction of tumors in tissues that can not have been exposed to
UVL.

In a recent paper Setlow et al. (41) described a defective
repair of gamma-ray-induced DNA damage in XP cells. Their results
indicate that some part of ionizing damage (induced under anoxic
conditions) mimics excision of UV damage in that the repair
patches are large. XP cells (complementation groups not indicated)
are defective in repairing this component of γ-ray induced damage.

These results indicate that in addition to pyrimidine dimers
also other molecular changes might be responsible for the clinical
defects in XP. In order to elucidate the lesion(s) resulting in
the neoplastic transformation of these cells, animal models for
xeroderma pigmentosum will be required (19). Those models would
facilitate the production of specific DNA lesions in vivo and the
demonstration of their carcinogenic activity.

THE DEFECTIVE STEP(S) IN DNA REPAIR IN XP

The defective enzyme or other proteins in the different XP
complementation groups have not yet been identified. The presence
of UDS and rejoining of DNA breaks (23) following exposure of XP

cells to X-rays and some alkylating agents (37) which do not require endonucleolytic excision has been interpreted as indicating
a defect in an early step in excision repair. A defective endonuclease which recognizes DNA lesions requiring long patch excision
repair would also explain the apparent lack in the excision of
pyrimidine dimers as well as the decreased levels of UDS following exposure to UVL and some carcinogenic agents. If so, the complementation analysis would indicate that at least 5 different
genes will be involved in the production and action of that enzyme.
This could be explained by assuming a multimeric composition of
the enzyme molecule and the action of other proteins required for
the incision process. Alternatively, if in mammalian cells the
repair systems act in a coordinated way, e.g. as part of a repair-
enzyme-complex a defect in one of the later steps in excision repair could block the entire repair process. This would also result
in decreased excision of the DNA lesions (20, 24).

By studying the excision of thymine dimers from UV irradiated
DNA treated with a dimer specific endonuclease from bacteriophage
T4 using extracts of mammalian cells, Cook et al. (12) observed
a normal dimer excision activity in extracts of XP cells. It was
concluded from their study that the hypothesis of a coordinated
enzyme complex for performing excision repair is not applicable to
excision of dimers in vitro. Moreover, all XP strains seem to perform the excision step in excision repair, ruling out the possibility of a defect in this step. These observations were supported
by experiments performed in Yoshio Okada's laboratory (45).
Addition of bacteriophage T4 endonuclease to the medium of UV exposed XP cells, which were pretreated with inactivated Sendai
virus, resulted in normal levels of UDS. The experiments of Cook
et al. (12) were performed on XP cells of the A and D group and on
two strains which were assigned to a complementation group (D and
E) based on residual UDS and clinical symptoms. The Japanese group
found similar effects of T4 endonuclease addition in A, B, C and
D group XP cells.

As shown by Mortelmans and coworkers (33), and presented at
this conference by Friedberg, extracts of XP cells possess endonucleolytic activity on UV irradiated and purified DNA but not on
chromatin preparations. These very recent results indicate that XP
cells might not be deficient in an endonuclease, but they might be
deficient in protein factors which are required for the action of
the repair enzymes. Moreover, Sutherland and coworkers presented
evidence for defective photoreactivation in XP (43), whereas
Lehmann and coworkers (27) found decreased levels of postreplication repair in excision deficient XP cell strains. Therefore, in
considering an association between defective DNA repair and carcinogenesis in XP, it has to be taken into account that the primary
genetic defect(s) in XP has (have) not yet been clarified.It has
still to be demonstrated that it concerns proteins which directly

play a role in DNA repair.

CORRELATION BETWEEN RESIDUAL UDS ACTIVITY AND SEVERITY OF THE DISEASE

Constant levels of residual UDS activity have been described in cells of patients of the same kindred (4). Examples of families with XP in the 5 complementation groups are presented in Table 2. These data indicate that exactly the same mutation results in comparable UDS levels in the cells of different patients in all complementation groups.

Although complementation between two different mutations in the same gene (intragenic complementation) cannot be ruled out, it seems likely to explain the results of cell fusion on the basis of intergenic complementation. Different genes will be involved in the different complementation groups. Within one complementation group different patients might have a mutation in the same gene but at different sites, possibly affecting the protein molecule in different ways. Table 3 presents the UDS levels found in strains belonging to the A and C group. At present these two groups contain the largest number of kindreds (Table 1). In earlier papers (25, 38) it has been reported that XP strains of the same complementation group have similar rates of UDS and that each complementation group has a characteristic rate. Fusion experiments with the XP8LO strain (48), having 36% residual UDS, assigned this strain to complementation group A (Table 3) and is therefore a clear exception of the rule. This observation stimulates a reevaluation of the different levels observed in complementation group C. In the Rotterdam series of patients these levels varied from about 8 to about 30% UDS compared with normal cells (Table 3).

TABLE 2

UNSCHEDULED DNA SYNTHESIS IN XERODERMA STRAINS OF THE SAME KINDRED

COMPLEMENTATION GROUP	XP STRAINS	UDS (% OF CONTROL)
A	XP12RO, XP25RO, XP26RO	< 5
C	XP4RO, XP21RO	10 - 20
	XP5RO, XP6RO, XP7RO	25 - 30
	XP2BE, XP8BE, XP9BE	15 - 25
D	XP5BE, XP6BE	2 - 10(25-55)
	XP2NE, XP3NE	10 - 20
E	XP2RO, XP3RO	40 - 50

TABLE 3

UNSCHEDULED DNA SYNTHESIS WITHIN THE SAME
COMPLEMENTATION GROUP

COMPLEMENTATION GROUP	XP STRAIN	UDS (% OF CONTROL)
A	XP12RO	< 5
	XP12BE	< 2
	XP1LO	< 2
	XP8LO	36
C	XP16RO	8
	XP9RO	10
	XP20RO	12
	XP4RO	15
	XP5RO	28

These levels of UDS have shown to be reproducible. It seems likely that these levels reflect differences in the molecular defect
of the protein molecule involved in complementation group C.

A strong argument in favour of a relationship between the
repair defect and carcinogenesis would be if the level of residual
UDS could be associated with the severity of the clinical symptoms
of the disease. Conflicting data have been published in this respect. From the Rotterdam (4) and the Japanese (44) series of
patients it was concluded that an inverse correlation exists between the level of repair DNA synthesis and the severity of the
disease. However, the careful clinical observations presented by
the NIH group (38) were in disagreement with the existance of
such a relationship. This discrepancy was based on the 25-55% residual activity found in the D group by the NIH investigators and
the data obtained with the patients XP12BE and XP1LO, who have 2%
of the normal repair rate and relatively mild symptoms of the disease (Table 4). Especially XP1LO is an interesting case as she
seems to be free of neurological abnormalities. In contrast with
the NIH data UDS experiments on two different D group kindreds
performed by the Rotterdam group (25) and by Huang and Vincent
(21) has revealed much lower levels of repair DNA synthesis (2-10%
for the BE strains in stead of 25-55). These values are not in conflict with an association between UDS level and severity of the
disease. Moreover they would also fit in the observations concerning the low survival of D group cells following UV exposure (3)
and the low host cell reactivation of UV irradiated adenovirus 2
in these cells (14). Therefore, it is tempting to consider the
XP12BE and the XP1LO as the only XP cases known so far being in
conflict with an inverse relationship between residual UDS and
severity of the disease. It has to be taken into account that the

TABLE 4

COMPARISON OF UNSCHEDULED DNA SYNTHESIS AND CLINICAL
MANIFESTATIONS IN DIFFERENT COMPLEMENTATION GROUPS OF
XERODERMA

COMPLEMEN-TATION GROUP	XP STRAIN	UDS (% OF CONTROL)	SKIN LESIONS	NEUROLOGICAL ABNORMALITIES
A	XP12RO	<5	+++	+++
	XP12BE	<2	++	+
	XP1LO	<2	+	-
	XP8LO	36	+	-
B	XP11BE	3 - 7	++	+++
C	XP16RO	5 - 10	++	-
	XP5RO	25 - 30	+	-
D	XP5BE	2 - 10(25-55)	++	++
E	XP2RO	40 - 50	+	-

expression of the clinical features of xeroderma will be highly
dependent on exogeneous factors which are difficult to be quantified, such as the amount of UV absorbed by the skin. This will obviously hamper the study of an association between UDS level and
expression of symptoms. Therefore the association indicated by the
large number of patients studied so far seems to be rather surprising and has to be considered with some caution.

XERODERMA PATIENTS HAVING NORMAL EXCISION REPAIR

The detection of XP patients having normal excision repair
(XP variants) (6, 10, 24) has been used as argument against a role
of DNA repair in carcinogenesis. The work of Lehmann and coworkers
(27) on postreplication repair in cells of these patients has solved this discrepancy. The mechanism of this repair process is not
yet fully understood. As in bacteria, newly synthesized DNA in
UV-irradiated mammalian cells is smaller than in unirradiated
cells, probably because of gaps in this DNA opposite to pyrimidine
dimers. On prolonged incubation of the cells the DNA attains a
high molecular weight similar to that from unirradiated cells.
The lengthening of the DNA might be the result of insertion of
new bases in the gaps (26) as well as of recombinational processes
between parental and daughter DNA strands (32). In XP variants the
time taken for the newly synthesized DNA to attain the normal
length was much longer than in normal cells. The conversion from
low- to high molecular weight DNA was found to be inhibited by
caffeine, which has little effect on normal cells. The biological

significance of this impairment of postreplication repair was also
indicated by the lower survival frequencies following UV exposure
of XP variant cells (30) and by the decreased host cell reactivation of UV irradiated adenovirus (15).

The detection of a DNA repair defect in the XP variants has
strongly emphasized the association between defective DNA repair
and carcinogenesis.

DNA REPAIR IN OTHER DISEASES IN MAN SHOWING PREDISPOSITION OF CANCER

The detection of a repair defect in XP has stimulated the
search for DNA repair defects in other diseases showing a genetically determined predisposition for cancer. Attractive candidates
were the diseases showing spontaneous chromosome breakage and a
relatively high incidence of cancer: Ataxia telangiectasia (the
Louis-Bar syndrome), Fanconi's anemia and Bloom's syndrome.

Ataxia telangiectasia (AT) patients show telangiectasia of
the conjunctiva and skin and cerebellar ataxia (loss of muscular
coordination). Sinopulmonary infections, immunological deficiencies, mental deficiency, progressive neurologic deterioration and
a predisposition to malignancies are also features of this syndrome.
The areas with the greatest exposure to the sun are the most affected. An enhanced level of spontaneous chromosome aberrations,
increased sensitivity to X-rays on receiving radiotherapy and increased chromosome aberrations induced by ionizing radiation in
leucocyte cultures have also been reported (18).

As demonstrated by Taylor and coworkers (46) AT cells show
an enhanced sensitivity to ionizing radiation in terms of cell
survival. In one of the symposia of this congress Dr. Paterson
discussed the work of his group on the characterization of the biochemical defect (35) in this disease. He presented evidence that
in a number of these patients repair DNA synthesis following γ-ray
exposure under anoxic conditions is inhibited to 40-60% of control
cells. By applying an in vitro enzymatic assay he showed that the
disappearance of a class of γ-lesions in the DNA proceeds at a four
times slower rate than in normal cells. Evidence was presented for
the existence of at least three different classes of AT patients.
Two classes were identified by a complementation test as performed
for xeroderma. A third class was distinguished on the basis of the
presence of patients having normal levels of repair replication.
In analogy with xeroderma these patients were referred to as Ataxia
variants. So far, a defect in a repair process in these patients
has not been demonstrated.

It is temping to consider Ataxia telangiectasia as a γ-ray
analogue of xeroderma. Again it presents a genetic disease in
man in which a DNA repair defect seems to be associated with increased malignancy.

Fanconi's anemia (FA) is an autosomal recessive disease
characterized by pancytopenia associated with diverse congenital
anomalies. Chromosomes from affected individuals exhibit structural lability and there is a strong predisposition to leukemia
and other cancers.

Evidence for a DNA repair defect in this disease was presented by Sasaki et al. (40). They found an increased susceptibility
of FA lymphocytes for DNA cross-linking agents in terms of chromosome aberrations. Further studies by Sasaki (39), in which the
chromosome breaking effects of mono- and difunctional mitomycins
were compared have indicated that FA cells have "a specific defect
in the repair of pre-aberration lesions induced by difunctional
mitomycin; the lesions are possibly DNA cross-links of the interstrand type". Moreover, Fujiwara et al. (17) found FA cells being
about 10 times as sensitive to mitomycin C cell killing as XP and
normal cells. Alkaline sucrose sedimentation studies of the DNA of
mitomycin C and psoralen and light treated FA cells indicated an
impairment of the excision of the DNA interstrand cross-links (17).
This work was performed on only one FA cell strain and has to be
extended to more patients in order to establish its reproducibi-
lity.

The finding of a dramatic increase in sister chromatid exchanges (SCE) in Bloom's syndrome (7) has stimulated the search
for a DNA repair defect in this disease. The repair test systems
applied on Bloom cells, including UDS and postreplication repair
following exposure to UVL has not revealed an abnormality so far.
An intriguing problem in this respect is the relationship between
SCE and DNA repair. Although this question has been subject to
many speculations (22) a definite proof for such a relationship
has not been presented. It is expected that the elucidation of the
biochemical defect in Bloom's syndrome will ultimately give the
answer.

MUTATION INDUCTION IN REPAIR DEFICIENT CELL STRAINS

In order to fit the concept of defective DNA repair in a model
for carcinogenesis, the somatic cell mutation hypothesis on the
origin of cancer, first suggested by Boveri (5) seems to be very
attractive. Inefficient or inaccurate repair of DNA damage might
result in mutations in critical regions of the genome that make
somatic cells malignant. This idea has encouraged several investigators to study the induction of mutations in repair deficient
cells and to compare the mutation frequencies with those obtained
with normal cells. As yet only mutation studies with XP cells
have been published, whereas others regarding AT and FA cells are
in progress. The system used in these studies is the induction of
forward mutations in an X-linked gene coding for the enzyme hypox-
anthine (guanine) phosphoribosyl transferase (HPRT) rendering the
cells resistant for 8-azaguanine. Maher and coworkers (29, 31)
and Simons (42) have demonstrated a much higher frequency of UV
induced mutations in excision deficient XP cells than in normal
cells. This increased frequency is correlated with the increased
sensitivity to UV in terms of cell survival and the decreased
capacity for excision repair of UV-induced lesions (Maher studied
complementation groups A and C, whereas Simons investigated the
D group with similar results). If the induced mutation frequencies
were plotted on the basis of cell survival, comparing mutation
induction at equal levels of cytotoxicity, the results obtained
with XP and normal cells were equal. This might indicate that the
excision repair process itself is "error-free". The critical event
might be the replication of DNA on a template still containing
non-excised UV induced DNA lesions.

Mutation experiments with XP variant cells (30) also revealed an increased frequency of mutations per unit of UV dose.
The frequency of mutations per lethal event in the XP variant
strain studied by Maher et al (30) was slightly higher than in
normal cells. Caffeine treatment increased the cytotoxic and
mutagenic effect of XP variant cells but did not increase its
cytotoxicity in normal cells. In the case of the XP variant one
might expect that the number of unexcised UV lesions at the time
of DNA replication will not differ significantly from that in normal cells, both have an efficient excision repair system. Moreover, the conversion from low- to high-molecular weight DNA, even
if this is delayed in the case of the variant, will be terminated
in both cell types before the next round of DNA replication will
start. It is therefore, difficult to understand how a difference
in rate of gap filling could be responsible for the increased
cytotoxic and mutagenic effects of UVL observed in the variants.
It seems likely to assume that the decreased gap filling is associated with another defect which is ultimately responsible for
both lethal and mutagenic events (30).

The finding of increased mutation frequencies in these repair
deficient cells of individuals showing increased susceptibility
to skin cancer provides support for the assumption that the increased frequency of neoplastic transformation is the result of
a higher incidence of mutational events at critical sites.

CONCLUSION

Evidence for an association between defective DNA repair and
neoplastic transformation comes mainly from the studies of two
genetic diseases in man: xeroderma pigmentosum and Ataxia telangiectasia. Although indirect evidence for a DNA repair defect
in these diseases is accumulating, a direct proof for impaired
DNA repair as primary genetic defect has not yet been presented.
Therefore, we have to consider the possibility of another defective mechanism in these XP and AT cells which in term might influence DNA repair. In that case defective DNA repair needs not
to be involved in the production of cancer in these patients.
An example of such a defective mechanism could be the immune deficiency observed in Ataxia telangiectasia.

If test systems become available for demonstrating the malignant state of human cells, they will facilitate a direct analysis
of the contribution of defective DNA repair to neoplastic transformation. Moreover, they might also allow genetic studies by
means of cell fusion, to elucidate the genetic changes that occur
in these cells during transformation.

ACKNOWLEDGEMENT

This work was supported by grants from The Netherlands
Organization for the Advancement of Pure Research (ZWO) and
Euratom, contract no. 123-74-IBJOC.

REFERENCES

1. Aaronson, S.A., and Lytle, C.D. (1970) Nature, 228, 359-361.
2. Abrahams, P.J. and van der Eb, A.J. (1976) Mutation Res.,
    35, 13-22.
3. Andrews, A.D., Barrett, S.F. and Robbins, J.H. (1976)
    The Lancet, ii 1318-1320.
4. Bootsma, D., Mulder, M.P., Pot, F. and Cohen, J.A. (1970)
    Mutation Res., 9, 507-516.
5. Boveri, T. (1929) The origin of malignant tumors (first
    published Jena, 1914), (Williams and Wilkins, Baltimore).

6. Burk, P.G., Lutzner, M.A., Clarke, D.D. and Robbins, J.H.
    (1971) J. Lab. Clin. Med., 77, 759-767.
7. Chaganti, R.S.K., Schonberg, S. and German, J. (1974) Proc.
    Nat. Acad. Sci., 71, 4508-4512.
8. Cleaver, J.E. (1968) Nature, 218, 652-656.
9. Cleaver, J.E. (1970) Int. J. Rad. Biol. 18, 557-565.
10. Cleaver, J.E. (1972) J. Invest. Dermatol., 58, 124-128.
11. Cleaver, J.E. and Trosko, J.E. (1970) Photochem. Photobiol.
    11, 547-550.
12. Cook, K., Friedberg, E.C. and Cleaver, J.E. (1975) Nature,
    256, 235-236.
13. Day III, R.S. (1974) Cancer Res., 34, 1965-1970.
14. Day III, R.S. (1974) Photochem. Photobiol., 19, 9-14.
15. Day III, R.S. (1975) Nature, 253, 748-749.
16. Fornace, A.J., Kohn, K.W. and Kann, H.E. (1976) Proc. Nat.
    Acad. Sci. USA, 73, 39-43.
17. Fujiwara, Y and Tatsumi, M. (1975) Biochem. Biophys. Res.
    Commun., 66, 592-598.
18. Harnden, D.G. (1974) Chromosomes and Cancer, edited by J.
    German (Wiley, New York), p. 619.
19. Hart, R.A. and Setlow, R.B. (1975) Molecular Mechanisms for
    repair of DNA, edited by P.C. Hanawalt and R.B. Setlow (New
    York: Plenum Press), p. 719.
20. Haynes, R.H. (1966) Radiation Res. Suppl. 6, 232.
21. Huang, P.C. and Vincent, R. (1975) Molecular Mechanisms for
    repair of DNA, edited by P.C. Hanawalt and R.B. Setlow (New
    York: Plenum Press), p. 729.
22. Kato, H. (1974) Exptl. Cell. Res., 85, 239-247.
23. Kleijer, W.J., Lohman, P.H.M., Mulder, M.P. and Bootsma, D.
    (1970) Mutation Res., 9, 517-523.
24. Kleijer, W.J., de Weerd-Kastelein, E.A., Sluyter, M.L.,
    Keijzer, W., de Wit, J. and Bootsma, D. (1973) Mutation Res.
    20, 417-428.
25. Kraemer, K.H., de Weerd-Kastelein, E.A., Robbins, J.H.,
    Keijzer, W., Barrett, S.F., Petinga, R.A. and Bootsma, D.
    (1976) Mutation Res., 33, 327-340.
26. Lehmann, A.R. (1972) J. Mol. Biol., 62, 319-337.
27. Lehmann, A.R., Kirk-Bell, S., Arlett, C.F., Paterson, M.C.
    Lohman, P.H.M., de Weerd-Kastelein, E.A. and Bootsma, D.
    (1975) Proc. Nat. Acad. Sci. USA, 72, 219-223.
28. Lytle, C.D., Aaronson, S.A. and Harvey, E. (1972) Int. J. Rad.
    Biol., 22, 159-165.
29. Maher, V.M. and Mc Cormick, J.J. (1976) Biology of Radiation
    Carcinogenesis, edited by J.M. Yuhas, R.W. Tennant and J.B.
    Regan (Raven, New York) p. 129.
30. Maher, V.M., Ouellett , L.M., Curren, R.D. and Mc Cormick, J.
    J. (1976) Nature, 261, 593-595.

31. Maher, V.M., Curren, R.D., Ouelette, L.M. and Mc Cormick, J.
    J. (1976) Fundamentals of Cancer Prevention, edited by W.
    Nakahara, S. Takayama and T. Sugimura (Plenum, New York), in
    the press.
32. Meneghini, R. and Hanawalt, P. (1976) Biochim. Biophys.Acta,
    425, 419-427.
33. Mortelmans, K., Friedberg, E.C., Slor, H., Thomas, G., and
    Cleaver, J.E. (1976). Proc. Nat. Acad. Sci. USA, in the press.
34. Paterson, M.C., Lohman, P.H.M. and Sluyter, M. L. (1973)
    Mutation Res., 19, 245-256.
35. Paterson, M.C., Smith, B.P., Knight, P.A. and Anderson, A.K.
    (1976). This Conference.
36. Rabson, A.S., Tyrrell, S.A. and Legallais, F.Y. (1969) Proc.
    Soc. Exp. Biol. Med., 132, 802-806.
37. Regan, J.D. and Setlow, R.B. (1974) Cancer Res., 34, 3318-
    3325.
38. Robbins, J.H., Kraemer, K.H., Lutzner, M.A., Festoff, B.W.
    and Coon, H.G. (1974) Ann. Intern. Med., 80, 221-248.
39. Sasaki, M.S. (1975) Nature, 257, 501-503.
40. Sasaki, M.S. and Tonomura, A. (1973) Cancer Res., 33, 1829-
    1836.
41. Setlow, R.B., Faulcon, F.M. and Regan, J.D. (1976) Int. J.
    Rad. Biol., 29, 125-136.
42. Simons, J.W.I.M. (1976) Personal Communication.
43. Sutherland, B.M., Rice, M. and Wagner, E.K. (1975) Proc.Nat.
    Acad. Sci. USA, 72, 103-107.
44. Takebe, H., (1976) Fundamentals of Cancer Prevention, edited
    by W. Nakahara, S. Takayama and R. Sugimura (Plenum, New York)
    in the press.
45. Tanaka, K., Sekiguchi, M. and Okada, Y. (1975) Proc. Nat.Acad.
    Sci. USA, 72, 4071-4075.
46. Taylor, A.M.R., Harnden, D.G., Arlett, C.F., Harcourt, S.A.
    Lehmann, A.R., Stevens, S., Bridges, B.A. (1975) Nature, 258,
    427-429.
47. De Weerd-Kastelein, E.A., Keijzer, W. and Bootsma, D. (1972)
    Nature New Biol., 238, 80-83.
48. De Weerd-Kastelein, E.A., Keijzer, W., Sabour, M. Parrington,
    J.M. and Bootsma, D. (1976) Mutation Res., in the press.

EXPERIMENTAL ULTRAVIOLET CARCINOGENESIS

P.D. Forbes, R.E. Davies, F. Urbach

Photobiology Program
The Skin and Cancer Hospital, Temple University Health
Sciences Center, Philadelphia, PA.   19140 USA

Cancer, it has been said, is an environmental problem.   As long as we live on the surface of the earth, we will have to deal with what must be the most ubiquitous carcinogen of all—sunlight. Soon after the recognition that ultraviolet radiation (UVR) could act as a carcinogen, a number of investigators attempted quantitative studies and made efforts to identify mechanisms of photocarcinogenesis. These aims were central to the investigations that Blum(1) and his colleagues performed at the National Institutes of Health during the 1940's and 1950's. The thrust of the Blum experiments was an effort to describe the relation of stimulus (UVR) to response (tumor formation and development) in quantitative terms. These data made a unique contribution to the cancer literature, and they remain an important component in the evolution of quantitative models of the carcinogenesis process. It is safe to predict that such models will continue to appear.

Carcinomas and sarcomas in experimental animals have been produced by UVR in a number of laboratories. The available leads in animal models for malignant melanomas have been only initially explored. Chemical carcinogen painting, followed much later by chronic UVR, produced melanomas in hairless mice (2). The staging and progression of melanomas have been followed in carcinogen-painted guinea pigs (3).

A number of factors influence skin cancer incidence rates: geographic, ethnic, occupational, etc. There is a growing awareness that human activities can influence cancer rates; obviously, anything that increases risk is cause for concern. Of the possible ways of increasing risk, two will be discussed here: increasing the effectiveness of the absorbed UVR, and increasing the amount

of absorbed UVR. This paper will review some of the evidence for
these kinds of influence on experimental UVR carcinogenesis.

INCREASED EFFECTIVENESS OF ABSORBED UVR

Carcinogens and Photosensitizers

Photosensitization and   chemical carcinogenesis are potentially
interactive processes, the results of which vary with a number of
recognized, and perhaps some unrecognized, experimental conditions.
Some polynuclear aromatic hydocarbon carcinogens are photodynamic
agents; the carcinogenic potential of some may be enhanced by UVR;
some are degraded by certain wavelengths and of these, some are
regenerated by other wavelengths. Thus, either enhancement, or no
net change, or inhibition of skin carcinogenesis may occur depending
on the carcinogen and wavelength and dose of light used.   The
available data have been reviewed by Blum (1), and more recently by
Emmett (4), Stenbäck (5), Davies et al (6,7) and Urbach et al (8,9).

A conceptually distinct problem concerns photoinduced carcinogenesis following administration of agents which are phototoxic, but
not in themselves carcinogenic. The potential for complex interactions would appear less, but the literature is limited and not
without controversy. In brief, there is ample evidence that 8-meth-
oxypsoralen (8MOP) applied topically can enhance photocarcinogenesis
in mice; there is considerably less evidence available for drawing a
conclusion on whether orally-administered 8MOP, or other photosensitizers by any route, can enhance photocarcinogenesis (10,11,12,
13). Two types of experiments involving P-UVA (oral 8MOP plus UVA,
i.e. $\lambda > 320$nm) are currently in progress in our laboratory.   The
studies involve repeated P-UVA exposures of mice, with or without a
short course of "priming" exposures to full-spectrum simulated sunlight.   Preliminary results suggest that at least a non-inflammatory
level of repeated P-UVA treatment alone produces few tumors, but that
P-UVA can enhance photocarcinogenesis induced by previous exposure to
UVR. One complicating factor is that even in unirradiated skin the
animals quickly develop a far greater tolerance for (oral) P-UVA
treatment than for topical 8MOP treatment.

Ionizing Radiation

A limited amount of information is available on carcinogenic
interaction of ionizing and non-ionizing radiation (14,15). We have
found that UVR photocarcinogenesis in mice is enhanced by
pre-exposure of skin to subcarcinogenic doses of β radiation.

Repair vs Cell Death

Epstein et al (16) and Zajdela and Latarjet (17) suggest that
the production of skin cancer by UVR is fostered by repair of DNA,
allowing the cell to survive, yet leaving in place or even favoring
subsequent errors in DNA replication, resulting in a greater likelihood of malignant change.   The weight of current opinion strongly
suggests   that   mutagenesis   and   carcinogenesis are consequenses of
errors in DNA repair (18,19); if this is so, then it can be expected
that factors influencing various repair systems will be found to influence the amount of response (skin cancer) to stimulus (a given
dose of UVR). A further consequence (though admittedly speculative)
may be that some psoralens, because of their DNA cross-linking capability, will be photocarcinogenic, while most photodynamic
compounds, causing either photooxidation or covalent DNA single
strand adducts, have less likelihood of causing permanent DNA change
and thus should be relatively less carcinogenic.   In any case,
identification of biologically important photoproducts will be an
important endeavor in the future (18,20).

Time-Dose Effects

If the total amount of UVR that a mouse could tolerate in a
lifetime of daily exposures were delivered all on one day, the result
would be a graphic demonstration of the fact that time-dose reciprocity does not apply to such indirect effects as epidermal survival.
Indeed, a large single UVR dose has little carcinogenic potential
(21). The extent to which time-dose reciprocity can be applied to
photocarcinogenesis is not known; consequently most models ignore the
possibility of departure from reciprocity.

What experimental evidence does exist indicates that dose deliv-
ery rates and "rest intervals" can significantly influence tumor
response (1). Some of our data indicate that a protracted UVR exposure is more effective than a brief, more intense exposure (the total
doses being equal). This result is unexpected, but assuming it can
be verified, an explanation may rest in mechanisms of DNA repair. It
is known that DNA repair is initiated immediately after injury and
progresses rapidly. Thus, long continued injury may lead to either
overloading of the repair systems, or injury so close to the onset of
cell division that repair has to occur postreplication; it is
believed that the repair mechanisms that operate postreplication may
be prone to error. In contrast, in high flux short duration exposures, more injury may occur in shorter period of time, but time for
repair is much longer, and so the more efficient excision repair
systems may predominate.

Within the context of environmental problems, the question of
time-dose reciprocity is inextricably involved with another reciprocity question, i.e., the additivity of spectral components in a

polychromatic source. This concept is expanded in the last section.

Immunologic Status

Several lines of evidence indicate that the immune status of the host and carcinogenesis are potentially interactive processes. Increased cancer risk in humans under prolonged immunosuppression following organ transplantation has been reported. Some investigators believe that chemical carcinogens depress the immune system of the host; others have provided evidence that immunosuppresive drugs can enhance chemical carcinogenesis. Nathanson et al (22) showed that rabbit anti-mouse, anti-lymphocyte serum (ALS) enhanced photocarcinogenesis. More recently, we have exposed hairless mice daily to UVR, treating three replicate subgroups with either ALS, or normal rabbit serum, or isotonic saline solution. Photocarcinogensis was enhanced by ALS, but inhibited by normal serum, as compared with the saline-injected controls (the differences became statistically significant by about 25 weeks; fig. 1). As administered, the ALS did not prolong survival of heterograft skin on these mice.

In a different system, Kripke and Fisher (23) have shown that UVR-induced skin tumors in C_3HF mice are highly antigenic, and are usually immunologically rejected when transplanted to normal nonirradiated syngeneic recipients. However, pre-irradiation of mice for periods of time too short to induce skin tumors in the lifetime of animals, made them susceptible to grafting with UVR-induced tumors. Another study shows that the skin, and not the eye, is the target organ for the UVR effect (24). This indicates that UVR irradiated mice are systemically altered in a way that prevents immunologic rejection of highly antigenic tumors, and that UVR can alter the host response against a tumor, in addition to initiating the neoplastic transformation.

INCREASED AMOUNT OF ABSORBED UVR

Changes in Lifestyle

As people become increasingly exposed to sunlight (geographically, anatomically, and temporally), the likelihood of long term changes, including neoplasia, increases. The association of actinic damage with basal cell carcinoma and squamous cell carcinoma is a familiar finding. It is somewhat more unexpected to find increasing evidence of such an association with malignant melanoma, in view of statements to the contrary in the older literature. There

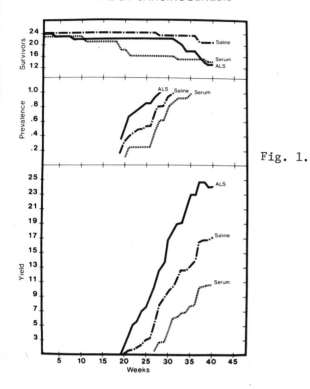

Fig. 1. Influence of chronic
treatment with antilymph-
ocyte serum (ALS) or
normal serum on UVR
photocarcinogenesis in
hairless mice. In terms
of yield (tumors/surviv-
ors) and prevalence
(affected/survivors),
both treatments are sig-
nificantly different from
saline injection at 27
weeks.

remains a paradox in the strong relationship of malignant melanoma to
exposure-latitude, but lack of strong concentration of melanomas on
exposed sites.   Even after corrections for improved diagnostic
techniques, the incidence and death rate from malignant melanoma is
rising rapidly in many developed societies (24,25).   If increased
exposure to sunlight is responsible for this rise, the effective
wavelengths remain a mystery; the available data do not any more
readily implicate the UVR spectrum than the visible or infrared. A
combination of factors is very likely involved (3).

Increase in Ambient UVR

Industrial, commercial, and medical applications of UVR can increase the average exposure of those directly involved, and to a
lesser extent, the general public.   Regulation of some types of exposure is made difficult by lack of reliable information on biological effectiveness of parts of the spectrum.

An entirely new and unanticipated problem arose with the recogninition that human activities could significantly modify the composition of the stratosphere (26,27). For many reasons, some of which are developed below, it became clear that the quantitative studies performed to date were not adequate for an extrapolation from hypothetical stratospheric modification to an alteration in human skin cancer risk. The human exposure experience has many variables which currently must be ignored in quantitative models. It is not even known in which direction some of the variables would influence the effectiveness of UVR exposure (26).

For present purposes we assume:

1. Various stratospheric processes may modify the ambient concentrations of O_3: the nature of such processes and their quantitative effects are not considered here.

2. Modification of ozone concentration will alter the quality and quantity of UVR reaching the ground. The nature of such changes is a function of the ozone absorption spectrum, although a series of moderator variables (scatter, angle, second-order processes, etc.) complicate the relationship.

3. Modification of UVR reaching earth will affect photosensitive biological processes. The specific process to be considered here is cutaneous carcinogenesis.

Most present analyses of photoinduced cutaneous carcinogenesis make the following additional assumptions:

1. The magnitude of the response can be related to the total delivered dose of effective radiation. This implies that the frequency, size and time distribution of components of the total dose are not significant to the overall effect. This assumption also implies that the effect of a polychromatic UVR source can be described in terms of the expected effects of its individual spectral components.

2. The dose of polychromatic energy can be represented, in biological effect calculations, by the sum of the doses of spectral components weighted for relative effectiveness by the appropriate action spectrum.

3. The action spectrum for carcinogenesis, which is not known, can be fairly approximated by the action spectrum for acute skin erythema (cf also 28).

Changes in ozone concentration will produce spectral changes largely confined to the 280-320 nm (UV-B) region, which is also the region of maximum effectiveness for many biological responses. The assumption of spectral additivity thus implies that longer wavelengths (320-400 nm: UV-A) need not be considered in assessing the effects of ozone changes. At most, this can only be a working assumption, pending availability of pertinent data.

An important implication of the total-dose assumption is that changes in the UV-B flux will have relatively little effect on the development of tumors in elderly people, who have already accumulated most of their lifetime doses. This suggests that they will be little affected by ozone changes and also that little can be done to protect them from further tumor formation. It further suggests that the effects of a change in ozone level will appear relatively slowly, over many years. There is some reason to question these conclusions: some clinical evidence suggests that progression of tumors to clinical status is affected by natural solar variations within a year, and there is also a clinical impression that the elderly can be benefited by protection from additional UVR exposure. Animal studies suggest that irradiation of pre-existing tumors has an effect on their development and aggressiveness (29).

It is probably fitting that the ozone problem illustrates the fact that distinguishing amount and effectiveness of UVR is convenient, but artificial. Depletion of ozone will assuredly lead to some increase in the amount of earth-level UVR; effects are not quantitatively predictable, precisely because the effectiveness of energy in this spectral region is so poorly known. Appropriate studies should allow considerably more accurate estimates of potential risks than are now available.

We are in the process of studying the carcinogenic effects of a simulated serial depletion of ozone effective thickness. The light source is a 6KW Xenon long arc water-cooled lamp. Quartz cylinders provide the only inherent filtration. Added filters are used to provide a series of cut-offs in the spectral region that is affected by ozone; the filters are various thicknesses of Schott WG320 material (fig 2). The curves are similar to those produced by ozone (fig 3), with those produced by the filter material slightly steeper (fig 4). The data thus far indicate a range of tumor response as a result of irradiating hairless mice through filters 0.55-3.00 mm thickness. Using this type of information, we will calculate change in effective energy as a function of change in ozone concentration.

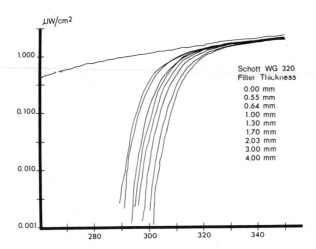

Fig. 2. UVR emission spectrum of long-arc
xenon lamp, with or without added filtration.

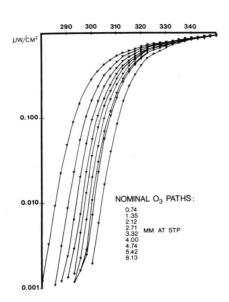

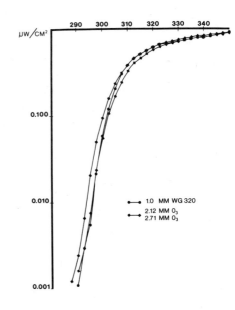

Fig. 3. UVR emmision spectrum
of the lamp as in Fig. 2, but
with varying thickness of ozone.

Fig. 4. Comparison of ozone
and WG320 filter material; the
latter has a slightly sharper
cut-off.

REFERENCES

1.   Blum H.F.: Carcinogenesis by Ultraviolet Light. Princeton U. Press, Princeton, 1959.
2.   Epstein J.H.:Ultraviolet carcinogenesis. Chapt. 8 In Vol. 5, Photophysiology (A.C. Giese, ed.), Academic Press, p. 235, 1970.
3.   Clark,W.C., M.J. Mastrangelo, A.M. Ainsworth, D. Berd, R.E. Bellet and E.A. Bernardino: Current Concepts of Biology of Human Cutaneous Malignant Melanoma. Adv. in Cancer Research, 1976 (in press).
4.   Emmett E.A.: Ultraviolet radiation as a cause of skin tumors. CRC Critical Reviews in Toxicology 2:211, 1973.
5.   Stenbäck F.: Studies on the modifying effect of ultraviolet radiation on chemical skin carcinogenesis. J. Invest. Derm. 64:253, 1975.
6.   Davies R.E., H.A. Dodge and L.H. DeShields: Alteration of the carcinogenic activity of DMBA by light. Proc. Am. Assn. for Cancer Res. 13:14, 1972.
7.   Davies R.E. and H.A. Dodge: Modification of chemical carcino-genesis by phototoxicity and photochemical decomposition of carcinogen. Proc. First Annual Meeting, Am. Soc. Photobiol, Sarasota, June 1972.
8.   Urbach F., P.D. Forbes, R.E. Davies and D. Berger: Cutaneous photobiology: past, present and future. J. Invest. Dermatol. 67:207.
9.   Urbach F., P.D. Forbes and R.E. Davies: Photocarcinogenesis: new light on an old problem. Sumbitted for publication.
10. Forbes P.D. and F. Urbach: Experimental modification of photo-carcinogenesis: III. Simulation of exposure to sunlight and fluorescent whitening agents. Fd. Cosmet. Toxicol. 13:343, 1975.
11. Forbes P.D., R.E. Davies and F. Urbach: Phototoxicity and photocarcinogenesis: Comparative effects of anthracene and 8-methoxypsoralen in the skin of mice. Fd. Cosmet. Toxicol. 14: in press 1976.
12. Griffin A.C., R.E. Hakim and J. Knox: The wavelength effect upon erythemal and carcinogenic response in psoralen treated mice. J. Invest. Derm. 31:289, 1958.
13. Langner A., H. Wolska, M. Jarzabek-Chorzelska and M. Pawinska: Dermal toxicity of 8-methoxypsoralen in hairless mice irradiat-ed with long UV. Joint meeting The Society for Investigative Dermatology, Inc. and the European Society for Dermatological Research, Amsterdam, Netherlands 1975.
14. Sanders C.L., R.H. Busch, J.E. Ballou and D.D. Mohlum (ed) Radionuclide Carcinogenesis. U.S.A.E.C. Conf. 720505 Nat. Tech. Inf. Serv. Springfield, Va. 22151 1973.
15. Bond V.P., S. Hellman, S.E. Order, H.D. Suit and H.R. Withers. (ed) Interactions of radiation and host immune defense mech-anisms in malignancy. Brookhaven National Laboratory 50418.

National Technical Information Service, Springfield, Va. 22151
1974.

16. Epstein W.L., D. Fukuyama and J.H. Epstein: Ultraviolet light,
    DNA repair and skin carcinogenesis in man. Fed. Proc. 30:
    1766, 1971.

17. Zajdela F. and R. Latarjet: The inhibiting effect of caffeine
    on the induction of skin cancer by ultraviolet light in the
    mouse. Compt Rendu Acad. Sci., Series D 277:1073, 1973.

18. Hanawalt P.C. and R.B. Setlow (Eds): Molecular mechanisms for
    repair of DNA, (Parts A & B) Plenum Press, New York and London
    1975.

19. Setlow R.B. and R.W. Hart: Direct evidence that damaged DNA
    results in neoplastic transformation: a fish story. Radiation
    Res. 59:73, 1974.

20. Grube D., R.D. Ley and R.J.M. Fry: Studies on the effect of
    psoralen and ultraviolet light. Photochem. Photobiol., in press,
    1976.

21. Hsu R., P.D. Forbes, L.C. Harber and E. Lakow: Induction of
    skin tumors in hairless mice following single exposure to
    ultraviolet radiation. Photochem. Photobiol. 21:88, 1975.

22. Nathanson R., P.D. Forbes and F. Urbach: Modification of photo-
    carcinogenesis by two immunosuppressive agents. Cancer Letters
    1:243, 1976.

23. Kripke M.L. and M.S. Fisher: Immunologic parameters of ultra-
    violet carcinogenesis. J.Nat. Cancer Inst. (In press) 1976.

24. Kripke M.L.: Target organ for a systemic effect of UV radia-
    tion. Photochem. Photobiol. (in press) 1976.

25. Lee J.A.H.: The trend of mortality from primary malignant
    tumors of the skin. J. Invest Derm 59:445, 1973.

26. Urbach F. (ed) Skin cancer and UV radiation. CIAP Monograph
    V. Chap. 7, part 2, p 7-43, Nat. Tech. Info. Serv., Springfield
    Va.   1975.

27. Davies R.E., C.A. Cole and P.D. Forbes: Effect of dichloro-
    difluoremethane (Halocarbon 12) on photochemical dissociation
    of ozone. Proc. Amer. Soc. Photobiol., p. 49, Fourth annual
    meeting, Denver, Co. 1976.

28. Setlow R.B.: The wavelengths in sunlight effective in producing
    skin cancer: a theoretical analysis. Proc. Nat. Acad. Sci.
    71:3363, 1974.

29. Forbes P.D.: Influence of continued exposure to ultraviolet
    light (UVL) on UVL induced tumors. Proc. Amer. Soc. Photobiol.,
    p.102, Second annual meeting, Vancouver BC 1974.

SUNLIGHT AND MELANOMAS

A. Wiskemann

Universitäts-Hautklinik Hamburg, Martinistr. 52

D 2000 Hamburg 20 West-Germany

SUMMARY

Melanoma incidence and mortality are influenced by melanin pigmentation as well as by latitude and exposure to sunlight. Studies by anatomical site give further information about the role of sunlight in the genesis of melanomas. Frequency, analysis by site, age and type of melanomas suggests that lentigo maligna melanoma of the face and superficial spreading melanoma of the lower limb in females are promoted by sunlight. Frequency on these sites also depends on complexion. Sunlight may work by realizing oncogenic viral information or by local immunosupression.

INTRODUCTION

Melanomas are one of the rarest but also one of the most malignant tumors of the skin. Nearly 6% of all skin cancers registered in East Germany were melanomas (1). The figures about the incidence and distribution of melanomas presented in this survey refer to people of different racial background, living in different latitudes with different sun exposure. Therefore, conclusions must be drawn very carefully.

INCIDENCE

Reliable figures of melanoma-incidence and mortality are available only from countries with specified cancer registry, based on obligatory reports. Thus, the figures from East Germany can be considered representative for Middle Europe. In that country in 1969 the incidence was in the same range as that of Hodgkins lymphoma, namely 2.7 per year per 100,000 inhabitants. Females were more frequently attacked than males (by a factor of 1,13) but mortality was less. The incidence in younger people is relatively high. It rises more slowly with age than does that of other skin cancers. The incidence has doubled in Norway from 1955 to 1969 (2). The same trend was observed in other countries. Obviously, this increase is due not only to improved diagnosis and certification.

MORTALITY

The mortality for both sexes was 1.6 per 100,000 in East Germany in 1969 (1). This is nearly 60% of the newly registered melanomas. In spite of improved treatment results, mortality has increased in the last decades (3).

ASSOCIATED FACTORS

The incidence of melanomas is influenced by endogenous and enviromental factors, the latter being more important. Family cases of maligmant melanoma have been observed (4). The inheritable liability to develop malignant melanoma among the citizens of Brisbane, Australia, has been estimated to be 11% (5). One inborn factor is the ability for enzymatic DNA repair. Xeroderma pigmentosum patients with defective DNA repair are pro-disposed to lentigo maligna and malignant melanoma. Another inborn factor is the ability to produce melanin which protects against solar damage. The percentage of patients with fair complexion, blue eyes and light hair has been found to be higher in melanoma patients than in a control group with or without carcinomas (6,7). Differences in racial pigmentation explain the low melanoma incidence in Negros, Indians and Japanese compared to the high incidence in Norway and Sweden.

SUN EXPOSURE HYPOTHESIS

Considering only the white skin of Caucasians, melanoma incidence rises with latitude and exposure to sun (8,9). In Queensland, Australia (10 - 30° S), the incidence of melanoma is three times and in Texas (30°N) twice that of Middle Europe (10). Analysis of melanoma incidence by latitude, age and sex shows that melanomas in Queensland develop ten to twenty years earlier than in East Germany. At both latitudes melanomas arise earlier in females than in males (1,10). Sun exposure is not only a question of light intensity but also of climate, habit and clothing. Estimated outdoor exposure was by far higher in melanoma patients, compared to a control group without skin cancer in New York (6). From all these findings it is strongly suggested that sunlight is associated with the genesis of melanomas (11). To get further information as to whether the action of sunlight might be direct or indirect, different localisation studies have been undertaken.

ANALYSIS BY ANATOMICAL SITE

In contrast to squamous cell carcinoma and basalcell skin cancer, melanomas are not mainly restricted to sunexposed skin

areas. They are distributed over the whole body surface but in
a different pattern for both sexes. From 1344 melanomas of the
German joint study "Malignes Melanom", 18% were located on
the head and neck (12). In this location, the ratio of males to
females is the same as that for the whole body. But in the remaining body surface, distribution differs with sex. Overrepre-
sentation occurs for males on the upper part of the trunk, especially on the back and in females on the lower part of the legs
(13). The high incidence of melanomas of the lower limb in
white females has been confirmed by many statistics (14), the
excess incidence being the greatest in women under 50 (15). The
question arises whether these distinctions in frequency by site
in both sexes are due to differences in sun exposure resulting
from clothing. Analysis of malignant melanomas in Norway, registered between 1955 - 1970, showed the same distribution for
melanoma incidence by site of origin, as mentioned above. Increase in incidence by calendar year had been the steepest on the
neck and trunk of the males and on the lower limbs of the females, born from 1900 to 1935 (2). Corresponding to the incidence
of melanomas on the lower limbs of white females ander 50, there
is a peak deathrate in women in middle age (15). The data are
in line with the increase of sun exposure to these areas by extended outdoor activities including sunbathing and changed clothing
habits since the end of World War I. The incidence and mortality rates of melanomas in the face rose with advancing age independent of sex (16). On the partially protected scalp and neck,
the trend was the opposite. The difference is probably due to
the development of lentigo maligna in the sun-exposed face of
older people (17). The effect of sun-exposure seems to be local.
There is no support for the hypothesis of a systemic effect, due
to materials released by the sun-exposed skin. Such a "solar
circulating factor" was suggested to be a melanocyte chalone or
an activated virus (11).

ANALYSIS BY SITE AND COMPLEXION
Patients of the German joint study group have been devided
into those with blue or grey eyes and light hair and those with
brown eyes and black or brown hair, respectively. In females
with light complexion melanomas on the sun-exposed lower
limbs are more frequent than in women with dark complexion.
On the trunk of females and males the ratio is opposite. Melanomas of the cheeks are very rare in fair skinned patients.

ANALYSIS BY TYPE OF MELANOMA
Classification of melanomas into Lentigo Maligna Mela-
noma (LMM), Superficial Spreading Melanoma (SSM) and Nodular
Melanoma (NM) has been adopted within the past decade (18).
Although LMM is the rarest type, about 50% of all the melano-
mas of the sun-exposed head and neck are LMM melanomas.
In the patients of the German joint study the percentage of LMM

was higher in females than in males (12). LMM has been found preferentially in older people of medium complexion. LMM in people with a fair complexion was exeptionally rare. LMM outside face and neck is also rare. Increase of LMM by age and multiple patches of Lentigo maligna in Xeroderma pigmentosum are in line with sunlight as an associated factor in the genesis of LMM. SSM prevails in the sun-exposed legs of females in middle age as well as in the sun-exposed chest and back of males (18). It is not clear whether the type of melanoma is favored by local conditions or whether sunlight or other factors play a part.

ANALYSIS BY PROGNOSIS

Prognosis of malignant melanoma is modified by stage, sex, type and grade of invasion, localization, onset and age. Taking stage and sex into account, the prognosis of female patients in the German joint study was somewhat better when melanomas were located on the sun-exposed areas of the body. This may be accounted for by the better prognosis of LMM on the face (19), but not by the mean level of melanomas of the sun-exposed skin, which are somewhat deeper than in the non-exposed skin. In men prognosis and grade of invasion were nearly the same in the sun-exposed skin as in covered skin. Prognosis and melanoma level in patients with fair complexion did not differ from those with medium complexion. Death rates from malignant melanomas of the lower limbs of females are higher than those of males (15). Thus prognosis is mainly determined by depth-level, type and site of melanoma, with the two latter dependent on sex distribution.

POSSIBLE MECHANISMS OF ACTION

In spinocellular carcinomas, sunlight acts directly by neoplastic transformation of the DNA of these cells (20). Irradiation of hairless mice by the sunburn spectrum causes squamous cell cancer, when sufficiently repeated (21). Localisation of squamous cell carcinomas on the sun-exposed skin of head and neck is dependent on the distribution of solar UV doses (22). In melanomas sunlight seems to act merely as a cofactor or as a promotor, which accelerates the ocurrence of the tumors. Blue naevi of hairless mice initiated by DMBA have changed to melanomas after UV irradiation (23). UV light alone did not produce melanomas. In 269 melanomas of the German joint study, located on the head and neck, the distribution did not correspond to the absorption pattern of UV light (24).

Xeroderma pigmentosum is an experiment of nature, which has enlightered our understanding of the role of sunlight in the transformation of melanocytes to freckles which are frequently followed by lentigo maligna and lentigo maligna melanoma. Freckles represent a clone of active melanocytes that might have undergone a UV induced mutation (25).

The developement of pigmented moles also seems to be promoted by sunlight. Moles on the most exposed parts of the body reach a peak of frequency sooner than on less exposed parts. Distribution pattern of moles is the same as that of malignant melanoms. In women the peak number of moles occurs between 20 - 29 years, and the peak incidence of melanomas between 30 and 50., Thus, the distribution of melanocytes and moles seems to account for the distribution of premalignant and malignant melanomas. In the dark-skinned population of Uganda, the majority of melanomas arise from pigmented spots of the foot soles (26), obviously not by the mutagenic effect of sunlight. But what is the mutagenic agent and how can sunlight work as a cocarcinogen? In my opinion, two suggestion deserve further investigation:

Research from the last decade has produced some evidence for the viral genesis of malignant melanoma. Oncogenic RNA viruses seem to be the transforming agent. Particles containing an single stranded high molecular weight RNA as well as reverse transcriptase have been found in hamster- and human melanomas. Oncogenic viral information can be realized by the additional action of UV light or by other enviromental cofactors (27,28,29).

Another possibility for the promotion of melanoma development by UV light is immunosupression. Like other tumors, melanomas are antigenic, stimulating a cell-mediated immune response for detection and destruction of the transformed cells. Depression of immunocytes by UV light may reduce immunoprotection of the host and thus raise the frequency of melanoma incidence. Suppression of immunocytes by sunlight, but not by the UVB lines of the mercury spectrum, is known from heliotherapy of mycosis fungoides.

But no matter in what way sunlight and melanomas may be related, fair skinned people have to be educated to avoid excessive sun exposure. The blue jeans fashion in both sexes as well as the development of better and better suncreens may be helpful to meet this challenge.

REFERENCES

1. Berndt,H. (1974) Arch. Geschw. Forschg. 44, 267-277.
2. Magnus,K. (1973) Cancer 32, 1275-1286.
3. Lee,J.A.H. and A.P.Carter (1970) J.Nat.Cancer Inst.45, 91-97.
4. Korn-Heydt,G.E. (1966) in: Handbuch der Haut- und Geschl. Krankht., hrsg.von J.Jadasson. Erg.Werk Bd.VII
   p.562-694,Springer, Berlin,Heidelberg,New York.
5. Wallace,D.C.,L.A.Exton u.G.R.C.MacLeod (1971) Cancer

27, 1262-1266.
6. Gellin, G. A. , A. W. Kopf and L. Garfinkel (1969) Arch. Derm.
   (Chicago) 99, 43-48.
7. Pack, G. T. , J. Davis and A. Oppenheim (1963) Ann. N. Y. Acad.
   Sci. 100, 719-742.
8. Doll, R. , J. Waterhouse and C. Muir (1970) Cancer Incidence
   in Five Continents Vol. II, Springer Verlag Berlin, Heidelberg,
   New York.
9. Holland, J. F. and E. Frei (1973) Cancer Medicine p. 288 ff.
   Lea and Febinger, Philadelphia.
10. Davis, N. C. , J. J. Herron and G. R. Mac Leod, Lancet 1966,
    2, 407-410.
11. Sunlight and Melanoma, Lancet 1971, 1, 172-173.
12. Jung, E. G. , A. Bersch und C. Köhler (1972) Arch. Derm.
    Forschg. 244, 195-200.
13. Hauss, H. und A. Proppe (1972) Arch. Derm. Forschg. 244,
    193-195.
14. Waterhouse, J. A. H. : Cancer Handbook of Epidermiology and
    Prognosis. Churchill Livingstone Edinburgh and London 1974.
15. Lee, J. A. H. (1970) J. Nat. Cancer Inst. 44, 257-261.
16. Lee, J. A. and H. J. Issenberg (1972) Brit. J. Cancer 26, 59-66.
17. Mc. Govern, V. J. (1966): Symposion on structure and control
    of the melanocyte. Springer Publishing Co. Inc. Berlin, Hei-
    delberg, New York.
18. Clark, W. H. jr. , L. From, E. A. Bernadino and M. C. Mihm
    (1969) Cancer Research 29, 705-726.
19. Thies, W. und U. Scherowsky (1972) Arch. klin. Forschg. 244,
    210-213.
20. Setlow, R. B. and R. W. Hart in: Proceeding 5. Internat. Congr.
    of Radiat. Research, edit by O. F. Nygaard, H. J. Adler and
    W. K. Sinclair. Academic Press Inc. New York 1975.
21. Epstein, J. H. and W. L. Epstein (1963) J. Invest. Derm. 41,
    463-473.
22. Urbach, F. in: The Biologic Effects of Ultraviolet Radiation
    ed. by F. Urbach, Pergamon Press New York 1969.
23. Epstein, J. H. , W. L. Epstein and T. Nakai (1967) J. Natl.
    Cancer Inst. 38, 19.-22.
24. Wiskemann, A. (1974) Hautarzt 25, 20-22.
25. Nicholls, E. M. Lancet 1968, 1, 72-73.
26. Levis, M. G. (1967) Brit. J. Cancer 21, 483.
27. Balda, B. R. , G. D. Birkmayer and O. Braun-Falco (1973)
    Arch. Derm. Forschg. 248, 229-236.
28. Birkmayer, G. D. , B. R. Balda and F. Müller (1974) Europ. J.
    Cancer 10, 419-424.
29. Balda, B. R. , R. Hehlmann, J. R. Cho and S. Spiegelman (1975)
    Proc. Nat. Acad. Sci. USA 72, 3697-3700.

CUTANEOUS CARCINOGENIC EFFECTS OF SUNLIGHT IN HUMANS

Thomas B. Fitzpatrick, M.D., Ph.D., Arthur J. Sober, M.D.,
Barbara J. Pearson, B.S., M.S., Robert Lew, Ph.D.

Department of Dermatology, Massachusetts General Hospital
Harvard Medical School
Boston, Massachusetts   02114

For normal Caucasians, malignant melanoma of the skin is the most
serious hazard of solar exposure, effecting people in their productive years. (age 25-60) (1)   Only two-thirds survive 5 years. (2)
Malignant melanoma is at present as common as Hodgkin's disease (3)
and is the leading cause of death of all skin disorders.(4) Many
investigators have implicated solar exposure in the etiology of
malignant melanoma. It is therefore, important for prevention,
to evaluate the evidence for this relationship.

The data supporting sunlight as a major or contributing cause of
melanoma, or non-melanoma* skin cancer takes four forms: I. difference in incidence or mortality with latitude, corresponding
generally to differences in solar exposure, (UV-B flux) (5) II. Anatomic locations of skin cancer with regard to difference in degrees
of protection by clothing. III. differences in incidence or mortality with the degree of pigmentation of the skin; i.e. black
people have a much lower incidence than Caucasians with lesions located mainly on the foot (7) while malignant melanoma occurs more
frequently in persons with fair complexions who tan poorly.(8) IV.
changes in incidence or mortality over time as related to changes
in exposure.(9) In the following discussion, these kinds of evidence will be compared in relation to non-melanoma and melanoma
skin cancer.

Non-Melanoma Skin Cancer

Epidemiologic data shows strong latitude dependence. Incidence is
directly dependent on intensity of UV-B flux.

Non-melanoma skin cancer occurs on locations of the body (face,neck)

*the term "non-melanoma" skin cancer includes basal cell and squamous
cell cancers.

receiving maximal exposure of sunlight. Over 90% of basal cell
carcinomas occur on the face and neck. (10) Squamous cell carcinomas
occur almost exclusively on the head and neck in a study by Urbach
et al.(11) Non-melanoma cancers also occur more frequently in persons with outdoor occupations who receive extensive solar exposure.
(e.g. sailors, farmers) Pigmented peoples rarely develop non-
melanoma skin cancer and the sites where these occur most frequently
are not those which receive maximum solar exposure.(7) Albino
blacks in Nigeria have a high incidence of non-melanoma skin cancer.
(13) These data suggest that non-melanoma skin cancer can be caused
by sunlight exposure as the principal etiologic agent.

Malignant Melanoma
Epidemiologic data also shows strong latitude dependence for malignant melanoma.(5) A higher incidence is seen in fair-skinned individuals and a lower incidence in pigmented peoples.(6,7,8,12) The
relationship of solar exposure to anatomic sites of tumor development is more complex for melanoma than for that seen in non-melanoma in that melanoma is not most frequent in the maximally exposed sites. In addition, a remarkable increase in incidence since
World War II has been noted, probably related to changes in exposure
behavior.(9) Epidemiologic data recently analyzed in the United
States (5) have confirmed that there is a variation with latitude
within the United States. A study of ocular melanomas (excluding
conjunctival lesions) showed no relationship between incidence
and latitude. (15) Melanoma is rare in blacks. The age adjusted
rate in 1970 for Caucasians was 4.6/100,000 in males and 4.4/100,000
in females while for blacks the rates were 0.9/100,000 in males
and 0.7/100,000 in females.(6) Previous studies (16,17,18) have
shown differences in location of primary melanoma in males and
females. Males and females both have a predominance of lesions
in the upper back, and in addition, females show a high incidence
on the lower legs. Magnus (19) in a study of 2,541 new patients
with primary melanoma observed in Norway during the period from
1955 to 1970 noted a striking increase in the total number of
melanomas from 100/yr. to 150/yr. and in the number of melanomas
on the trunk in males and females and on the lower legs in females,
while the incidence on the face and feet remained the same.   He
also noted a steep increase in incidence starting in adolescence
and a levelling off at late middle age. In addition, the increase
in incidence was largest for those persons born between 1900-1935.
He attributed these changes in the pattern of location of melanomas
to "a transition in fashions and habits from World War I to post
World War II."

These observations motivated our group to determine the precise
anatomic location on humanoid contoured life-size models to
delineate the relation of the tumor origin to anatomic prominences
and clothing patterns. This data is derived from an extensive
computerized data base collected by the Melanoma Clinical Cooper-

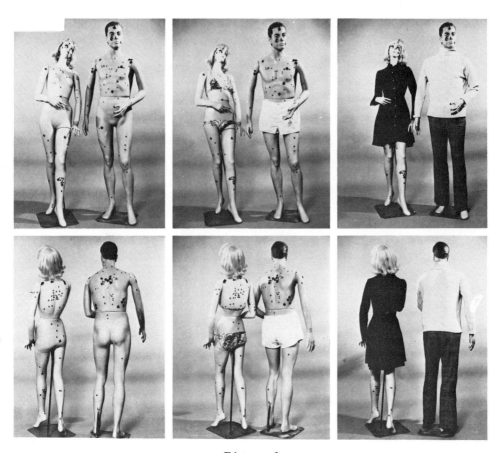

Figure 1.

Table 1 Distribution of cutaneous malignant melanoma by site
         and sex in 534 cases

	Males	Females
Back	35%	24%
Chest and Abdomen	22%	8.2%
Lower legs	2.1%	19.7%
Other sites	41%	48.1%

ative Group, (Harvard Medical School, New York University, Temple University, and the University of California at San Francisco). A total of 534 primary melanomas of all three types were included. (19) Inasmuch as dress styles affect the skin exposure in males and females, life-size mannekins were used to determine the differences of sites in males and females.

There is a heavy density of primary melanomas on totally exposed (face) and intermittently exposed areas (trunk) and a low density of melanomas on the totally unexposed areas (covered parts of the breasts in females, bathing trunk of males and panty area of females). (Fig.1,Table 1) In addition, there is a striking disparity in the number of melanomas on the lower legs of females and males. This increased number of melanomas on the lower leg in females might be related to the degree of exposure of the skin surfaces below the knee in women since 1917 when the skirt length was raised from ankle length to knee. The shift from opaque to translucent stockings has also increased the solar exposure of the lower extremity of females.

From these data, two hypotheses can be offered: 1. the increased incidence on torso-legs, while the face-hand have remained constant is reflected by the already maximal rates for face-hand, while the torso-legs now receiving more exposure are "catching-up" to their full potential. 2. the predominance of primary melanomas on the intermittently light exposed areas suggests an interaction of a pre-existing susceptible lesion target with sunlight; this increased susceptibility could be genetic or acquired. (virus or exposure to chemicals)

There appear to be persons at risk for developing melanoma based on genetic control of their melanocytes to supply protective shielding melanin pigment. In a recent study of 343 melanoma patients and 150 age-sex matched controls a substantially greater percentage of melanoma patients gave a history of developing painful sunburn and poor or no tanning than the control group. Both groups need to be expanded and the determination of sun reaction quantified before these conclusions can be firmly established. Melanoma can no longer be considered a rare disease. In the U.S.A. alone, there are 8,400 new cases each year and this figure is increasing by 5 to 10% each year. (8) Based on the current survival rates, this means approximately 2,800 of these new patients will die within 5 years. The evidence for sunlight-derived UV-B as a major factor in the causation of melanoma is sufficiently strong that we should not permit changes in the stratosphere that would increase UV-B flux at the earth's surface.

Acknowledgement: Mr. Larry Cherkas prepared the photographs.

References

1. Sober, A.J., Fitzpatrick, T.B., Mihm, M.C., Jr., et al: Primary cutaneous malignant melanoma: Recognition and Management. American Academy of Dermatology, Independent Study Series, Volume 1, 1976.

2. Cutler, S.J., Myers, M.H., Green, S.B.: Trends in survival rates of patients with cancer. N.Eng. J. Med. 293:122-124, 1975.

3. Cancer Statistics, 1976. American Cancer Society Professional Education Publication. 1976, 0. 22.

4. Odland, G.F.: The Skin: A description of the external organ and its common afflictions. University of Washington, School of Medicine, Seattle, 1971. p.89.

5. Environmental Impacts of Stratospheric Flight, National Academy of Science. Washington, D.C. 1975.

6. Third National Cancer Survey 1969-1971: Incidence Data. National Cancer Institute Monograph 41, 1975.

7. Fleming, I.D., Barnawell, J.R., Burlison et al: Skin cancer in black patients. Cancer 35:600-605, 1975.

8. Gellin, G.A., Kopf, A.W., Garfinkel, L.: Malignant melanoma: a controlled study of possibly associated factors. Arch. Dermatol. 99:43-48, 1969.

9. Magnus, K.: Incidence of malignant melanoma of the skin in Norway 1955-1970. Cancer 32:1275-1286, 1973.

10. Brodkin, R.H., Kopf, A.W., Andrade, R.: Basal cell epithelioma and elastosis: A comparison of distribution. The Biologic Effects of Ultraviolet Radiation: with Emphasis on the Skin. F. Urbach, ed. Pergamon, New York. 1969 pp. 581-618.

11. Urbach, F., Davies, R.E., Forbes, P.D.: Ultraviolet radiation and skin cancer in man. In Montagna W. and Dobson, R.L. eds. Advances in Biology of Skin. Vol. 7 Carcinogenesis. Pergamon Press, New York 1966. pp. 195-214.

12.. White, J.E., Strudwick, W.J., Ricketts, W.N. et al: Cancer in the skin of negroes - a review of 31 cases. J.A.M.A. 178: 845-847, 1961.

13. Okoro, A.M.: Albinism in Nigeria: A clinical and social study. Brit. J. Derm. 92:485-492, 1975.

14. Levin, M.L., Haenszel, W., Carroll, J.E. et al: Cancer
     Incidence in Urban and Rural Areas of New York State. J. Nat.
     Cancer Inst. 24:1243-2157, 1960.

15, Scotto, J., Fraumeni, J.F., Lee, J.A.H.: Melanomas of the eye
     and other noncutaneous sites: Epidemiologic Aspects. J. Nat.
     Cancer Inst. 56:489-491, 1976.

16. Petersen, N.C., Bodenham, D.C., Lloyd, O.C.: Malignant mela-
     nomas of the skin - a study of the origin, development, aetiology,
     spread, treatment and prognosis. Brit. J. Plast. Surg. 15:49-94,
     1962.

17. Anchev, N., Popov, I., Ikonopisov, R.L.: Epidemiology of
     malignant melanoma in Bulgaria. In Della Porta, G. and Muhlbach,
     D., eds. Structure and Control of the Melanocyte, New York,
     Springer-Verlag, 1966, pp. 286-292.

18. Elwood, J.M., Lee, J.A.H.; Recent data in the epidemiology of
     malignant melanoma. Seminars in Oncology 2:149-154, 1975.

19. Melanoma Clinical Cooperative Group, unpublished observations,
     1976.

ROLE OF DNA REPAIR IN PHYSICAL, CHEMICAL AND VIRAL CARCINOGENESIS

Antonio Caputo

Regina Elena Institute for Cancer Research, Rome

291, Viale Regina Elena, Rome - Italy

It is generally assumed that a great variety of carcinogen-induced changes to DNA lead to one type of DNA repair synthesis which can be detected by an unscheduled uptake of labelled DNA precursors. Thus, by measuring DNA repair synthesis data can be obtained about DNA damages without defining the type of carcinogen-DNA bondage and the type of alterations. There is evidence to suggest that the removal of carcinogen-induced products occurs at various rates after the initial damage. Even if extensive dose-response relationships have been worked only in few cases, it can be assumed that the rate of excision is dose-dependent and follows a general kinetics. In the case of UV-irradiation, according to the sensibility of the various cell lines, the rate of excision in vitro is dependent on the dose of irradiation and gradually drops toward a plateau reached, even at the highest doses, within 24 hours. Similarly, when cultured human fibroblasts are exposed to one low dose of 4NOO, the study of the kinetics of unscheduled incorporation of thymidine indicates that the repair has occurred in the first few hours after the treatment. If a second dose of carcinogen is given the cells do not respond at a normal level, yet when repair of the first dose is completed further DNA damages are effectively repaired.

It is well known that the number of tumors macroscopically evaluable in an animal after the administration of a given carcinogen is very small if related to the number of cells that have been exposed to the carcinogen. For example, after the intravenous administration of methylnitrosourea, which does not require activation, the carcinogen is present at about the same concentration throughout all tissues and organs. Nevertheless, only few tumors are later detected, and sometimes only one. Among the millions of cells interacting with the ultimate carcinogen, the rare event of malignant

transformation is regulated by a steady-state equilibrium in which
unrepaired carcinogen-induced damages to DNA prevail on repaired
ones, thus determining the fixation of the transformed state.
Recently, Goth and Rajewsky (1) have reported that after a single
injection of N -ethylnitrosourea, a very powerful carcinogen which
specifically induces brain tumors, the rate of disappearance from
the brain of O^6-ethylguanine was very much slower than that from
the liver. This finding gives further confirmation of the biological importance of alkylation in the O^6 position of guanine for carcinogenesis and strongly supports a differential rate of DNA repair
in vivo between target and non-target organs.
By using a combined in vivo and in vitro system, Stich and coworkers (2) have confirmed the validity of this assumption for its generalization to the organotropic action of carcinogens. It can be
assumed that the precarcinogens are organ-specifically activated
to ultimate carcinogens in those tissues in which they induce damages to DNA and these damages in turn initiate a repair process,
on whose capability depends the fixation of the transformed state.
Although the general assumption that all chemical carcinogens produce DNA alterations which are followed by a DNA repair process,
is up to date still an assumption, not fully supported by experimental evidences, a good correlation has been found after the exposure of cultured mammalian cells to carcinogens of diverse molecular structure, to precarcinogens following their metabolic activation, and mutagens.

Repair of damages induced to DNA by alkylating agents and nitroso
carcinogens in the whole animal in vivo has been evaluated by
Farber's (3) group. Single and double strand breaks and the time
of repair have been documented in liver cells DNA after treatment
of the rat with different hepato and non-hepato carcinogens. Double
strand breaks are slowly repaired, in contrast single strand breaks
are mostly, if not completely, repaired within a short period of
time ranging from 4 to 72 hours. In addition, although both hepato
and non-hepato carcinogens induces single strand breaks to liver
DNA, hepato carcinogens fall at one end of the spectrum in that the
damages are slowly repaired. On the contrary, the damages caused by
non-hepato carcinogens are repaired rapidly.
The more comprehensive analysis on the role of repair processes in
carcinogenesis implies to consider the problem of virus-induced
tumors. Evidence has been obtained that oncogenic viral DNA resembles the cell genomic DNA and its organization. The type of structure sequenced by cy osine-rich and thymine-rich clusters is essential for the replication cycle of DNA polyoma and human papil-
loma viruses for supplying peculiar informations coding for non-
structural proteins, including enzymes for activation of DNA precursors, nucleases, polymerases and polyligases, and structural
proteins. The early messenger RNA molecules are transcribed from
genes adjacent to poly-dA-dT sequences. The late messenger which
transcribes informations for capsid proteins, viral antigens and
enzyme machinery for the synthesis of host's genomic DNA, might

therefore contain poly-A sequences. It is implied that the cells
able to be transformed by oncogenic viruses are those having their
DNA in a conformation suitable for offering the necessary counter-
parted poly-dG-dC sequences. But, over all, the crucial point for
the integration is the interaction of the complementary chromosomal and viral sequences, after that a specific endonuclease has incised the bonds adjacent to the poly-dG-dC sequence. Such type of
endonuclease has been recognized in Adenovirus types 2 and 12, and
in virions of Rous sarcoma virus. Like oncogenic DNA viruses, RNA
tumor viruses also act by stimulating cellular activity. In addition, one of the most striking feature in the mode of replication
of RNA tumor viruses is the requirement for DNA synthesis. The duplication of the virus, as well as the neoplastic transformation
it causes, can be prevented by known inhibitors of DNA synthesis.
A property common to many chemical carcinogens applied in vivo, as
well as oncogenic viruses applied to cell cultures is the causation
of an elevated level of DNA synthesis and cell proliferation in the
respective target cell systems. This elevated rate of DNA replication and cell division, following the exposure to chemical carcinogens or oncogenic viruses, increases the frequency of the "fixation" of the induced alterations in informational DNA molecules
and facilitate the phenotypic expression of malignant properties
in a sufficient high proportion of initiated cells. The recombinational repair mechanism could be common mean for the mutated state
of genomic DNA and the neoplastic transformation caused by a virus:
the loss and/or the defective repair determines the occurrence of
gaps in DNA which, according to the phases of cell cycle, persist
in various degrees in newly synthetized daughter strands. The gaps
are filled by a recombinational mechanism with pieces from parental
strand of the other DNA molecules. The recombinational process is
error-prone and therefore viral DNA may be incorporated and later
be present in the cell transformed by oncogenic virus.
In trying to answer the question "What is the sequence of metabolic
events occurring between the exposure to a given carcinogen and the
burst of tumor cells proliferation?" we shall proceed backward,
that is, we shall start from cell proliferation and thence retrace
our steps back to the early chemical signals that, from a remote
distance, control cell proliferation. The first step in this jour-
ney is the initial attack to DNA and how significant the damage is.
Significancy means here the capacity of modifying the genomic informations which code for the molecular system controlling the normality of cell proliferation. Does the initial alteration of a carcinogen preferentially occur with target cells in a specific phase
of the cell cycle or the initiation occurs with equal probability
independently on the cycle? If there are cell cycle phases with
higher sensitivity, do characteristic differences exist in relation to different types of carcinogen?

Beware of making a fixed scheme for a general rule, repair processes
must be viewed as a possibly basic mechanism in the contest of what

we currently believe about carcinogenesis. It is likely that when
the oncogenic effectors interact with DNA in the prereplicative
phase of the cycle, the exposed conformation of DNA molecules and
the lower repairability of the induced damages will more extensively modify essential genes, according to significant alterations of
the templating and/or priming activity of DNA.

Several attempts have been made to define conditions necessary for
the fixation of the transformed state. Recently, Terzaghi and Little
(4) have shown a high degree of correlation between the repair interval leading to both maximal survival and transformation following X-irradiation, thus confirming previous observations on the
transformation obtained with methylcholanthrene and SV40 virus. The
picture then emerges that if these is an interval after which a
cell does not divide, it will repair the damages which would eventually lead to malignant transformation.

The question arises whether the transformation involves direct induction of critical lesion or whether the error is inserted in genomic DNA during repair. For the moment we shall simply conclude
by reminding that transformation frequency and repair of potentially
lethal damages are maximal in G_1 and therefore the hypothesis of
error insertion is strongly supported. If the cell does insert the
error which determines the transformed state, and ultimately leads
to malignancy, it may also be able to go back over and delete the
inserted error from genomic DNA molecule.

Before concluding, it is important to take one further point: that
all carcinogens produce DNA damages followed by an induced repair
synthesis is an assumption to be proved on a larger scale, according to the variety of their nature. An other unanswered question is
whether there are compounds which elicit DNA repair but lack a carcinogenic effect. In addition, aspects such as the exact chemical
nature of the carcinogen-induced damages, the enzymes specifically
involved in the various types of repair and the rate limiting
steps need to be substantially clarified for an understanding of
the whole matter.

Although the many open questions, DNA repair is a short-term test
available for the detection of possible cancer-causing agents. Hopefully, the evaluation of DNA repair as prescreening test for environmental chemicals appears, at present, one of the most promising approaches in assessing the carcinogenic hazard to man.

References

1 - Goth R. and Rajewsky M. - Proc.Nat.Acad.Sci.U.S. 71, 639-643,
    (1974)

2 - Stich H.F., San R.H.C., Lam P.P.S., Koropatnick D.J., Lo L.W.
    and Laishes B.A. - I.A.R.A. Scint.Publ. 12, 617-637, (1976)
3 - Farber E. and Sarma D.R.S. - In "Control Processes in Neoplasia",
    Editors: Mehlam M. and Hanson R., Academic Press, N.Y., pag
    173-185, (1974)
4 - Terzaghi M. and Little J.B. - Nature, 253, 548-549, (1975)

SYMPOSIUM XI

LIGHT AND DEVELOPMENT

CONTROL OF CHLOROPLAST DEVELOPMENT AND CHLOROPHYLL SYNTHESIS BY PHYTOCHROME

H. MOHR and H. KASEMIR

Biological Institute II/University of
Freiburg
7800 Freiburg/Schaenzlestr. 9-11
West Germany

INTRODUCTION

In higher plants the development of mature chloroplasts takes place only in the light. The developing photosynthetic apparatus is thus a useful model system to investigate the control of development by an exogenous factor, light. We have studied over the years the development of the photosynthetic apparatus in the cotyledons of the mustard seedling (Sinapis alba L.), a representative of the dicotyledonous plants [1]. As long as the seedling develops in complete darkness, the cotyledons function as storage organs. They do not grow or develop significantly. However, when the seedling is illuminated with white light of sufficiently high irradiance, the cotyledons are transformed rapidly into photosynthetic organs, very similar in internal structure and in function to a normal photosynthetically active leaf. The control by phytochrome of plastogenesis has been analyzed on the level of fine structure, with regard to protochlorophyll synthesis, chlorophyll accumulation, and formation of Calvin cycle enzymes. A major problem has been whether the multiple control by phytochrome of the development of the plastid compartment is the consequence of a single initial action of Pfr.

THE PHYTOCHROME SYSTEM IN THE MUSTARD COTYLEDONS

The photochromic sensor pigment phytochrome is a bluish chromoprotein [2]. In the following, we use the

phytochrome model as given in Fig. 1. This model describes quantitatively the properties of the phytochrome
system as it occurs in the cotyledons (and in the hypocotylar hook) of the mustard seedling during the period
of our experimentation.

An important property of the phytochrome system is
that it develops a steady state under continuous light
[4]. Based on the model in Fig. 1 any change of "total
phytochrome" ([Pr] + [Pfr]) can be described by the
equation

$$\frac{d\ [Ptot]}{dt} = {}^{o}k_{s} - {}^{1}k_{d}\ [Pfr]$$

Under steady state conditions, i.e. "no change of Ptot",
the rate of Pr synthesis is equal to the rate of Pfr
destruction, ${}^{o}k_{s} = {}^{1}k_{d}\ [Pfr]$. Therefore

$$[Pfr]\ \text{steady state} = {}^{o}k_{s}/{}^{1}k_{d}$$

In words: the steady state concentration of the effector
molecule Pfr is only a function of the rate constants for
Pr synthesis and for Pfr destruction, which are both
light-independent. This means that the steady state concentration of the effector molecule Pfr does not depend
on the wavelength of the light incident on the system.
The only prerequisite is that the incident light is absorbed by both phytochrome forms to an extent sufficient
to establish a steady state. This property of the phytochrome system offers the opportunity to run the steady
state of the phytochrome system at a wavelength which
does not cause significant chlorophyll synthesis or
photosynthesis. We have been using a high irradiance
($3.5\ Wm^{-2}$) far-red light source which is equivalent, as
far as the phytochrome system is concerned, to the wavelength 718 nm.

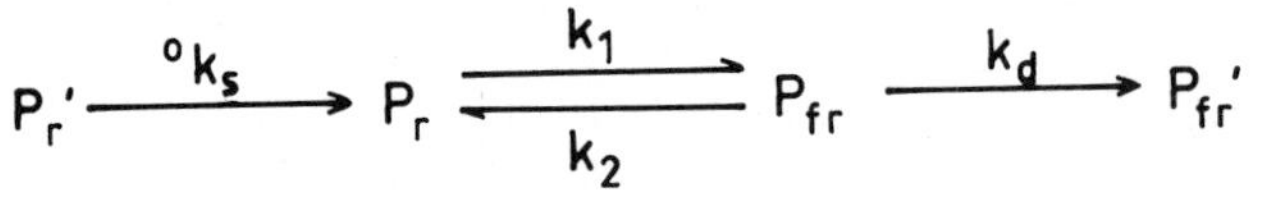

$$P_{r}' \xrightarrow{{}^{o}k_{s}} P_{r} \underset{k_{2}}{\overset{k_{1}}{\rightleftharpoons}} P_{fr} \xrightarrow{k_{d}} P_{fr}'$$

Fig. 1. A model of the phytochrome system as it occurs
in mustard seedling cotyledons and hypocotylar hook.The
symbols ${}^{o}k_{s}$, k_{1}, k_{2}, k_{d} represent the rate constants of
de novo synthesis of Pr, back and forth phototransfor-
mation and Pfr destruction respectively. Only k_{1} and
k_{2} are light dependent. The half-life of Pfr in the
mustard cotyledons is 45 min at 25° C (after [3]).

INDUCTION OF RIBULOSEBISPHOSPHATE CARBOXYLASE
(CARBOXYLASE) BY PHYTOCHROME

Induction by phytochrome of enzyme levels, including Carboxylase in developing plastids is well known [5]. Figure 2 shows the time course of Carboxylase and NADP-dependent glyceraldehydephosphate dehydrogenase (GPD) in the mustard seedling cotyledons. In the dark grown mustard seedling, the level of Carboxylase and GPD in the cotyledons remains low; however both enzymes can be induced by phytochrome, operationally, by continuous far-red light (Fig. 2). Both enzymes cannot be detected before 36 h after sowing. Figure 2 shows that for the induction of the Calvin cycle enzymes it does

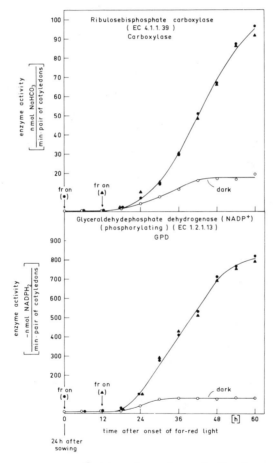

Fig. 2. Time course of the levels of Carboxylase and GPD in the cotyledons of the mustard seedling in the dark and under continuous far-red light. Onset of far-red light (fr) at 24 or 36 h after sowing (after [5]).

not matter whether Pfr functions in the seedling from
24 h after sowing or only from 36 h after sowing onwards.
This indicates the action of endogenous regulatory factors which prevent enzyme induction by phytochrome before
42 h after sowing irrespective of light treatment.Other
enzymes in the mustard cotyledons, not related to photosynthesis, can be induced by phytochrome much earlier.
As an example, phenylalanine ammonia-lyase (PAL), the
key enzyme of phenylpropanoid biogenesis in plants, becomes inducible by Pfr at 28 h after sowing [6]. There
is obviously a temporal pattern of inducibility of enzymes in the mustard cotyledons. The specification of
this pattern is not influenced by phytochrome [7].

We have studied the problem whether the phytochrome-
mediated accumulation of Carboxylase in the cotyledons

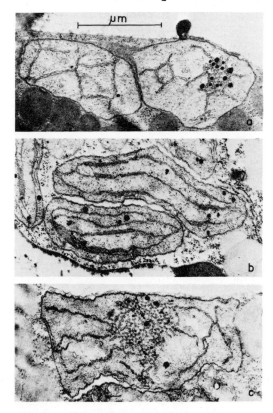

Fig. 3. Representative sections through plastids in
palysade parenchyma cells of cotyledons of mustard
seedling grown in the dark (a), under continuous white
light (b) or under continuous far-red light (c). Onset
of light at time zero (time of sowing of the seeds).
Fixation was performed 48 h after sowing (after [8]).

of the mustard seedling is related to size, ultrastructure, or organization of the plastid compartment. It turned out that under different light conditions (e.g. continuous far-red or continuous white light) which lead to con- spicuously different plastids (Fig. 3), the time course of the Carboxylase level remains precisely the same (Fig. 4). It is concluded that the onset and the rate of Carboxylase synthesis is not related to the organization- al state of the plastid compartment as discernible under the electron microscope.

CONTROL OF CAPACITY OF CHLOROPLAST BIOGENESIS BY PHYTOCHROME

The biosynthetic pathway of chlorophyll (Chl) syn- thesis is known in principle. We are interested in the sites of "regulation", i.e. where the decision is made over the flux in the biosynthetic channel. A photoreac- tion which occurs directly in the pathway leading to

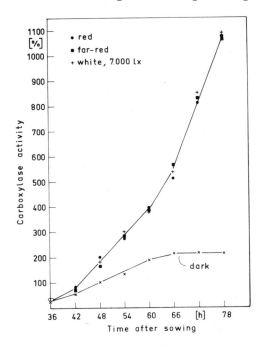

Fig. 4. Time courses of the levels of Carboxylase in the cotyledons of the mustard seedling in the dark and under continuous red, far-red, or white light. Onset of light at time zero (time of sowing of the seeds). The enzyme level in the dark at 48 h is taken as a reference point (100 per cent) (after [8]).

Chl-a has been known for more than 20 years: the protochlorophyll —> chlorophyll(ide) a photoconversion. The
light absorption is performed by the protochlorophyll-
holochrome (PChl) [9]. We have measured the "capacity"
of the mustard cotyledons to synthesize Chl-a. By "capacity" we designate the maximum flux in the biosynthetic
channel leading to Chl-a. In order to saturate the PChl
—> Chl-a photoconversion, the mustard seedlings are
illuminated with white fluorescent light at an illuminance
of 7.000 lx. Figure 5 shows that the "capacity" depends
on the developmental state (age) of the seedling and that
it is strongly increased by phytochrome, operationally,
continuous far-red light. The time course of the "capacity" under continuous <u>white</u> light (which permits Chl
synthesis and photosynthesis) deviates only slightly
from the time course of the "capacity" under continuous
standard far-red light (which does not permit significant Chl accumulation). It is concluded that the capacity
for Chl-a synthesis is determined by phytochrome irrespective of whether or not Chl-a is actually being synthesized. In continuous darkness or in continuous light, we
do not see any manifestation of an endogenous circadian
rhythm. However, when we turn off the light after 12 h
the physiological clock [11] comes into play (Fig. 6).
It manifests itself in a strong overshoot and in the conspicuous circadian oscillation of the "capacity". The induction (or release) of the physiological clock is caused
by a light —> dark transition. The photoreceptor involved
is phytochrome [10]. Experiments to localize the site of
regulation of the "capacity" have led to the result that

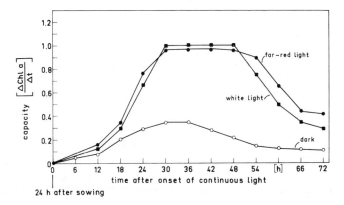

Fig. 5. Time course of the "capacity" of the chlorophyll
a synthesizing pathway in the mustard seedling cotyledons in continuous darkness, continuous far-red or continuous white light (after [10]).

both phytochrome and the physiological clock operate at
the level of 5-amino-levulinate (ALA) synthesis. It is
obvious from Fig. 7 that the capacity of the ALA-forming
system and the capacity of the Chl-a synthesizing system
almost coincide under continuous far-red light as well
as under conditions where the physiological clock plays
a major part. It is concluded that phytochrome and the
physiological clock determine the capacity of Chl-a
biogenesis through the capacity of ALA synthesis.

CONTROL OF CHLOROPLAST DEVELOPMENT BY PHYTOCHROME

Phytochrome exerts a strong influence on the growth
of the plastids (see Fig. 3) and on the appearance of
the prolamellar body [12]. Results obtained recently by
C. Girnth indicate that a pretreatment with phytochrome
even controls the rate of grana formation in white light.
A pretreatment of the mustard seedling with brief red
light pulses (which establish a high level of Pfr) strongly accelerates grana formation in white light. The red
light effect is far-red reversible, in accordance with
the established operational criteria for the involvement
of phytochrome in a response.

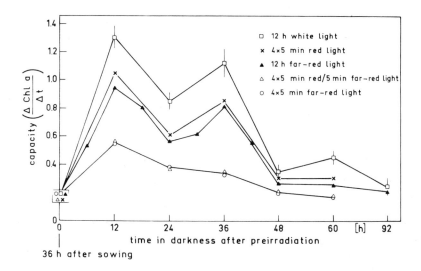

Fig. 6. Time course of the "capacity" of the chlorophyll
a synthesizing pathway in the mustard seedling cotyledons in darkness following a 12 h light pretreatment
(between 24 and 36 h after sowing). The light pulses
(5 min each) were given at 24, 28, 32 and 36 h after
sowing (after [10]).

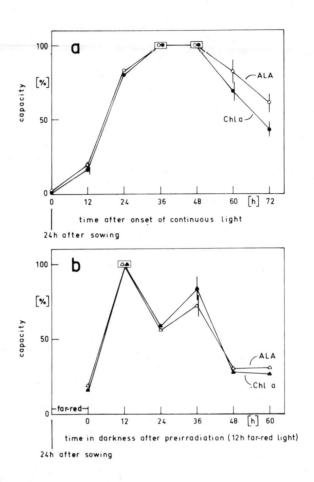

Fig.7. Time course of the "capacities" for Chlorophyll a (Chl a) synthesis and 5-aminolevulinate (ALA) synthesis in the mustard seedling cotyledons in continuous far-red light (a) and in darkness (b) following a 12 h pretreatment with far-red light (between 24 and 36 h) after sowing (after [10]).

REFERENCES

1 Mohr, H. (1972) Lectures on Photomorphogenesis. Springer, Berlin-Heidelberg-New York
2 Mitrakos, K. and Shropshire, W (eds.) (1972) Phytochrome. Academic Press, New York
3 Oelze-Karow, H., Schäfer, E. and Mohr, H. (1976) Photochem.Photobiol.23, 55-59
4 Schäfer, E. and Mohr, H. (1974) J.Math.Biol. 1, 9-15
5 Brüning, K., Drumm, H. and Mohr, H. (1975) Biochem. Physiol.Pflanzen 168,141-156
6 Peter, K. and Mohr, H. (1974) Z.Naturforsch.29c,222-228
7 Mohr, H. (1976) Biologie in unserer Zeit, in press
8 Frosch, S., Bergfeld, R. and Mohr, H. (1976) Planta, in press
9 Smith, J.H.C. (1960) In: Comparative Biochemistry of photoreactive Systems (M.B.Allen, ed.).Academic Press, New York
10 Gehring, H. and Mohr, H. (1976) Planta , submitted for publication
11 Bünning, E. (1973) The Physiological Clock, 3rd ed. Springer, Berlin-Heidelberg-New York
12 Kasemir, H., Bergfeld, R. and Mohr, H. (1975) Photochem.Photobiol. 21, 111-120

PROGRESS TOWARD AN UNDERSTANDING OF THE MOLECULAR MODE OF ACTION

OF PHYTOCHROME

Lee H. Pratt

Department of Biology, Vanderbilt University

Nashville, Tennessee 37235, U.S.A.

Summary. Evidence concerning the possible existence of a
receptor for Pfr (the physiologically active form of phytochrome),
with which Pfr would bind before its morphogenic action may be expressed, is summarized. Two different associations between Pfr and
particulate material are described. The most widely examined interaction, studied with extracts of Cucurbita pepo, appears to be artifactual. The other association has not yet been as well characterized, but possesses features which make it a candidate for a Pfr-
receptor interaction.

INTRODUCTION

The gradual elucidation of the morphogenically active plant
chromoprotein phytochrome during the 1940's and 1950's by the "Belts-
ville Group" has become a classic example of a successful photobiological investigation (1). Physiological experiments led to the
early prediction that a single pigment should exist in two different
forms to explain (a) the induction by red light of morphogenic
responses such as inhibition of stem elongation and promotion of
seed germination, (b) the reversal of these red-induced responses
by far-red light and (c) the repeated photoreversibility of these
responses (1). This prediction in turn led to the design and use
of a dual wavelength spectrophotometer which permitted the detection
(1), isolation (1) and eventual purification (2) of phytochrome.

Physiological experiments have also resulted in a variety of
hypotheses regarding the primary reaction leading from Pfr, the
physiologically active (1) far-red-absorbing form of phytochrome,
to observed responses. Among such hypotheses are that phytochrome

511

functions as an enzyme (1), that it directly alters the expression
of gene activity (3), or that it modulates one or more membrane
properties (4).

In considering the mode of action of phytochrome, it is perhaps
trivial to note that the most useful hypotheses would be those derived from physiological responses which most closely follow in
time the photoconversion of Pr, the physiologically inactive red-
absorbing form of phytochrome, to Pfr. A consideration of such
appropriately rapid responses, typified by bioelectric potential
changes which occur within 10 to 15 s following light treatment (5),
leads to the present, most widely accepted hypothesis, which suggests
that Pfr functions by modulating one or more membrane activities.
The early suggestion that Pfr functions as an enzyme was discarded
long ago and, as discussed in detail elsewhere (4), one should now
consider phytochrome-mediated alteration of gene expression to be a
secondary, rather than primary, response to Pfr. With respect to
the generally accepted hypothesis concerning membrane involvement,
many experiments, especially those related to leaflet movements and
associated K^+ fluxes (4), further indicate that one should anticipate
possible Pfr effects on ion pumps.

Since molecular studies are needed to elucidate definitively
the mode of action of phytochrome, and since our understanding of
phytochrome has undergone considerable change in recent years, I
will begin here by briefly summarizing the molecular properties of
the pigment. I will then present a simplified pair of readily test-
able alternative hypotheses followed by a summary of recent data
related to these hypotheses. It is hoped that the arguments and
data presented here will lead to investigation of phytochrome action
at a molecular level comparable in effort to the physiological
studies which have brought us to our present state of knowledge.

SUMMARY OF THE MOLECULAR PROPERTIES OF PHYTOCHROME

Phytochrome is readily degraded by neutral proteases found in
crude plant extracts yielding a relatively stable, photoreversible
chromopeptide (6). Thus, many fundamental studies on phytochrome
either utilized preparations now known to have been degraded or,
perhaps worse, preparations of unknown status (2). Native phytochrome has a subunit weight of 120,000 daltons and is apparently a
dimer while the proteolytically degraded chromopeptide is most often
a 60,000 dalton monomer (2). The number of open chain tetrapyrrole
chromophores per subunit is as yet unresolved. Both the protein (2)
and chromophore (7) moieties change as a function of phototransformation between the two forms. The molecule is highly soluble in
aqueous media indicating that it would probably not be found in a
highly lipophilic environment such as the interior of a membrane.

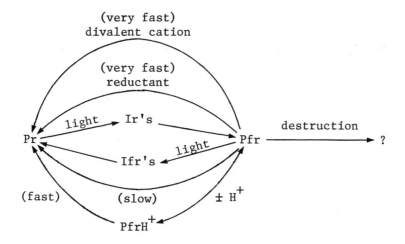

Fig. 1.   Schematic representation of phytochrome reactions as described in the text.

Phytochrome is synthesized by the plant as Pr and, as depicted in Fig. 1, upon absorption of light becomes transformed to Pfr with a quantum yield of 0.17 (8).   Absorption of light by Pfr results in the reverse transformation with an identical quantum yield.   Intermediates (Ir's and Ifr's) are involved in both transformation pathways (9) and the transformations occur in highly purified solutions in the absence of any added cofactor or activator.   Pfr is known to revert directly to Pr by a slow, temperature dependent, nonphotochemical process (2).   In vitro, reversion may proceed less slowly through a protonated intermediate (10), or rapidly in the presence of catalytic amounts of either reductant (11) or divalent cation (12,13).   Thus, control in situ of the molecular environment of phytochrome may be one mechanism by which a plant could regulate its endogenous Pfr level.   Pfr, in addition to causing morphogenic responses, also undergoes destruction, a process resulting in a loss both of photoreversibility and of antigenically detectable material (14) leading to the suggestion that it involves a general degradation of the protein moiety.

Pr has an absorption maximum near 665 nm both in vitro and in vivo while Pfr has a maximum at 724 nm in vitro and about 735 nm in vivo (2,13).   Since the absorption spectra of Pr and Pfr overlap, light produces a photoequilibrium between the two forms, consisting in red light (665 nm) of 75% Pfr, 25% Pr for undegraded phytochrome in vitro (8).   These latter values are different from those obtained for 60,000 dalton phytochrome (81% Pfr, 19% Pr) and this difference may be significant for some calculations, for example those related to threshold control of lipoxygenase activity (15).

ALTERNATIVE HYPOTHESES CONCERNING THE PRIMARY REACTION OF Pfr

Models have been presented which serve formally to explain phytochrome action (16-19) and to account for so-called phytochrome
paradoxes (4). However, some of these models involve several independent assumptions and parameters of freedom which make them difficult if not impossible to test as a whole. Thus, one must select
simpler questions for experimental analysis. The first question
to be addressed is whether a receptor, analogous to the situation
with many hormones, exists for Pfr (Fig. 2). That is, does photochemically produced Pfr directly modulate some as yet unknown process
(e.g., an ion pump) represented by A→B (Fig. 2a) or must Pfr first
bind with a receptor, R (Fig. 2b)?

DIRECT EVIDENCE CONCERNING A POSSIBLE RECEPTOR FOR Pfr

Phytochrome is readily extracted from fully dark grown tissue
and is not generally found associated in appreciable quantity with
particulate subcellular fractions except as noted below. The observation that Pr, as synthesized in etiolated plants, is a soluble protein is reinforced by both light and electron microscope level immunocytochemical labelling experiments which indicate that Pr is generally dispersed throughout the cytosol and not uniquely associated
with any identifiable organelle or structure (21). Reports, which
will be discussed below, that phytochrome associates with particulate
subcellular fractions after conversion to Pfr, led Mackenzie (21) to
repeat these immunocytochemical experiments with light-treated
plants. Mackenzie found that while Pr was generally dispersed
throughout the cytosol of oat (<u>Avena sativa</u> L., cv. Garry) coleoptile
parenchyma cells (Fig. 3a), Pfr produced by red light became associated with discrete areas about 1 μm in size (Fig. 3b) which do not
possess readily identifiable morphology. Following phototransformation back to Pr, no immediate change in phytochrome localization was

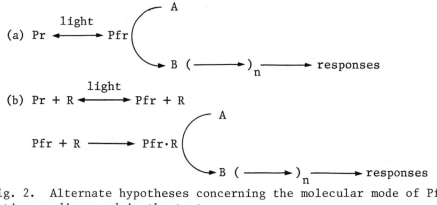

Fig. 2. Alternate hypotheses concerning the molecular mode of Pfr
action as discussed in the text.

observed (Fig. 3c) but upon subsequent dark incubation at 24°C phytochrome distribution returned (Fig. 3d) to that observed in the dark control (Fig. 3a). Although this 'relaxation' process was relatively slow, Mackenzie has since observed (personal communication) that the initial red-induced redistribution is quite rapid, with a significant change being observed in less than 1 min at 3°C.

As alluded to above, substantial amounts of phytochrome are not found associated with particulate material in crude plant extracts unless it has first been converted to Pfr. The initial report on red-induced phytochrome pelletability (22) was followed by numerous investigations which led to claims that Pfr was bound to plasma

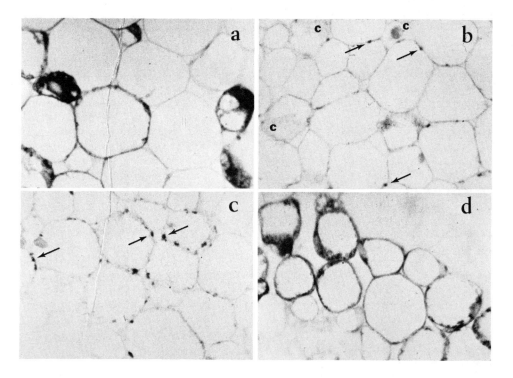

Fig. 3. Bright-field micrographs showing phytochrome-specific immunocytochemical stain in dark-grown 3-day-old Garry oat coleoptile parenchyma cells. All tissue was fixed at 0°C immediately following the indicated treatment. (a) dark control; (b) 8 min 'red' at 25°C; (c) 8 min 'red' followed immediately by 5 min far-red at 25°C; (d) as (c) except that a 2 h dark period at 24°C followed the irradiations and preceded fixation at 0°C. Arrows identify areas of discrete phytochrome-associated stain. Note unstained cytoplasm, c, in (b) which indicates phytochrome is no longer uniformly distributed throughout the cytosol. Magnification = 365x. (After Ref. 21)

Table I. Effect of light on pelletability of phytochrome from Garry oats. Except in the one case noted, light treatments were given to intact tissue. Pelletable phytochrome = units in pellet/ (units in pellet + units in supernatant)·100 $\pm$ average deviation from a minimum of 3 experiments. All treatments and incubations at 0°C. (After Refs. 26, 27)

Treatment	Pelletable phytochrome
Dark control	7.8 $\pm$ 1.0
3 min red in vitro	7.8 $\pm$ 1.2
5 s red	63 $\pm$ 4
60 s far-red	8.4 $\pm$ 0.6
5 s red followed immediately by 60 s far-red	9.7 $\pm$ 0.6
5 s red, 30 min dark, 60 s far-red	62 $\pm$ 1

membrane and/or rough endoplasmic reticulum and to the tacit assumption that this binding involved a specific receptor (23,24). However, as Quail (25) has shown, much or all of the observed binding in most of these experiments was a nonspecific, electrostatic interaction between Pfr and 31S ribonucleoprotein particles which is probably artifactual. Concurrent with Quail's work, other data led to the suggestion that two types of interaction between Pfr and pelletable material were being observed (23) and these suggestions have been substantiated by more recent data (26,27).

The two types of interaction just referred to differ in at least four significant characteristics: (a) binding to 31S particles occurs in vitro (23) while the other type of interaction is observed only following incubation of Pfr in situ (26,27); (b) in vitro binding is observed in the absence of a divalent cation (25) while the in vivo-induced association is not (27); (c) in vitro binding is readily reversed by increasing ionic strength while the in vivo-induced association is not (27); and (d) in vitro binding has a much greater pH dependence than the in vivo-induced association (25,27). Furthermore, in vitro binding is found with only 2 of 11 plants examined (including Cucurbita pepo L., the organism used in most studies) while the other type of association has been observed in all 11 so far tested (including C. pepo) (27). Thus, since the in vitro binding to 31S particles may well be an artifact as argued by Quail (25), one must reevaluate the significance of most of the binding studies using C. pepo.

Since the in vivo-induced association has not yet been well characterized, I shall summarize here some data, obtained with the same oat cultivar used for the immunocytochemical studies, which indicate that this second type of association may yet represent the

hypothesized interaction between Pfr and a membrane-bound receptor.
A brief irradiation of oat tissue prior to extraction increases the
proportion of phytochrome found in a 15 min, 20,000g pellet by about
8-fold (Table). The composition of the extraction buffer is relatively unimportant as long as it contains a divalent cation. As
pointed out above, irradiation of the crude extract has no effect
(Table). Red-induced pelletability is almost completely reversed
by far-red light unless a dark period is interposed between the
two light exposures (Table). The kinetics of the increase in pelletability, which may be determined by closely controlling this
dark interval (Fig. 4a), indicate that the process follows first
order kinetics, has a half-life of 40 s at $0.5°C$, is strongly temperature dependent, and goes virtually to completion within 5 to 10 s
at $22°C$ (27). The kinetics of this process at $0.5°C$ correspond well
with the immunocytochemical observations reported above.

     While extraction immediately after far-red irradiation of pel-
letable phytochrome does not result in any reversibility (Table), if
a dark interval follows the far-red irradiation and precedes extraction, the red-induced pelletability is clearly reversible (Fig. 4b).
Again, the process follows first order kinetics (although the time
course is much slower) and is strongly temperature dependent (27).
At $25°C$ the half-life is 25 min while it is 100 min at both $13°C$ and
$3°C$. The reversal kinetics observed at $25°C$ again agree well with
the immunocytochemical observations. As for enhanced pelletability,
reversal is not observed in vitro (27).

     An interpretation of the immunocytochemical and pelletability

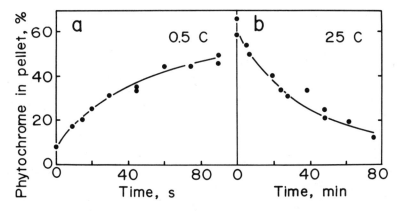

Fig. 4. Percentage of phytochrome which is pelletable in extracts
of oat shoots (a) as a function of time in darkness between a 5 s
irradiation with red light and a 10 s homogenization, all at $0.5°C$,
and (b) as a function of time at $25°C$ between a 5 min red, 10 min
far-red irradiation sequence and extraction. (After Ref. 27)

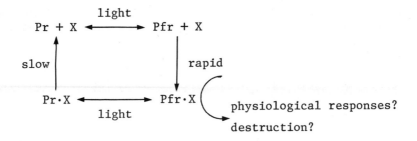

Fig. 5. Summary of observations as discussed in the text.

data, which may be two methods of observing the same phenomenon, is given in Fig. 5. As indicated, a simple and favored interpretation, especially with regard to the immunocytochemical data, is that the interaction represents a binding process following the photoconversion of Pr to Pfr (Pfr + X→Pfr·X). However, enhanced pelletability may also result from a Pfr-induced change in the system which leads either to stabilization of a pre-existing association or to the possibility that the association may be created at the instant of extraction (but not later [27]). Upon phototransformation of Pfr·X to Pr·X, phytochrome slowly returns to the nonpelletable condition. Whether Pr and X are unaltered by the cycle as indicated here is not known. While one may thus conclude that the in vivo-induced association described here is consistent with the possibility that a receptor, possibly membrane-bound, exists for Pfr much more must be done before any conclusions may be made regarding the nature of 'X,' its possible multiplicity, or whether X in Fig. 5 is the same as R in Fig. 2b.

    Acknowledgements. Preparation of this manuscript and the author's research program have been supported by grants from the National Science Foundation (GB-17057, PCM75-19125) and the Vanderbilt University Research Council.

REFERENCES

1. Hendricks, S. B., in Giese, A. C. (ed.), Photophysiology, Vol. I, p. 305, Acad. Press, New York (1964).
2. Briggs, W. R., and Rice, H. V., Annu. Rev. Plant Physiol., 23, 293 (1972).
3. Mohr, H., Photochem. Photobiol., 5, 469 (1966).
4. Satter, R. L., and Galston, A. W., in Goodwin, T. W. (ed.), Chemistry and Biochemistry of Plant Pigments, Acad. Press, New York, p. 680 (1976).

5. Newman, I. A., in Bielski, R. L., Ferguson, A. R., and Cresswell, M. M. (eds.), Mechanisms of Plant Growth, Bull. 12, The Royal Society of New Zealand, Wellington, p. 355 (1974).

6. Gardner, G., Pike, C. S., Rice, H. V., and Briggs, W. R., Plant Physiol., 48, 686 (1971).

7. Grombein, S., Rüdiger, W., and Zimmerman, H., Hoppe-Seyler's Z. Physiol. Chem., 356, 1709 (1975).

8. Pratt, L. H., Photochem. Photobiol., 22, 33 (1975).

9. Kendrick, R. E., these proceedings.

10. Anderson, G. R., Jenner, E. L., and Mumford, F. E., Biochemistry, 8, 1182 (1969).

11. Mumford, F. E., and Jenner, E. L., Biochemistry, 10, 98 (1971).

12. Negbi, M., Hopkins, D. W., and Briggs, W. R., Plant Physiol., 56, 157 (1975).

13. Pratt, L. H., and Cundiff, S. C., Photochem. Photobiol., 21, 91 (1975).

14. Pratt, L. H., Kidd, G. H., and Coleman, R. A., Biochim. Biophys. Acta, 365, 93 (1974).

15. Oelze-Karow, H., and Mohr, H., Photochem. Photobiol., 18, 319 (1973).

16. Borthwick, H. A., Hendricks, S. B., Schneider, M. J., Taylorson, R. B., and Toole, V. K., Proc. Natl. Acad. Sci., 64, 479 (1969).

17. Hartmann, K. M., Photochem. Photobiol., 5, 349 (1966).

18. Mancinelli, A. L., and Rabino, I., Plant Physiol., 56, 351 (1975).

19. Schäfer, E., J. Mathematical Biol., 2, 41 (1975).

20. Coleman, R. A., and Pratt, L. H., Planta, 121, 119 (1974).

21. Mackenzie, J. M., Jr., Coleman, R. A., Briggs, W. R., and Pratt, L. H., Proc. Natl. Acad. Sci., 72, 799 (1975).

22. Quail, P. H., Marmé, D., and Schäfer, E., Nature New Biol., 245, 189 (1973).

23. Marmé, D., J. Supramolecular Structure, 2, 751 (1974).

24. Schäfer, E., in Zimmerman, V. and Dainty, J. (eds.), Membrane Transport in Plants, Springer Verlag, Berlin p. 435, (1975).

25. Quail, P. H., Planta, 123, 235 (1975).

26. Grombein, S., Rüdiger, W., Pratt, L., and Marmé, D., Plant Science Letters, 5, 275 (1975).

27. Pratt, L. H., and Marmé, D., Plant Physiol, in press.

PHOTOTRANSFORMATIONS OF PHYTOCHROME

Richard E. Kendrick

Department of Plant Biology, The University

Newcastle upon Tyne, NE1 7RU, U.K.

SUMMARY

The phototransformations of phytochrome involve several intermediates. The initial photoreaction of the chromophore is followed by a series of dark relaxation states of the chromophore and
apoprotein. The nature of these intermediates is outlined and they
are discussed in relationship to our knowledge of the tetrapyrrole
chromophore.

INTRODUCTION

Phytochrome is the photoreversible chromoprotein which controls many aspects of plant growth and development (4,8,18,19,35,
38). It exists in two forms: Pr, thermostable, absorbance maximum
in the red, and Pfr, less stable, absorbance maximum in the far-
red and physiologically the active form

$$\text{Pr} \underset{h\nu}{\overset{h\nu}{\rightleftarrows}} \text{Pfr} \longrightarrow \text{Physiological Action}$$

Phytochrome has been extracted and purified from several species and is a chromoprotein with a molecular weight of 120,000
daltons (12). The chromophore has been shown to be an open-chain
tetrapyrrole similar to the chromophore of the algal photosynthetic
pigment c-phycocyanin (4,37,38). The absorption spectra of Pr and
Pfr overlap strongly in the red region of the spectrum and light in
this region results in a photostationary equilibrium between Pr and
Pfr (6,15). Red light leads to a photostationary equilibrium (Pfr/

521

P total) of 0.80, whereas far-red light nearly completely maintains phytochrome in its Pr form (Pfr/P total = c 0.05). Since it is impossible to obtain a pure sample of Pfr, analysis of phototransformation spectra of Pfr must be cautious since they involve a component of 20% Pr. However, it is possible to study the behaviour of a sample of pure Pr.

Although phototransformations are first order in both directions (6,11) in 1966 it was demonstrated that photoconversion of phytochrome was more complex involving several intermediates in both pathways (25). It was concluded that each pathway consisted of an initial photoreaction followed by a series of dark relaxation reactions (25,38).

$$\text{Pr} \; \underset{d}{\overset{h\nu}{\rightleftarrows}} \; \underset{d}{\overset{d}{\rightleftarrows}} \; \underset{d}{\overset{d}{\rightleftarrows}} \; \underset{h\nu}{\overset{d}{\rightleftarrows}} \; \text{Pfr}$$

METHODS

Several techniques have been used to investigate phytochrome intermediates (19,25). Initially flash spectroscopy (27) and low temperature techniques (9,26,31,39-44) were used to demonstrate the existence of intermediates. More recently intermediates have been studied under conditions of pigment cycling (2,3,10,11,20-23) and in dehydrated tissue, where photoconversion is not possible (1, 16,17,24,25,46,47,49). A system of intermediate terminology based on that used for the visual pigment Rhodopsin has been adopted (25, 46). The first stable photoproducts of Pr and Pfr are called lumi-R and lumi-F respectively. The dark relaxation intermediates of these photoproducts are designated by the prefix meta (meta-Ra, meta-Rb, etc. being relaxations of lumi-R and meta Fa, meta-Fb, etc. being relaxations of lumi-F). The intermediates studied by these techniques are transients at physiological temperatures. Their spectral characteristics have been observed to vary *in vitro* and *in vivo* as well as with temperature making comparative analysis of results difficult (25,46).

RESULTS

The Nature of lumi-F and its Relaxations

The first stable photoproduct of Pr, called lumi-F, has been detected by flash spectroscopy (27) and low temperature studies (9,23,26,31) *in vitro*, and low temperature studies (39-44) and dehydration (16,24,46) *in vivo*. Difference spectra for the formation of lumi-R show a sharp absorbance maxima at about 698 nm. Figure 1

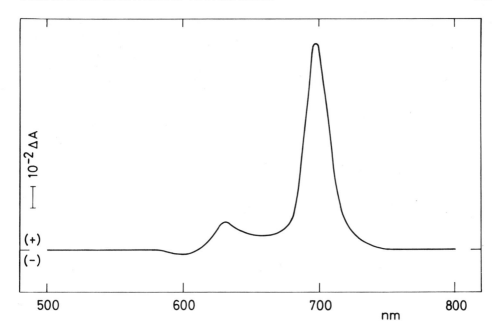

Figure 1. Difference spectrum for the phototransformation of Pr to lumi-R in epicotyl hooks of peas at 77K.

shows a typical difference spectrum for this reaction *in vivo* at 77K. At 77K lumi-R is thermostable, but can undergo photoconversion back to Pr (22,46). Actinic red light at this temperature results in the formation of a photostationary equilibrium between Pr and lumi-R. At higher temperatures lumi-R is unstable in darkness and either reverts to Pr or undergoes a series of relaxation reactions to form Pfr. At physiological temperatures lumi-R is just a transient in photoconversion, but even at these temperatures can be stabilized to some extent by dehydration (16, 24). In freeze-dried tissue at room temperature no Pfr is formed from Pr, but a mixture of lumi-R and meta-Ra which reverts to Pfr in a few minutes of darkness (46). In hydrated tissue above 203K the first relaxation intermediate of lumi-R has been called meta-Ra (25,44,46). Flash spectroscopy (27) and photoconversion at 228K *in vitro* (9) and 254K *in vivo* (44) suggest that lumi-R transforms into meta-Ra with an absorbance maximum in difference spectra of about 710 nm. Actinic light can photoconvert meta-Ra back to Pr at 254K *in vivo* (23,44).

Both flash spectroscopy (27) and low temperature studies (9)

in vitro as well as investigations under conditions of pigment cycling *in vitro* and *in vivo* (20,21,23) suggest that a weakly abosor-
bing intermediate directly precedes Pfr. This intermediate is called meta-Rb. Figure 2 shows a difference spectrum for the reactions
of intermediates which form Pfr in darkness after the pigment *in
vitro* has been cycled under high intensity incandescent light.
Since the meta-Rb to Pfr reaction is the slowest in the Pr to Pfr
and Pfr to Pr pathways, it is this intermediate which accumulates
in greatest concentrations when the pigment cycles and Figure 2 is
essentially a difference spectrum for the meta-Rb to Pfr relaxation
reaction.

The Nature of lumi-F and its Relaxations

The first photoproduct of Pfr has been defined as lumi-F (25,
46). At 77K *in vivo* (39,44) photoconversion results in a stable
product with a difference spectrum maximum at 650 nm (Fig. 3).
Flash spectroscopy provides evidence for a similar intermediate

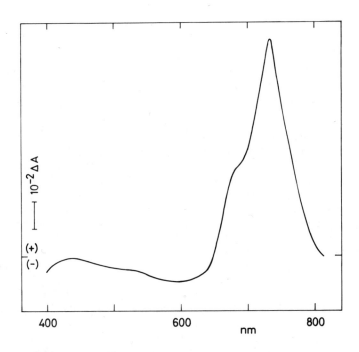

Figure 2.   Difference spectrum for the dark relaxation reactions
of the intermediates (principally meta-Rb forming Pfr) accumulating during irradiation at $0^{\circ}C$ with high-intensity incandescent
light.   Phytochrome in glycerol-buffer (3:1).

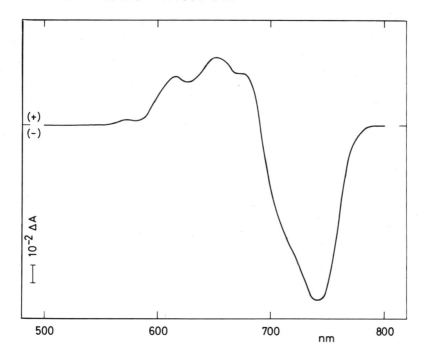

Figure 3. Difference spectrum for the phototransformation of Pfr
to intermediates in epicotyl hooks of peas at 77K.

in vitro at 0°C (27). This product can at least partially be photoconverted back to Pfr at 77K, a reaction seen more obviously at
higher temperatures (22). A similar situation is seen in dehydrated tissue at 0°C, where a stable photoproduct is formed from Pfr,
which can be photoconverted back to Pfr (16,24,46). There is also
evidence for the dark reversion of this product to Pfr at temperatures above 77K (44,46). Recent data (45) suggests that this
stable photoproduct at 77K is not the first photoproduct of Pfr but
a mixture of relaxation intermediates of lumi-F. It now seems
likely that the true lumi-F has an absorbance maximum of about 720
nm. At 77K actinic far-red light given to Pfr produces lumi-F
which is itself photoconverted to meta-Fa. Only by carefully designed experiments is it possible to detect lumi-F. The situation
concerning relaxation intermediates of lumi-F is complex (25). At
higher temperatures at least two are detected in dehydrated tissue
(46). The absorbance bands are all in the red region of the spectrum and have not yet been completely resolved. Here they are called
meta-Fa$_1$, meta-Fa$_2$, meta-Fa$_3$ etc. Since they occur in dehydrated
tissue it is assumed that they represent chemical events restricted

to the chromophore and apoprotein. All these meta-Fa intermediates are relatively stable at 0°C in dehydrated tissue, but slowly revert to Pfr in darkness (46). A further intermediate meta-Fb which is relatively bleached compared to other intermediates has been proposed from data with hydrated tissue (22,44). Formation of this intermediate and its conversion to Pfr requires an aqueous matrix (46).

DISCUSSION

Figure 4 shows a summary of the phototransformations of phytochrome based on the data available. Similarities can be seen in both pathways but so far there is no evidence to support the idea of intermediates common to each pathway. At physiological temperatures these intermediates are presumed to be only transient states, except for those directly preceding the slowest dark reactions in the cycle. This has been demonstrated for the intermediate meta-Rb, which significantly accumulates under conditions of pigment cycling and forms Pfr in darkness (23).

What can be said about the nature of the changes involved during these photoreactions and relaxations? The phytochrome chromophore is a straight chain tetrapyrrole (33,37) and recently the work of Rüdiger and his co-workers has presented a detailed structure for the points of attachment to the apoprotein (14,33,34,36). The difference in chromophore structure between Pr and Pfr lies in the double bond connecting the first two rings of the tetrapyrrole. During photoconversion from Pr to Pfr it is lost and Pfr has a

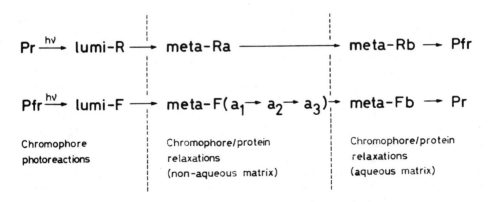

Figure 4. Scheme proposed for the phototransformations of phytochrome.

shortened conjugated system (14). The apoprotein has a strong influence on the absorbance band of the chromophore, shifting the
band some 105 nm in the case of Pfr, although the effect is less
in the case of Pr (14). These large influences of apoprotein on
the absorption band make it difficult to relate absorption properties of intermediates with definite chromophore structures. The
Pr and Pfr chromophores conform to extended configurations of the
tetrapyrrole chromophore whereas the relatively bleached meta-Rb
conforms to a tightly folded form (5). This suggests that large
three-dimensional changes occur between apoprotein and chromophore
during the intermediate relaxations. Although during photoconversion the overall change in shape of the Pr and Pfr molecules is
small (13,48) the active centre must be significantly altered. The
different reactivity of Pr and Pfr (4,13,32,38) must demonstrate
changes in this local region of the molecule.

On the basis of results with dehydrated tissue it appears that
the formation of lumi-R and lumi-F, and their relaxation to meta-
Ra and meta-(Fa$_1$, Fa$_2$, Fa$_3$) are chemical events restricted to the
chromophore and apoprotein which do not require an aqueous matrix.
It is only the subsequent steps via meta-Rb to Pfr and meta-Fb to
Pr that involve conformational interaction of the apoprotein and
chromophore which requires water. It is these reactions which are
slowest in the whole cycle Pr to Pfr to Pr. The substitution of
D$_2$O for water slows down the formation of Pfr from meta-Rb (23).
A similar slowing down is observed in the presence of high glycerol
and sucrose concentrations (3,10,23) due to increased viscosity or
dehydration. Interestingly, this reaction is increased by reducing
agents *in vitro* (23) and under anaerobic conditions *in vivo* (21).

The Pfr form of phytochrome is meta stable *in vitro* at 0°C (23,
28-30) and in many tissues *in vivo* at physiological temperatures (7,
8) reverting to the stable Pr form. This reaction is strongly influenced by the molecular environment. It is favoured by reducing
conditions and it has been postulated to occur via a protonated form
of Pfr, called PfrH (29). It has been speculated that this may be
a common intermediate with the photochemical pathway, but so far no
evidence to support this hypothesis is available (38,46).

It is hoped that study of phytochrome intermediates will provide information about the chemical and biophysical events occurring
at the active centre of the molecule. Although intermediates themselves have not yet been found to be physiologically active, the
maintenance of a significant proportion of the phytochrome under
conditions of pigment cycling as found in natural daylight must be
of physiological significance. In the special case of seeds, where
phytochrome is present naturally in a dehydrated state, a knowledge
of the intermediate reactions observed under such conditions has
enabled germination behaviour to be understood (17).

ACKNOWLEDGEMENTS

I would like to thank: W. Rüdiger, H. Scheer, and C.J.P. Spruit, for providing unpublished material; The Royal Society for financial support.

REFERENCES

1. Balangé, A.P. Physiol. Veg., 12, 95 (1974).
2. Briggs, W.R. and Fork, D.C. Plant Physiol., 44, 1081 (1969).
3. Briggs, W.R. and Fork, D.C. Plant Physiol., 44, 1089 (1969).
4. Briggs, W.R. and Rice, H.V. Ann. Rev. Plant Physiol., 23, 293 (1972).
5. Burke, M.J., Pratt, D.C. and Moscowitz, A. Biochemistry, 11, 4025 (1972).
6. Butler, W.L., Hendricks, S.B. and Siegelman, H.W. Photochem.
P Photobiol., 3, 521 (1964).
7. Butler, W.L. and Lane, H.C. Plant Physiol., 40, 13 (1965).
8. Butler, W.L., Lane, H.C. and Siegelman, H.W. Plant Physiol. 38, 514 (1963).
9. Cross, D.R., Linschitz, H.,Kasche, V. and Tenenbaum, J. Proc. Nat. Acad. Sci. U.S.A., 61, 1095 (1968).
10. Everett,M.S. and Briggs, W.R. Plant Physiol., 48, 679 (1970).
11. Gardner, G. and Briggs, W.R. Photochem. Photobiol. 19, 267 (1974).
12. Gardner, G., Pike, C.S., Rice, H.V. and Briggs, W.R. Plant Phytol., 48, 686 (1971).
13. Gardner, G.m Thompson, W.F. and Briggs, W.R. Planta, 117, 367 (1974).
14. Grombein, S., Rüdiger, W. and Zimmermann, H. Hoppe-Seyler's Z. Physiol. Chem., 356, 1709 (1975).
15. Hanke, J., Hartmann, K.M. and Mohr, H. Planta, 86, 253 (1969).
16. Kendrick, R.E. Nature, Lond., 250, 159 (1974).
17. Kendrick, R.E. Sci. Prog., Oxf., 63, 347 (1976).
18. Kendrick, R.E. and Frankland, B. Phytochrome and Plant growth. Study in Biology 68. London: Edward Arnold (1976).
19. Kendrick, R.E. and Smith, H. In: Chemistry and Biochemistry of Plant Pigments (Ed. Goodwin) 2nd ed., volume 2. London and New York: Academic Press p. 334 (1976).
20. Kendrick, R.E. and Spruit, C.J.P. Nature New Biol., 237, 281 (1972).
21. Kendrick, R.E. and Spruit, C.J.P. Photochem. Photobiol., 18, 139 (1973).
22. Kendrick, R.E. and Spruit, C.J.P. Photochem. Photobiol., 18, 153 (1973).
23. Kendrick, R.E. and Spruit, C.J.P. Plant Physiol,, 52, 327 (1973).
24. Kendrick, R.E. and Spruit, C.J.P. Planta, 120, 265 (1974).
25. Kendrick, R.E. and Spruit, C.J.P. In: Light and Plant

Development (Ed. H. Smith). Proceedings of the 22nd Easter
School in Agricultural Science, University of Nottingham.
London: Butterworth (1976).
26. Kroes, H.H. Meded. Landbouwhogesch. Wageningen, 70, 1 (1970).
27. Linschitz, H., Kasche, V., Butler, W.L. and Siegelman, H.W.
    J. Biol. Chem., 241, 3395 (1966).
28. Mumford, F.E. Biochemistry 5, 322 (1966).
29. Mumford, F.E. and Jenner, E.L. Biochemistry, 10, 98 (1971).
30. Pike, C.S. and Briggs, W.R. Plant Physiol., 49, 514 (1972).
31. Pratt, L.H. and Butler, W.L. Photochem., Photobiol., 8, 477
    (1968).
32. Roux, S.J. Biochemistry 11, 1930 (1972).
33. Rüdiger, W. In: Phytochrome (Ed. Mitrakos and Shropshire).
    London and New York: Academic Press p.129 (1972).
34. Rüdiger, W. and Correll, D.L. Liebigs Ann. Chem., 723, 208
    (1969).
35. Satter, R.L. and Galston, A.W. In: Chemistry and Biochemistry
    of Plant Pigments (Ed., Goodwin) 2nd ed., volume 1. London and
    New York: Academic Press p.680 (1976).
36. Scheer, H. Z. Naturforsch C. (in press).
37. Siegelman, H.W., Turner, B.C. and Hendricks, S.B. Plant Phys-
    iol., 41, 1289 (1966).
38. Smith, H. and Kendrick, R.E. In: Chemistry and Biochemistry of
    Plant Pigments (Ed. Goodwin) 2nd ed., volume 1. London and New
    York: Academic Press p.377 (1976).
39. Spruit, C.J.P. In: Currents in Photosynthesis (Eds. Thomas,
    J.B. and Goedheer, J.C.). Rotterdam: Donker p.67 (1966).
40. Spruit, C.J.P. Meded. Landbouwhogesch. Wageningen, 66, 1 (1966).
41. Spruit, C.J.P. Biochim. Biophys. Acta, 120, 186 (1966).
42. Spruit, C.J.P. Biochim. Biophys. Acta, 120, 454 (1966).
43. Spruit, C.J.P. Meded. Landbouwhogesch. Wageningen, 71, 1 (1971).
44. Spruit, C.J.P. and Kendrick, R.E. Photochem. Photobiol., 18,
    145 (1973).
45. Spruit, C.J.P. and Kendrick, R.E. Unpublished data.
46. Spruit, C.J.P., Kendrick, R.E. and Cooke, R.J. Planta 127, 121
    (1975).
47. Tobin, E.M. and Briggs, W.R. Plant Physiol., 44, 148 (1969).
48. Tobin, E.M. and Briggs, W.R. Photochem. Photobiol., 18, 487
    (1973).
49. Tobin, E.M., Briggs, W.R. and Brown, P.K. Photochem. Photobiol.,
    18, 497 (1973).

BIOGENESIS OF CHLOROPLAST MEMBRANES IN ALGAE

I. Ohad, S. Bar-Nun, D. Cahen, J. Gershoni, M. Gurevitz,

F. Kretzer, R. Schantz* and S. Schohat

Department of Biological Chemistry, The Hebrew University of
Jerusalem, Jerusalem, Israel
* Institute of Botany, University of Strasbourg, France.

INTRODUCTION

The intracellular origin of the polypeptides forming the
photosynthetic membranes in Chlamydomonas reinhardi has been
studied extensively. Based on results obtained with conditioned
mutants in which the protein synthesis process by the cytoplasmic
or chloroplast ribosomes was alternatively blocked by specific
inhibitors it was proposed that both subcellular compartments
contribute polypeptides essential for the formation of the membrane
and development of its photosynthetic function (1). This concept
is also supported by experimental results obtained with synchronous
cultures of wild type cells and mutants in which the synthesis of
specific membrane components is impaired (1-3). In this presentation we would like to summarize recent data obtained in our
laboratory regarding the origin and assembly of polypepttides
required for the formation of photosystems I and II and the formation of chlorophyll-protein complexes CPI and CPII in Chlamydomonas
reinhardi and Euglena gracilis.

MATERIALS AND METHODS

Euglena gracilis (Z) was kindly supplied by Dr. M. Lefort-Tran,
Laboratory of cytophysiology of photosynthesis, Gif-sur Yvette,
France; the T_4 mutant of Chlamydomonas reinhardi was obtained by
Dr. P. Bennoun and previously characterized (4); the work with this
mutant was carried out in cooperation with him; the y-1 mutant of
Chlamydomonas was described before (1). Cell growth, preparation of
membranes and experimenta procedures were as previously described (5).

531

Analysis of polypeptides, composition of membranes and identification of protein-chlorophyll complexes and their composition were carried out by SDS-polyacrylamide gel-electrophoresis (6). For additional experimental details regarding measurements of fluorescence induction, quantum yield and flash yield, see ref. (7).

RESULTS AND DISCUSSION

1. Synthesis and Role of Membrane Polypeptides of Cytoplasmic Origin in Chlamydomonas reinhardi

Chloroplast membranes are specifically synthesized when dark grown y-1 mutant cells are exposed to the light under non-dividing conditions (greening) (1). When the greening is carried out in the presence of chloramphenicol (CAP), chloroplast membranes containing significant amounts of chlorophyll are formed However, such membranes do not contain active PSI or PSII reaction centers and exhibit only the light harvesting chlorophyll-protein complex CPII (1, 8). Analysis of the polypeptide composition of such membranes reveals the abundant presence of polypeptides 22 KD, 24 KD, 28 KD and 32 KD as well as additional polypeptides (see below). When the greening is carried out in the presence of radioactive precursors the above mentioned polypeptides are highly and specifically labelled in the presence of CAP (9, 10), which prevents protein synthesis by the chloroplastic ribosomes. These polypeptides are not synthesized in presence of cycloheximide (CHI) which blocks protein synthesis by cytoplasmic ribosomes. Chloroplast membranes are not formed and chlorophyll is not synthe-sized when the synthesis of these polypeptides is inhibited (1). Polypeptide 28 KD was found to be present in small amounts in isolated active preparations of PSI and enriched in similar preparation of PSII (10) and it was suggested that its presence is a prerequisite for the formation of the membrane structure. Isolation and analysis of the light harvesting chlorophyll protein complex CPII from Chlamydomonas reinhardi membranes discloses that this complex contains chlorophyll a and b in equal amounts and polypeptides 28 KD, 24 KD and 22 KD (8, 11). The synthesis of chlorophyll and that of the latter two polypeptides is light-dependent (1, 9). During the growth of the y-1 mutant in the dark the polypeptides and chlorophyll are diluted out through cell division concomitantly with the loss of the CPII complex, while the loss of the 28 KD polypeptides is much less prnounced (8). Apparently the 28 KD polypeptide continues to be synthesized in the dark. Its presence alone is probably not sufficient for the preservation of a detectable CPII complex. No specific role can be assigned yet to the 32 KD polypeptide, which on gradient polyacrylamide gels can be further resolved into two distinct bands (2, 4). This polypeptide is readily digested by trypsin acting upon intact membranes without loss of reaction center activity (12). However, following digestion, the sensitivity of PSII to DCMU and

ability to transfer electrons from PSII to PSI are lost (12). All
the above-mentioned polypeptides are synthesized in the presence
of rifampicin which prevents the transcription process in the
chloroplast of <u>Chlamydomonas</u> (13). Based on the above data it
is suggested that the 32, 28, 24 and 22 KD polypeptides are products
of cytoplasmic transcription and translation.

2. Synthesis and Role of Membrane Polypeptides of Chloroplastic Origin in <u>Chlamydomonas reinhardi</u>

        When the y-1 mutant cells greened in the presence of CAP
(see 1. above) are washed free of the drug and further incubated
in the presence of CHI, photosynthetic activity is reestablished.
Labelling of the cells with radioactive precursors and analysis
of the radioactive pattern of chloroplast membrane polypeptides
shows that several minor membrane polypeptides are synthesized in
the presence of this drug. These are polypeptides in the range
of 65-67 KD, 40-50 KD and 26 KD (10). Isolated active PSI membrane
preparations contain the high molecular weight polypeptides as well
as polypeptides in the range of 50 KD while active PSII preparations contain polypeptides in the range of 42-47 KD and 26 KD (10,
14). These are missing from preparations obtained from inactive
membranes formed in the presence of CAP (14). In the thermosensi-
tive nuclear mutant T_4 polypeptide 47 and 44 KD are absent from
membranes obtained from cells grown at 37°C but present in membranes
from cells grown at 25°C (4, 15). The missing polypeptides are
specifically synthesized de novo and inserted into the membrane in
cells grown at 37°C when these are incubated at 25°C under non-
dividing conditions (15). Cells in which these two polypeptides
are missing do not exhibit PSII activity neither when using H_2O nor
DPC as electron doner. The polypeptides are synthesized de novo
and inserted into the membranes in cells grown at 38°C when these
are incubated at 25°C under non-dividing conditions for 7-10 hrs
(15). H_2O splitting and DPC oxidation activities are resumed
following the synthesis and integration of the 47 and 44 KD polypeptides into the membranes. The synthesis of both is specifically
inhibited by CAP while that of the 47 KD is also sensitive to
rifampicine (15). Membranes in which only the 44 KD polypeptide
had been inserted, regain the ability to oxidise DPC but cannot
use H_2O as electron donor (15). The synthesis of both polypeptides
is light independent. However, when their synthesis occurs in
the dark their proper integration into the membrane is impaired.
In this case only DPC oxidation can be detected and the peptides
as well as the activity are easily lost from the membrane following
temperature shock (38°C, 1 hr) and treatment with mild detergents.
Proper integration of the dark synthesized polypeptides can be
obtained in the absence of additional protein synthesis if photosynthetic electron flow through PSII is permitted for a short time
(15). Based on these data it is suggested that polypeptides 44
and 47 KD are required for establishing the active reaction center

and the H_2O-splitting activity of PSII respectively. Both are translated in the chloroplast. While the 47 KD polypeptide appears to be coded by the chloroplast genome (rifampicine sensitive) it is possible that the 44 KD polypeptide is a nuclear gene product. This is suggested by the fact that the mutation of T_4 is nuclear (4). The synthesis of this polypeptide is resistent to rifampicine but sensitive to α-amanitine, an inhibitor of nuclear transcription in eukaryotic cells. Due to the mutation in the T_4 cells, the primary sequence of this polypeptide has been apparently altered, thus rendering its proper tertiary structure unstable at 37°C. As a result, the polypeptide is lost from the membranes at the non-permissive temperature if not properly integrated during growth at 37°C or repair at 25°C in absence of electron flow through PSII. The improper integration or lack of this polypeptide from the membrane might have a pleotropic effect, resulting in the loss of the 47 KD polypeptide required for the H_2O-splitting activity (15).

The 65-67 KD polypeptides are required for the formation of the chlorophyll-protein complex CPI (8), considered to be the P_{700} component of PSI (16). This complex contains chlorophyll a and its formation requires concommitant synthesis of both chlorophyll and the polypeptides 65-67 KD (8). In the T_4 cells grown at 38°C and lacking the 47 and 44 KD polypeptides the CPI complex is not detectable. However, these cells exhibit a normal PSI activity and have normal amounts of the 65-67 KD polypeptides in the membranes (15).

Thus, it appears that polypeptides 47 and 44 KD might play a role in the organization of the membrane and might contribute to the stabilization of the CPI complex towards SDS extraction.

3. Synthesis and Role of Membrane Polypeptides of Cytoplasmic and Chloroplastic Origin in Euglena gracilis

Dark grown Euglena cells exposed to a short illumination followed by further incubation in the dark respond by a fast greening upon subsequent illumination (17). When the initial illumination period is extended for 6-8 hrs, sufficient chlorophyll ($\sim$ 2 $\mu g/10^7$ cells) and membranes are formed to allow detection of photosynthetic activity of both PSI and PSII.

If the cells are now transferred to the dark for an additional period of 8-10 hrs, synthesis of certain membrane polypeptides continues, resulting in a significant increase in PSI and PSII-activity in absence of synthesis of chlorophyll. The increase in PSI activity during the dark incubation is not inhibited by CAP while PSII activity decays following addition of the drug, indicating that under these experimental conditions polypeptides required for PSII activity of chloroplastic origin are synthesized and turn over (18). The loss of activity is more pronounced for H_2O-splitting activity than for DPC oxidation (11). Addition of CHI to the dark incubated cells results in a partial inhibition of further PSII development and has no effect on that of PSI. When

cells in which the PSII activity has been significantly lost in
the dark in the presence of CAP are re-exposed to the light, repair
of PSII activity ensues, concomittantly with chlorophyll synthesis
(18). Chlorophyll synthesis however is not required and can be
prevented by addition of CHI without effect on the repair of the
PSII activity. This activity, as expected, is not repaired during
the light period if CAP is not removed. Thus, using the above
described experimental procedure it is possible to resolve in time
the synthesis of membrane components required from PSI and PSII
activity and that of chlorophyll. By appropriate labelling of
the cells with [^{35}S] in presence or absence of CHI and CAP, it is
possible to identify the polypeptides synthesized under different
conditions and correlate their presence with the development of
photosynthetic function. The results of such experiments show
that membrane polypeptides, in the molecular weight range of
78-80 KD of chloroplastic origin, whose synthesis is light independent, are associated with development of PSII activity; polypeptides
of cytoplasmic translation (35 KD, 33 KD, 23 KD), whose synthesis
is light dependent, might be involved in binding chlorophyll into
a complex which might be similar to the chlorophyll-protein light
harvesting complex CPII.
        Thus, like in Chlamydomonas also in Euglena, the chloro-
plast contributes polypeptides required for the development of
PSII activity while the synthesis of several polypeptides of cytoplasmic origin is light dependent.

4. Assembly of Pre-existing and Newly Synthesized Components into
Active Photosynthetic Units

        Measurements of overall electron flow can detect
functional units, but fails to detect presence of only partially
formed electron transfer chains or unconnected components. On the
other hand, electrophoretic analysis can detect presence of
polypeptides, but cannot give much information on whether they are
properly integrated into functional units. However, combination
of the above techniques in conjunction with spectrophotometric
measurements of light induced changes in the redox state of
components such as cytochromes or P_{700}, and measurements of partial
reaction and fluorescence induction, quantum yield, flash yield
and apparent size of the photosynthetic units permits to obtain
more information on the presence of the electron chain constituents
and the degree of their interconnection. Based on this approach,
it was demonstrated that residual membranes present in dark grown
y-1 Chlamydomonas cells contain reaction centers unconnected to
the rest of the chain, as well as chlorophyll which apparently
cannot transfer light absorbed energy to the reaction center (7).
During the initial phase of the greening process, the PSII units
became connected and form   active chains.   This is accompanied
by a drastic reduction in the apparent size of the photosynthetic
unit, increase in quantum yield and reduction of intrinsic

fluorescence (7). The decay of PSII activity in the presence of
CAP in dark incubated Euglena cells described above (see Section 3)
seems to be at least partially due to disconnection between the
PSII units and the plastoquinone pool in addition to the degradation
of the H$_2$O-splitting activity (19). The connection is re-established
during the repair of PSII activity following synthesis of chloroplastic
membrane proteins (see above, Section 3).

        The formation of the photosynthetic membranes can thus be des-
cribed as a process of reorganization of pre-existing units or   comp-
nents, and integration of newly formed ones into a growing membrane (1).

5. Role of Light in the Regulation of Membrane Polypeptide Synthesis

        It has been previously reported that in Chlamydomonas reinhardi
y-1 synthesis of membrane polypeptides of chloroplastic origin is
inhibited in absence of concommitant or previous synthesis of membrane
polypeptides of cytoplasmic origin which is light controlled (1).
The control of 70s ribosomal activity by 80s ribosomal products seems
to be specific for the synthesis of membrane polypeptides. Synthesis
of soluble proteins by 70s ribosomes (CAP sensitive) can proceed in
presence of CHI in Chlamydomonas for extended periods of time (5-10
hrs) (20). The above translation control is not complete. Small
amounts of certain membrane polypeptides synthesized by 70s ribosomes appear to continue and be formed in the dark grown y-1 cells
at a reduced rate such as 65-67 KD, and 40-50 KD polypeptides (11)
while the synthesis of the 26 KD is completely blocked.

        Once a certain amount of polypeptides of cytoplasmic origin
have been synthesized and incorporated into the membrane, the synthesis
of the chloroplast made polypeptides can continue for prolonged
periods of time (3-4 hrs in Chlamydomonas, up to 10 hrs in Euglena)
in the dark (1, 18). The effect of light on the synthesis of membrane
polypeptides of cytoplasmic origin in the y-1 mutant of Chlamydomonas
has been ascribed to the accumulation of protochlorophyll precursor(s)
which can be converted into chlorophyll in the dark in this mutant
and might cause a repression of the synthesis of the required mRNA
(1). This concept is based on circumstantial evidence and requires
further and more direct proof.

ABBREVIATIONS: DCMU, 3-(3,4 dichlorophenyl)-1,2 dimethylurea; DPC,
    1,5 diphenylcarbazide; SDS, sodium dodecylsulphate; PSI, PSII,
    photosystems I and II.

REFERENCES

1. Ohad, I. in: A. Tzagoloff, ed., "Membrane Biogenesis, Mitochondria,
    Chloroplasts, and Bacteria", Plenum Press (1975) New York, p.279.

2. Bourguignon, L.Y.W. and Palade, G.E., J.Cell Biol. 69, 327 (1976).

3. Chua, N.H., Matlin, K. and Bennoun, P., J.Cell Biol. 67, 361 (1975).

4. Chua, N.H. and Bennoun, P., Proc. Nat. Acad. Sci. U.S. 72,
    2175 (1975).

5. Eytan, G. and Ohad, I., J. Biol. Chem. 245, 4297 (1970).

6. Laemmli, U.K., Nature 227, 680 (1970).

7. Cahen, D., Malkin, S., Shochat, S. and Ohad, I., Plant
    Physiol. in Press (1976).

8. Bar-Nun, S., Schantz, R. and Ohad, I., Biochim. Biophys.
    Acta, in press.

9. Hoober, J.K., J. Biol. Chem. 245, 4327 (1970).

10. Bar-Nun, S. and Ohad, I. in: M. Avron, ed., "Proceedings of
    the Third International Congress on Photosynthesis", Elsevier
    (1974) Amsterdam, p. 1627.

11. Schantz, R., Bar-Nun, S. and Ohad, I., Plant Physiol.
    in press.

12. Regitz, G. and Ohad, I., J. Biol. Chem. 251, 247 (1976).

13. Armstrong, J.J., Surzycki, S.J., Moll, B. and Levine, R.P.,
    Biochemistry 10, 692 (1971).

14. Bar-Nun, S. and Ohad, I., Plant Physiol., in press.

15. Kretzer, F., Ohad, I. and Bennoun, P. in: Th. Bücher, ed.
    "Symposium on Genetics and Biogenesis of Chloroplasts and
    Mitochondria", Elsevier (1976) Amsterdam, in press.

16. Shiozawa, J.A., Alberte, R.S. and Thomber, J.P., Arch.
    Biochem. Biophys. 165, 388 (1974).

17. Klein, S., Schiff, J.A. and Holowinsky, A.W., Developmental
    Biology, 28, 253 (1972).

18. Gurevitz, M., Kratz, H. and Ohad, I., Israel J. Med. Sci.
    in press (1976).

19. Gurevitz, M., Cahen, D. et al., in preparation.

20. Gershoni, J.M.and Ohad, I., in "Proceedings of Advance Course
    Nuclei Acids and Protein Synthesis in Plants", organizers
    Weil, J. and Bogorard, L., Strasbourg, France, July 1976.
    In press.

LIGHT-INDUCED CYTOCHROME REDUCTION IN

NEUROSPORA CRASSA MEMBRANE FRACTIONS

Robert D. Brain and Winslow R. Briggs
Department of Plant Biology
Carnegie Institution of Washington
Stanford, California 94305 U. S. A.

SUMMARY

Neurospora crassa mycelium harvested from liquid culture was homogenized and the homogenate fractionated by differential centrifugation. Particulate fractions enriched for mitochondria, presumptive plasma membrane, and endoplasmic reticulum respectively were obtained. All of the fractions contained a cytochrome system which became reduced on irradiation with blue light, and became reoxidized upon subsequent darkness. The distribution of this activity matched closely that of a sodium-dependent ATPase, a putative plasma membrane marker, but was markedly different from the distribution either of mitochondrial or endoplasmic reticulum marker enzymes. The light reaction could only be observed when samples were partially anaerobic. The light-minus-darkness difference spectrum, dark decay kinetics, and requirement for partial anaerobicity all closely resembled the same three parameters reported in the literature for intact mycelium of Neurospora (1). Preliminary investigations with a mutant thought to be deficient in b-type cytochromes revealed reduced light sensitivity in a test for light-suppression of circadian formation of conidia. It seems possible that a flavin-cytochrome pigment system could serve as the photoreceptor complex for the in vivo photoresponse.

INTRODUCTION

Munoz and Butler (1) recently reported a blue-light-inducible reduction of a b-type cytochrome in partially anaerobic mycelium of Neurospora crassa. The cytochrome became reoxidized in

539

darkness at room temperature in about 30 sec. The light-minus-
dark difference spectrum showed a characteristic shift in the
Soret region and the appearance of an alpha band near 560 nm.
Substantial loss of absorbance near 450 nm was consistent with
the photoreduction of a flavin or flavoprotein. The action spectrum for the absorbance changes was consistent with the absorption spectrum of a flavoprotein, suggesting that it was a flavoprotein component which was the photoreceptor for the process
rather than the cytochrome itself. The Neurospora response described by Muñoz and Butler (1) was thus in marked contrast to
one reported by Poff and Butler for Dictyostelium (2,3,4). In
the latter case, the reaction to blue light was an oxidation, not
a reduction, and the photoreceptor was a porphyrin itself rather
than an associated flavin moiety.

Briggs (5) reported briefly on the isolation of a membrane
fraction from corn coleoptiles (Zea mays L.) which showed a blue-
light-induced cytochrome reduction under partially anaerobic conditions, and Schmidt and Butler (6) subsequently reported blue-
light-inducible cytochrome reduction in the total pelletable fraction from a Neurospora homogenate. The latter authors made no
attempt to separate various particulate components, and reported
light-induced reduction of cytochrome c and cytochrome oxidase as
the commonest reaction, with observation of reduction of cytochrome b in some samples.

The purpose of the present work was two-fold: first to iso-
late and if possible characterize the specific membrane fraction
showing the light-inducible cytochrome b reduction; and second,
to explore any possible link between the system and known photoresponses of intact Neurospora. Schmidt and Butler (7) showed
that flavins in solution could mediate photoreduction of cytochrome c, an observation which raises the possibility that the
photoreaction may be nonspecific and therefore not necessarily
physiologically significant. Lipson and Presti (8) found light-
induced cytochrome reduction in phototropically impaired mutants
of Phycomyces as well as in a cultured mammalian cell line, Hela.
These experiments also suggested that the observed photoreduction
might be non-specific and have nothing to do with photoreception
in vivo. It is clearly essential to demonstrate some direct connection between light-induced cytochrome reduction and some in
vivo photoresponse before concluding that a flavin-cytochrome is
actually involved in photoreception in vivo.

METHODS

A carotenoidless strain of <u>Neurospora</u> <u>crassa</u> (albino-timex) was obtained from the Fungal Genetics Stock Center, California State University, Humboldt, Arcadia, California. A poky mutant (mi-1) was crossed with the albino-timex strain for investigation of the possible role of <u>b</u>-type cytochromes in photoreception (see below). Brain <u>et</u> <u>al</u>. describe growing conditions, harvesting, and fractionating techniques in detail elsewhere (9), so they will be mentioned only briefly here. Following growth in liquid culture in darkness mycelia were harvested and homogenized in a Braun Homogenizer, in an extraction buffer consisting of 250 mM sucrose, 100 mM N-morpholinepropanesulfonic acid (MOPS), 14 mM 2-mercaptoethanol,3 mM EDTA, and 0.1 mM $MgCl_2$, titrated to pH 7.4 with KOH. All operations were done under safelights (red or green) which were ineffective in inducing physiological responses in <u>Neurospora</u>. The homogenate was filtered and then centrifuged successively at 2,000, 9,000, 20,000, and 50,000 x g for 10, 15, 30, or 75 minutes respectively. The pellets are designated 2KP, 9KP, 20KP and 50KP. The pellets were resuspended in a small volume of buffer similar to that used for extraction, but adjusted to pH 7.0 and without 2-mercaptoethanol. Assays for cytochrome <u>c</u> oxidase, NADH-dependent cytochrome <u>c</u> reductase, and Na^+-dependent ATPase as markers for mitochondria, endoplasmic reticulum, and putative plasma membrane respectively are those reported by Brain <u>et</u> <u>al</u>. (9), as are spectrophotometric methods for measuring light-induced absorbance changes.

RESULTS

Irradiation of the resuspended pellets with blue light caused an increase in absorbance at 423 nm, a decrease at 450 nm, and a small increase near 560 nm. Thus the light-minus-dark difference spectrum was similar to that reported by Muñoz and Butler (1) for the intact mycelium. Time course measurements showed that the decay half time was near 30 sec. at room temperature, again consistent with the <u>in</u> <u>vivo</u> observations (1). Photoactivity was found in all three resuspended pellets, but its specific activity on a per mg protein basis was twice as high in the 20KP as in either of the other two pellets. By contrast, cytochrome <u>c</u> oxidase activity was twice as high in the 9KP, and the NADH-dependent cytochrome <u>c</u> reductase was almost entirely in the 50KP. The distribution of photoactivity parallelled the distribution of the Na^+-dependent ATPase exactly, however.

Regardless of which pellet was examined, it was always necessary to wait several minutes before a light-inducible cytochrome

reduction could be observed. At the very start of an experiment,
absorption spectra indicated all of the cytochromes to be fully
reduced, and light-induced absorbance changes were absent. After
a few minutes, as verified by absorption spectra, the system
became partially reduced, and thereafter repeatable light-induced
cytochrome reduction could be observed. Addition of dithionite
produced complete reduction of all cytochrome present, again verified by absorption spectra, and eliminated the light-inducible
changes. Thus the condition necessary for obtaining the cytochrome reduction in light was an intermediate redox state. Hence
the isolated membrane fraction shared yet another property with
the in vivo system studied by Muñoz and Butler (1): it required
partial anaerobicity to show light-inducible cytochrome reduction.

Sargent et al. (10) showed that the circadian production of
conidia by the timex mutant of Neurospora crassa could be suppressed by low intensity white light, with an intensity of 0.4
ergs cm^{-2} sec^{-1} being close to threshold. The poky mutant of
Neurospora is well known to be deficient in b-type cytochrome (11).
If a b-type cytochrome were involved in photoreception for suppression of the circadian conidial production, then one would
expect a double mutant, poky-timex, to show reduced sensitivity
to light, and a higher intensity required for suppression. Preliminary experiments showed that an intensity of white light of
210 ergs cm^{-2} sec^{-1} was insufficient to suppress formation of
bands of conidia, though higher intensities were effective. Hence
the poky-timex mutant does indeed show impaired photosensitivity.
Since the action spectrum for this system (12) is similar to that
obtained by Muñoz and Butler for cytochrome photoreduction in
Neurospora in vivo (1), it is possible that the flavin-cytochrome
b complex is indeed part of the photoreceptor machinery.

DISCUSSION

A particulate system showing light-inducible cytochrome reduction has been isolated from Neurospora mycelium. Its properties are very similar to those of the in vivo system studied by
Muñoz and Butler (1): both have dark decay half lives of approximately 30 sec at room temperature; both have light-minus-dark
difference spectra showing positive peaks near 425 nm and 560 nm
and a negative peak at 450 nm; and both can only be seen under
conditions of partial anaerobicity. The system described here is
clearly not associated either with mitochondria or endoplasmic
reticulum. If, as has been suggested by Scarborough (13), the
Na^{+}-dependent ATPase is a marker for plasma membrane, then the
present system is probably also in the plasma membrane. Unlike
the results of Schmidt and Butler (6), the present work showed no
evidence for reduction of cytochrome c or cytochrome c oxidase

on illumination. The alpha band which appeared on illumination was clearly near 560 nm, rather than 550 nm. Hence though differential centrifugation is hardly the best way to purify membrane fractions, it is sufficient to show that the system is not mitochondrial.

There are at least two properties which one might expect of a pigment system if it is to be considered the legitimate photoreceptor for blue-light-induced processes in higher plants and fungi. First, it should be ubiquitous. The present system has been described in several fungi (see 14 for review) and has also been reported from corn coleoptiles (5,9,15), and hence is reasonably widely distributed. Second, however, and more important, one must demonstrate in some way an association between a defective component of the proposed photoreceptor system and a concomitant decrease in photosensitivity. Lipson and Presti (8) looked at Phycomyces mutants with altered phototropic sensitivity, but found unaltered light-inducible cytochrome reduction. However, the phototropic lesion need not necessarily involve the pigment itself in the Phycomyces mutants, but could involve a step further along the reaction chain. In the present work, the approach was the opposite: examine photosensitivity of a Neurospora mutant known to be deficient in a suspected component of the pigment system, in this case, a b-type cytochrome. Preliminary experiments do indeed reveal impaired photosensitivity, with manyfold higher intensities of light required to produce a given photoaction in the mutant than in the wild type.

It must be cautioned that the present authors have not yet rigorously characterized the mutant in terms of its cytochrome content, in comparison with wild type, nor have they verified that its membrane preparations show an appropriately lower amount of cytochrome photoreduction. Any firm conclusions concerning the role of the flavin-cytochrome complex in in vivo photoreception must await such investigations.

The photoreduction seen under partially anaerobic conditions cannot itself represent the exact mechanism of photoreception in vivo in any case. Clearly fully aerobic Neurospora (and for that matter corn coleoptiles) are photosensitive. Partially anaerobic conditions simply provide a convenient tool for observing the system and following it through purification. The alteration of electron flow through the system could indeed be a possible consequence of photoreception, but this suggestion remains to be verified.

The results discussed briefly in this paper will be presented in detail elsewhere (9).

REFERENCES

1. Muñoz, V. and Butler, W. L. (1975) Plant Physiol., 55, 421.
2. Poff, K. L., Butler, W. L. and Loomis, W. F. Jr. (1973) Proc. Nat. Acad. Sci. 70, 813.
3. Poff, K. L., Loomis, W. F. Jr. and Butler, W. L. (1974) J. Biol. Chem. 249, 2164.
4. Poff, K. L. and Butler, W. L. (1974) Photochem. Photobiol. 20, 241.
5. Briggs, W. R. (1975) Abstracts, 3rd Annual Meeting, Am. Soc. Photobiol., Louisville, Ky., 54.
6. Schmidt, W. and Butler, W. L. (1976) Photochem. Photobiol. 24, 77.
7. Schmidt, W. and Butler, W. L. (1976) Photochem. Photobiol. 24, 71.
8. Lipson, E. D., and Presti, D. (1976) Photochem. Photobiol., in press.
9. Brain, R. D., Freeberg, J., Weiss, C. V. and Briggs, W. R. (1977) Submitted to Plant Physiol.
10. Sargent, M. L., Briggs, W. R. and Woodward, D. O. (1966) Plant Physiol. 41, 1342.
11. Mitchell, M. B., Mitchell, H. K. and Tissieres, A. (1953) Proc. Nat. Acad. Sci. 39, 606.
12. Sargent, M. L. and Briggs, W. R. (1967) Plant Physiol. 42, 1504.
13. Scarborough, G. A. (1976) Proc. Nat. Acad. Sci. 73, 1485.
14. Briggs, W. R. (1976) In Light and Plant Development, H. Smith, Ed., Butterworths, London, in press.
15. Briggs, W. R., Freeberg, J. and Weiss, C. V. (1976). Carnegie Inst. Washington Year Book 75, in press.

SYMPOSIUM XII

LIGHT-INDUCED DEGENERATION OF SKIN:
CHRONIC ACTINIC DERMATOSIS

STUDIES ON THE PATHOMECHANICS OF CHRONIC ACTINIC

DERMATOSIS

F. SERRI,* A. TOSTI,** D. CERIMELE,* M.L. FAZZINI,**
S. VILLARDITA,**, G. COMPAGNO**

*Dept. of Dermatology, University Medical School, Pavia, It.
**Dept. of Dermatology, University Medical School, Palermo,
Italy

The structural alteration in chronic actinic dermatosis
consists of dermal as well of epidermal changes. In the
dermis a double order of changes are evident:
1) regressive metamorphosis (elastosis of the middle dermis, including disappearance of the argyrophilic reticulum and blood vessels,
2) fibrosis of the uppermost dermis with flattening of
the dermoepidermal interface,since the dermal papillae
disappear.
It may be interesting to show some aspects of the alterations which are not apparent with conventional histology: we will consider at first the behaviour of the
elastotic tissue observed by means of ultrasoft X-rays
(1). In the contact microradiographs taken at tension
ranging from 600 and 1500 Volts with an open X-ray tube
the X-rays absorption in the elastotic areas is clearly
lowered with an irregular distribution giving a spongy
appearance to the elastotic colloid. As absorption, in
the case of X-rays, can be converted in terms of concentration of dry mass, the images strongly suggest that
the regressive metamorphosis resulting in elastotic colloid is connected to a severe decrease of proteic concentration of mass. Another kind of information is given by the "absorption images" by contact photomicro –

graphy in the U.V. spectrum (2). When observed at 260 nm
the absorption by elastotic tissue is lowered and this
may reflect a variation in the chemistry, rather than
in the mass of the tissue, connected with the altera –
tion. When explored with infrared "false colour" photo-
micrography of fresh sections (3), the elastotic areas
fail to yield the prevailing infrared response of the
normal connective and record orange instead of cyan or
red, that is, the ability to absorb the visible radiations is lowered in the altered tissue. Also the typical anisotropic properties of collagen are lost when
elastosis occurs. On the contrary, birefringence is
strong in the uppermost dermis, where fibrosis is present. Severe alterations of the surface stereomorpholo-
gy of the elastotic connective are also evident in tissue
sections observed by means of scanning electronmicro –
scope (4, 5, 6).
In the chronically sun exposed skin biochemical inve –
stigations (7, 8, 9) show increase of elastin,non fibrous
proteins and hexosamines, as compared to the unexposed
skin; total collagen is slightly decreased, soluble collagen increased and insoluble collagen decreased. The
elastin of exposed skin seems to be qualitatively altered;the amino acid analysis has shown that polar amino-
acid residues (aspartic and glutamic acid,lysine and
arginine) increase and tyrosine decreases. Phenylalanine does not vary and the ratio tyrosine/phenyl-alanine
changes from 1 to 0,5. In sun exposed skin hydroxypro-
line decreases and total hexosamine increase. The in –
crease in total hexosamines in exposed skin is thought
to be due to an increase in acid glycosaminoglycans,
more particularly to hyaluronic acid than to chondroitin-
sulfate B. The separate estimation of glucosamine,which
is indicative of the hyaluronic acid content, and of galactosamine which is thought to be indicative of chon-
droitin-sulfate B, shows that in the exposed skin glucosamine increases, while galactosamine decreases, the
ratio glucosamine/galactosamine is very close to that
found in the skin of the fetus and remote from that
found in unexposed skin of old people. The amount of
hexosamine in exposed skin was also found greater than
it would have been expected from the known values of
hyaluronic acid. Therefore was thought that a fraction

of the glucosamine content of the skin could be due to
glycoproteins and that it was not correct to take glucosamine estimation as an estimation of hyaluronic acid
and total hexosamines estimation as an estimation of
acid glycosaminoglycans. Sialic acid is commonly found
in glycoproteins and its amount was used to estimate
the glycoprotein content. Thus it was observed that the
sialic acid content of the skin decreases greatly, from
the fetus to old age, but a larger quantity of sialic
acid was found in exposed than in unexposed skin (10,11);
this was interpreted as being due to an increase of glycoproteins in the exposed skin.
Let us consider the chronically sun exposed epidermis
under a morphometrical viewpoint. Measurements referred
to 10.000 μ^2 of keratinized surface were carried out
comparatively on biopsy specimens of exposed and unexposed skin. The measurements included: tissue-volume,
surface/undersurface ratio, number of cell population
in the stem and expanding compartments, cell volumes.
The method, consisting of the use of photomicrographic
maps, is reported in previous works (12, 13). The results
showed a reduction in the cell population in both the
stem and expanding compartments, as well as a decreased
tissue volume and a lowered surface/underface ratio. In
spite of this, a high rate of regenerative activity was
observed, as shown by the important colchicine effect
in epidermis overlying severe chronic actinic dermatosis (14). Also the determination of DNA content by Feul-
gen micro-spectrophotometry (15) performed on chronically photoexposed skin showed a high percent distribution
of the hyperploid and tetraploid individual DNA contents
associated to the stem compartment. This can be assumed
as evidence of an active regeneration in the tissue. Hypoplasia and high mitotic rate strongly suggest that
the dynamics of the cell-turnover differs from the one
of normal epidermis. In the latter, mitotic activity is
focal and episodic, yielding clusters of new born cells;
in the former the mitoses are diffusely spread and are
likely to occurr continuously on the basal layer.
A thorough study of the activity of the enzymes responsible of the main metabolic transformations in the human epidermis was planned with quantitative histochemi-
cal methods. This research was undertaken with two main

goals; first of all in order to know if the chronic damage following to a very prolonged sun exposure could
affect some basic metabolism processes like glycolysis,
the tricarboxylic acid cycle and pentose phosphate pathway; secondly to investigate if alterations similar to
those found in the skin of Prosimians treated with chemical carcinogens (16) or in the neoplasma of the human
skin (17) could be shown keeping in mind that sun exposed human skin can be considered a precancerous state,
due to the high frequency of epitheliomas arising on
this skin. Fourteen enzymes representative of the glycolytic pathway, the tricarboxylic acid cicle, the pentose phosphate pathway have been studied, in the epi —
dermis of the exposed and unexposed skin of two groups
of healthy human subjects, one of young people, the o-
ther of old farmers. The enzymes activities were assayed by microfluorometric methods (18) as adapted to
the skin by Hershey (19) and al . In the group of young
people the following enzymes had higher activity in the
exposed skin: phosphofructokinase, glyceraldehyde-3-
-phosphate dehydrogenase, enolase, glucose-6-phosphate
dehydrogenase, 6-phosphogluconic-dehydrogenase, aspartate transaminase. The enhancement of the activity of
glyceraldehyde-3-phosphate dehydrogenase is statistically significant, the enhancement of all other enzymes
is highly significant. In the group of old people the
epidermis of exposed skin,compared to the epidermis of
the unexposed skin, had enhanced activity of enolase,
glucose-6-phosphate dehydrogenase and 6-phosphogluconid
dehydrogenase, while the activity of alfa-glycerophosphate dehydrogenase, isocitric-dehydrogenase, malic enzyme, alanine transaminase, and beta-hydroxylacyl dehydrogenase was depressed. The diminution of the activity
of alfa-glycerophosphate dehydrogenase, isocitric dehydrogenase and malic enzyme is statistically significant;
the diminution of alanine transaminase and betahydroxy-
acyl dehydrogenase is highly significant. The comparison
between the results in exposed skin of the young and
old groups showed that the activity of almost all the
enzymes was depressed in the epidermis of the old group.
The decrease of alfa glycerophosphate dehydrogenase,
isocitric dehydrogenase, malic dehydrogenase, glucose-
6-phosphate dehydrogenase and beta-hydroxylacyldehydro-

genase was statistically significant; the decrease of
enolase, lactic dehydrogenase, alanine and aspartate
transaminase was highly significant. Our results show
that in exposed skin there is a lowering of the activity of most enzymes; this lowering affects in different
degree the different metabolic pathways. The activity
of both transaminases is highly depressed, the enzymes
of the glycolytic pathway, the Krebs cycle and the fatty
acid metabolism show a smaller depression, while the
enzymes of the pentose phosphate shunt are only slightly
effected. (This last result could be explained by the
fact that, even in old age, the epidermal cells show
high mitotic activity and therefore have a high requirement of pentose sugars which are produced through the
pentose phosphate pathway). Our results could fit with
the hypothesis that the chronic sun damage of the human
skin and neoplasma that develops on such skin, are two
related but different processes.

We have been till here concerned with structural and
biochemical changes induced by chronic over-exposition
of skin to sunlight. Let us consider now in what extent
the mechanical properties of skin are affected in chronic actinic dermatosis. At this purpose a ballistometric
investigation has been carried out. The ballistometry
is based on the "drop impact" of a body onto a stationary, solid surface. That is, a collision in one dimension is provoked by allowing a hard body to drop from a
given height onto the surface to be tested. The amount
of the energy absorbed by the material subjected to the
impact is proportional to the difference between the
height of fall of the impacting body and the height of
the rebound. Thus, in a collision between a weighing $\underline{P}$
body falling freely onto a stationary surface from the
height $\underline{H}$ and rebounding to the height $\underline{H'}$, the energy of
the impacting body is $\underline{P.H}$ while the energy returned by
the surface is $\underline{P.H'}$. The ratio:

$$e = \frac{P.H'}{P.H}$$

represents the coefficient of restitution and in percentage indicates the % rebound energy. The method is reported in previous work. For this presentation the tests

were carried out on 12 skin areas of 46 subjects ranging
from 8 to 80 years. The coefficient of restitution e of
the skin tested underwent variations related both to the
age and to the different skin areas. The highest values
of e were present in very young subjects, while a pro –
gressive, significant decrease occurred with aging. As
a rule, the decrease of the % rebound energy in the older subjects was more evident in the sun exposed skin
areas than in the unexposed ones. The lowest values occurred in severe chronic actinic dermatosis. Obvious
differences of e according to the various skin regions
were also present, some skin areas, as the dorsum of the
hand and the forehead producing the lowest values. No
significant differences were recorded in symmetrical
areas. On the basis of the results, there is evidence
of a significant progressive decrease of the % rebound
energy connected with aging in the sun-exposed skin.

REFERENCES

1) Tosti A.: Arch. Dermatol. 89, 131, 1964
2) Tosti A.,Villardita S.,Fazzini M.L.: J. Invest.Der-
   matol. 57, 166, 1971
3) Tosti A.,Fazzini M.L.,Villardita S.,Scalici R.:Ital.
   Gen.Rev.Dermatol. 9: 7, 1969
4) Tosti A.,Villardita S.: Chron. Dermat. 2: 23, 1971
5) Dawber R.,Schuster S.: Brit. J. Derm. 84, 130, 1971
6) Tosti A.,Villardita S.,Fazzini M.L.:Proc. XIV Int.
   Congr. Dermat. Padua-Venice, May,1972
7) Cerimele D.,Pisanu G.: Boll. S.I.B.S. XLVI,107,1970
8) Cerimele D.,Serri F.: Research Progress.Organic Biological Medicinal Chemistry, vol.3,part II,623,1972
9) Cerimele D.,Yamosama S.,Serri F.:Brit.J.Derm. 87, 149
   1972
10) Giannetti A.,Rabbiosi G.: Proc.XIII Int.Congr. Dermat.
    1050, 1969
11) Rabbiosi G.,Giannetti A.: Min.Dermat. 44, 387, 1969
12) Tosti A.,Scerrato R.,Fazzini M.L.:Ann.It.Derm. 14,
    185, 1959
13) Fazzini M.L.: Ann.It.Derm.Clin.Sperim.16,235,1962
14) Tosti A.: Proc.Symp. Farmitalia, 41, 1967
15) Tosti A.,Fazzini M.L.,Aricò M.:Ann.It.Derm.Clin.Sperim.
    19, 125, 1965

16) Adachi K.,Yamasawa S.,Montagna W.: J.Natl.Cancer
    Inst. 42, 61, 1969
17) Cerimele D.,Gerna Torsellini M.,Serri F.:Pigment Cell;
    Karger, 1976 (in press)
18) Lowry O.H.: J. Histochem. Cytochem. 1, 420, 1953
19) Hershey F.B.,Lewis C.Jr.,Murphy J.,Schiff T.: J. Hi-
    stochem.Cytochem. 8, 41, 1960
20) Tosti A.,Villardita S., Fazzini M.L.,Compagno G.:Proc.
    Symp. on measurements of Physical properties of skin;
    Miami, March 1976, J. Invest.Dermatol.,in press.

CLINICAL PATTERN OF THE ACTINIC CHRONIC DERMATOSIS

A. PUISSANT

Gabrielle NOURY-DUPERRAT; Liliane DIDIERJEAN

Clinique Dermatologique (Pr.B.Duperrat), Hôpital
Saint-Louis, 2 Place du Dr Fournier - 75475 PARIS 10

Actinic chronic dermatosis include skin diseases the
common feature of which is that they are circumscribed on skin
areas chronicaly exposed to sunlight.

The most frequent are lesions of the elastic dermal
tissue; they are called actinic elastosis or solar elastosis. Very
frequent from the age of 40 among farmers and seamen, they can
be observed more and more often and earlier and earlier on fair-
skin subjects because of a pronounced taste among our contempo-
raries for a free life in the sun.

We shall describe in this chapter the following clinical
pattern of actinic elastosis : 1/ " cutis rhomboïdalis nuchae;
2/ " citrin skin "; 3/ " diffuse elastoma " ; 4/ " nodular elastoïdosis with cysts and comedones " ; 5/ " cutis punctata linearis
colli ", or " erythrosis interfollicularis colli ".

" Cutis rhomboïdalis nuchae " was described by
Jadassohn in 1925; in France it is called "peau losangique" . The
skin of the nape is thicker, yellowish or yellow-brown and sometimes erythematous; it is rough and consists in deep lines for -
ming a rhomboïdal network of an irregular pattern; it is often
dotted with a multitude of comedones of various sizes. This pattern often extends, though less markedly, to the sides of the neck,
to the forehead and sometimes to the upper part of the chest.

" Citrin skin " was described by Milian in 1921. It consists in areas of varying sizes where the skin is white or yellow,
wrinkled, finely cross-lined as a lemon skin. In these areas, the
teguments are very loose with sometimes dilatation of the follicular
pores and presence of comedones.

" Diffuse elastoma " described by Dubreuilh in 1892
consists, on the face and the neck, in marked areas where the
skin is yellowish, old ivory with an uneven surface, looking
thick, fattened. When located on the sides of the neck, this elastoma is not to be mistaken for " pseudoxanthoma elasticum ", which
is a genetic disorder, affecting not only exposed areas but also
retina and media and intima of the blood vessels. Elastomas of
limited extend are sometimes found on sun exposed skin areas of
young subjects : often tumoral-like and erythematous, they can
be mistaken, before a pathologic examination, for lymphocytoma,
discoïd tumidus lupus, erythematosus, sarcoïdosis, and chiefly
basal-cell epithelioma.

" Nodular elasteïdosis with cysts and comedones ",
described by Favre in 1931, was the subject of his pupil Racouchot's thesis in 1937. It is almost always circumscribed on the
face, around the eyes, on the forehead, the temples and the
cheeks. It associated a very dense pattern of comedones and nodules; the comedones are of varying sizes, deeply anchored in
the skin, and hard to extract. The nodules are more or less pro-
tuding, whitish or pink, or translucid, with or without comedones.
This state is associated with an actinic skin elastosis. Nodular
elastoïdosis is seldom observed on the nape or on the posterior
face of the ears or in another area periodically exposed to sunlight.

We think that the " cutis punctata linaeris colli " described by Evan-Paz and Jaquer in 1963 is the equivalent of
"Miescher's erythrosis interfollicularis colli". Very common, it
starts appearing from the age of 35, on the sides of the neck. The
area of the neck shadowed by the chin is not affected. Lesions
consist in numberless small skin-coloured or more often yellowish
follicular papules, needle head size; they occur in multiple sinuous
lines pattern; between the papules, the skin is more or less inten-
sely red; it keeps its smoothness; there is no functionnal signs .
P. Temine (1) has shown the importance of dermal elastic fibers
alteration; the fibers are either scarces, or thickened, or broken
up around hypertrophic sebaceous glands; the acuteness of these

alterations can make the diagnosis difficult with " pseudo-xanthoma elasticum " .

　　　　　All the above lesions are the consequence of an increasing elastic tissue, much more, important in areas exposed to
sunlight, the alterations of which results in a more or less com -
plete degenerescence and a loss of skin elasticity; the initial lesion
would be at the level of fibroblasts, the irradiation of which de -
viate synthesis of collagene towards that of elastine. A. Klingman
(2) studied the relationship between actinic elastosis and pigmentation; the observation of an area of chronic vitiligo on a 40 years old
black patient nape enhances the importance of the protection given
by the pigmentation ; the unpigmented area is "rhomboïdalis " as
the normally pigmented area shows no sign of actinic elastosis;
the early occurence of actinic elastosis on albino subjects con -
firms that fact.

　　　　　" Colloïd milium " , also called " pseudo-colloïd milium " , starts between the age of 15 and 50. The eruption develops over a few years and then stabilisies. It is symetrical and
circumscribed on exposed areas, mainly the upper third of the
face, around the orbits, the back of the hand, the nape, the sides
of the neck, the pavilion of the ears; it consists in micro-papules,
1 to 2 mm. in diameter, yellow, often translucent, in more or
less coalescent croups; there is no functionnal sign; when the papule is cut open, a gelatinous matter appears. K. Hashimoto and
his collaborators have recently shown (3) that this colloïdal matter
consists in structural glycoproteïnes synthetised in too large quantities by the dermic fibroblats deteriorated by sunlight.

　　　　　The role that sunlight plays in the constitution of "lupus
erythematosus" lesions is amply demonstrated by clinical observation and by the fact that clinical and histological lesion of this
disease is produced by repeated U.V. irradiation on clinically
uninvolved patients skin (4). According to M. Landry and W. M.
Sam (5), and to J. Thivolet and coll. (6), an antibody-antigene
complex would be located within the basal membrane zone where
it would absorb a certain quantity of U.V. rays; this would result
in lesion of surrounding basal epidermal cells and of basal lamina
which becomes antigenic. In chronical discoïd "lupus erythema -
tosus (L. E. C.), besides histological signs which are typical of this
disease, it often exists an important dermal elastosis, which does
not happen in systemic " lupus erythematosus " .

In acne rosacea, actinic elastosis in common; association with non specific inflammatory infiltrates sometimes renders difficult histological diagnosis between L.E.C. and acne rosacea. To our opinion, this actinic elastosis is one of the main factors entertaining acne rosacea.

Hyperpigmented i. e. freckles and hypopigmented macules can be observed, on the skin of rather young subjects, when chronically exposed to sunlight and even more if the sub - jects are red headed with a fair complexion. They can be observed from age of 25 to 30 years. Hypopigmented macules could be similar to " idiopathic guttate hypomelanisis ", the initial lesion of which consists in an alteration of the stratum corneum which is unfolded like a basket net (7).

Colomb's spontaneous starries like-scars always appear on aging skin, atrophy of which is sometimes increased by corticotherapy. They are almost only circumscribed on the back of the hand and the forearm. Sunlight, on top of the aging, plays at least an worsening role. This can be said about Bateman's senile purpura (8). All those lesions are often associated with other solar degenerative changes.

" Hydroa vacciniforme " described by Bazin in 1860 appears during the first months or years of life; it suddenly starts with first spring sun; afterwards, it appears again every year. It is circumscribed on the nose, cheeks, ears, back of the fingers and the hands, and sometimes on other exposed areas. It is characterized by the development of painful umbilicated bullae, 5 to 10 mm in diameter. They quickly become hemorragic, developp into a necrosis, and leave permanent scars. The beginning of "Möller's hydroa aestivalis " is similar but it does not leave any scars. Erythropoïetic protoporphyria and trytophan metabolism defect must be excluded. Etiology of hydroa vacciniforme is unknown.

Actinic granuloma was described by T. Allen in 1966 under the name of "annular granuloma-like reaction to sunlight". In 1975, J.P. O'Brien (9) studied 18 cases of this disease. It can be observed mainly on fair-skinned subjects. It is located on the face, and rather often appears during the week following a strong sunburn. It consists in one or more eruptive elements; each of which in turn consists of papules distributed in an annular configuration sometimes incomplete; the annuli

diameter is 0, 5 to 2 mm; the papule are skin-coloured or lighty
erythematous or amber. As the dimension of the annulus increase
very slowly, the center of the lesion heals with sometimes a light
atrophy. There is no functionnal sign. Each element develops over
several months or years. This macrophagic granuloma seems to
tend towards resorption of actinic elastosis; it sould therefore
be an peculiar repair of sunlight affected derma.

        " Actinic porokeratosis " was described by
M.E. Chernosky in 1966 (10). It appears only on caucasian
adults and in circumscribed sun exposed areas. The eruptive
elements are multiple; their shape is geographical with strongly
defined edges; they are of variable sizes, from 5 to 10 mm in
diameter, lightly depressed, skin-coloured or pinky-brown; their
cornified wireloop edge is slighty darker and can be more easily
felt than seen, which allows to make the diagnostis. The lesions
are increased by sunlight. It is a peculiar anomaly of keratinisa-
tion; it is inherited as an autosomal dominance, but it appears
lately under the action of sunlight.

   In " acrokerato-elastoïdosis " described by O.G. Costa in
1954 (11), E. Kocsard (12) incriminated, at least in the cases he
has observed in Australia, a combined action of occupational
micro-traumas and solar rays.

        " Actinic reticuloïd " was described by F.A.
Ive, I.E. Magnus and collaborator in 1969 (13). It consists in a
chronic erythemato-papular eruption, beginning on exposed areas.
It is worsened by each exposure to sunlight but improves when
the patient keeps away from it. It corresponds to very large dermal infiltrates which are similar to the lymphoma; light-tests on
clinically uninvilved patients skin reproduces these infiltrates.
The unending and very painful evolution always appears to remain
benign.

                    S U M M A R Y

        Actinic chronic dermatosis implies a group of
changes in the skin of the exposed areas :

1/ dermal elastosis as wrinkles, " cutis rhomboïdales nuchae"
(Jadassohn), " peau citrinee " (Milian), " nodular skin elastoï-

-dosis with cysts and comedones " (Favre et Racouchot), " elastoma diffusum senilis " (Dubreuilh), " erythrosis interfollicularis
colli " (Leder); those clinical patterns, in some patients have to
be differentiated from " pseudo-xanthoma elasticum " (Darier)
and from "pseudo-milium colloïdale "; we have also to debate
about dermal elastosis in "lupus erythematosus chronicus " and
in " acne rosacea ";

2/ hypermelanotic et hypomelanotic macules as freckles and
"hypomelanose en goutte idiopathique " (Bazex et Dupre), and
perhaps as " pseudo-cicatrices stellaires " (Colomb);

3/ telangiectasia and purpura senilis (Bateman);

4/ light is also implicated in chronic dermatosis as "hydroa vacciniforme " (Bazin) and " hydroa aestivale " , " actinic granuloma ", " disseminated superficial actinic porokeratosis ", and
"actinic reticuloïd ";

        We do not study here " keratosis senilis ",
" xeroderma pigmentosum ", "porphyria ", which overlap
other reports.

REFERENCES

1. TEMINE P., COULIER L., XIIè Congrès de l'
Association des Dermatologistes et Syphiligra-
phes de Langue Française - Paris, 1965 -
Edit. Masson et Cie, Paris, 409, (1960)
2. KLIGMAN A.M., Fitzpatrick, University of Tokyo Press,
157, (1974)
3. HASHIMATO K., KATZMAN R.L., KANG A.H., KANZAKI T.
Arch. Dermatol., 111, 49, (1975)
4. EPSTEIN J.H., TUFFANELLI D.L., DUBOIS E.L., Arch.
Dermatol., 91, 483, (1965)
5. LANDRY M., SAMS W.M. Jr., J. Clin. Invest., 52, 1871,
(1973)
6. THIVOLET J., BEYVIN A.J., MOT J.L., Lyon Med. 227,
241, (1972)
7. BAZEX A., DUPRE A., CHRISTOL B., Rev. Med. Toulouse,
4, 557, (1968)
8. COLOMB D., PINCON J.A., MARTAUD J., Ann. Derm.

Syph., 94, 3, 273, (1967)
9. O'BRIEN J.P., Arch. Dermatol., 111, 460, (1975)
10. CHERNOSKY M.E., FREEMAN R.G., Arch. Derm., 96, 611, (1967)
11. COSTA A.G., Ann. Derm. Syph., (Paris), 83, 146, (1956)
12. KOCSARD E., Dermatologica (Bâle), 131, 169, (1965)
13. IVE F.A., MAGNUS I.A., WARIN R.P., WILSON JONES E., Brit. J. Derm., 81, 469, (1969)

DISTRIBUTION AND GENERAL FACTORS CAUSING

CHRONIC ACTINIC DERMATOSIS

J.D. EVERALL

Skin Department

Royal Marsden Hospital, London SW3

INTRODUCTION

Solar irradiation is essential for life as we know it on this
planet, but it is also harmful and potentially lethal to living
matter. Since the skin organ, the outer covering of the body
is the front line of defence in man's ever changing environment,
it is not surprising that solar radiation has a role in the
production and exacerbation of many conditions and diseases of
the skin, as illustrated in Figure 1.

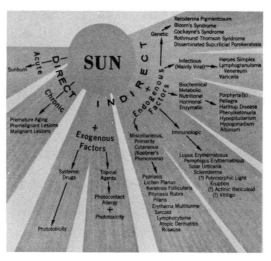

Figure 1. Sunlight related disorders (Willis JAMA Aug. 23 1971)

One could spend many hours discussing the effect of the sun on
human skin, however this paper will be confined to a discussion
of the distribution and general factors involved in the production
of chronic actinic dermatosis, by which we mean those changes
produced in Caucasian skin by repeated exposures to doses of
ultraviolet radiation, which are insufficient of themselves to
produce the acute reaction, which we refer to as sunburn.

CLINICAL DESCRIPTION

The changes of chronic actinic dermatosis occur on the parts
directly exposed to ultraviolet irradiation, i.e. face and
forehead, neck and 'V' area of chest, the backs of the hands and
the extensor surface of the forearms. There is diffuse atrophy,
xeroderma and decreased elasticity of the skin of the face and
forehead, and in the later stages patchy yellow plaques (often
referred to as Dubreuilh's elastoma, particularly in the
circumoral and circumorbicular areas, or occurring as a single
plaque on the nose.

In extensively actinically damaged skin of the periorbicular
region especially of elderly males, keratin cysts and giant
comedones may occur, these appearances are referred to as
nodular elastoidosis or the Favre-Racouchot syndrome. Occasionally
atrophy, dyskeratotic desquamation and telangiectasiae may occur
on the lower lip.

On the sides of the neck and often the post-auricular sulci,
actinic damage results in the appearance of perifollicular
yellowish papules with surrounding atrophy bounded by telangiectatic dermal vessels in a reticular pattern. These changes I
refer to as perifollicular elastosis and reticular atrophy. This
pattern of actinic damage may also occur in the skin of the 'V'
of the neck and chest.

The changes in the exposed skin of the back of the neck are of a
more pronounced pattern. The skin becomes diffusely thickened
and leathery with furrowing and exaggeration of the normal skin
marks. These changes are commonly referred to as cutis
rhomboidalis nuchae, but I prefer the term furrowed elastosis.

Actinic damage of the backs of the hands, extensor surface of
the forearms and arms, results in a diffuse atrophy, xeroderma
and inelasticity. Rarely are there the yellowish plaques seen
so characteristically on the face.

Guttate, scarred, depigmented lesions are frequently seen on
the forearms, hands and face.

The above changes are accompanied by a diffuse yellowish-brown pigmentation at all sites.

Increasing damage results in the appearance of irregularly shaped solar lentigines on the face, forehead 'V' of the neck and chest, and the affected areas of the forearms and hands. Solar lentigines on the face and less commonly on the arms and hands, may undergo malignant change with the development of a lentigo maligna (Hutchinson's melanotic freckle).

In contradistinction to the reticular pattern of telangiectatic vessels occurring on the sides of the neck, the vascular changes in the remaining actinically damaged sites, take the form of patchy ill-defined telangiectasiae, frequently over wide areas, and often being most prominent on the trauma-prone sites of the hands and forearms. The latter are the sites of the purpura seen in the advanced stage of actinic damage.

A variety of skin tumours is often seen in actinically damaged skin. Those individuals prone to develop basal cell papillomas of the trunk, frequently develop flat and papillomatous basal cell papillomas (seborrhoeic keratoses) of the exposed parts of the face, forehead, sides and front of the neck and 'V' of the chest. These individuals tend also to have plane warts of the backs of the hands and squamous papillomas of the forearms (often referred to as acanthomas).

Eventually premalignant lesions in the form of actinic keratoses occur at all sites - except in my experience the nape of the neck, and are most frequently seen on the backs of the hands, the forehead and helix of the ears.

Actinically damaged skin is more prone to the development of basal cell and squamous cell carcinomas.

HISTOLOGY

Histologically the major changes appear in the dermis rather than the epidermis and skin appendages. Freeman 1962 (1) showed a marginal thinning of the epidermis resulting from a flattening of the rete ridges, while the suprapapillary minimal thickness remains constant, and the stratum corneum becomes a little thicker than skin of a comparable age with little actinic damage. My own work (2) estimating the thickness of the epidermis by weight, suggests that when correction for shrinkage is made, there is in fact a slight thickening of the whole epidermis in severely damaged skin. Many abnormal cells are present in the epidermis and the melanocytes show great variation in size, distribution and tyrosinase content.

There is a progressive degeneration of the papillary and
subpapillary zone of the dermis, vascular ectasia develop, acid
mucopolysaccharides and abnormal appearing fibrocytes accumulate,
there is loss of total collagen but an increase in the soluble
fraction, and a marked increase of elastic fibres, which in
advanced damage become thickened, curled and more branched with
eventual replacement of the dermis, and disorganisation of the
connective tissue into amorphous masses.

Undoubtedly there are changes in the dermal lymphatics, and these
changes may account for the late metastasis of squamous cell
carcinoma arising from actinically damaged skin, except at the
mucocutaneous junction. However I know of no good study which
has described the changes in the lymphatics. Chemotactic
mechanisms are also probably defective, and we know from studies
of purpuric areas that there is faulty macrophage function.

CAUSES

Epidemiological surveys in Australia, America and Ireland, and
work on the induction of elastosis in experimental animals, all
indicate that the most important factor in the development of
chronic actinic dermatoses is undoubtedly repeated exposure to
solar irradiation, within the wavelengths 290-320 nanometers, i.e.
the sunburn spectrum. Johnson et al (3) suggested that the
connective tissue change, was the result of photochemically
indued alterations in fibroblast function by these rays, rather
than degradation of connective tissue elements. In support of
this, Epstein et al (4) reported direct injury of human dermal
connective tissue cellular DNA in vivo, within a few minutes after
irradiation with wavelengths shorter than 320 nanometers. Mackie
and McGovern (5) have suggested that the longer wavelengths
(i.e. 320-400 nm of U.V-A) have also a significant influence and
intensify the damaged produced by U.V-B.

The sunburning effectiveness of the short wave fraction varies
with latitude, season and climate, as illustrated in Figure 2.
This short wave end of the sunlight spectrum, also varies in
intensity and spectral distribution with movements of the skin, and
the state of the sky during any given day. The dominant factor
determining the amount of sunburning ultraviolet which penetrates
the atmosphere, is the angular altitude of the sun above the
horizon. This governs the path length of direct sunlight. For a
given solar altitude on a clear day, the ultraviolet is generally
within 10% of the mean value. The sunburning effectiveness falls
to half the strength from overhead sun when the solar altitude is
about 50° (+5), and one quarter about 35°. Exposure during the
hours of 10 a.m. and 3 p.m. is therefore the most harmful, and
exposure before 9 a.m. or after 3 p.m. the least harmful. Figure 3.

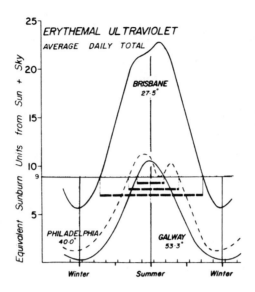

Figure 2. Variation of erythemal ultraviolet light with latitude, season and climate (Robertson, private communication).

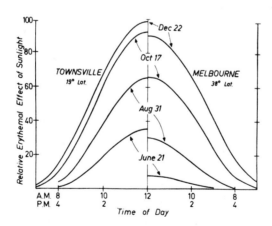

Figure 3. Erythemal effect of sunlight at Townsville and Melbourne at various times during the day and during each season of the year (Robertson, private communication).

A considerable amount of sunburning ultraviolet comes in all
directions from the blue sky. The relative amount in sunlight
and skylight varies with the solar altitude, and the direction of
view. When the sun is fairly high, skylight is roughly equivalent
in effectiveness to sunlight. Light clouds dotted about the sky
scatter as much U.V-B as the blue sky which they hide, and thus
neither increase or decrease the sunburn fraction. However a
cloud over the sun decreases the direct sunlight considerably, and
may increase skylight slightly. A fine weather cloudy sky may
still maintain 70-80% of the clear sky U.V-B, heavier cloud reduces
it to about 50%, while rain clouds may only leave 20% or less.
Sunburn during cloudy conditions invariably follows prolonged
exposure, while absorption of infra-red by the clouds leaves one
unaware of the intensity.

Because of the skylight contribution, it is often only of limited
value to shelter from direct sunlight by keeping in the shadows,
however in the shade of a beach umbrella sunburning rays are
about half as strong as in open sun, and the shade of a building,
tree or brim of a hat, may cut down the sunburning effectiveness
to as little as 25% of the original value.

Normally the ground around us scatters back only a few per cent
of the U.V-B which falls on it. Fresh snow, however, effectively
replaces the ground by another sky. Dry sand may rise to being
one quarter as effective as the lower sky. Water reflects no
more than 5% (except at very low angles of incidence), and
transmits well to give 50% at 50 cm. depth. Swimming in a pool
therefore provides little reduction in exposure.

Increase in altitude adds 3 to 4% to the sunburning strength of
open daylight for each 1000 feet. Direct sunlight increases more
rapidly and skylight less rapidly.

Dust and the ozone content of the upper atmosphere help to reduce
the fraction of harmful ultraviolet rays reaching the ground,
these factors have however less effect on the intensity of sunburn
radiation than the solar altitude. Measurements at Philadelphia
show that a persistent industrial atmosphere can reduce ultraviolet penetration significantly, the effect being greatest when
the humidity is high.

There is good epidemiological evidence that regular continuity
of exposure, gives the greatest acceleration to actinic change
in persons with tolerant skin, which is exposed to 1, 2 or more
hours daily for 4 to 5 days a week. There is little experimental
evidence that long periods at very low intensity may permit some
recovery. Recovery of skin is slow, so that the effect may not
be significant, unless there are a few months with negligible
exposure. The fraction of the year in which regular exposure

occurs is therefore important in determining actinic damage.

An individual's occupation will obviously influence the degree and frequency of daily exposure, and is in many cases the over-riding factor determining frequency of exposure and hence actinic damage. Indeed the role of occupation in actinic damage has been long recognised, hence the names Sailor's skin and Farmer's neck, previously applied to actinically damaged skin.

GENETIC

Chronic actinic dermatosis appears mainly in skins which are sufficiently sheltered by pigment, to be able to accept longer hours of exposure, while there is no uncomfortable deterrent such as sunburn. Very fair skinned non-pigmenters tend to burn so easily, that they learn to limit their exposure at an early stage. Unfortunately in pigmenters, loss of collagen and disorderly increase in elastic fibres in the dermis progresses significantly before any change in superficial appearance is recognised, somewhat too late.

In chronic actinic damage, unfortunately there is no magic threshold as with sunburn, separating the acceptable from the unacceptable. All exposed skin ages more rapidly and more severely than unexposed skin, and owing to the popular cult of sunbathing and the scantier clothes of the 20th century, actinic dermatosis will no doubt in the future be seen with increasing frequency over a wider area of the skin organ.

PREVENTION

Clearly from the above facts, it can be seen that the most effective simple measures to prevent the development of chronic actinic dermatosis, with its attendant danger of skin cancer, must be a reduction in the time and frequency of exposure, and for those individuals whose occupation necessitates exposure, a re-organisation as far as possible of the time of day of exposure, and the seeking of as much shade and protective cover as possible.

REFERENCES

1. Freeman, R.S., Cockerell, E.G., Armstrong, J. and Knox S.M. 1962. Sunlight is a factor influencing the thickness of the epidermis. J.I.D. 39, 295.

2. Whitton, J.T. and Everall, J.D. 1958. The thickness of the epidermis. Br. J. Dermatol. 1973, 89, 467.

3. Johnson, B.E., Daniels, F., Jr. and Magues, I.A., Response
    of human skin to ultraviolet light in photophysiology.
    Vol. IV (Ed. A.C. Giese) (New York Press, 1968) pp 139-202.

4. Epstein, W.L., Fukuyana, K. and Epstein J.H.   Early effects
    of ultraviolet light on DNA synthesis in human skin in vivo.
    Arch. Dermatol. 100, 84-89, 1969.

5. Mackie and McGovern.   The mechanism of solar carcinogenesis
    a study of the role of collagen degeneration of the dermis
    in the production of skin cancer.   A.M.A. Arch. Dermat. 78.
    218-244.

SYMPOSIUM XIII

ENVIRONMENT-SPACE INTERACTIONS:
PHOTOBIOLOGICAL IMPLICATIONS

Prof. H.F. BLUM

Skin and Cancer Hospital
Temple University, Health Sciences Centre
Philadelphia, U.S.A.

It is 35 years since Hugh Grady, John Kirby-Smith and I carried out at the National Cancer Institute a long series of experiments on induction of cancers in mice with repeated doses of
ultraviolet radiation. The experiments showed that the effect of
repeated doses accumulate, this being expressed in terms of the
square of the number of doses. The results are summarized in
slide 1, where the data are plotted on probability paper. At B,
incidence of cancers is plotted against the logarithm of the
square of number of doses. The points fall along straight lines,
showing that the effects accumulate.

Cancers grow by increase in size of the cell population and
by increase of rate of cell duplication. These two things cannot
be measured simultaneously, so we face uncertainty. By assigning
equal weight to each of these factors the data can be brought together as in Figure 1 at A, where the points fall along the same
line. This probabilistic dodge permits analysis of the data
quantitatively, but does not resolve the uncertainty. We are
faced with indeterminacy, which may be compared with Heisenbergs
indeterminacy regarding the position and direction of an electron
within an atom.

Within the last few years epidemeological studies, and
measure of the carcinogenic radiation in sunlight have shown that
the effects of repeated exposures to sunlight accumulate, this is

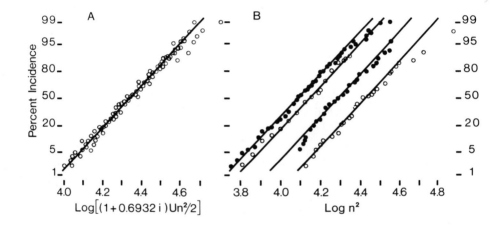

Figure 1.

shown in Figure 2 where data for four regions of the United States are plotted in the same way as the data for mice. The points fall along parallel straight lines. But the lines do not follow the expected order with regard to latitude, measured sunlight or sex, and cannot be brought to the same line as in figure 1A. We again face uncertainty, which is better shown by plotting as in Figure 3, where the stippled area indicates uncertainty of current predictions about sunlight and skin cancer. More measurements covering a wider range of latitude may reduce the uncertainty, but cannot eliminate it from our predictions.

Various factors : genetics, human behaviour, kind of skin, variation of intensity of sunlight, etc., could all affect the position of the curves in slide 2; but cannot be evaluated quantitatively because of uncertainty.

These studies on skin cancer cannot be directly applied to other aspects of the sunlight factor in ecology. It would seem that most effects of sunlight on plants and animals are not cumulative. But in others they may be, as for example in plankton, where Caldwell has shown that sunlight may kill organisms, and also decrease rate of reproduction.

Possibility of inherent indeterminacy should be kept in mind in the study of all effects of sunlight.

Reference

The argument presented here appears in more extended form in my paper in the next issue of "Photochemistry and Photobiology", where figures and further references are given.

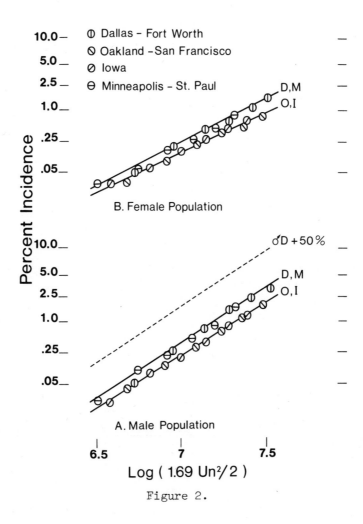

Figure 2.

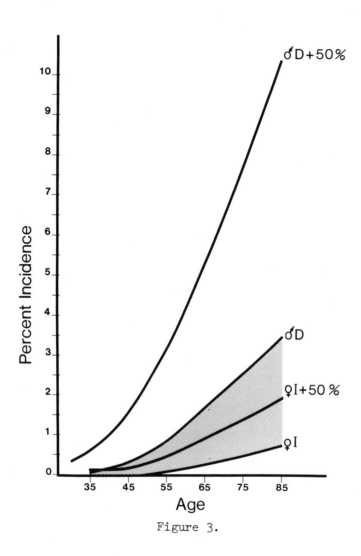

Figure 3.

The Stratospheric Photochemistry of Chlorine Compounds and
Its Influence on the Ozone Layer

F. S. Rowland

Department of Chemistry, University of California

Irvine, California 92717 USA

Introduction

Most gaseous chemical species released at the surface of the earth are removed from the atmosphere rather rapidly, and consequently are not important as stratospheric pollutants. Among the effective tropospheric removal processes or sinks are rainout and washout, especially for water-soluble species; biological interactions; and decomposition by absorption of visible (400-700 nm) or near ultraviolet (300-400 nm) solar radiation. Chemical reaction, either with surface materials or with common tropospheric gaseous species (for example OH radicals), can also be very effective in other cases. However, whenever a molecular species is reasonably inert toward these tropospheric interactions, the molecules can survive long enough to penetrate into the stratosphere and become potential stratospheric pollutants. As such molecules rise higher and higher into the stratosphere, they become exposed to shorter and shorter wavelengths of ultraviolet light, and eventually, at high enough altitudes, all diatomic or polyatomic molecules can be photochemically decomposed by intense solar ultraviolet radiation.

Most of the molecules which often penetrate into the stratosphere from below are rather inert chemically, or they would already have been removed by tropospheric processes. Consequently, solar ultraviolet photochemistry often plays the dominant role in their stratospheric removal. Sometimes the photochemical reaction itself is the removal process, as in (1) for CCl_2F_2.[1, 2]

$$CCl_2F_2 + h\nu\,(<220\ nm) \longrightarrow CClF_2 + Cl \qquad (1)$$

In others, as with CH_4 and N_2O, some important processes involve attack on the molecule by reactive species of photochemical origin, as in (2) and (3). The OH radical of (2) can be formed by photolysis

$$CH_4 + OH \longrightarrow H_2O + CH_3 \qquad (2)$$

$$N_2O + O(^1D) \rightarrow 2NO \qquad (3a)$$
$$\;\rightarrow N_2 + O_2$$

of H_2O, while the excited singlet O atom in (3) is usually formed by the photolysis of ozone, as in (4).

$$O_3 + h\nu \xrightarrow{\lambda<310\ nm} O_2 + O(^1D) \qquad (4)$$

Another class of potential stratospheric pollutants also exists: the effluents from man's stratospheric technologies-aircraft, rockets, etc. Since these molecules are deposited directly into the stratosphere, they can be either chemically reactive (e. g. the free radicals NO and NO_2 from supersonic transport aircraft) or relatively inactive (e. g. HCl from the proposed space shuttle). Although some of these molecular species may be quite reactive chemically, photochemical processes often play important roles here as well.

Stratospheric Ultraviolet Radiation

The solar radiation striking the top of the earth's atmosphere has its peak intensity in the visible wavelength near 500 nm, but also contains much more energetic wavelengths ranging down to 120 nm and below. However, the radiation with <300 nm is removed by interactions high in the earth's atmosphere. Most of the radiation below 242 nm is removed by dissociative absorption in O_2, as in (5), with the formation of O atoms. These O atoms normally

$$O_2 + h\nu\,(\lambda<242\ nm) \longrightarrow O + O \qquad (5)$$

react with additional O_2 to form O_3, ozone, in the three-body reaction (6), and ozone is capable of absorbing with dissociation

$$O + O_2 + M \longrightarrow O_3 + M \qquad (6)$$

radiation with wavelengths as long as 1100 nm. The ozone absorption is sufficiently strong in the ultraviolet that radiation with $\lambda < 290$ nm does not penetrate to the surface of the earth. The absorption of radiation by ozone in (7) is followed by the combination reaction (6), which converts the extra energy into heat for

$$O_3 + h\nu \longrightarrow O + O_2 \tag{7}$$

the third body, M, and is the primary source of heat for the stratosphere. Basically, since the stratosphere is defined as an atmospheric region of rising temperature (approximately from 210 K at 10 km to 275 K at 50 km), reactions (6) and (7) determine its actual location, and the temperature gradient as well. [3-5]

Molecules which can absorb radiation with > 290 nm normally have rather short atmospheric lifetimes and are removed in a few days or less by such photodecomposition processes. Molecules which are transparent from 290-800 nm, however, are not subject to direct photodecomposition until they have risen in the atmosphere above most of the O_2 and O_3 which remove u.v. radiation by reactions (5) and (7). As such molecules rise upward, the next important wavelength band is between 190-230 nm, radiation which penetrates more deeply into the stratosphere than that on either side of this band. This radiation becomes appreciable above 20 km, reaching high intensity by 35 km altitude, i.e. in the mid-stratosphere. Otherwise inert polyatomic molecules will usually be photolyzed between 20-35 km by this 190-230 nm radiation, while a very few molecular types which are still transparent for $\lambda > 190$ nm must rise even higher until dissociated by the Lyman α radiation at 121 nm or other very short wavelenth radiation. The molecular N_2 of the atmosphere is actually only decomposed at altitudes well above 100 km, while CF_4 is transparent even to Lyman-α radiation.

Important Photochemistry of the Natural Stratosphere

Ozone is a very reactive chemical and its formation by reactions (5) to (7) is balanced by several destruction mechanisms, including the direct reaction with O as shown in (8). This reaction

$$O + O_3 \longrightarrow O_2 + O_2 \tag{8}$$

not only removes one molecule of ozone, but it also removes an O atom which would otherwise have formed O_3 by reaction (6).

The net reaction of (8) is thus the removal of two ozone-equivalents. Catalytic reaction sequences also exist which have the same result as (8), of which the most important in the natural stratosphere is the NO_x chain carried by NO and NO_2, as in (9) and (10). [2-5] The NO_x which causes this chain does not

$$NO + O_3 \longrightarrow NO_2 + O_2 \qquad (9)$$

$$NO_2 + O \longrightarrow NO + O_2 \qquad (10)$$

$$\text{sum:} \quad O + O_3 \longrightarrow O_2 + O_2$$

enter the stratosphere as NO or NO_2, both of which are chemically reactive and are easily removed by tropospheric reactions. Instead, the prime source of natural stratospheric NO_x is the decomposition in the stratosphere of the much more inert molecule, N_2O, by the reaction shown in equation (3a). Most N_2O which reaches the stratosphere is decomposed by direct photolysis, equation (11) and additional N_2O is converted to N_2 and not

$$N_2O + h\nu \longrightarrow N_2 + O \qquad (11)$$

NO, as in (3b). Only about 5% of the N_2O decomposed in the stratosphere is converted into NO, with about 95% converted to the chemically inert N_2.

Other catalytic chains based on the HO_x radicals (H, OH, HO_2) also exist in the natural stratosphere, and an estimate of the relative importance of the various contributions is given in Table 1. [5] In such calculations the apparent very large rate of destruction of O_3 by (7) is not included because the usual fate of the O atom released in (7) is the reformation of O_3 by (6) with no net permanent loss of O_3. For the purposes of Table 1 the only reactions of ozone which are counted are those that lead to a net change in total "odd oxygen," i.e. O_3 or O atoms. Reactions (6) and (7) do not involve any net change in odd oxygen, but merely the change from one form to the other.

Effects of Stratospheric Pollutants

Concern about the possible contamination of the stratosphere by man's activities is centered upon materials whose introduction there can appreciably alter the overall stratospheric

TABLE 1

Global Rates of Formation and Removal of Ozone
in the Natural Stratosphere
(Units in 10^{29} molecules O_3/second for entire earth)

[Reference 5]

Formation:		Removal:		Percent
Photolysis of O_2	500	NO_x Catalysis	250-350	50-70
		HO_x Catalysis	60	12
		Direct $O + O_3$	90	18
		Other Catalysis	0-100	0-20
		Diffusion to troposphere	6	

behavior. [2-6]   Generally, the possible influences of man
upon the stratosphere which are under prime consideration are:
(a) addition of reactive chemicals which are able to catalyze ozone
removal by chain-catalytic processes such as the NO_x chain; (b)
addition of materials which either are particulate already, or
which can react in the stratosphere to form particles.

The natural aerosol layer in the stratosphere around 20 km
altitude is composed largely of particles of sulfuric acid, and one
of its properties is that it scatters incoming solar radiation.   The
chief source of conern here from man's activities arises from the
possibility that this scattering process might be altered if additional
particulate matter were added, with consequences for the radiation
balance of the earth, and in turn possible climatic effects. [3, 4]

Since the rates of formation and removal of ozone shown in
Table 1 are very large (about 4, 000 metric tons per second, or
3300 megatons per year), the primary concern is felt for chemicals
with the possibility of long catalytic chains in the stratosphere.
These chains invariably involve free radical catalysts, and in
principle any potential source of stratospheric free radicals is
an area of concern.   The two potential consequences of an alteration
in the ozone balance of the stratosphere which have been most considered are (a) the increased penetration ot the surface of ultraviolet radiation with diminished quantitites of ozone overhead;
and (b) alterations of the stratospheric temperature structure
through changes in the amount or even just the distribution with

altitude of stratospheric ozone. [1-6] The natural variations in ozone concentration (daily, yearly, perhaps longer cycles) are such that perturbations in the average level are difficult to detect until they are quite large, perhaps 5%. Smaller depletions in the average level are nevertheless quite important because all of the fluctuations would merely be superimposed on a lower base.

Solar ultraviolet radiation with $\lambda < 290$ nm is so strongly absorbed by ozone that it is all removed in the stratosphere and a decrease of 5-10% in stratospheric ozone would not permit any detectable increase in the amount of such radiation at the earth's surface. However, radiation in the 290-320 nm region is only partially absorbed now by ozone, and decreased amounts of O_3 would permit an increase in this radiation at the surface. Radiation in the 290-320 nm range is designated as ultraviolet-B, or u.v.-B, or because of its biological effect on human skin, "sunburn ultraviolet". A decrease of 10% in the average ozone concentration in the stratosphere would lead to an increase of about 20% in the penetration to the surface of u.v.-B radiation. (The percentage increase varies with the wavelength, and this calculation gives the overall increase weighting the solar intensities at each wavelength and the erythemal sensitivity of human skin to each wavelength).

Extensive studies have shown that u.v.-B radiation is the primary cause of human skin cancer, and estimates of the relationship between decreased ozone concentrations and increased incidence of skin cancer are primarily based upon the increased incidence of such skin cancer at latitudes nearer the equator. U.S. estimates of the potential effects of ozone depletion on the incidence of skin cancer generally have been that a 20% increase in u.v.-B would lead to a 30-50% increase in skin cancer incidence, i.e. that 10% decrease in average ozone concentration would lead to a 30-50% increase in skin cancer. [7]

Since all biological species on earth have developed for hundreds of thousands of years under solar irradiation conditions not very different from the present (i.e. absence of u.v. radiation with $\lambda < 290$ nm because of shielding by ozone in the atmosphere), it would not be surprising to find that many other biological species might also be sensitive to an increase in u.v.-B radiation. However, most biological experiments on u.v. sensitivity have been carried out with 254 nm mercury resonance lamps, and relatively few with u.v.-B radiation. Therefore, there is an almost complete absence of detailed scientific estimates concerning the potential effects of ozone depletion on other biological species. Such data are urgently needed, and will certainly be sought in the next decade. [2, 4, 6]

The potential meteorological effects of a change in amount or distribution of stratospheric ozone is also very difficult to assess. Meteorology is so complex that it is not possible to state now whether a change in stratospheric temperature structure, following a change in ozone concentrations in the upper stratosphere, would have any appreciable effect on tropospheric climate or if it did what the effects would be. The development of satisfactory models of climatic behavior is needed before any accurate estimate of the potential climatic consequences of ozone depletion is possible. The time scale for such successful research probably involves at least 10-20 years. [2-4]

Potential Sources of Stratospheric Pollutants

Stratospheric pollutants can either be introduced directly by man through activities in the stratosphere itself, or by the release at or near the surface of materials which are inert enough to reach the stratosphere, there to be converted into reactive species by solar irradiation. Direct stratospheric introduction can occur through the operation of aircraft there, or by its penetration by rockets on their way out of the atmosphere.

Any engine which exposes air to high temperatures must convert some of the normal N_2, O_2 mixture into NO gas; the operation of such an engine in the stratosphere will release the NO effluent there; and the NO can then serve as an additional source of NO_x-catalysis of ozone removal by reactions (9) and (10). The overall chain length of the NO_x-chain from release to atmospheric removal is strongly dependent upon the altitude at which the NO_x is introduced into the stratosphere, since the final termination of the chain occurs with the downward diffusion of the NO_x into the troposphere where it can be removed by rainout, etc. The relative effects of the introduction of NO_x at several altitudes are shown in Table 2, illustrating clearly why the operation of the proposed Boeing SST (flight altitude, 19.5 km) and the actual aircraft of the Concorde/Tupolev classes (16.5 km) are quite different from that of the present subsonic, stratospheric jet aircraft. [4] The effect on stratospheric ozone is a cumulative one with roughly proportional effects from lesser numbers of aircraft.

TABLE 2

Relative Total Chain Length Versus Altitude of Injection of NO at Different Stratospheric Altitudes

Altitude, km	Relative Chain Length	Estimated Accuracy
10.5	5	Factor of 10
13.5	40	Factor of 10
16.5	110	Factor of 3
19.5	200	Factor of 2

Concern has also been expressed over the releases in the stratosphere of sulfur contained in jet fuel as an impurity because of the possibility of its being converted to stratospheric sulfate aerosol.[3] The task of removing sulfur from supersonic jet fuel, while not easy, seems considerably easier than the redesign of jet engines such that the high-temperature conversion of $N_2 + O_2$ into NO is drastically reduced.

The total volume of effluent gases introduced into the stratosphere by rockets has been negligible in the past in comparison to the rates of formation and destruction of global ozone, and the observed effects have been confined to local areas. The proposed operation in the future of the U.S. space shuttle, however, anticipates weekly flights of large rockets, and the potential for introduction of contaminants, especially HCl from NH_4ClO_4 used in the first stage booster, is not entirely negligible.[2]

Chlorofluorocarbons

The world production of chlorofluorocarbon molecules was approximately 10^6 tons in 1975, including 5×10^5 tons of CCl_2F_2 and 3×10^5 tons of CCl_3F. [1, 2, 6] Most of this material is manufactured for use in applications involving release to the atmosphere, either quickly (aerosol propellants, solvents, plastic foams) or later (refrigeration). No important tropospheric removal processes have been discovered for either CCl_2F_2 or CCl_3F, and the primary fate for each is believed to be stratospheric photolysis, as in (1) and (11). The Cl atoms released by

$$CCl_3F \;+\; h\nu(\lambda < 230 \text{ nm}) \longrightarrow Cl \;+\; CCl_2F \qquad (11)$$

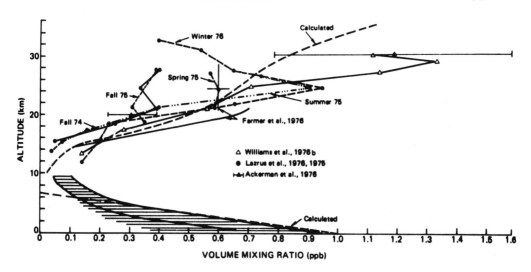

Figure 1. (Reference 2). Measured and Calculated Vertical Distributions of HCl versus Altitude. Measurements with balloon-borne filters (Lazrus et al.) and stratospheric infrared.

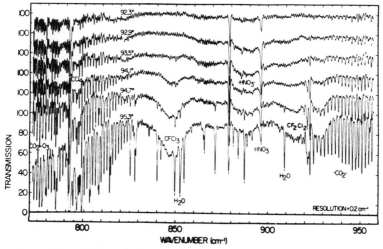

Figure 2. Solar Infrared Spectra taken by D. Murcray et al., University of Denver, September, 1975. (See references 10 and 11). Zenith angles of sun as marked. The minimum altitude of the tangential path for 92.3° is 24.8 km, and for 95.3° is 2.5 km. Absorption positions are shown for HNO_3, H_2O, CO_2, CCl_3F and CCl_2F_2.

this reaction can then initiate the ClO_x-chain catalysis of reactions (12) and (13), with the net result of removal of two ozone-equiva-

$$Cl + O_3 \longrightarrow ClO + O_2 \qquad (12)$$

$$ClO + O \longrightarrow Cl + O_2 \qquad (13)$$

lents (one O_3, one O atom) just as in the natural NO_x cycle. The general understanding of chlorine chemistry in the stratosphere includes a number of other reactions summarized by the equilibria of (14), as well as other reactions of lesser importance[1, 2, 8].

$$HCl \xleftarrow[CH_4,\, H_2,\, HO_2]{OH} Cl \xleftarrow[O,\, NO]{O_3} ClO \xleftarrow[h\nu]{NO_2} ClONO_2 \qquad (14)$$

As shown in Figure 1, extensive stratospheric measurements have been made for HCl [2] and both free radicals, Cl and ClO have recently been measured as well. [9] Restrictive upper limits have been placed on the concentration of chlorine nitrate by stratospheric infrared observations. [10] Such observations are illustrated in Figure 2 which shows both CCl_3F and CCl_2F_2 in stratospheric spectra. Similar spectra made seven years earlier showed about 2.5 times less of both CCl_3F and CCl_2F_2.[11] The continued release to the atmosphere of CCl_2F_2 and CCl_3F at the 1973 rate would lead, in an estimate made in 1974, to an average ozone depletion of 7-13% in steady state.[12] After extensive additional measurement and calculation, the current best estimates fall at the lower end of this earlier estimate. [2]

The 1974 prediction of the altitude profile for CCl_3F in the stratosphere[1] as the result of the combined influence of upward stratospheric diffusion and of solar ultraviolet irradiation in the 190-230 nm band has been amply confirmed by measurement of the concentrations of CCl_3F in the stratospheric samples captured by balloon, as shown in Figure 3. [13, 14] The measurements with CCl_2F_2 also confirm its photolysis in the stratosphere, and at higher altitudes than for CCl_3F because of its lower absorption cross-sections in the 190-230 nm region. The altitude for peak photodecomposition of CCl_2F_2 is about 32 km and for CCl_3F about 27 km. The calculated atmospheric lifetimes of CCl_2F_2 and CCl_3F are in the 40-70 year and 70-150 year range, respectively.

The measured atmospheric lifetime of CCl_3F is now >30 years, from close comparison of the amounts already released with the amounts actually found in the atmosphere.[15] Similar comparisons with CCl_2F_2 are made uncertain by the slow release of material used in refrigeration.

Many other chlorofluorocarbon molecules are also in technological use, and can generally be divided into two categories: (a) saturated chlorofluorocarbon molecules not containing hydrogen; and (b) all others. Those molecules which contain C-H bonds, such as the widely used $CHClF_2$ (10^5 tons in 1974), are primarily removed by reaction with OH in the troposphere, as in (15) and are

$$CHClF_2 + OH \longrightarrow H_2O + CClF_2 \qquad (15)$$

lesser stratospheric hazards. Unsaturated molecules such as $CCl_2=CCl_2$, are also removed by reaction in the troposphere. However, compounds such as $CClF_3$, $CClF_2CClF_2$, etc., [16] also do not appear to have tropospheric removal processes, and are potentially as hazardous to stratospheric ozone as are the much more widely used CCl_2F_2 and CCl_3F if widely used.

Compounds containing bromine and not containing hydrogen (for example $CBrF_3$ and $CBrF_2CBrF_2$) are also potential stratospheric hazards [21] since bromine atoms can participate in a BrO_x—catalytic chain very similar to the ClO_x-chain. On the other hand, the fluorine atoms eventually released from CCl_2F_2 and CCl_3F are not a serious stratospheric problem because the FO_x-catalytic chain is rapidly terminated by reaction to form HF, and this molecule is not attacked by OH. Thus, HCl and HBr are only temporary terminations for the ClO_x and BrO_x chains, to be restarted by OH attack until the whole chain diffuses to the troposphere and is removed by rainout, etc. Once formed, however, HF is essentially a permanent sink for F, and no further catalytic removal of ozone by the FO_x chain occurs.

The ClO_x-chain characteristically attacks ozone most effectively between 30-50 km (i. e. upper stratosphere) and is predicted to reduce O_3 concentrations much more seriously at these altitudes which influence the stratospheric temperature gradient. Again, the potential climatic alterations from this severe change in O_3 gradient can not now be calculated. One final climatic worry accompanies the accumulation of fluorocarbons in the atmosphere through a strong "greenhouse" effect similar to that from CO_2 released

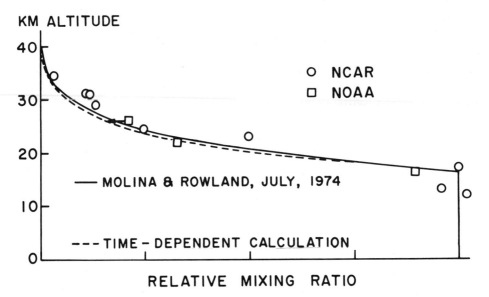

Figure 3.   Comparison of Altitude Profiles for CCl₃F.   Solid line, prediction (Ref. 1).   Measurements: NCAR (13), NOAA (14)

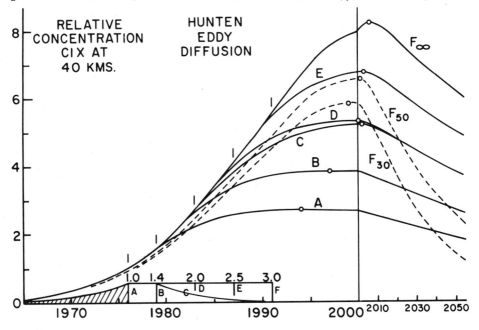

Figure 4.   Predicted Growth of ClX for Various Assumptions of Future Patterns of Atmospheric Release.   Release Patterns are marked A to F.   Solid lines: growth patterns for ClX with no tropospheric sinks.   Dotted lines: 30-year and 50-year sinks.

by burning fossil fuel. Both CCl_3F and CCl_2F_2 are much more effective in retaining the earth's infrared radiation per molecule than CO_2 because the strong absorptions (e.g. the C-F stretching vibration) of the chlorofluorocarbons fall in nearly transparent infrared regions, while those of CO_2 of course are nearly opaque already from the 330 parts per million of CO_2 already in the atmosphere. [2, 17]

Time Scales for Atmospheric Recovery from Chlorofluorocarbon Pollution

The motion of molecules throughout the troposphere is quite rapid, and the chlorofluorocarbon gases are quickly mixed both horizontally, and vertically to the tropopause. Mixing into the southern hemisphere requires about 2 years. However, above the tropopause the motions are very much slower and the peak concentrations of ClX (i.e. Cl-containing decomposition products) are reached a decade or more after further release at ground level has ceased. (See Figure 4). [1, 2, 15] Furthermore, only a small fraction of the chlorofluorocarbons are exposed to solar u.v. in the 190-230 nm band at any one time, with the consequence that many decades are required on the average before the molecules are actually decomposed by absorption of solar u.v. radiation. For this reason, the ClX concentration in the stratosphere will decline only very slowly (Figure 4), with appreciable chlorofluorocarbon pollution remaining throughout the 21st century.

Finally, great effort has been expended in the past several years to determine whether important removal processes exist in the troposphere for CCl_3F and CCl_2F_2. The searches for specific tropospheric sinks have all resulted in failure, with the exception of a minor sink requiring several hundred years for removal by dissolution in the ocean. [1, 2] If there were a number of undetected tropospheric sinks leading to combined removal rates averaging 30 or 50 years then the expected effects on CCl_3F concentrations are shown in Figure 4, as F_{30} and F_{50}, as compared to the stratospheric sink alone, F_∞ [15] The effects of such cumulative processes, if discovered, can be seen to be of relatively little importance for the first 25-50 years after termination of further release of chlorofluorocarbons to the atmosphere, with increasing importance in the latter part of the 21st century. However, the most recent comparison of the atmospheric burden of CCl_3F with the amount already released indicates no "missing" CCl_3F that would be indicative of as yet undetected tropospheric sinks. [15] Some

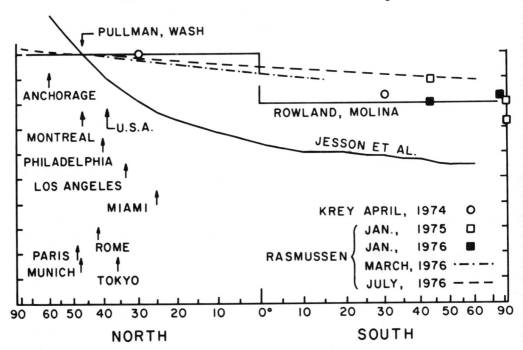

Figure 5. Measured Interhemispheric Gradients for CCl_3F
Compared with Latitude Gradients used in Modeling for the
Presence of Tropospheric Sinks. See references 15 and 19.
Arrows indicate measurement base of R. Rasmussen in Pullman,
Washington (Washington State University) and latitudes of major
cities.

other such comparisons of atmospheric burden versus atmospheric release have been interpreted to indicate enough "missing" CCl_3F to indicate tropospheric lifetimes as short as 10-15 years. [10] The major difference between these estimates has been in the estimation of the tropospheric concentrations of CCl_3F in the tropics and in the southern hemisphere. The prediction of relatively small CCl_3F gradients with latitude (two years or less to the southern hemisphere, Rowland and Molina, ref. 15) have been strongly supported by the latest world-wide CCl_3F measurements (See Figure 5) in contrast to the very slow north-south mixing even within the northern hemisphere (postulated by Jesson et al, [19]. The "missing" CCl_3F of the latter estimates is thus essentially an artifact of a poor model for the tropospheric north-south mixing of CCl_3F.

At the present stage, then, the chlorofluorocarbon molecules have been shown to be present in the stratosphere, along with the decomposition products including the actual chain-reacting species Cl and ClO. All of the major calculations indicate an appreciable reduction of total ozone with continued long-term use and accumulation of CCl_3F and CCl_2F_2 in the atmosphere, and much more substantial percentage ozone losses in the 35-50 km range. No important alternative removal processes have either been found or are now likely to be found. As a consequence of this overall situation, several regulatory agencies in the United States have announced that strong restrictions on the use of chlorofluorocarbons can be expected in the U.S. during 1978.

References

1. "Chlorofluoromethanes in the Environment", F. S. Rowland and Mario J. Molina, Rev. Geophys. and Space Physics, 13, 1 (1975); "Stratospheric Sink for Chlorofluoromethanes: Chlorine Atom-Catalysed Destruction of Ozone," M. J. Molina and F. S. Rowland, Nature, 249, 810 (1974).

2. "Halocarbons: Effects on Stratospheric Ozone", U.S. National Academy of Sciences, Washington, D.C., 1976.

3. Climatic Impact Assessment Program Monographs. Six volumes. U.S. Department of Transportation, DOT-TST-75-51, Washington, D.C., 1975. See especially Vol. 1, "The Natural Stratosphere of 1974."

4. "Environmental Impact of Stratospheric Flight. Biological
    and Climatic Effects of Aircraft Emissions in the Strato-
    sphere". Climatic Impact Committee, National Research
    Council, National Academy of Sciences, National Academy of
    Engineering, Washington, D. C. 1975.

5. "Global Ozone Balance in the Natural Stratosphere", H. S.
    Johnston, Rev. Geophys. and Space Phys., 13, 637 (1975).

6. "Fluorocarbons and the Environment", Report of Federal
    Task Force on Inadvertent Modification of the Stratosphere
    (IMOS), June, 1975.

7. "Measurements of Ultraviolet Radiation in the United States
    and Comparisons with Skin Cancer Data", J. Scott, T. R.
    Fears and G. B. Gori, National Cancer Institute, DHEW No.
    (NIH) 76-1029 , 1976.

8. "Stratospheric Formation and Photolysis of Chlorine Nitrate",
    F. S. Rowland, John E. Spencer and Mario J. Molina, J. Phys.
    Chem., 80, 2711 (1976); "Estimated Relative Abundance of
    Chlorine Nitrate among Stratospheric Chlorine Compounds",
    ibid., 80, 2713 (1976).

9. "Free Chlorine in the Stratosphere: an In Situ Study of Cl
    and ClO", J. G. Anderson, J. J. Margitan and D. H. Stedman,
    preprint, February, 1977.

10. "Upper Limit for Stratospheric $ClONO_2$ from Balloon-Borne
    Infrared Measurements", D. G. Murcray, A. Goldman, W. J.
    Williams, F. H. Murcray, F. S. Bonomo, C. M. Bradford,
    G. R. Cook, P. L. Hanst and M. J. Molina, preprint, Mar. 1977.

11. "Detection of Fluorocarbons in the Stratosphere", D. G.
    Murcray, F. S. Bonomo, J. N. Brooks, A. Goldman, F. H.
    Murcray and W. J. Williams, Geophys. Res. Letters, 2,
    109 (1975).

12. F. S. Rowland and M. J. Molina, paper presented at the
    American Chemical Society Meeting, Atlantic City, N. J.,
    September 1974.

1 3. "Stratospheric Profiles of CCl_3F and CCl_2F_2", L.E. Heidt, R. Lueb, W. Pollock and D.H. Ehhalt, Geophys. Res. Letters, $\underline{2}$, 445 (1975).

14. "Measurements of Stratospheric $CFCl_3$, CF_2Cl_2 and N_2O", A.L. Schmeltekopf, P.D. Goldan, W.R. Henderson, W.J. Harrop, T.L. Thompson, F.C. Fehsenfeld, H.I. Schiff, P.J. Crutzen, I.S.A. Isaksen, Geophys. Res. Letters, $\underline{2}$, 393 (1975).

15. "Estimated Future Atmospheric Concentrations of CCl_3F (Fluorocarbon-11) for Various Hypothetical Tropospheric Removal Rates," F.S. Rowland and M.J. Molina, J. Phys. Chem., $\underline{80}$, 2049 (1976).

16. "Stratospheric Photodissociation of Several Saturated Perhalo Chlorofluorocarbon Compounds in Current Technological Use (Fluorocarbons-13, 113, 114, 115)," C.C. Chou, R.J. Milstein, W.S. Smith, H. Vera Ruiz, M.J. Molina and F.S. Rowland, J. Phys. Chem. (in press.)

17. "Greenhouse Effect Due to Chlorofluorocarbons: Climatic Implications", V. Ramanathan, Science, $\underline{190}$, 50 (1975).

18. "Production of Atmospheric Nitrous Oxide by Combustion", R.F. Weiss and H. Craig, Geophys. Res. Letts., $\underline{3}$, 751 (1976); and unpublished work by H. Craig, R.F. Weiss and W.L. Dowd.

19. "The Fluorocarbon-Ozone Theory II: Tropospheric Lifetimes: An Estimate of the Tropospheric Lifetime of CCl_3F", J.P. Jesson, P. Meakin, and L.C. Glasgow, preprint published in "Chlorofluorocarbon Effects and Regulations", U.S. Senate Hearings on Dec. 15, 1976.

THE EFFECTS OF SOLAR UV-B RADIATION (280-315 nm) ON HIGHER PLANTS:

IMPLICATIONS OF STRATOSPHERIC OZONE REDUCTION

Martyn M. Caldwell

Dept. Range Science and Ecology Center

Utah State University, Logan, Utah 84322 U.S.A.

Summary

For each 1% reduction in the atmospheric ozone column there would be a predictable increase of approximately 2% in solar UV-B radiation weighted for biological effectiveness, UV-B_{BE}. Yet, physiological manifestations reflecting the net unrepaired damage to higher plants would not necessarily increase as a linear function of the UV-B_{BE} enhancement. Recent evidence shows that for impairment of photosynthetic capacity, the response to UV-B_{BE} was quite nonlinear. There was, however, no apparent threshold of UV-B_{BE} to which the plants must be exposed before photosynthetic damage ensued and reciprocity in the response to UV-B applied over a considerable period of time. If the accumulative mode of damage to the photosynthetic apparatus is generally applicable to higher plants, even moderately UV-B resistant species may be affected by small increases in UV-B radiation, if such species possess long-lived leaves which are exposed to full sunlight. Because higher plant species appear to be differentially sensitive to UV-B radiation, the ecological significance of ozone reduction may lie more in changes in the competitive balance of species rather than in reduced primary productivity per se. The differential UV-B sensitivity of higher plant species may be related to the optical properties of tissues covering physiological targets and the efficacy of repair mechanisms.

Introduction

Higher plants have evolved to be efficient collectors of solar radiation which is of necessity an effective photosynthetic

597

energy capture. However, this also maximizes the exposure of
aerial plant parts to solar UV radiation. Of particular potential
detriment are wavelengths shorter than 315 nm since nucleic acids
and proteins have an appreciable absorption cross section in this
waveband. The absorption cross section of ozone also becomes
appreciable only at wavelengths shorter than 315 nm. With
decreasing wavelength, the ozone absorption coefficient increases
exponentially (Fig. 1). Even though the average total atmospheric
ozone column is only on the order of 3 mm if condensed to standard
temperature and pressure (commonly denoted as 0.300 atm·cm), it
is sufficient to effectively attenuate the solar UV flux to less
than 10^{-3}W·m^{-2}·nm^{-1} at wavelengths below 295 nm. If the atmospheric ozone column were reduced to only 10% of this thickness,
calculations using the model of Green et al., (1) suggest that UV
irradiance at ground level, with the sun directly overhead, would
still not include wavelengths shorter than 280 nm at intensities
greater than 10^{-3}W·m^{-2}·nm^{-1}. Thus, any consideration of biological
effects of increased solar UV flux resulting from reduced atmospheric ozone must be confined to the waveband between 280 and 315
nm (commonly designated as UV-B) and for moderate ozone reduction
only the 290 to 315-nm waveband would be of concern.

With changes in optical thickness of the atmospheric ozone
column there is a predictable change in ground-level UV-B irradiance
as illustrated in Fig. 2. As would be expected from the ozone
absorption coefficient, a change in ozone thickness is reflected
much more in spectral irradiance at shorter wavelengths. The
greater representation of shorter wavelengths with reduced ozone
should be of biological importance if the biological effectiveness
of UV-B radiation corresponds to the action spectra for effects
as mediated by nucleic acid and protein chromophores as shown in
Fig. 1. To accommodate this relationship, a useful expression of
biologically effective UV-B irradiance, UV-B$_{BE}$, can be defined as

$$UV\text{-}B_{BE} = \int_{280}^{315} I_\lambda \cdot E_\lambda \cdot d\lambda \qquad (1)$$

where I_λ is the spectral irradiance at wavelength λ, and E_λ is the
relative biological effectiveness at wavelength λ as defined by
Caldwell (4). This effectiveness term closely approximates the
action spectra for biological effects involving nucleic acids and
proteins as shown in Fig. 1 in the 280 to 315-nm waveband. This
function was, however, derived from a range of action spectra for
both higher and lower plants involving several physiological
responses as shown in Fig. 3. Except in the case of 3I, the
chromophores were considered to be nucleic acid or protein compounds. Despite the fact that action spectra can be modified by
wavelength-dependent filtration through plant tissues covering
physiological targets, the correspondence of this sundry of action

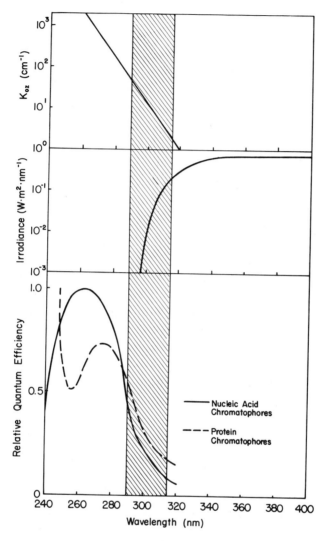

Figure 1. Ozone extinction coefficient adapted from Green et
al., (1); the solar ultraviolet flux (direct sun plus diffuse
sky radiation) at ground level for conditions of 0.300 atm·cm
ozone thickness and a solar angle of 30° from the zenith taken
from calculations using the model of Green et al. (1) and measurements of Bener (2); generalized action spectra for biological
responses to UV irradiation involving protein (dashed line) and
nucleic acid (solid line) chromophores. Adapted from Giese (3).
The hatched area designates that portion of the spectrum which
is involved in the stratospheric ozone depletion question.

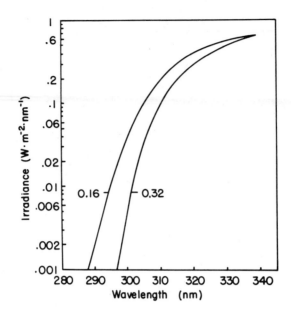

Figure 2. Global (direct sun plus sky) spectral irradiance
calculated for a solar angle of 30° from the zenith and atmospheric
ozone concentrations of 0.32 and 0.16 atm·cm from the model of
Green et al. (1).

spectra is reasonably good. In the case of 3I, which deviates
most from the generalized spectrum, the chromophore was considered
to be a plant hormone, indoleacetic acid. Although the number of
available action spectra for physiological phenomena of higher
plants in the UV-B is quite limited, some generality of this E_λ
relationship still seems reasonable. When the effect of decreasing
atmospheric ozone concentrations on UV spectral irradiance is
calculated (1) and integrated with respect to biological effectiveness according to equation 1, the net effect is an approximate 2%
increase in UV-B_{BE} for each 1% decrease of the atmospheric ozone
column.

Although UV photobiology has progressed far during the last
few decades, this research has only occasionally involved higher
plants and has been primarily concerned with the effects of 254-nm
radiation. Nearly monochromatic radiation of this wavelength can
be easily generated by low-pressure mercury lamps and this
conveniently falls within the waveband of maximum inactivation cross
section of nucleic acids. Although much of what has been learned
(particularly the elucidation of various radiation repair mechanisms)
may also likely apply to the response of higher plants to UV-B

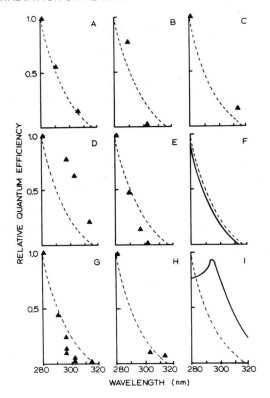

Figure 3. A generalized plant UV–B action spectrum relative to
280 nm (dashed line) as it corresponds to action spectra data
for: A, mutation of liverwort spores, *Sphaerocarpus donnellii*
(5); B, mutation in the fungus *Trichophyton mentagrophytes* (6);
C, inhibition of photosynthesis in *Chlorella pyrenoidosa* (7); D,
cessation of cytoplasmic streaming in epidermal cells of *Allium
cepa* (8); E, frequency of "endosperm deficiencies" in maize due
to chromosomal deletion or mutation (9); F, germidical action
spectrum (10); G, epidermal cell damage in *Oxyria digyna* leaves
(11); H, induction of chromosomal aberrations in *Tradescantia
paludosa* pollen (12); I, base curvature of *Avena coleoptiles*
(13). All of the action spectra have been expressed in terms of
relative quantum efficiency which denotes either relative quantum
effectiveness (spectra A, B, E, F, H, and I) or relative quantum
sensitivity (spectra D, D, and G). From Caldwell (4).

radiation, a simple extrapolation or prediction of ecological
consequences of ozone depletion cannot be made from this information.
Not only are there appreciable differences in the quantitative
responses to UV-B radiation as opposed to 254-nm radiation as is
indicated by the action spectra for physiological responses, but

there may also be qualitative differences in the patterns of response as has been demonstrated by Nachtwey (14) for phytoplankton. Furthermore, most higher plants have apparently evolved defense mechanisms to minimize the UV-B radiation insult. These include absorption of UV-B radiation in outer tissue layers and molecular repair mechanisms such as photoreactivation and dark repair systems. Thus, of ultimate biological interest is the net unrepaired UV-B-induced damage that would result in nature if the ozone layer were diminished.

Plant Response to UV-B Radiation

Although the relationship between ozone reduction and UV-B_{BE} has been calculated to show an approximate 2% increase in UV-B_{BE} per 1% decrease in ozone thickness, the physiological responses of plants reflecting the net unrepaired radiation damage is not necessarily a linear function of UV-B_{BE}. Thus, the question of the basic dose-response relationships must be addressed. Several related questions are also pertinent: Is there a threshold level of UV-B_{BE} to which plants must be exposed before physiological manifestations of unrepaired damage appear? Do plants respond to the total accumulated exposure of UV-B_{BE} or are the physiological manifestations better correlated with the level of UV-B_{BE} irradiance during rather brief periods of time such as during critical developmental stages.

Although a variety of physiological manifestations of UV irradiation have been reported in the literature for both higher and lower plants (4), two effects which are of particular ecological import are reductions in photosynthetic capacity and curtailment of leaf expansion, both of which would contribute to an impaired carbon economy of the plant. Studies recently completed by W. B. Sisson in our laboratory with *Rumex patientia* exposed to various levels of polychromatic UV-B irradiance corresponding to solar UV-B irradiance at various concentrations of atmospheric ozone, provide some indication of the pattern of response to UV irradiation. For impairment of photosynthesis, response to UV-B_{BE} was quite nonlinear and there did not appear to be a particular threshold of UV-B_{BE} to which the plant must be exposed before photosynthetic damage ensued. Instead, reduction of photosynthetic capacity appeared to be only a function of the total accumulated exposure to UV-B_{BE} and even UV-B irradiance corresponding to that under present-day ozone conditions resulted in some impairment of photosynthesis relative to control plants exposed to no UV-B radiation. Reciprocity in the response to UV-B_{BE} applied over a considerable period of time (5 to 45 days) for individual leaves exposed to 3 levels of UV-B_{BE} irradiance. Thus, damage to the photosynthetic system appears to be accumulative over a considerable portion of the ontogeny of individual leaves and is apparently only a function

of the total UV-B$_{BE}$ to which the leaf is exposed. In terms of
leaf enlargement, reciprocity did not apply over long periods of
time. Instead, UV-B$_{BE}$ was most effective in curtailment of leaf
expansion during only a period of two or three days during early
developmental stages. The response was, however, a function of
the level of UV-B$_{BE}$ irradiance during this short period of time.
The reduction of leaf expansion rates appears to be due to effects
of UV radiation apart from the simple limitation of photosynthate
supply (15).

This study suggests solar UV-B irradiance, even under present-
day conditions, may be a significant ecological factor for *Rumex
patientia*. This species occurs in open habitats exposed to full
insolation and there is a minimal degree of self-shading among
leaves of the same plant. Individual leaves of this species are
active for periods of time sufficient to accumulate appreciable
exposure to solar UV-B radiation. The degree to which other higher
plant species would correspond to this pattern of reduction in
growth and photosynthesis as a function of UV-B irradiation is
difficult to predict. *Rumex patientia* was selected in this study
because it is both sensitive to UV-B radiation and exposed to full
sunlight in nature. Many other higher plant species are, however,
much less sensitive to UV-B radiation (16). Nevertheless, if the
accumulative mode of damage to the photosynthetic apparatus is
generally applicable to higher plants, even moderately UV-B
resistant species may be affected by UV-B radiation in the event
of ozone reduction if such species possess long-lived leaves which
are exposed to full sunlight.

The effects of UV-B irradiance on photosynthetic capacity have
been shown to be associated with the nonstomatal component of
photosynthesis (15). That is, reduction of photosynthesis was not
due to a simple increase in the diffusion resistance to CO_2 entering
the leaf through the stomates. Instead, the damage appeared to be
associated with the photosynthetic apparatus itself and related
studies have shown that electron transport associated with photosystem II may be most affected. Since stomatal diffusion resistance
did not change, the capacity for transpirational water loss would
remain the same; therefore, the water use efficiency of photosynthesis (photosynthesis/transpiration ratio) would also become less
favorable for a sensitive plant such as *Rumex patientia* exposed to
UV-B radiation.

Differential UV-B Sensitivity of Higher Plant Species

Another complicating factor in the assessment of the potential
impact of reduced stratospheric ozone is the apparent differential
sensitivity of higher plants to UV-B radiation. Short-term studies
conducted during the past few years indicated that higher plant

species vary considerably in their sensitivity to UV-B radiation
in terms of parameters such as reduction in biomass accumulation
and height growth (16). The basis for this differential
sensitivity is not well understood. No striking correlations
between sensitivity and taxonomic affinity, life form, or ecological
status seem to have emerged from the studies to date.

Coping with present-day solar UV-B radiation appears to involve
avoidance of the radiation by reflection or absorption of the UV-B
radiation by tissues covering physiological targets. A second
major facet of resistance likely involves the molecular radiation
repair mechanisms such as photoreactivation and dark repair systems.

Transmittance through the epidermis of most higher plant
leaves is on the order of 80 or 90% in the visible part of the
spectrum and typically decreases to less than 10% in the UV-B
portion of the spectrum (17, 11). Although the epidermis itself
must be considered a physiological target of some importance, this
selective absorbance of UV-B radiation by the epidermis undoubtedly
reduces the impact of UV-B radiation on mesophyll tissues.

Photoreactivation has been demonstrated in a variety of higher
plants for several physiological manifestations of UV irradiation
at 254 nm (18, 4). A few studies also have demonstrated that some
UV-B induced damage can be photoreactivated (11). Dark repair
systems in higher plant tissues have also been recently shown (19).
It is quite likely that these molecular repair systems such as
photoreactivation are of survival value to higher plants under
present-day ozone conditions rather than being simply a vestigial
element lingering from earlier stages of the evolution of plant
life during that time when the earth's ozone layer was only just
forming.

The importance of photoreactivation for plants growing under
a normal solar irradiation regime was implied in earlier experiments
where leaves of several alpine plant species were exposed to 297-
and 302-nm radiation with intensities controlled to closely
simulate UV-B_{BE} on a summer day at alpine elevations (11). During
and following this irradiation, the leaves were not exposed to any
other wavelengths. Some of these leaves did exhibit severe lesions
indicating UV-B-induced damage which had never been observed under
field conditions. The absence of such lesions in nature might
suggest the importance of photoreactivation of UV-B-induced damage
as an effective protective mechanism. Photoreactivability of
these UV-B-induced lesions was also shown in these studies.

The degree to which variations in optical properties of
plants, such as leaf epidermal UV-B transmittance, and the
efficacy of repair systems contribute to the apparent differential
sensitivity of higher plants has not yet been elucidated. Nor
have extensive studies been undertaken to evaluate the acclimation

potential of individual plants to accommodate increased intensities of UV-B irradiance. This acclimation might be elicited by prior exposure to UV-B radiation or perhaps other environmental factors which might lead to a heightened tolerance of UV-B radiation.

Ecological Implications of Ozone Reduction

Although there is little doubt as to the actinic potential of solar UV-B radiation for at least sensitive higher plant species, the probable ecological impact of increased UV-B radiation resulting from decreased mean atmospheric ozone concentrations is difficult to assess. Because higher plant species do appear to be differentially sensitive to UV-B radiation, the ecological significance of ozone reduction may lie more in changes in the competitive balance of species rather than in reduced primary productivity *per se*. The distribution of plants in nonagricultural ecosystems, and even to a certain extent in intensively-managed situations, is often not immediately limited by the direct effects of physical environmental factors such as temperature extremes, moisture limitations, nutrient supply, etc. Though such factors set limits on the potential distribution of species, the prevalence of species and their actual limits are dictated more by the outcome of competition with neighboring species.

With ozone reduction, even small depressions of photosynthesis, growth, and water use efficiency of photosynthesis caused by exposure of sensitive species to increased UV-B radiation may under some circumstances shift the competitive balance in favor of more UV-B-resistant species. Recent experiments in our laboratory by F. M. Fox suggest that the competitive balance between species can be changed with even modest supplements of UV-B radiation.

Although it has never been documented or perhaps even much contemplated, some shifts in competitive balance between higher plant species may be taking place during the course of natural fluctuations in atmospheric ozone. Long-term records of atmospheric ozone concentrations are unfortunately very limited. Nevertheless, such records bear some witness to changes in mean annual concentration of about 10% which appear to follow an approximate 10- to 12-year cycle (20, 21). These changes would result in roughly a 20% change in UV-B_{BE} over periods of half a dozen years which might result in some shifts in competitive balance of species. Though evidence for changes over longer periods of time do not exist, they might be expected. For example, if the periods of low ozone during the 10- to 12-year variation are indeed linked to periods of low sunspot activity, with appropriate lags, as has been suggested by Christie (21), periods when there was an extended lull in sunspot activity such as has been suggested to have

occurred during the latter part of the 17th century (22) may have
resulted in an extended period of low ozone concentrations.
Unfortunately, a retrospective surmise as to the biological
consequences of such extended periods of low ozone concentration,
or even during the course of the 10- to 12-year fluctuations
has not been attempted and would undoubtedly be difficult, if not
impossible.

The natural fluctuations of ozone concentration, particularly
over the longer term, do, however, suggest that higher plants,
and ecosystems in general, must display a certain resiliency
and ability to accommodate some change in UV-B flux.

The magnitude of change over the 10- to 12-year cycle, i.e,
10%, is at least equal to the ozone reduction anticipated from
several man-induced depletions of atmospheric ozone. Nevertheless,
the anthropogenic reduction of mean atmospheric ozone would be
superimposed on natural fluctuations. Therefore, periods of low
ozone concentrations in the course of natural variations, if
compounded by a man-induced reduction in mean ozone concentration,
could result in exposure of ecosystems to UV-B flux not yet
experienced in particular regions. A concerted effort to place
some perspective on the ecological consequences of such UV-B flux
intensifications is urgent in the face of current trends in
stratospheric flight and the release into the troposphere of agents
which can be calculated to have potential ultimate consequences
for the global stratospheric ozone layer.

REFERENCES

1. Green, A.E.S., Sawada, T., and Shettle, E.P., Photochem.
   Photobiol., 19, 251 (1974).

2. Bener, P., Technical Report, U.S. Army, London, Contract No.
   DAJA 37-68-C-1017, 1972.

3. Giese, A. C., in "Photophysiology" (A.C. Giese, ed.), Vol. II,
   p. 203. Academic Press, N.Y., 1964.

4. Caldwell, M. M., in "Photophysiology" (A.C. Giese, ed.), Vol.
   VI, p. 131. Academic Press, N.Y., 1971.

5. Knapp, E., Reuss, A., Riesse, O., and Schreiber, H.,
   Naturwissenshaften, 27, 304 (1939).

6. Hollaender, A., and Emmons, C. W., Cold Spring Harbor Symp.
   Quant. Biol., 9, 179 (1941).

7. Bell, L. N., and Merinova, G. L., Biofizika, 6, 21 (1961).

8. Glubrecht, H., Z. Naturforsch. B, 8, 17 (1953).

9. Stadler, L.J., and Uber, F. M., Genetics, 27, 84 (1942).

10. Luckiesch, M., Applications of Germicidal, Erythemal, and
    Infrared Energy, Van Nostrand-Reinhold, Princeton, New
    Jersey.

11. Caldwell, M.M., Ecol. Monogr., 38, 243 (1968).

12. Kirby-Smith, J.S., and Craig, D.L., Genetics, 42, 176 (1957).

13. Curry, G.M., Thimann, K.V., and Ray, P.M., Physiol. Plant.,
    9, 429 (1956).

14. Nachtwey, D.S., in "Impacts of Climatic Change on the Bio-
    sphere: Part 1 Ultraviolet Radiation Effects" (D. S. Nachtwey,
    M.M. Caldwell and R.H. Biggs, eds), p 3-50. U.S. Dept.
    Transportation, Wash., D.C., 1975.

15. Sisson, W.B., and Caldwell, M.M., Plant Physiol., (in press)
    (1976).

16. Biggs, R.H., Sisson, W.B., and Caldwell, M.M., in "Impacts
    of Climatic Change on the Biosphere: Part 1 Ultraviolet
    Radiation Effects" (D. S. Nachtwey, M.M. Caldwell and R.H.
    Biggs, eds), p 4-34. U.S. Dept. Transportation, Wash, D.C.,
    1975.

17. Lautenschlager-Fleury, D., Ber. Schweiz. Bot. Ges., 65, 343
    (1955).

18. Cline, M.G., Conner, G.I., and Salisbury, F.B., Plant Physiol.,
    44, 1674 (1969).

19. Howland, G.P., Nature, 254, 160 (1975).

20. Dütsch, H.U., Can. J. Chem. 52, 1491 (1974).

21. Christie, A.D., Pure Appl. Geophys., 106, 1000 (1973).

22. Eddy, J.A., Science, 192, 1189 (1976).

EPIDEMIOLOGY OF MALIGNANT MELANOMA OF THE SKIN IN
NORWAY WITH SPECIAL REFERENCE TO THE EFFECT OF SOLAR
RADIATION

Knut Magnus

The Cancer Registry of Norway

The Norwegian Radium Hospital, Oslo 3, Norway

SUMMARY

During the 20 year period 1955-1974 the incidence
rate of malignant melanoma in Norway was more than
tripled for both sexes. A marked north - south gradient
was observed, the incidence in Southern Norway being
almost three times larger than that in the Northern
part of the country. Incidence variations in time and
space were most pronounced for malignant melanoma of
the male neck-trunk and the female lower limb.
It is concluded that the increase in incidence in
the main must be real and that exposure to sunlight is
an important factor in the etiology of malignant
melanoma of the skin. The effect of exposure seems to
be local and not systemic.

INTRODUCTION

Malignant melanomas are malignant tumours of nevo-
cytic or melanocytic origin, mostly located on the
skin. Few other types of cancer disclose such remarkable time trends and geographical variations in
incidence and mortality, and marked differences are
observed in the topographic distribution of the tumours
according to sex and age. The variety of epidemiologic
contrasts indicates that exogenic factors may be
important in the etiology of this disease.
Observations on the epidemiology of malignant melanoma
seem to be quite consistent within as well as between countries. A rapid rise in incidence and

609

mortality has been reported from many parts of the world. McGovern and Brown (1969)[1] and Lee and Carter (1970)[2] in their analyses of data from Australia, England and Wales, and the United States have found an increase in the rate between 50-100% from 1950-1965, and they conclude that the rise is probably a real increase in incidence rather than an improvement in certification.

The exceptionally high mortality in Australia was first emphasized by Lancaster (1956)[3] who also demonstrated the increase in rates with increasing proximity to the equator. The highest incidence was reported in the tropical and subtropical areas of Queensland. In the United States, a geographical pattern similar to that in Australia has been observed. Dorn and Cutler (1959)[4] found a ratio of 3 between incidence rates in southern and northern areas and Haenszel (1963)[5] reported about the same ratio for 4 southern versus 4 northern cities.

The epidemiologic evidence summarized above indicates that exposure to sunlight may be an important factor in the etiology of malignant melanoma. The association between ultraviolet light and cancer was suggested more than 100 years ago, and epidermal tumors were produced by mercury-arc radiation in 1928. The hypothesis that exposure to sunlight was associated with human malignant melanoma was first proposed by McGovern (1952)[6] and Lancaster (1956)[3]. The pathogenesis is not well known but the degree of pigmentation of the skin is apparently of significance. In clinical studies Lancaster and Nelson (1957)[7] and Gellin et al. (1969)[8], when comparing cases and controls, showed that patients with malignant melanoma tend to have lighter complexions and to spend more time outdoors.

Not all investigators accept exposure to sunlight as an important factor in the etiology of malignant melanoma. The main argument against the hypothesis is that the topographic distribution of the primary site does not correlate well with the degree of exposure to sunshine (Blum, 1959[9]; Dorn and Cutler, 1959[4]; Gellin et al., 1969[8]). An indirect effect has been proposed by Lee and Merrill (1970)[10] who hold a "solar circulating factor" responsible for the high incidence at unexposed sites.

The present study may elucidate these problems because of the large and complete material available and the special geographical conditions of Norway. Exposure to sunlight differs greatly throughout the country. The lattitude ranges from 58-72° and climatic conditions differ from South to North as well as from coastal to inland areas.

MATERIAL AND METHODS

The material is based on reports submitted to the
Cancer Registry of Norway, and comprises all new cases
of malignant melanoma of the skin in the total Norwegian population 1955-1970. A total of 2541 cases was
registered. This material has been supplemented, in a
few graphic illustrations, with data for the years
1971-1974.

Cancer registration in Norway is based on compulsory
notification and it aims to register all recognized
cases of cancer among the total population of the
country. Reports are required from all hospital departments and all institutes of patology.

All cases were confirmed histologically. The cases of
lentigo maligna, although reported to the registry, are
not included in the present material.

RESULTS

Total malignant skin melanoma

Time trends   The number of new cases of malignant
melanoma of the skin in Norway has increased from less
than 100 a year to nearly 400 during the period 1955-
1974. A certain rise was to be expected as a result of
the changes in the size and age structure of the population, but as seen from Figure 1, the incidence rate,
even after age-adjustment, has risen by about a factor
of 3 in the course of these 20 years. The trend is
rather similar for both sexes.

Fig. 1. Total age adjusted incidence rates of malignant
melanoma of the skin in Norway 1955–1974.

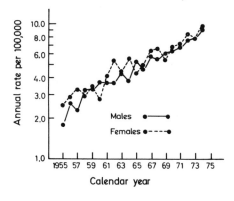

Age-specific rates   The curves of the age-specific
incidence rates shown in Figure 2 are unlike those for
most other forms of cancer. There is a steep increase
starting in adolescence, a levelling-off through middle
age followed by a new rise during the late years of life.

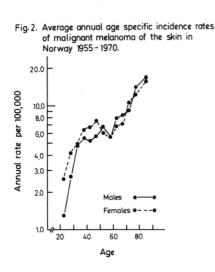

Fig. 2. Average annual age specific incidence rates
of malignant melanoma of the skin in
Norway 1955-1970.

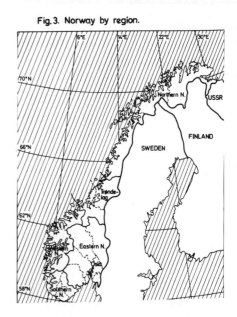

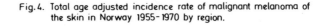

Fig. 3. Norway by region.

Geographical variations   The map (Fig. 3) shows the
situation of the capital and the 5 regions for which
incidence data have been given. Large geographical
variarions are seen (Fig. 4). The highest rates are
observed in the capital, its surrounding counties

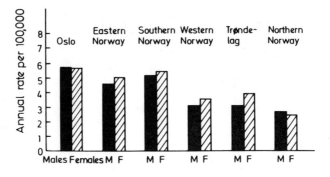

Fig. 4. Total age adjusted incidence rate of malignant melanoma of
the skin in Norway 1955-1970 by region.

(Eastern region), and in the southern part. In Northern Norway the rate is reduced to between one half and one third. Intermediate rates are observed in Trøndelag and in the Western region.

Analysis by anatomical site

The material has been classified into the following groups according to the primary site of the malignant melanoma: face (including scalp and external ear); neck-trunk; upper limb; foot; lower limb (excluding foot); and "other and unspecified site" (including anus, scrotum and mamma).

The total age-adjusted rates by site are given in Figure 5. The graph discloses striking contrasts by sex. Male preponderance is observed for the incidence of melanoma of the neck-trunk, whereas the reverse is found for the lower limb. For the latter site the rate among females is more than 3 times higher than among males. An excess incidence among females is also observed for the upper limb, whereas for the face and the foot there is no significant sex differential.

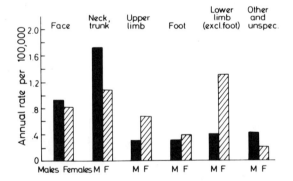

Fig. 5. Total age adjusted incidence rate of malignant melanoma of the skin in Norway 1955-1970 by anatomical site.

Time trends

Time trends in incidence for the various sites are shown in Figure 6. The sites for which variations with calendar time seem to be negligible have been selected in the upper part of the figure. A slight upward trend may be present for the face, whereas no systematic variation can be observed for the foot. As a contrast to these two sites the results are shown for the neck-trunk and the lower limb (lower part of Fig. 6). During the 15-years a manifold increase in the incidence rate is observed for malignant melanoma of these sites, particularly for neck-trunk among males, and lower limb among females.

Age-specific rates

Contrasting patterns are also seen in the curves of age-specific rates (Fig. 7). The concurrence between face/foot and neck-trunk/lower limb

as regards time trends, is also depicted on this graph.
A consistent and steep increase of the age-specific
rates with age is shown for the face and the foot
(upper part), whereas the curves for neck-trunk and
lower limb reach their maximum at middle age (lower
part). Similar contrasts in the patterns of age-specific rates for individual sites have been demonstrated
by Lee and Yongchaiyudha (1971) [11]

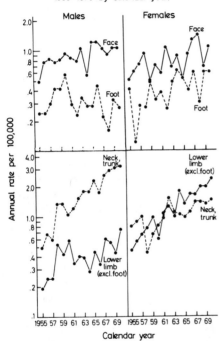

Fig.6. Total age-adjusted incidence rates of
malignant melanoma of various anatomical sites of the skin, in Norway
1955-1970 by calendar year.

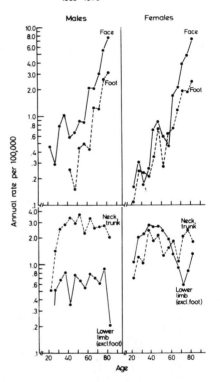

Fig.7. Average annual age-specific incidence
rates of malignant melanoma of various
anatomical sites of the skin, in Norway
1955-1970.

Geographical variations   The incidence of malignant
melanoma for all sites differed distinctly by region
(Fig. 4). When analyzed by site the geographical variations form a more complex picture (Fig. 8). The decrease in incidence from the southern to the northern
parts of the country is clearly demonstrated for the
neck-trunk and the lower limb. The results for the face
and the foot again deviate from this pattern. The face
is relatively rare as the primary site for malignant
melanoma in Oslo. Only the most northern part of the
country reveals lower incidence rates. It seems worth

noticing that for the foot the rates appear to be
stable in the five regions.

DISCUSSION

The rate of increase in the recorded incidence of
malignant melanoma of the skin in Norway during the
last 20 years has been even larger than for cancer of
the lung (Cancer Registry of Norway, 1972)[16].Extrapola-
tion of the observed trends for this period suggests
that the age-adjusted incidence rate of malignant
melanoma is likely to increase 5-6 times during the
next 50 years.

These predictions indicate the relative importance
of malignant melanoma in the future field of cancer.
But how valid are these estimates? An observed increase
in the recorded incidence of a chronic disease should
initially be viewed with scepticism and particularly so
as a basis for extrapolation. One is always faced with
the problem of whether the recorded increase is real or
only an artefact.

The diagnosis of malignant melanoma is based on
histopathological examination, and it might be questioned whether the criteria used by pathologists for
the diagnosis of a malignant melanoma had become less
strict over the study period.

A growing tendency in the population to have "suspicious moles" excised, or an increasing inclination by
the surgeon to submit specimens for histologic examination might also lead to a fictitious rise in incidence
over a certain period. However, great emphasis should
be laid on the difference in time trends for malignant
melanomas of the various anatomical sites. For the face
and the foot, little or no increase in incidence has
been observed. For the neck-trunk among males and the
lower limb among females, a spectacular increase has
taken place. It is difficult to see how cosmetic considerations, a growing awareness of "suspicious moles",
or any diagnostic factor could produce such a pattern.
It seems reasonable to postulare, therefore, that an
ever increasing number of moles recognized as malignant melanoma has appeared on certain sites during the
last decade or two in Norway.

But how fatal are these tumors? Is the rise in incidence accompanied by a corresponding rise in mortality? This is shown in Figure 9 which is based on the
official mortality statistics of Norway, as supplied
by the Central Bureau of Statistics of Norway (1961,
1964, 1974) [12, 13, 14]. Results are shown for total skin
melanoma, as data for individual sites are not available.

There is no doubt about the statistical significance of
the increase, but the slopes are less steep than those
for incidence (Fig. 1), particularly among females.
Could this discrepancy in slopes of incidence rates and
mortality rates be due to a lower degree of malignancy
of the excess melanomas during the later part of the

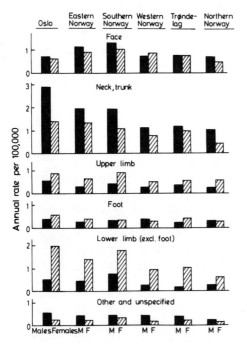

Fig.8. Total age adjusted incidence rate of malignant
melanoma of the skin in Norway 1955-1970
by region and anatomical site.

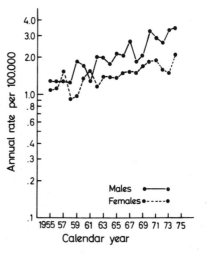

Fig.9. Total age adjusted mortality rate of malignant
melanoma of the skin in Norway 1955 - 1974
by calendar year.

study period? In order to evaluate this problem survival
data for cases diagnosed 1955-1963 and 1964-1971 are
given separately in Table 1. As one might expect from
the mortality data shown in Figure 9 there is a rise in
survival from the first to the second period. However,
the rise in survival is far from prominent for the
sites where the excess melanomas have occurred - the
neck-trunk and the lower limb. The largest progress in
survival is noticed for melanomas of the face among
females, the incidence of which has remained practically
constant. These results seem to point to earlier diagnosis or therapeutic progress rather than a lower degree
of malignancy of the tumors in the later part of the
study period.
How do the epidemiologic patterns fit the hypothesis

that solar radiation is a main factor in the etiology
of malignant melanoma? The degree of human exposure to
sunlight depends on the duration of exposure, the surface exposed, and the intensity of the radiation. The
industrialization and urbanization of Norway during
this century has led to a shift from outdoor to indoor
occupation for many people. On the other hand, economic progress and increasing leisure have created new
possibilities for outdoor sports and recreation. Long
vacations, weekends, and easy transportation have
facilitated access to open-air areas. The net result of
increased indoor occupation and increased open-air leisure as far as sun exposure is concerned is difficult
to evaluate. Less ambiguous is the trend in clothing
habits. There has been an almost consistent trend away
from the total concealment of the human body at the
beginning of this century towards the present mini-
models. The naked skin is no longer taboo - quite the
contrary.

Table 1. Malignant melanoma in Norway (1955-1971): Five
year observed survival rate of localized cases
diagnosed 1955-1963 and 1964-1971, according
to sex and anatomical site

| | Males | | Females | |
Site	1955-63	1964-71	1955-63	1964-71
Face	0.49	0.53	0.55	0.70
Neck-trunk	0.52	0.59	0.65	0.70
Upper limb	0.62	0.72	0.73	0.85
Foot	0.48	0.35	0.62	0.63
Lower limb	0.68	0.64	0.84	0.88
Total	0.55	0.58	0.69	0.78

The attitude to suntan has changed radically. The
idol of the interesting pale and buttoned-up lady has
been transformed into the sporty, suntanned woman. Suntan has become a symbol of mental and physical health.
People now select for their holidays those places where
they can be guaranteed sunshine, as for instance Southern Europe. On the basis of these changing habits and
attitudes one would expect the increase in exposure to
be most prominent for the previously concealed parts
of the human body. Accepting the hypothesis that solar
radiation is a main etiologic facor in malignant melanoma it is therefore not surprising that no or very
little change in incidence has taken place for the face
and the foot, whereas an excessive increase is observed

for the neck-trunk and the lower limb. The particularly
steep rise for the neck-trunk among males and for the
lower limb among females may be associated with typical
masculine and feminine habits of clothing. Outdoor-
working men frequently wear long trousers and bare
torso, whereas bare legs are a very common feature in
the everyday life of women during the spring and
summer.

The north-south gradient in incidence of malignant
melanoma is also in line with the "sun exposure hypothesis". This may not be due to the north-south gradient
in the amount and intensity of sunlight alone, but to
the increase in mean temperature with decreasing latitude as well. Sun bathing may, even during mid-summer,
seem far less tempting in Northern Norway than in the
southern parts of the country because of climatic conditions.

The face and the foot represent the two extremes as
far as sun exposure is concerned. Regardless of clothing habits and temperature the face will be exposed
during outdoor activities. Protection of the scalp and
face from sunlight is much rarer in Scandinavia than
in more southerly countries. The exceptionally low
incidence of malignant melanoma of the face in Oslo
may therefore be due to predominantly indoor occupations and activities, or to a low intensity of ultraviolet radiation because of absorption by atmospheric
pollution. The foot is probably sunexposed only to a
very slight degree, irrespective of residence and
occupation, and it fits the "sun exposure hypothesis"
in that no geographical variations can be detected.

The present data indicate that exposure to the sun
has a local effect, as the increase in incidence of
malignant melanoma is most marked for those sites
where an increase in exposure has taken place. There
is no support in favor of the hypothesis of a systemic
effect. The absence of a North-South gradient and an
increase in incidence of malignant melanomas of the
eye and other non-cutaneous sites seems to contradict
this hypothesis.[15][16][17]

REFERENCES

1. McGovern, V.J. and Brown, M.M.L. (1969): In: The
    Nature of Melanoma, p. 113. Charles C. Thomas,
    Springfield, Ill.
2. Lee, J.A.H. and Carter, A.P.(1970):J.nat.Cancer
    Inst., 45,91.
3. Lancaster, H.O. (1956): Aust.med.J., 1, 1082.
4. Dorn, H.F. and Cutler, S.J. (1965):Publ.Hlth
    Monogr., 56,60.

5. Haenszel, W. (1963):Nat. Cancer Inst.Monogr.10,225
6. McGovern, V.J.(1952): Aust.med.J., 1,134
7. Lancaster, H.O.and Nelson,J.(1957):Aust.med.J.,1,452
8. Gellin,A.G.,Kopf,A.W. and Garfinkel,L.(1969):Arch.
    Dermatol., 99,43.
9. Blum,H.F.(1959):Carcinogenesis by Ultraviolet Light,
    p.293.Princeton University Press,Princeton, N.J.

10. Lee,J.A.H.and Merrill,J.M.(1970):Aust.med.J.,846
11. Lee,J.A.H.and Yongchaiyudha,S.(1971):J.nat.Cancer
    Inst.,47,253.
12. Central Bureau of Statistics of Norway(1961):Medical
    Statistical Reports, 1955-1961,Oslo.
13. Central Bureau of Statistics of Norway(1964):Health
    Statistics,1962-1964,Oslo.
14. Central Bureau of Statistics of Norway(1970):Causes
    of Death, 1965-1974, Oslo.
15. Scotto,J.,Fraumeni,J.F.,and Lee,J.A.H.(1976):J.nat.
    Cancer Inst.,56,489.
16. Cancer Registry of Norway(1972):Trends in Cancer
    Incidence in Norway 1955-1967.Universitetsforlaget.
17. Ericsson,J.L.E. and Ringertz,N.(1972):Läkartidningen,
    60/50,5907.

ESTIMATING THE INCREASE IN SKIN CANCER CAUSED BY INCREASES IN
ULTRAVIOLET RADIATION

Elizabeth L. Scott and Marcella A. Wells

Statistical Laboratory, University of California

Berkeley, California 94720, USA

SUMMARY

There is substantial evidence that ultraviolet radiation
causes skin cancer. Any impact to the environment that will increase ultraviolet radiation, such as depletion of the stratospheric ozone shield, is expected to produce an increase in skin
cancer. The usual method of estimating the increase in skin cancer involves estimating the relation between the occurrence of skin
cancer at different localities and the intensity of ultraviolet
radiation at the same localities. We study the influence of the
difficulties in estimating the relationship: (i) the observations
of skin cancer, and also of ultraviolet radiation, are uncertain
and incomplete, (ii) the carcinogenesis of skin cancer is unclear
and possibly complicated by a complex ultraviolet repair mechanism,
(iii) there are other associations that may be interacting.

A second method is to study the relation between skin cancer
and ultraviolet flux by studying the increase in skin cancer with
age at each of several localities (separately for malignant melanoma and for nonmelanoma). This method is complicated by the
increase in melanoma rates through the years, especially in the
younger population, and by the under-reporting of nonmelanoma.
We need a better understanding of the mechanism of the origin and
growth of skin cancer and of its metastasis. A stochastic model
corresponding to the present experimental observations is suggested.

Estimates of the expected increases in skin cancer due to
increased ultraviolet flux are provided under several models of
action, both simple and synergistic, with estimates of the uncertainty resulting from sampling variability and, roughly, the

imprecision of the models.

INTRODUCTION

We only summarize the evidence that ultraviolet radiation (uv) is a causal factor for human skin cancer, since there have been extensive reviews recently by Emmett (1), Epstein (2) Urbach, Rose, and Bonnem (3), Urbach (4), and others. We use the presentation in the preliminary report of the Panel to Review Statistics of Skin Cancer (5), from which much of this paper continues. The evidence for a causal relationship is epidemiological, clinical, and experimental and consists of six principal categories: (i) the association of skin cancer with exposed areas of the skin (the face, neck, arms and hands, and especially for women, the legs.); (ii) the association with protection against ultraviolet radiation (among races with dark skin, in which the pigment filters out uv radiation, there is very little skin cancer while fair-skinned persons, especially Celtics, are much more susceptible); (iii) the association with amount of exposure to the sun (such as the fact that Caucasians who spend more time outdoors are more likely to have skin cancer); (iv) the association with the intensity of solar radiation was noticed early (towards the equator there is an increase in the prevalence of skin cancer among Causcasians and also an increase in the amount of solar radiation and in the intensity of uv radiation); (v) the association with ultraviolet radiation in laboratory studies (skin cancer can be produced in mice with repeated doses of uv but not with visual wavelengths alone): (vi) the association with poor ability to repair DNA damaged by uv radiation (persons with xeroderma pigmentosum, which indicates an inherited defect in the repair of DNA damage caused by uv radiation, almost always develop skin cancer which usually causes death at an early age).

The two main types of skin cancer are malignant melanoma, the less common but more deadly type, and nonmelanoma, the most common of all cancers among humans. For both types of skin cancer, for both sexes, and for all age groups there is a strong association between uv flux and the incidence rates for skin cancer. The association between uv flux and mortality is better determined (because the data are better) but less pronounced than for incidence, as shown in Figure 1. We can use parallel lines to approximate the relation between the logarithm of the rates and uv flux for each age group (we show only two age groups in Figure 1 to decrease confusion) and each sex with a constant difference due to sex. This is the reason that we have used log rates in all our main computations; other scales do not fit the observations so well and the results are quite insenstive to scale for moderate changes in uv flux and thus for moderate depletion of the stratospheric ozone. Not surprisingly, the two types of cancer and the

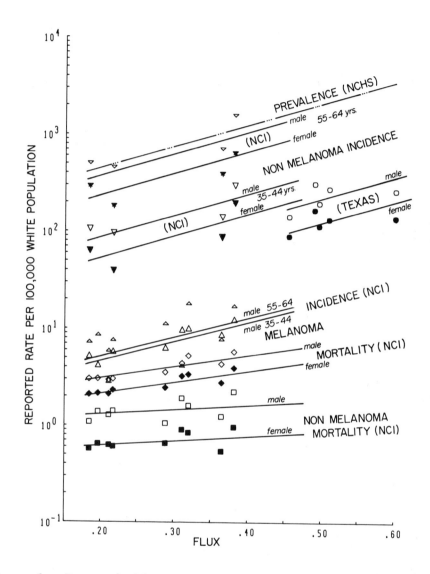

Figure 1. Reported skin cancer per 100,00 white population as a function of increasing weighted effective ultraviolet flux in 10^{16} photons. Points are shown for the 9 TNCS localities.

different kinds of rates show different slopes for the relation
between log rate and uv flux.

We have shown all the data on incidence that are available
for recent years for the United States. In general, it is not
possible to combine data from different surveys due largely to
differences in definitions. This problem is reflected in the
observations collected in the Third National Cancer Survey (6)
of the National Cancer Institute (NCI) compared to those (Texas)
collected by the M. D. Anderson Hospital survey (7) of MacDonald.
Although the levels are not consistent, the same slope can be
used to approximate the relationship between log incidence of
nonmelanoma and flux in each set of data. We use the data with
the widest base in flux, the NCI data, to estimate the relationship
between skin cancer rates and flux, and propose to use the estimate
to predict the relationship for other surveys, including those
in other countries such as the excellent set of data presented by
Magnus (8) at this Symposium, so as to have an independent check
on the relation.

It is important to note that the rates plotted in Figure 1 are
reported crude rates among white. All nonmelanoma rates are difficult to observe; its incidence rates were measured in a special
study (9) in only four localities out of nine localities of the Third
National Cancer Survey. Nonmelanoma is grossly under-reported.
Age-adjusted mortality rates averaged over the period 1950-1969
are available from Mason and McKay (10) for every county in the
United States. We have plotted them only for the same nine
localities for comparison. Melanoma is increasing and so we have
multiplied the rates by 2.3 to adjust them to a standard of 1970,
roughly consistent with the period of the other surveys shown. The
nonmelanoma mortality rates are very uncertain. Preliminary information on prevalence from the Health and Nutrition Examination
Survey (HANES) was kindly supplied by McDowell (11).

The observations illustrated in Figure 1 are used to estimate
how much skin cancer will increase if the uv flux is increased by
a decrease in the protective stratospheric ozone layer or for some
other reason. For any given increase in flux, we can read (or
calculate) from the lines drawn in Figure 1 what the corresponding
percentage increase in each rate would be. All that is needed is
to estimate the slope of each set of lines because this slope
provides the increase in log rate, and thus of percentage increase
in rate, corresponding to any assigned increase in uv flux. This
idea is essentially what was used in our preliminary report (5)
and in other studies (McDonald (12), Urbach, Davies, and Berger (13),
van der Leun and Daniels (14), Green and Mo (15), Scotto, Fears, and
Gori (16)). The difficulties in obtaining good estimates of the
increase in skin cancer using this method arise from the nature of
the observations, and the lack of clear understanding of the mech-

anisms of the origin of skin cancer.

DIFFICULTIES IN ESTIMATING THE INCREASE IN SKIN CANCER DUE TO INCREASING UV

The uncertainties and incompleteness in the observations of skin cancer rate have been described. Data on uv flux is just beginning to be available in a few places for a short period of time, not enough to average out the strong diurnal and seasonal fluctuations and the changes from year to year. We have used the solar irradiance outside the stratospheric ozone layer from Thekaekara (17), multiplied these values at each wavelength by the transmittance throught the stratospheric ozone layer and the clear atmosphere kindly computed for us by Venkateswaran and St. John (18) for 20 wavelengths in the range 297 to 320 nm for a series of values of the depletion of the stratosphere, then multiplied by the transmittance through the layer of clouds and murk at the earth's surface times the transmittance through an individual's clothing and other protection into his skin. Every term in this product, which we call weighted effective flux, is uncertain, especially the sensitivity of the skin as a function of wavelength, the so-called action spectrum. We have used the computations of Setlow (19) corresponding to the damage to DNA of E. coli by radiation, reflecting the evidence that skin cancer arises from the faulty repair of uv damage to DNA. We have used a series of other action spectra, including a standard erythema spectrum, because the resulting estimated increases in skin cancer depend, by at least a factor of two, on the choice made for this sensitivity curve. It is important that animal experiments be extended to study the interactive effect of uv dose-rate-schedule-wavelength in causing skin cancer lesions and in promoting tumor growth.

The weighted effective flux is not well approximated by simple latitude, an assumption made in many studies. This flux is a sharply peaked function which shifts with changes in any of the factors going into it, being especially sensitive to changes in the action spectrum of the skin and to changes in the intensity of the flux at the ground, in particular, the increases in the uv range due to ozone depletion. We have checked that 20 wavelengths are more than enough to use when estimating the increase in skin cancer rates due to increasing uv flux; in fact, only 4 wavelengths carefully handled will give the same results as are obtained with 20 for each locality considered under moderate ozone depletion, contrary to the complaint made by Green (20) about our earlier study (5).

Just which function of skin cancer rate should be considered as dependent on which function of uv flux is not known. The carcinogenesis of skin cancer is uncertain and there may enter a complex uv repair mechanism. As we have noted above, it is important to extend the study of the interactive effect of uv dose-rate-schedule-wavelength on the mechanism of skin cancer origin and growth in laboratory animals and bacteria to try to get an understanding that will extend to humans. In principle, the relationships can be estimated for humans from the data on skin cancer by sex and age for different localities where the uv flux is known. However, as we have just seen, the available observations are unreliable and incomplete. Curve fitting applied to the observations shown in Figure 1 suggests using the logarithm of the rate as a linear function of the total flux because this best approximates the data, and this is what we have assumed for our main calculations. We have also used other functions of the rate, including the rate itself, and other functions of the flux, such as the maximum flux during the year or the flux with a queuing model of uv repair. In general, the estimated increases in skin cancer for a moderate depletion of the ozone (such as 10%) are not much affected by the choice of functions to be used. However, there are exceptions, in particular, the repair model with queuing means less cancer than otherwise for low doses but little change at higher doses when the system is surfeit. This distorts the estimates of skin cancer increase. More biological information is needed.

Not all persons at the same locality receive the same dose of uv since there is a large variation in the amount of exposure to uv with different styles of clothing and different behavior. This means that simple estimates of the slopes of the lines in Figure 1, such as estimates by least squares as used in our preliminary study and elsewhere, will be biased. The variability in flux as well as in rates produces a statistical problem known as "both variables subject to variation" and the estimated slope, when the variability of the flux is ignored, will be too small. We have computed what the underestimate of slope will be using the survey organized by Urbach (21) to study the variability in exposure at the same locality in the two localities, Philadelphia in the United States and Galway in Ireland, for which Urbach has collected observations. Our computations indicate that the slope estimated by least squares should be multiplied by 1.3 to correct for this bias.

The deviations from the lines drawn in Figure 1 are very systematic; any one locality tends to have all of its observations above the corresponding lines or all below. We interpret this consistency in the deviations as an indication that other factors besides uv flux are affecting skin cancer rates in these localities. One locality has a warm humid climate so that the population tends to remove clothing, to spend time on the beaches and so forth, while another locality has a different micro-climate, perhaps an extremely hot, dry climate that forces the population to protect itself from the sun or to spend the middle of the day indoors. We have carried out a series of computations testing what other variables, in addition to uv flux, should be used and estimating the resultant effect of uv flux in the presence of these other variables. In these computations, we have considered total solar radiation, both simply and as a synergistic variable as suggested by the laboratory experiments of Forbes (22), two measures of micro-climate, urbanicity, access to medical care, access to melanoma clinics, proportion of the population that is very fair skinned, and so forth. The results do depend to some extent on whether other variables besides uv flux are used, as shown in Table 1 below.

There are further difficulties in estimating the relationship between the logarithm of the skin cancer rates and the uv flux at the same locality. One is the use of logarithms since we do not want the results as increases in log rates but as increases in rates. This leads us to take the inverse function with results in bias in the estimated increase in rates. We have corrected for this bias, following Neyman and Scott (23). Another statistical difficulty is the spurious correlation introduced by using rates -- the skin cancer observations come to us as rates with the white population as the denominator, while many of our predictor variables are also rates, such as the proportion of very fair-skinned population. The result is a distortion of the estimated relation between skin cancer and flux. By repeating our computations on subsets of the data where the population is rather constant, we find that the estimated relations between skin cancer and uv flux on skin cancer rates is slight. The forthcoming observations from the HANES study and new mortality counts from the National Center for Health Statistics (24) will allow us to make estimates free of this spurious correlation.

SECOND METHOD FOR STUDYING THE RELATION BETWEEN SKIN CANCER AND ULTRAVIOLET FLUX

The increase of skin cancer with age at each of several localities provides a second method for studying the relation between skin cancer and uv flux. The method is illustrated schematically in the left panel of Figure 2, which shows the increase in the log of the age-specific rates with increasing age at two localities, Dallas in the south of the United States and Minneapolis in the north. The solid line is for males and the dashed line for females, with the rates for Dallas consistently above those for Minneapolis. This difference is to be used to estimate the increase in skin cancer with an increase in flux corresponding to the increase from Minneapolis to Dallas. The increase in skin cancer with increasing age compared to the increasing dose of uv flux received through the years provides another estimate of increase in skin cancer with increasing uv flux. The estimates should be consistent. We need to consider an additional complication. Melanoma rates are increasing through the years, especially among the young and in sites which were previously protected from sunlight such as the legs of women.

As noted by Magnus (8) and others, we need to use cohort studies, that is, to study those individuals born in a given time period separately from the individuals born in the next time period, and so forth. The plot in the right panel of Figure 2 illustrates the idea: at I are plotted the skin cancer rates for the persons aged 25-34 years in 1950, at II are plotted the same persons in 1960, now aged 35-44, and at III are plotted the same persons again for 1970 when their ages are 45-54. All these persons were born between 1916 and 1925. Another plot is considered for the persons born between 1926 and 1935, and so forth. There is thus a series of data sets to study how skin cancer rates increase with increasing uv flux; each cohort study will exhibit a component of age and a component of locality, as well as site and sex, which when rescaled to flux and allowing for the increase in melanoma should provide consistent estimates. However, the rescaling to flux requires a better understanding of the mechanism of the origin of skin cancer and of tumor growth and of metastasis, and how these vary with increasing uv dose. The simple cumulative theory (the one-hit model) provides inconsistent results, with a much larger slope from the second method than from the first and thus a much larger value for the estimated increase in skin cancer due to increasing ultraviolet radiation. The experiments with mice of Blum (25) and of Forbes and Urbach (26) suggest a more complicated interactive effect of dose, rate, schedule, and wavelength in the carcinogenesis of skin cancer and its growth.

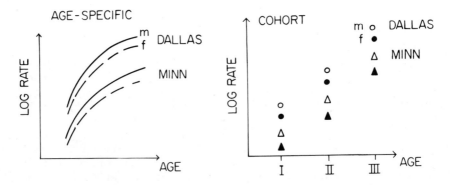

Figure 2. Schematic drawing of the increase in log rate of skin cancer with increasing age, for males and for females, at two diffrent localities which have different uv flux. Left panel-age-specific rates increasing with age; right panel - cohort study of persons born in the same time period observed every ten years for skin cancer rate.

STOCHASTIC MODEL OF CARCINOGENESIS OF SKIN CANCER
AND OF TUMOR GROWTH WITH UV RADIATION

We cannot perform experiments on men but there are experiments on mice. We can try to build a stochastic model that will "explain" the observations in the experiments on mice and that can be correlated with models of bacterial experiments, and then try to extend the models to man utilizing the better understanding of the process. What we describe here is the result of a cooperative seminar in the Statistical Laboratory with Darden, Hammerstrom, Preisler, Sid and Troya and we are indebted to our colleagues for allowing us to describe this joint study. The model is an extension of earlier two-stage models such as that of Neyman and Scott (27), with a cooperative analysis using the experiments already carried out or now being made by Forbes (26).

A typical experiment includes the following data: each of 48 mice receives the same dose-rate-schedule-wavelength treatment of uv radiation. Other groups receive slightly different treatments. Once a week each animal is observed and any lesion is mapped for location and its major and minor diameters are measured and recorded. We propose to use these diameters from week to week, the times to tumors of preassigned size, and the joint occurrence of tumors of size k_1, k_2, say, to construct a stochastic model of carcinogenesis and tumor growth in these mice as it depends on dose-rate-schedule-wavelength interaction.

Suppose we consider the scheme illustrated in Figure 3. We assume that there are many normal cells (essentially infinitely many), each of which has the instaneous probability $\gamma_1(t)$ of suffering mutation, or a mutation-like change, to a state which is often called <u>promoted</u>. Let $X_1(t)$ be the number of promoted cells at time t. Then $X_1(t)$ is an instaneous inhomogeneous Poisson $\gamma_1(t)$ process that undergoes an independent birth and death pro-

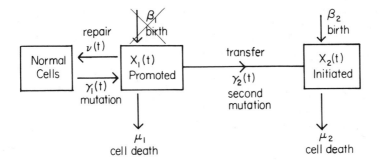

Figure 3 - Stochastic model for the carcinogenesis of skin cancer

cess with parameters β_1 for birth and μ_1 assumed constant in
time as a first approximation. In addition there is a rate $\nu_1(t)$
of repair returning the mutant to normality. Each cell in the
promoted state is subject to a second mutation-like change to the
initiated state with probability $\gamma_2(t)$. Let $X_2(t)$ equal the
number of cells in the initiated state at time t. Then $X_2(t)$
is an inhomogeneous Poisson $X_1(t)\gamma_2(t)$ process that undergoes
an independent birth-death process (β_2, μ_2). Neither $X_1(t)$ nor
$X_2(t)$ is observable. If $\beta_1 = 0$, then we can write the solution
for the probability generating function of $X_2(t)$ and trace the
fate of each mutation; otherwise, it is very difficult. Suppose
we put $\beta_1 = 0$ as a first approximation.

Let $N(t)$ denote the number of first-order mutations, that
is, promotions that occur in the time interval $[0,t]$. Let
$X_{1i}(t) = 1$ if the i-th first-order mutation is in the promoted
state at time t, and let it be zero otherwise (if has not yet
occurred, is repaired, initiated, or dead). Let $X_{2i}(t)$ equal
the number of cells in the promoted state that are descended from
the i-th mutation. We handle the problem as a diffusion process.
The solution is known explicitly for the joint distribution that
simultaneously $N(t) = n$ and that each of the $X_{2j}(t)$ takes on a
preassigned value x_j for $j = 1,\ldots,n$.

Now consider the number of tumors of a specified size. Let
$Y_1(t)$, $Y_2(t),\ldots,Y_k(t),\ldots$ be the number of tumors with exactly
1 cell, 2 cells,$\ldots,k$ cells,$\ldots$, respectively, at time t. If
$k \geq k_0$ assumed known roughly, then $Y_k(t)$ corresponds to an observable tumor; otherwise the tumor is too small to notice. In order to
construct the counts, introduce $Z_{ki}(t)$ equal unity if the i-th promoted cell gave rise to a tumor of k cells. Then $Y_k(t)$ is the
sum of the $Z_{ki}(t)$ summed for $i = 1,\ldots,N(t)$, for each $k = 0, 1,\ldots$.
We recall that $N(t)$ is the number of promotions up to time t and
that it is a Poisson process with parameter equal to the integral of
$\gamma_1(\tau)$ from 0 to t. Now we can compute the expected value of $Y_k(t)$
which is the same as its variance under this model. We can also
compute the joint distribution that there will be exactly n_1 tumors
of size 1 cell, and simultaneously exactly n_2 tumors of size 2
cells, and so on over the total number of tumors, some of them in-
visible.

If we observe each visible tumor over a period of weeks, assuming that its two diameters provide an approximation of the number of
cells, we are taking observations on a birth and death process for
which we have transition probabilities in terms of β_2, μ_2. If we
assume that the tumor grows like a diffusion process (can assume instead that grows only on outer shell, etc.), then it is easy to write
the explicit solution as a Bessel function.

Suppose that we observe a tumor at times $t_1, t_2, \cdots, t_m$ and record the size to be $s_1, s_2, \cdots, s_m$. Then we can estimate the rates of birth and of death, β_2 and μ_2 for this tumor by regression if we assume that s_i is a constant times the number of cells in the tumor at time t_i. We can now study whether different tumors (the first, the second, etc.) have the same parameters β_2 and μ_2 for their birth and death processes. We can also check whether different mice have the same parameters or nor. We can study the growth process of the tumor by doing the computations for successive blocks of weeks to see whether it is "growing" at a constant rate or is growing only on the circumference, and so forth. We will carry out the computations separately for each tumor on each mouse, and study the variability of the results.

We can also study the two mutation rates since the output of this stochastic model is the product of two time-dependent Poisson processes whose parameters involve the mutation rates. For example, suppose that $\gamma_1(t) = aD(t)$ and $\gamma_2(t) = c_1 + c_2D(t)$ as simple cases of any polynomials in $D(t)$ which can be used. Here, $D(t)$ is the instantaneous dose, known, while a, c_1, c_2 are unknown constants to be estimated. The c_1 term is to account for those transfers from promotion to initiated that take place without more uv (perhaps through a chemical) and c_2 is to account for the promoted cells initiated by a second bit of uv damage. We now can obtain the distribution of $Y_k(t)$, the number of tumors of size k cells, and then get a, c_1, c_2 and $\mu_1 + \nu$. By using different length time intervals we can determine whether the linear relation $\gamma_2(t) = c_1 + c_2D$ fits the observations or whether D^2 and higher order terms must be included. However, since $Y_k(t)$ is usually 0 or 1, that is, there are no tumors or perhaps one with exactly k cells in it at a specified time t, we want to find a better method for estimating $\gamma_1(t)$ and $\gamma_2(t)$.

The observations of T_1, the waiting time to the first tumor of size at least k_0 cells, provides this possibility. The empirical distributions of the observed T_1 over the 48 mice are hazard functions, which can be fitted to find $\gamma_1(t)$ and $\gamma_2(t)$. We can also use T_j, the time to the j-th tumor. The estimates will be further improved if we extend our computations to the joint distribution of $Y_{k1}(t)$, $Y_{k2}(t)$.

Skin cancer provides an excellent opportunity to study tumor origin and growth: the lesions can be observed without sacrificing the animals, and they can be followed over time and their growth studied. This is a wonderful opportunity to study details of the stochastic process with changes in the treatment, for example, to study the repair rate $\nu(t)$ as a queuing process under heavier and heavier doses of uv radiation.

PRELIMINARY ESTIMATES OF THE INCREASE IN SKIN CANCER DUE TO INCREASING UV RADIATION CAUSED BY REDUCING OZONE

We have estimated the increase in skin cancer rates (mortality, incidence, and prevalence) due to increasing uv radiation under many sets of assumptions about the underlying processes. The results are too extensive to present in full and, in many cases, the resulting changes in our estimates are small. Therefore, we present results for two types of basic models, one set has uv flux as the only predictor for skin cancer and the other set corresponds to a linear combination of predictor variables with a possible interaction (synergism) between uv flux and general solar radiation (the insolence, almost none of which is uv) in causing cancer.

The estimated percentage increase in skin cancer depends on the characteristics of the locality. We have calculated the increase for each of the localities occurring in one of the surveys in the United States, for 75 localities in all. For brevity, the results for only two localities, Minneapolis-St. Paul and Dallas-Ft. Worth, are shown in Table 1. The results can be given for any specified ozone depletion; we restrict attention to the case of 10% ozone depletion (which is approximately 20% increase in uv flux). We have also computed a 90% confidence interval for each estimated increase but there is not enough space to table these.

Although the estimates are not so well determined as one would wish, the results in Table 1 show that there is a strong effect of uv flux on skin cancer, and that this effect persists even when other possible predictors are taken into account. It turns out that the numerical size of the percentage increase in melanoma, taking into account other possible predictors, is greater than that for nonmelanoma for several data sets. Part of the reason is that the data are more reliable for melanoma, as we noted earlier. Our results are still preliminary but they indicate that uv flux is the most important factor in both types of skin cancer and that other predictors, such as temperature and pigmentation, also enter.

ACKNOWLEDGEMENTS

The authors are indebted to the members of the Panel to Review Statistics in Skin Cancer, Peter Bloomfield, Philip Cole, Richard A. Craig, Harley A. Hanes, and Richard B. Setlow, for continuing help and advice. We appreciate the cooperation of our colleagues in the Statistical Laboratory, University of California, Berkeley who participated in the Skin Cancer Seminar and many more who helped in the analysis.

This paper was prepared with the partial support of the National Institutes of Health research grants No. ESO1299-13 and NO1-ES-6-2128

Table 1. ESTIMATED PERCENTAGE INCREASE IN SKIN CANCER CORRESPOND-
ING TO 10% OZONE REDUCTION (ABOUT 20% INCREASE IN UV FLUX) ACCORD-
ING TO SEVERAL ASSUMPTIONS OF CAUSE OF CANCER.

Disease	Rate	Data Source	No. of Localities	Possible Causes	Estimated Increase Minn–StP	Estimated Increase Dall–FtW
Skin cancer	Incid.	SNCS Auerbach	10	F only	62%	142%
				F,S,FS	34	38
Skin can. + severe keratosis	Preval.	HANES McDowell	first 35	F only	44	95
				F,S,FS	42	90
Nonmel.	Incid.	TNCS Scotto	4	F only	35	79
	Mort.	Mason & McKay	9 in TNCS	F only	5*	10*
			75 in HANES	F only	3*	5*
				F,S,FS	5*	− 8*
				F,S,K_1,K_2,M,P	5*	10*
Melanoma	Incid.	TNCS Cutler	9	F only	30	66
	Mort.	Mason & McKay	9 in TNCS	F only	17	36
			75 in HANES	F only	8	16
				F,S,FS	9	− 3*
				F,K_1,K_2	9	17
				F,S,K_1,K_2,M,P	7	13

F = uv flux, S = solar radiation, K_1 = microclimate very hot,
K_2 = microclimate hot and humid, M = access to melanoma clinics,
P = differential pigmentation among whites,
* = confidence interval covers zero.

REFERENCES

1. Emmett, E. A. (1973) in CRC Critical Reviews in Toxicology 2, 211-255.
2. Epstein, J. H. (1970) in Photophysiology, A. C. Giese, ed. (Academic Press, N.Y.) 5, 235-273.
3. Urbach, F., Rose, D. B., and Bonnem, M. (1972) in Environment and Cancer: A Collection of Papers (Williams and Wilkins, Baltimore, Md.) 355-371.

4. Urbach, F. (1975) in Impacts of Climatic Change on the Biosphere, A. J. Grobecker, ed. (Washington, D.C.) CIAP Monograph 5, 7-13 to 7-28.

5. Scott, E. L., Bloomfield, P., Cole, P., Craig, R. A., Haynes, H. A., Setlow, R. B., Straf, M. L., and Woolsey, T. D. (1975) in Environmental Impact of Stratospheric Flight (Nat. Acad. Sci., Washington, D.C.) 177-221.

6. Cutler, S. J. and Young, J. L., Jr. (1975) Third National Cancer Survey: Incidence Data (Nat. Canc. Inst., Bethesda, Md.) Monograph 41.

7. Macdonald, E. J. (1974) M. D. Anderson Hospital, manuscript.

8. Magnus, K. (1973) Cancer 32, 1275-1286.

9. Scotto, J., Kopf, A. W., and Urbach, F. (1974) Cancer 34, 1333-1338.

10. Mason, T. J. and McKay, F. W. (1974) U.S. Cancer Mortality by Country: 1950-1969, DHEW No. (NIH) 74-615. (U.S. Govt. Printing Office, Washington, D.C.).

11. McDowell, A. (1974) Nat. Cent. Health Statist., personal communication.

12. McDonald, J. E. (1971) statement submitted at hearings before House Subcommittee on Transportation Appropriations, March 2, 1971.

13. Urbach, F., Davies, R. E., and Berger, D. (1975) in Impacts of Climatic Change on the Biosphere, A. J. Grobecker, ed. (Washington, D.C.) CIAP Monograph 5, 7-42 to 7-61.

14. van der Leun, J. C. and Daniels, F., Jr. (1975) ibid. 7-107 to 7-124.

15. Green, A. E. S. and Mo, T. (1975) ibid. 7-127 to 7-140.

16. Scotto, J., Fears, T. R., and Gori, G. B. (1975) Measurements of Ultraviolet Radiation in the United States and Comparisons with Skin Cancer Data (Nat. Canc. Inst., Bethesda, Md.) DHEW No. (NIH) 76-1029.

17. Thekaekara, M. P. (1974) Applied Optics 13, 518-522.

18. Venkateswaran, S. V. and St. John, D. E. (1976) Transmittance of Solar Ultraviolet Radiation at 20 Wavelengths from 297 to 320 nm for 74 Localities, preprint.

19. Setlow, R. B. (1974) Proc. Nat. Acad. Sci. USA 71, 3363-3366.

20. Green, A. E. S. (1975) Additional Material for the Record, House Subcommittee on Environment and Atmosphere, July 31, 427.

21. Urbach, F., Epstein, J. H., and Forbes, P. D. (1974) in Sunlight and Man, T. B. Fitzpatrick, et al., eds. (Univ. Tokyo Press, Tokyo) 259-283.

22. Forbes, P. D. (1973) Amer. Soc. Photobiol. Abs. (Sarasota, Fla.) 136.

23. Neyman, J. and Scott, E. L. (1960) Annals Math. Statist. 31, 643-655.

24. Patterson, J. E. (1976) Standardized micro-data tape transcripts, 1968-1974, Nat. Cent. Health Statist., Rockville, Md.

25. Blum, H.F. (1959) Carcinogenesis by Ultraviolet Light (Princeton Univ. Press, Princeton, N.J.).

26. Forbes, P. D. and Urbach, F. (1975) Fd. Cosmet. Toxicol. 13, 335-346.
27. Neyman, J. and Scott, E. L. (1967) Proc. Fifth Berkeley Symposium on Math. Statist. and Prob. (Univ. Calif. Press, Berkeley, Calif.) 4, 745-776.

SYMPOSIUM XIV

VISION

MSP MEASUREMENTS OF ROD AND CONE VISUAL PIGMENTS IN A RETINA (<u>SCARDINIUS ERYTHROPHTHALMUS</u>) THAT CAN BE EITHER VITAMIN A_1- OR VITAMIN A_2-BASED

E. R. Loew and H. J. A. Dartnall

Medical Research Council's Vision Unit

University of Sussex, Brighton, England

SUMMARY

MSP measurements on populations of the Rudd, <u>Scardinius erythrophthalmus</u>, show that the outer segments of both rods and cones have variable absorbance spectra. It is known from work on extracts that the variability of rod pigment is due to the presence of mixed A_1- and A_2-based pigments in different proportions. Since the λ_{max} of individual "red" and "green" cones correlate with the λ_{max} of nearby rods, it is concluded that the "red" and "green" cones similarly contain mixtures of A_1- and A_2-based pigments.

INTRODUCTION

On extraction with digitonin solution, the Rudd retina yields both A_1- and A_2-based visual pigments[1]. Proportions of these two rod pigments depend on several factors, such as daylength[1], habitat[2], the age of the fish[3] and the retinal location[4,5]. All our knowledge of this mixed visual pigment system has come from analysis of extracts or from retinal patch measurements and relates only to the rod pigments. This study attempts to answer two questions: first, what spectral classes of cone pigments occur (e.g. red-, green- and blue-sensitive)?; second, do the cones, as do the rods, contain mixtures of A_1 and A_2 pigments and, if so, is the mixture ratio the same for rods and cones from the same retina?

METHODS

The Rudd were taken from two stock populations maintained in the laboratory.   One population had been kept in continuous darkness for two years, the other under a regime of alternate 12-hr artificial light and 12-hr darkness.

Techniques of retinal isolation and preparation for MSP measurements on the Liebman instrument at Sussex were as previously described(6,7).   Since the ratio of A_1- to A_2-based pigment in rods is known to vary in different retinal regions(4,5), each retina was cut into 1.0mm^2 pieces under infra-red viewing and visual cells from each piece were examined separately on the MSP.   In this way, regional differences in pigment composition were avoided. Rods from a particular piece of retina rarely varied by more than 3 nm in λ_{max}, while those from different pieces of the same retina could vary by as much as 10 nm.   Smooth curves were drawn through the middle of the noise bands of the MSP records for unbleached rods and cones and likewise through the accompanying baseline records.   From these data, absorbance spectra of the various receptors were constructed and estimates of λ_{max} made as described below.

RESULTS

In the case of rods, the highest spectral locations of the absorbance maximum were clustered around 540 nm and the lowest around 517 nm.   Those rods with maxima around 540 nm contained pure A_2 pigment (P540$_2$),for exposure to red light did not cause a progressive shift of absorbance to shorter wavelengths during bleaching.   The rods with lowest placed maxima, when similarly bleached, however, were found to contain substantial amounts of P540$_2$, showing that the pure A_1 rod pigment must lie below 517 nm.   Using the Dartnall/Lythgoe formula(8),

$$\lambda_{max}(A_1) \quad = \quad 0.6 \ \lambda_{max}(A_2) \quad + \quad 186$$

it is calculated that the P540$_2$ should pair with a P510$_1$. Values for the λ_{max} of the rod records were therefore obtained in the following way.   A series of standard spectra for mixtures of P540$_2$ and P510$_1$ in various proportions were constructed from the appropriate nomograms, and from them a chart was prepared to relate pigment composition with the wavelength at which various percentages of maximum absorbance were attained.   With this chart ten values of λ_{max} were obtained for each original record and the mean value (maximum standard deviation $\pm$ 3 nm) was used.

The "green" cones had λ_{max} clustered around 523 nm at the longwave extreme and 507 nm at the shortwave extreme.   By analogy with the rod records we assume that the 523-nm cones contained a pure A_2-based pigment.   From the above formula, a $P523_2$ porphyropsin should pair with a $P500_1$ rhodopsin and the analyses of the records were carried out as before, but from a chart prepared from the nomogram spectra of those pigments.

The "red" cones had λ_{max} ranging from around 620 nm down to around 575 nm.   By analogy with the rod records we assume that the 620-nm cones contained a pure A_2-based pigment.   From the formula again, the A_1-based pigment pairing with a 620 nm 'cyanopsin' is a 562 nm 'iodopsin'.   The records with λ_{max} between 620 and 575 nm were then analysed in a similar way, this time using a chart prepared from data for pure cyanopsin(9) and pure iodopsin(10).

The above spectral assignments are based on values of over 20 spectra from each of the "red" and "green" cones.   In addition, a small number of "blue" cones were observed in pieces of peripheral retina and found to have λ_{max} between 450 and 460 nm.   Further analysis of these was not attempted, however, in view of the fact that only six records were obtained.

Scholes(11) identified five cone types from sections of Rudd retina.   These were principal and accessory members of double cones, free principal cones, single cones and oblique cones.   In our preparations we could identify three of these types; the principal, the accessory, and the oblique.   Since the double cones were usually separated we could not distinguish positively between principal members of doubles and free principals, nor between accessory members of doubles and singles. This makes the statement, in the next paragraph, about the spectral assignments to cone types only a probable one, for it is conceivable that we missed the free principals and singles, though we think that unlikely.

We provisionally assign the five cone types to three spectral regions, as did Scholes.   The principal and free principal cones absorb maximally between 575 and 625 nm, the accessory and single cones between 507 and 527 nm, and the oblique cones between 450 and 460 nm.

Does the variation observed in cones of the same type but from different fish result from variation in the proportions of A_1- and A_2-based visual pigments, as is the case with rods? Because of their small size it is not practicable to subject the cones to partial bleaching tests as can be done with rods. It was, there-fore, decided to test whether there existed a correlation between the λ_{max} of cones and those of nearby rods. The spectra of 11 rods and a number of cones of each type from a particular piece of retina were measured. With each piece of retina, the λ_{max} value for the 11 rods were cal-culated and averaged, as were those for the cones of each class. The average λ_{max} values were then plotted on a graph relating the λ_{max} of the cones (abscissa) to the λ_{max} of the rods (ordinate). This graph, embodying data from pieces of retina from 20 different fish, is shown in Fig. 1. The correlation between the λ_{max} of both "red" and "green" cones and that of nearby rods strongly sug-gests that the variation stems from variation in the A_1/A_2 ratios, as in the rods.

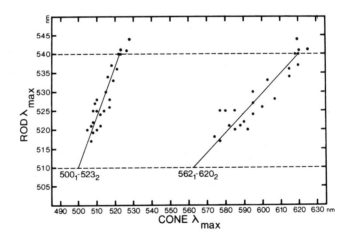

Fig. 1. λ_{max} of "green" and "red" cones versus λ_{max} of nearby rods. The filled circles are data from individual pieces of retina of 20 fish kept on 12-hr days for almost 2 yr. The broken lines are drawn at values of rod λ_{max} corresponding to pure A_1 (510 nm) and pure A_2 (540 nm) pigments. The slanting lines connect the pigment pairs positions for the "green" (500_1 - 523_2) and "red" (562_1 - 620_2) cones(12).

DISCUSSION

The present evidence that the cones of the Rudd contain mixtures of A_1- and A_2-based visual pigments, the ratios of which are identical or nearly so to those in nearby rods, implies that both types of photoreceptors draw their chromophoric groups from a common pool. As suggested by Bridges(2) it is likely that this pool is supplied by the pigment epithelium. This being the case, those factors already known to affect the A_1/A_2 ratio in rods (daylength, habitat, age, etc.) could similarly affect the visual pigment composition of the cones.

REFERENCES

1. Dartnall, H.J.A., Lander, M.R. & Munz, F.W. (1961) In: Progress in Photobiology (Ed. by Christensen, B. Chr. & Buchmann, B.), pp. 203 - 213, Elsevier, Amsterdam.
2. Bridges, C.D.B. (1972)   In: Handbook of Sensory Physiology, VII/1. (Ed. by Dartnall, H.J.A.), pp. 417 - 480, Springer, Berlin.
3. Bridges, C.D.B. & Yoshikami, S. (1970) Vision Res. 10 1315 - 1332
4. Muntz, W.R.A. & Northmore, D.P.M. (1971)   Vision Res. 11, 551 - 561
5. Denton, E.J., Muntz, W.R.A. & Northmore, D.P.M. (1971)   J. mar. biol. Ass. U.K. 51, 905 - 915
6. Bowmaker, J.K., Loew, E.R. & Liebman, P.A. (1975) Vision Res. 15, 997 - 1003
7. Loew, E.R. (1975)   J. Physiol. Lond. 251, 47P
8. Dartnall,H.J.A. & Lythgoe, J.N. (1965) Vision Res. 5 81 - 100
9. Wald, G., Brown, P.K. & Smith, P.H. (1953) Science 118, 505 - 508
10. Wald, G., Brown, P.K. & Smith, P.H. (1955) J. gen. Physiol. 38, 623 - 681
11. Scholes, J.H. (1975)   Phil. Trans. R. Soc. Lond.B. 270, 61 - 118
12. Loew, E.R. & Dartnall, H.J.A. (1976) Vision Res. 16 891 - 896

THE IONOCHROMIC BEHAVIOR OF GECKO VISUAL PIGMENTS

Frederick Crescitelli

Department of Biology

University of California, Los Angeles

The lizards of the family Gekkonidae have been found to possess two photolabile, vitamin A_1-based pigments within their visual cells. In the Tokay gecko (Gekko gekko) these two pigments have spectral absorbances maximal at 521nm and 467nm. They have been detected and studied by the methods of microspectrophotometry (1,2) and of extraction (3).

In response to certain physical and chemical treatments the two pigments appear to be significantly different. When solubilized in digitonin, pigment-521 shifts its spectral absorbance hypso-chromically as the temperature is raised from about 2°C to about 25°C (4). This thermometric or thermochromic effect is a reversible action for the pigment absorbance returns to its original spectral position on recooling. Raised to above 25°C pigment-521 begins to denature irreversibly. This special temperature sensitivity is a property of the solubilized pigment-521, for in situ the spectral absorbance is the same (521nm) at 25°C and at 5°C. This means that action spectra can only be compared with extract spectral absorbances providing the latter are measured at the low temperature. In contrast to pigment-521, the second pigment, present as only 8% of the total photopigment density in extracts, is less sensitive to temperature and will survive temperature treatments that completely destroy pigment-521.

Solubilized pigment-521 also shifts its absorbance to shorter wavelengths on adding an -SH poison such as p-chloromercuribenzoate or p-mercuriphenyl sulfonate (5). This, too, is reversible, for on adding dithiothreitol (Cleland's reagent) to the poisoned pig-ment the original spectrum is quickly restored. Furthermore, the prior addition of Cleland's reagent to an extract, itself without

effect, prevents the hypsochromic response to the subsequently
added mercurial poison. This thiochromic action is interpreted
to mean that -SH groups are involved in the structure of the
chromophoric group of pigment-521.

The most intriguing effect discovered for pigment-521 is an
ionochromic action that may be called the chloride effect. This
effect has appeared in a number of different types of experiments
conducted with extracts of the gecko retina. In one type, two
groups of retinas are extracted with digitonin dissolved in two
different buffers: veronal-HCl and Tris maleate. Prior to extraction the retinas are hardened in 4% potassium alum and thoroughly washed with twice-distilled water and with the buffer made
up in the twice-distilled water. The spectral absorbance curves
for these two extracts are significantly different in spectral
position, though similar in form. For the pigment in veronal-HCl
buffer the spectral maximum is at 521nm which is the normal position,
i.e., the spectral location for the pigment in situ. For the pig-

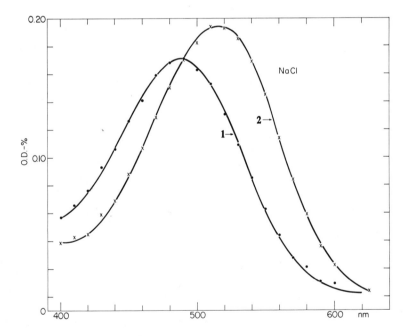

Fig. 1: The chloride effect. Curve 1: the pigment dissolved in
1% digitonin with Tris-maleate buffer (pH 7.2). Curve 2: after
adding NaCl to a concentration of 7.96 x 10^{-3}M. Temperature:
3.6°C.

ment in the Tris buffer the maximum is at about 485nm. Initially,
I interpreted this to signify the presence of two different photopigments each extracted differently. This is not correct, however,
for on adding a chloride to the pigment in the Tris buffer, the
spectrum at 485nm quickly shifts bathochromically toward 521nm.
This chloride effect is a very specific response to the anion for
the same action is seen with NaCl, LiCl, KCl, RbCl, CaCl$_2$, MgCl$_2$,
BeCl$_2$, CdCl$_2$, and LaCl$_3$. Even the organic chloride, choline
chloride, gives the same shift in spectrum. Other anions, e.g.,
sulfate, phosphate, bromate, iodide, fluoride, thiocyanate fail in
evoking a bathochromic response with the pigment in the hypsochromic, chloride-deficient state. Only with bromide ions have I
succeeded in obtaining a chloride effect. The nature of the effect
suggests that the size of the anion is critical. The small fluoride
ion and the large iodide ion fail to elicit the spectral shift
while chloride and bromide, intermediate in size to the other two
halides, are both effective.

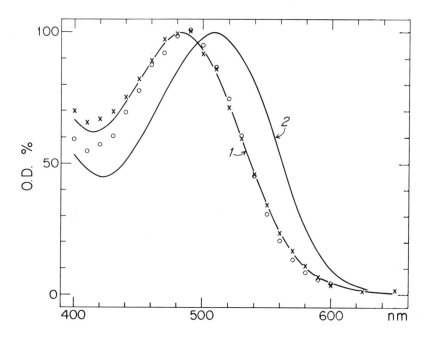

Fig. 2. Specificity of the effect. Curve 1: full line, pigment
in 2% digitonin, borate buffer (pH 8.8); X, after adding NaF to
7.4×10^{-3}M; O, after adding NaI to 7.70×10^{-3}M. Curve 2: after
adding NaBr to 1.03×10^{-1}M. Temperature: 3.3°C.

The chloride ionochromic response is a quantitative effect related to the chloride concentration. Chloride added to make a concentration of about 3×10^{-5}M is sufficient to produce a measurable shift in spectrum. Increases above this value, up to about 5×10^{-3}M, result in progressively greater shifts. Above 5×10^{-3}M there is little or no further shift. The maximum position to which the pigment is shifted is the normal position at 521nm.

The chloride effect is an easily reversible one for on diluting an extract (which already has been chloride-shifted by about 5 to 10 nm) and thus diminishing the chloride concentration, the spectrum shifts back toward its original hypsochromic, chloride-deficient state. It needs to be pointed out that this ionochromic behavior with chloride or bromide is not a general one. Frog rhodopsin,

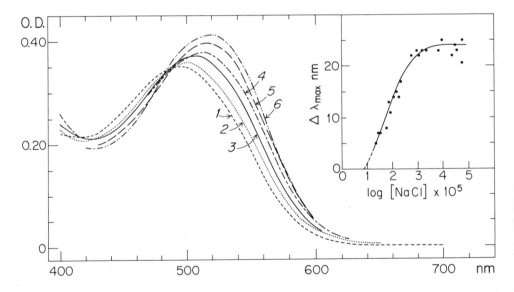

Fig. 3. Chloride effect, quantitative action. Curve 1: pigment in 2% digitonin, borate buffer (pH 8.8). Curves 2 to 6: adding NaCl to increasing concentrations, i.e. 2.70×10^{-4}, 8.25×10^{-4}, 1.81×10^{-3}, 1.04×10^{-2}, 3.75×10^{-1}M. Inset: bathochromic shift of λ_{max} as function of log NaCl molarity. Temperature, $3.6°$C.

extracted and treated in the same manner, shows no evidence of an
ionochromic reaction. It is not in an hypsochromic state when
extracted and it remains at its normal position of 502nm even after
adding NaCl. Insofar as it has been possible to test, pigment-467
of Gekko gekko, present in the same extract along with pigment-521,
fails to show a chloride effect. Why it has been possible to prepare pigment-521 in an ion deficient state, and not frog rhodopsin,
and what, if any, is the normal ion in the cell that adjusts the
position at 521nm, are questions not answerable at this writing.

The mechanism of action of this chloride effect cannot be
precisely defined with the present state of knowledge. We know
nothing of the composition or the structure of pigment-521 and how
it differs from cattle rhodopsin, for example. If we assume that
it is like other visual pigments in having 11-cis retinal linked
to a free amino group as a protonated Schiff base (6) then we can
argue by analogy from what is known about cattle rhodopsin, the most
thoroughly studied visual pigment. The high specificity of the
chloride effect, the response to such low concentrations of this
anion, the rapidity and easy reversibility of the effect, and its
occurrence at low temperatures (2°-5^oC) all argue against the usual
ion effects on proteins, such as the lyotropic effects, the ionic
strength effects, the chaotropic actions. For the moment, at least,
an interpretation that seems reasonable is that of an electrostatic
effect by the chloride or bromide ions at a very specific site in
the opsin. If we again assume the existence of a secondary stabil-
izing charge interaction between the positive charge of the polyene
chain of retinal and specific charged negative groups of the opsin
in the region of the chromophore (6,7) then we can postulate a
specific action on this charge interaction by the chloride ions
located at specific sites on the opsin. We must assume that for
the case of pigment-521, and not for frog rhodopsin, essential ions
are removed in the extraction procedure that displace the normal
spectrum to its deficient, hypsochromic state. The addition of
chloride simply restores the charge-stabilizing electrostatic
effect. The action of chloride could be a direct effect on the
secondary charge field or it could be an indirect effect that
involves alteration of the conformation of the opsin in or near the
region of the chromophore.

REFERENCES

1. P. A. Liebman, Handbook of Sensory Physiology, VII/1 Photochemistry of Vision, ed. H. J. A. Dartnall, Springer-Verlag,
   Berlin-Heidelberg, 1972, 481.
2. F. Crescitelli, E. Loew & H. J. A. Dartnall, Fed. Proc., 35,
   1976, 562.
3. F. Crescitelli, J. Gener. Physiol., 47, 1963, 33.
4. F. Crescitelli, Vision Res., 14, 1974, 243.

5. F. Crescitelli, Vision Res., $\underline{15}$, 1975, 743.
6. E. W. Abrahamson & J. R. Wiesemfeld, Handbook of Sensory
    Physiology, VII/1 Photochemistry of Vision. ed. H. J. A.
    Dartnall, Springer-Verlag, Berlin-Heidelberg, 1972, 69.
7. R. Hubbard, Visual Problems of Colour, Symp. #8, National
    Physical Laboratory, Her Majesty's Stationery Office, London,
    $\underline{I}$, 1958, 151.

ACKNOWLEDGEMENT

     Aided by a Research Grant from the Division of Research Grants
and Fellowships, National Eye Institute, National Institutes of
Health, U.S. Public Health Service.

MEMBRANE ADAPTATIONS OF VISUAL PHOTORECEPTORS FOR THE ANALYSIS OF PLANE-POLARIZED LIGHT

Timothy H. Goldsmith

Department of Biology, Yale University

New Haven, Connecticut, 06520, U.S.A.

The ability to analyze plane-polarized light is a remarkable feature of the visual system of many insects and other arthropods. It seems remarkable because it is foreign to the direct experience of most observers, being appreciated by the human eye only with the aid of extraocular devices such as polaroids, or only weakly through the entoptic phenomenon known as Haidinger's brushes. The sensory basis resides in the receptors, but not in a profoundly different chemistry of visual pigments. It arises first, in the way the pigment molecules are organized within the photoreceptor membranes, and second, in the manner the membranes are assembled in the receptor cells. Of course there is also an appropriate neural circuitry, but this lies beyond the scope of the present discussion.

In the popular contemporary view of cell membranes (1) globular proteins are inserted into a lipid bilayer whose hydrocarbon chains are in a fluid state. The depth of insertion of each protein molecule depends on its surface distribution of polar and non-polar groups. Although this distribution of hydrophylic and hydrophobic surfaces imposes restraints on tilting motions of the protein, the molecules are able to diffuse laterally and to rotate around molecular axes that are normal to the plane of the lipid bilayer. In fact, the disk membranes of amphibian rod outer segments have provided one of the clearest examples of these kinds of molecular motion (2, 3, 4, 5).

The transition moments for photoexcitation of the vertebrate visual pigment rhodopsin lie nearly coplanar with the disk membranes (6, 7, 8, 9), and, as shown in Fig. 1, it is possible to resolve the absorption vectors into three mutually perpendicular

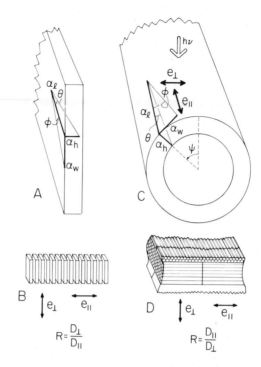

Fig. 1 Comparison of the photoreceptor membranes of vertebrate
outer segments (A and B) and arthropod microvilli (C and D), modified from (10) and (11).

   A. Vertebrate disk membrane. The absorbance of rhodopsin can
be resolved into three mutually perpendicular coefficients, as described in the text. Because of molecular rotations, $\alpha_1 = \alpha_w$.

   B. The rod or cone outer segment consists of a stack of membranes, which exhibit dichroism from the side. Absorbance with
$e_\perp$ is greater than with $e_\parallel$ (axes referred to the stack). Dichroic
ratio, $R = D_\perp / D_\parallel$.

   C. The photoreceptor membranes of arthropods are microvilli.
In crayfish the transition moments are aligned ± 50° with the
microvillar axes; i.e., $\alpha_1 > \alpha_w$.

   D. The microvilli of crayfish are assembled in bands or layers
about 25 µm in diameter and 5 µm thick. All the microvilli in one
layer are parallel to each other and at right angles to the microvilli of adjacent layers. This diagram shows one complete band
and parts of the two contiguous layers. Isolated organelles can
be viewed from the side and a small measuring beam placed in a
single layer. In measuring dichroic ratio the plane of polarization of the measuring beam is referred to the microvillar axes, as
shown. In vivo the rhabdom is surrounded by seven retinular cells,
but these detach when the tissue is disrupted. From (15).

absorption coefficients (10, 11). Because of Brownian rotation about molecular axes perpendicular to the membrane, the coefficients α_w and α_1 are equal. The coefficient α_h arises because of a net tilt of about 20° into the membrane (6).

In each rod outer segment approximately 1000 disks occur together in a stack, and in vivo the light penetrates along the axis of the outer segment, perpendicular to the planes of the disks. As $\alpha_1 = \alpha_w$, there is no dichroism, or variation in strength of absorption with plane of polarization. With axially incident light, vertebrate rods and cones are therefore not differentially sensitive to plane of polarization.

If an outer segment is detached from the retina and viewed from the side, it exhibits dichroism (12). Absorption of light is greater for light polarized perpendicular to the axis of the rod, $e_\perp$, and less for $e_\parallel$. The dichroic ratio, which is normally 4 or 5, provides the basis for calculating the tilt of the transition mement out of the plane of the membrane (6, 7, 8, 9).

Recent work has indicated much less freedom of motion in the rhabdomeric photoreceptor membranes of the crayfish (Orconectes, Procambarus) (13, 14, 15). These membranes are microvilli rather than plane sheets, and in a few species one can examine small masses of mutually parallel microvilli by the technique of microspectrophotometry. Microbeams, laterally incident on individual isolated rhabdoms, can be placed within a single stack of microvilli, such as the one illustrated in Fig. 1 D, and absorption measured with the e-vector either parallel or perpendicular to the microvillar axes. These studies have produced several interesting results.

First, the amount of linear dichroism exhibited by these micorvilli is too great to be consistent with a random distribution of chromophores in the tangent planes of the microvilli. A number of years ago Moody and Parriss (16) showed that such a random array should give a dichroic ratio of 2, or slightly less if form dichroism is significant (10, 11). Although the first measured values in crustacea seemed consistent with this model (17), more recent measurements have yielded larger values (18). Some tendency of the chromophores to axial alignment with the microvilli implies that Brownian rotation does not occur freely. In recent experiments we have sought to test for this latter possibility by looking for photoinduced dichroism.

The general conditions influencing dichroism and photodichroism are summarized in Figs. 2 and 3. The arguments, developed here in qualitative terms for plane sheets, are in principle readily extended to microvilli and expressed quantitatively (15). Clearly a random orientation of chromophores does not exhibit dichroism,

Dichroism in Plane Sheets

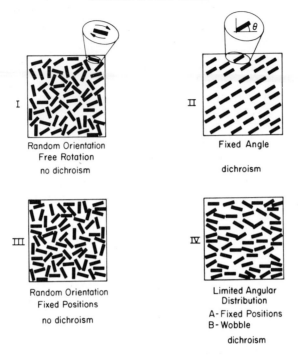

Fig. 2   Conditions for dichroism in plane sheets.   See the text.

whether or not the individual molecules are free to rotate on axes
normal to the plane of the sheet (Fig. 2, I) or are locked in
position (III). Dichroic absorption by the array requires some
net alignment of the chromophores, either perfect (II) or imperfect (IV).   (In these drawings the plane of polarization of the
measuring beam is assumed to be either parallel or perpendicular
to the edges of the square sheets, and the angle θ in II is assumed
to be different from 45°).   Whether or not imperfect alignment (IV)
is accompanied by limited molecular rotation ("wobble"), the array
of chromophores is still dichroic.

Photoinduced dichroism is the change in dichroic ratio produced
by bleaching some of the chromophores with plane-polarized light.
The dichroic ratio of the part of the population that survives is
then compared with the initial dichroism; specifically, the ratio
of the initial and final dichroic ratios provides a measure of
photodichroism.

The conditions necessary for photodichroism are not identical

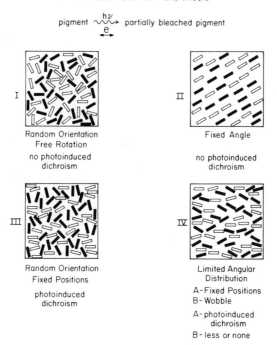

Fig. 3 Conditions for photoinduced dichroism in plane sheets. See the text.

to those required for dichroism (Fig. 3). Time-random orientation caused by free Brownian rotation is inconsistent with photodichroism as well as dichroism, as long as the time required for a molecule to rotate is short with respect to the duration of the bleaching exposure. Under such conditions all molecules will have an equal chance of absorbing the plane-polarized bleaching light (I). On the other hand, when all chromophores share the same fixed alignment (II) there is no photodichroism, even though the ensemble is, in general, strongly dichroic. Regardless of how many of the molecules have been bleached and at what rate, those that remain have the same dichroic ratio as the initial population.

The condition that yields maximal photodichroism is shown in Fig. 2, III: random orientation of the population, with individual molecules locked in place and unable to rotate. For either plane of polarization of the bleaching light, there will be some chromo-

phores with orthogonally oriented absorption vectors and a high
probability of being spared. The population initially displays a
dichroic ratio of unity, and an increasing dichroic ratio as
bleaching progresses.

A limited angular distribution of chromophores may give high
initial dichroism but only small values of photoinduced dichroism
(IV). The more restricted the distribution, the less photodichroism, as can be readily seen in the limiting case illustrated
in II. If the molecules "wobble", the photodichroism will be
less than if the molecules were rigidly fixed in place. (Compare
with Fig. 3, I as the limiting case).

These considerations have been extended to isolated rhabdoms
and a model developed for comparison with results of the following
experiments. Partial bleaches of otherwise stable metarhodopsin
were rapidly achieved through a combination of light, alkaline pH,
and aldehyde fixatives. The aldehydes have the additional effect
of preventing gross deterioration of the rhabdoms. Based on work
with other membranes, the bifunctional reagent glutaraldehyde
should block both rotational and translational diffusion, but
modest amounts of formaldehyde should not (2, 3, 19). Photoinduced
dichroism occurs in rhabdoms after either treatment, but it is
larger in glutaraldehyde than formaldehyde. Two conclusions can
be drawn from this experimental result: (i) the presence of photodichroism means that not all rotational positions of the chromophores occur with equal frequency, and (ii) the greater photodichroism in glutaraldehyde indicates that this reagent damps some
molecular motions that occur in the presence of formaldehyde. A
quantitative comparison of results with the mathematical model
yields a bonus: even in glutaraldehyde the amount of photodichroism
is less than predicted if the absorption vectors were to occupy all
angles in the tangent planes of the microvilli. This provides
another argument that free rotation is absent, an argument that
does not require the supposition that formaldehyde fixation proceeds
without cross-linking. Quantitatively, the absorption vectors seem
to lie within ±50° of the microvillar axes (15).

Finally, experiments to detect the translational diffusion of
metarhodopsin along the lengths of the microvilli have yielded
negative results. Local bleaches are not followed by a redistribution of remaining pigment, and there is thus no evidence for
movement of pigment over distances of 15 μm in times of about 20
min (14, 15).

These results indicate that there are significant differences
in the molecular architecture of vertebrate rod disk and arthropod
rhabdomeric membrane. What the bases for these differences are
remain to be discovered; the effect, however, is to make the rhabdomeric membranes a useful component in a neural system for anal-

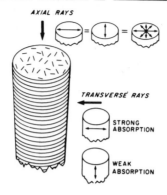

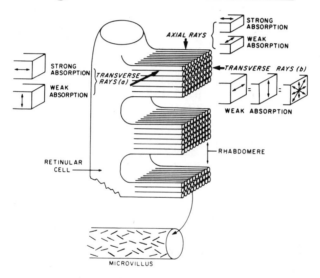

Fig. 4 Comparison of vertebrate and crustacean photoreceptors. In each case the axial rays show the direction along which light is incident on the receptors in vivo, whereas the transverse rays indicate how dichroism can be measured in the laboratory by microspectrophotometry. Because of molecular rotation, vertebrate outer segments have no dichroism for axial rays. In crustacea (and other arthropods), the receptors exhibit dichroic absorption of axial rays. This is because i) the absorption vectors of the rhodopsin molecules are aligned within ±50° of the microvillar axes, and ii) all the microvilli from a given photoreceptor cell are approximately parallel to each other. The sensitivity of the cell is therefore a function of the plane of polarization of the stimulating light. Modified from (20).

yzing plane-polarized light, where in vivo the light is incident
normal to the microvilli and the dichroic ratio of the microvilli
determines how much plane-polarized light will be absorbed (Fig. 4).

Supported by USPHS grant EY00222.

REFERENCES

1. Singer, S.J. and Nicolson, G.L. (1972) Science 175, 720-731.
2. Brown, P.K. (1972) Nature New Biology 236, 35-38.
3. Cone, R.A. (1972) Nature New Biology 236, 39-43.
4. Liebman, P.A. and Entine, G. (1974) Science 185, 457-459.
5. Poo, M. and Cone, R.A. (1974) Nature 247, 438-441.
6. Liebman, P.A. (1962) Biophys. J. 2, 161-178.
7. Wald, G., Brown, P.K., and Gibbons, I.R. (1963) J. Opt. Soc.
    Am. 53, 20-35.
8. Harosi, F.I. (1975) J. Gen. Physiol. 66, 357-382.
9. Harosi, F.I. and MacNichol, E.F., Jr. (1974) J. Gen. Physiol.
    63, 279-304.
10. Laughlin, S.B., Menzel, R., and Snyder, A.W. (1975) in
    "Photoreceptor Optics", pp. 237-259, A.W. Snyder and R. Menzel,
    Eds., Springer Verlag, Heidelberg.
11. Snyder, A.W. and Laughlin, S.B. (1975) J. Comp. Physiol. 100,
    101-116.
12. Denton, E.J. (1959) Proc. Roy. Soc. B. 150, 78-94.
13. Goldsmith, T.H. and Wehner, R. (1975) Biol. Bull. 149, 427-428.
14. Wehner, R. and Goldsmith, T.H. (1975) Biol. Bull. 149, 450.
15. Goldsmith, T.H. and Wehner, R. (1976) (Submitted for publication).
16. Moody, M.F. and Parriss, J.R. (1961) Z. vergl. Physiol. 44,
    268-291.
17. Waterman, T.H., Fernandez, H.R., and Goldsmith, T.H. (1969)
    J. Gen. Physiol. 54, 415-432.
18. Goldsmith, T.H. (1975) in "Photoreceptor Optics", pp. 382-409,
    A.W. Snyder and R. Menzel, Eds., Springer Verlag, Heidelberg.
19. Edidin, M., Zagyansky, Y. and Lardner, T.J. (1976) Science
    191, 466-468.
20. Goldsmith, T.H. (1973) in "Comparative Physiology", pp. 577-
    632, C.L. Prosser, Ed., Saunders.

DYNAMIC ASPECTS OF THE MOLECULAR ARCHITECTURE OF PHOTORECEPTOR MEMBRANES

L.GIULIO

Istituto di Fisiologia veterinaria,Università di Torino

v.Campana,I6. 10125 Torino,Italy

The purpose of future studies on the fine structure of the light--trapping device of photoreceptors,the so-called visual cells outer segment,is,as WORTHINGTON says,mainly to "provide a more detailed picture so that correlation between structure and function can became a reality"(1974). The cooperative use of very different probing techniques(birefringence and dichroism analysis,transmission and scanning electron microscopy,specific staining techniques,freeze-etching and freeze-fracture methods,low-angle X-ray diffraction,microspectrophotometry,fluorescent probes,gel electrophoresis, radioactive tracer detection,small-angle neutron scattering,etc.) has provided in the last few years a valuable sum of information on the way the various molecular kinds may be assembled in the photoreceptor membranes. From this point of view the vertebrate rod outer segment(ROS) is the most extensively studied and best understood object:the main purpose of this short lecture will be to try to give a simplified (perhaps oversimplified) picture of the dynamic aspects of the molecular architecture of photoreceptor membranes of the ROS.

1) Birefringence

The frog ROS,in unbleached condition,exhibits a positive uniaxial birefringence($\Delta n = n_e - n_0$) of strength ca. 1.0×10^{-3}(i.e. a 6 nm retardation $-\Gamma-$ for a thickness of specimen ca. 6μ)which rises,after bleaching,to a value about 1.7×10^{-3} (Γ = 10 nm). This difference(SCHMIDT,I938;as SCHMIDT wrote," die Doppelbrechung

des Stäbchens als positiv einachsig mit längs vorlaufender opti-
scher Achse bezeichnet",loc.cit.;LIEBMAN,1975)is still unexplained.
According to WEALE(1971),this change follows an exponential(half-
-time = 46 s);at any rate the final value occurs after a time > 4
minutes. WEALE suggests that the light exposure shortens(of 200-
500 Å)the ROS. It will be noted,in fact,that the total birefringence of ROS is made up of three major components,i.e.

$$\Delta n_T = \Delta n_I + \Delta n_F + \Delta n_C$$

were Δn_T is total, Δn_I is intrinsic, Δn_F is form and Δn_C is
chromatic birefringence. The Δn_I of frog ROS achieves ca. +
6 x 10^{-3} , the Δn_F ca. -4 x10^{-3}(LIEBMAN,1975:resolution of net
birefringence into form and intrinsic components).
According to WIENER's formula the Δn_F refers to a structure made
by a stack of lamellae interleaved with spaces of different n
(e.g. n_l= 1.475; n_s = 1.365):

$$\Delta n_F = \frac{-f_s \, f_\ell \, (n_\ell^2 - n_s^2)^2}{(f_\ell \, n_s^2 + f_s \, n_\ell^2)}$$

where f_l =0.37 and f_s =0.63. An interesting LIEBMAN's observation
(1975) is the axial gradient of birefringence in Rana pipiens rods.
The change of birefringence from dark to light adaptation could
be,at first sight,a simple consequence of the photolytic process:
the rhodopsin,in fact,would affect more strongly the transverse
refractive index(i.e. the ordinary ray)and the shift toward the
shorter λ of the last photoproducts could cause a Δ n increase
(for the relationships between birefringence and dichroism see
WEALE,1971 a & b,and BACCI,in press).
To a first approximation,notwithstanding this,the slow rise of Δ n
is independent of measuring wave-length: it is present when measurements are effected with orange to red light(practically not absorbed by the rod pigment:WEALE,1970)or with light of 671 nm λ
(FACELLO et al.,in press). According to LIEBMAN et al.(1974)the
rate of rise,following a bleaching saturating flash,is very low,
the stady state conditions being reached only after 8 minutes(
the slow achromatic increase in birefringence occurs with a half-
-time of about 20 minutes at + 20 °C).
LIEBMAN et al.(1974) were able to detect a flash-induced birefringence transient(BRT),corresponding to a 3-4 % decrease in the net
birefringence. The possible sources of BRT are,in the Aa's opinion,a "loss of crystallinity of the lipids of the ROS membranes",
perhaps correlated with the metarhodopsin II formation(whose half-

-time in an ARRHENIUS plot fits that of BRT). The BRT could, therefore,"be due to the disorientation of only one phospholipid molecule per rhodopsin bleached". Here the employment of an averager
technique(e.g. summation mode) with very low repetitive flashes
becomes crucial.

However, no simple statements or implications can be made about the
results of this biophysical technique. Because the well-known relationships between the retina and the pigment epitelium during
the recovery(regeneration of rhodopsin by frog), FACELLO et al.(in
press) have measured the ROS $_{671\ nm}$ Δn_T at predetermined time intervals following a set of bleaching flashes delivered to the my-
driatic eyes of the intact animals. In this way the Δn_T increase
(latency ca. 20 minutes; peak about 40 minutes) and then the return
to the normal values(within 65 minutes) present a time course which
overlaps exactly the curve showing the progressive regeneration
of rhodopsin in the frog intact eye(PESKIN, 1942). The course of
events, summarizing, following a saturating flash of light, could be
as follows:

net(dark) $\begin{cases} \text{light} \\ \text{LIEBMAN's et} \end{cases}$ Increase of   Peak of   Decrease of
birefringence al.BRT.Loss Δn_T.Laten= Δn_T: Δn_T to a
          of Δn_T.Latency   cy= 20 min.   40 min.   normal value:
          of ms range.   65 minutes.

It is likely that in the sequence of related events the retinol
plays a significant rôle(the retinol becomes aligned to the rod
long axis, i.e. to the hydrocarbon chains long axes). The rôle of
this substance, at the membrane level, is now firmly established
(retinol and retinoic acid as not competitive inhibitors to cholesterol biosynthesis: ESKELSON et al., 1970; PRODOUZ & NAVARI,
1975; altering action of the retinol in the packing of lipid bilayers of cell membranes: BLOUGH, 1963; the retinol alters the membrane
permeability by forming micelle regions into the phospholipid bilayers: LUCY, 1969; etc.).

2) Dynamic Aspects

     To explain the lack of photodichroism in axially illuminated
photoreceptors(HAGINS & JENNINGS, 1959; GIULIO & MESSINA, 1966; PAK &
HELMRICH, 1968)one could suppose that the visual pigment molecules
possess some rotational freedom in multilayered structures of photoreceptor cells; the "closely packed" pigment molecules could present a "dynamic" disorder at their absorption oscillators direct-

ion level,disorder which is presumably due to continous brownian
motion. KITAIGORODSKIJ(1966) talks about of crystal rotational,or
gas crystalline state(rotational crystal): some degree of rotational freedom(perhaps restricted to only one degree) coexists with a
spatial distribution of the single molecules severely fixed(this
signifies that the pigment molecules maintain tho electric dipoles
of the chromophores confined to the plane of the lamellae:dichroic
properties of ROS when illuminated from the side. Small-angle X-ray
diffraction studies suggest that the pigment molecules at the ROS
disk membranes level behave like a "planar liquid",the rhodopsin
molecules possessing a considerable freedom of movement which is
temperature dependent(WORTHINGTON,1971). Additional information
about this argument can be achieved by freezing or fixing the retina,manipulations that probably hinder the rotational freedom of
molecules:in this way(experiments after glutaraldehyde fixation:
BROWN,1972;STRACKEE,1971),long-lasting photodichroism may be induced in the tissue itself.
The failure to detect the photodichroism in earlier contributions
was evidently due to the very poor time resolution of the technique used. With a suitable apparatus for detection of a transient
photodichroism with a time resolution of 40 ns,and capable of detecting changes of the order IO^{-3} -10^{-1} absorbance units,CONE(1971,
1972)was able to measure(when frog ROS rhodopsin was partially
bleached with a flash of plane-polarized light)a maximum dichroic
ratio somewhat greater than 2 but less than 3. This dichroic ratio decayed with a half-time of 3.0 $\pm$ 1.5μs at + 20 °C;from the
half-time for the decay rate it could be estimated a rotational
relaxation time(ϱ_{o}) for rhodopsin in fresh dissected retinas of
4-50μs at + 10 °C(ca. = 20μs at + 20 °C).
Using the well-known equation:

$$\varrho_{o} = 4\,\pi\,\eta\,r^{3}\big/\kappa\,T$$

where η = poise(CGS),T = °K; k = 1.38 x 10^{-16} ,and r = molecular
radius in cm,CONE estimates the viscosity of the medium surrounding the pigment molecules to be around 2 poise(0.7-6 poise),as-
suming spherical molecules with a radius around 2.2 - 2.8 x 10^{-7}
cm. According to us,information about the existence of a brownian
motion of pigment particles,and the range of numerical value of
ϱ_{o} ,can be achieved indirectly by recording a quantity related
to the absorption probability after a transient,selective effect
of a plane-polarized intense light pulse delivered to the "sample"

(GIULIO & FERRARO,in press).
It is not the purpose of this lecture to review exhaustively the
achievements in the field of theoretical analysis of the brownian
motion of molecules,above all those of biological interest. They
have been reviewed elsewhere(EINSTEIN,1906;PERRIN,1914;CHANDRASE-
KHAR,1943;WEBER,1953 & 1973;DALE-FAVRO,1960;DARTNALL,1972;EHREN-
BERG & RIGLER,1972;RIGLER & EHRENBERG,1973;EDIDIN,1974).
From the well-known EINSTEIN's equation(1906),giving the average
small rotation of a spherical particle about an instantaneous axis
during time t,as a function of the absolute temperature and of the
frictional coefficient of the rotation for the sphere:

$$\overline{\alpha}^2 = \kappa T t \Big/ 4 \pi \eta r^3$$

The rotation of the particle itself may be characterized in terms
of a single constant $\varrho_0 = 4\pi\eta r^3/kT$,which has the dimensions of
a time. This quantity,called rotational relaxation time,may be defined in general,following DEBYE(1929)as the time after which a
direction bound to the rotating molecule has accomplished on the
average an angle ϑ with its starting position such that $\overline{\cos\vartheta}$ =
e^{-1} . The modern theory of the brownian motion starts with the so-
-called LANGEVIN's equation:

$$du/dt = -\beta u + A(t)$$

where u denotes the velocity of the particle;- β u represents a dynamic friction experienced by the particle,and A(t),independent of
u,is the "fluctuating part...characteristic of the brownian motion"
itself(CHANDRASEKHAR,1943).
Regarding the frictional term $-\beta$ u it is assumed that this is governed by STOKE's law which states that the frictional force dece-
lerating a spherical particle of effective radius $\pm$ r and mass m
is given by:

$$6\pi r \eta \; u/m$$

where η denotes the coefficient of viscosity of the surrounding
medium.
According to the theory of translational brownian motion for spherical particles,the translational diffusion coefficient is given
by:

$$D = kT/6\pi r\eta$$

(where r is the radius of the particle,k is the BOLTZMANN constant,
T the absolute temperature, η the viscosity of the "solvent" in
which the particle moves) POO & CONE(1974) were able to estmate
the rate of arrival of "bleached" rhodopsin on the "unbleached"

side of a amphibian ROS disk membrane: a uniform distribution is
exponentially approached,with a $t^{1/2}$ of 35 s in frog diskal membranes. From calculations,D is estmated to be $5 \pm 1.5 \times 10^{-9}$ cm^2
s^{-1} at + 20 °C(η = 2 poise).

The lipid environment of visual pigment molecules must therefore
be,according to the quoted papers,highly fluid. The rotational
freedom of the visual pigment molecules can be explained(according to TRAUBLE & SACKMANN,1973) in another way,i.e.in terms of the
appearance and disappearance of vacancies in the lipid bilayers of
the membrane in position next to local protrusion of the protein
(the last molecule exchanging position with a vacant lattice site).
X-ray diffraction studies indicate that the average distance between adjacent molecules of rhodopsin in the disk membranes is ca.
70 Å (BLAUROCK & WILKINS,1972). The average time τ between collisions with another rhodopsin molecule can then be measured as:

$$\tau = s^2 / 4 D$$

(where s is the distance travelled between collisions; s,owing to
the above mentioned value,equals 25 Å assuming a diameter for
rhodopsin molecule of ca. 45 Å).

The time between collisions(τ) is then 3.9 μs at + 20 °C(POO &
CONE,loc.cit.);this signifies a collision frequency in the range
of 10^5–10^6 s^{-1} .

A possible consequence of multiple collisions by thermal motion
might be,according to DE VRIES(1956),the "activation" of the photochemical substance. The probability that a sufficient quantity
of energy(or more) is concentrated "on the trigger" at some time,
is given by the BOLTZMANN's factor   exp($-E/k$ T). When the number
of new energetic configurations in one second is f,the number of
excitations will be,per second, f exp($-E/k$ T). The time for a vibration is ca. 10^{-13} s(compared with the electron rearrangement
time of ca. 10^{-15}s): substituting the factor 10^{13} for f and since
one ROS contains ca. 3×10^9 molecules(frog retina: PIRENNE,1962)
we obtain 3×10^{22} exp($-E/k$ T). This figure indicates the probability the rod becomes excited per second. The coincidence-mechanism will be able to suppress several of the spurious excitations.
(According to BRAY & WHITE(1966) the "rate of a reaction must depend on the frequency of collisions between the atoms or molecules concerned and it is useful to consider how a quantitative expression of this may be derived". The expression for reaction rate,
for unit concentration of reactants,may be written:

$$V = PZ \, e^{-E_{act}/RT}$$

where V is the velocity of reaction,P the probability factor(which arises from the fact that the reactivity of a molecule varies from one part to another) Z the collision number, E^{act} the energy of activation,R the gas constant,and T the absolute temperature). The brownian motion can also play a hypothetic positive rôle when the electromagnetic radiation,falling on the photoreceptors,has not enough energy to dissociate the pigment molecules,i.e. when $h\nu < E$,providing the missing quantity of energy W . The probability,according to DE VRIES(loc.cit.)that the rhodopsin molecule has this amount of energy(if no more than one degree of freedom plays a part) equals $\exp(-W/k\,T)$; more sophisticated expression is needed if more degrees of freedom have to be considered,namely(DE VRIES,loc.cit.):

$$W = \sum_{r=0}^{S-1} \frac{1}{r}! \left(W/\kappa T\right)^{r} e^{-W/\kappa T}$$

The co-operative action between electronic and vibrational ener= gies refers to well-known FRANCK-CONDON principle.

The basic problem of the coupling photon-absorption/signal- -generation at single ROS level is evidently linked to the archi- tecture of photoreceptor membranes,described in dynamic molecular terms;from a still unknown photointermediate of the pigment pho- tolysis arises,as wrote HECHT SHLAER and PIRENNE in a celebrated paper of 1942,the "initial impetus for a visual act".

References

BACCI,V.:in press.
BLAUROCK,A.E.,& WILKINS,M.H.F.:Nature,236,313(1972).
BLOUGH,H.A.:Virology,19,349(1963).
BRAY,H.G.,& WHITE,K.:Kinetics and Thermodynamics in Biochemistry,
    II ed.,Academic Press,N.York,(1966).
BROWN,P.K.:Nature(N.B.),236,35(1972).
CHANDRASEKHAR,S.:Revs.Mod.Phys.,15,1(1943).
CONE,R.A.:Biophys.Soc.Abstracts,TLAM-E 2(1971).
CONE,R.A.:Nature(N.B.),236,39(1972).
DALE FAVRO,L.:Phys.Rev.,119,53(1960).
DARTNALL,H.J.A.:in Handbook of Sensory Physiology,Photochemistry
    of Vision(ed.by H.J.A.Dartnall),Springer,Berlin(1972,p.122).
DEBYE,P.:Polar Molecules,Chemical Catalog Co. N.York (1929).
DE VRIES,Hl.:Progress in Biophysics,6,207(1956).

EDIDIN,M.:Ann.Rev.Biophys.Bioengng.$\underline{3}$,179(1974).

EHRENBERG,M.&RIGLER,R.:Chem.Phys.Letters,$\underline{14}$,539(1972).

EINSTEIN,A.:Ann.Physik,$\underline{19}$,371(1906);in Investigations on the Theory of the Brownian Movement,ed. by R.Furth,Dover Publ.,N.York (1926).

ESKELSON,C.D.,JACOBI,H.P.,& CAZEE,C.R.:Physiol.Chem.Phys.,$\underline{2}$,135 (1970).

FACELLO,C.,FERRARO,M.,LAUDATO,M.T.,PAROLA,I.,& GIULIO,L.:in press.

GIULIO,L. & FERRARO,M.: in press.

GIULIO,L. & MESSINA,F.:III° Convegno Naz.Biofisica & Biologia Molecolare,p.3^a,Roma(1966).

HAGINS,W.A. & JENNINGS,W.H.:Trans.Fraday Soc.,$\underline{27}$,180(1959).

HECHT,S.,SHLAER,S. & PIRENNE,M.H.:J.gen.Physiol.,$\underline{25}$,819(1942).

KITAIGORODSKIJ,A.I.:Poriadoc i Besporiadoc v Mire Atomov. Nauka, Moskwa(1966).

LIEBMAN,P.A.:in Photoreceptor Optics,ed. by A.W. Snyder & R.Menzel,p.199,Springer,Berlin(1975).

LIEBMAN,P.A.,JAGGER,W.S.,KAPLAN,M.W. & BARGOOT,F.G.: Nature,$\underline{251}$, 31(1974).

PAK,W.L.,& HELMRICH,H.G.:Vision Res.,$\underline{8}$,585(1968).

PERRIN,J.:Les Atomes,Alcan,s.l.(1914).

PESKIN,J.C.: J.gen.Physiol.,$\underline{26}$,27(1942).

PIRENNE,M.H.: Vision and the Eye,Chapman & Hall,London(1967).

POO,M. & CONE,R.A.: Nature,$\underline{247}$,438(1974).

PRODOUZ,K.,NAVARI,R.M.:Nutr.Rep.Int.,$\underline{11}$,17(1975).

RAZI-NAQVI,K.,GONZALEZ-RODRIGUEZ,J.,CHERRY,R.J. & CHAPMAN,D.: Nature(N.B.)$\underline{245}$,249(1973).

SCHMIDT,W.J.:Kolloid Ztschr.,$\underline{85}$,137(1938).

STRACKEE,L.: Biophys.J.,$\underline{11}$,728(1971).

TRAUBLE,H. & SACKMANN,E.:Nature,$\underline{245}$,210(1973).

WEALE,R.A.:J.Physiol.,$\underline{210}$,28 P(1970).

WEALE,R.A.:Experientia,$\underline{27}$,403(1971).

WEALE,R.A.:Vision Res.,$\underline{11}$,1373(1971 a).

WEALE,R.A.:Vision Res.,$\underline{11}$,1387(1971 b).

WEBER,G.:in Advances in Protein Chemistry,$\underline{8}$,415(1953).

WEBER,G.: in Fluorescence Techniques in Cell Biology,ed. by A.A. Thaer & M. Sernetz,p.5,Springer,Berlin(1973).

WORTHINGTON,C.R.:Fed.Proc.,$\underline{30}$,57(1971).

WORTHINGTON,C.R.:Ann.Rev.Biophys.Bioengng.,$\underline{3}$,53(1974).

THE PHOTOCHEMISTRY OF RHODOPSIN EXCITED IN THE 280nm (γ) BAND.

Tchiya Rosenfeld and Michael Ottolenghi

Department of Physical Chemistry
The Hebrew University of Jerusalem
Israel

SUMMARY

Photochemical experiments using continuous and pulsed Nd laser sources are carried out at room temperature, exciting bovine rhodopsin in its γ (280 nm) ultraviolet band. It is shown that light absorption within this band leads to the same sequence of photochemical transformations (with bathorhodopsin as the primary product) as characterizing excitation within the main visible (α) and near uv (β) bands of the pigment. From a quantum yield analysis, it is concluded that energy transfer, from excited tryptophanes to the retinal moiety, accounts for most of the photochemical activity in the γ-band. The effect may yield information relevant to the geometrical distribution of tryptophanes around the retynilic chromophore of visual pigments.

INTRODUCTION

The room-temperature photobleaching of bovine or frog rhodopsins excited in their α (498,502 nm), β (330 nm) and γ (280 nm) bands, have been extensively investigated using continuous excitation methods (1-3). The application of laser flash spectroscopy has recently shown that bathorhodopsin, the primary photoproduct at 77°K, is also generated at room temperature following excitation in both α (4-6,8) and β (7) absorption bands. A recent analysis (6) has also proved that, as the sole primary photoproduct, bathorhodopsin is the precursor of the subsequent thermal intermediates, not only at low temperatures, but also in the room temperature sequence. Moreover, during the laser pulse, bathorhodopsin is photochemically converted to either rhodopsin

or isorhodopsin (the pigment with a 9-cis chromophore), in analogy
to the low-temperature behaviour.

Excitation of frog rhodopsin within its γ band was found to
be associated with bleaching of the pigment, yielding all-trans
retinal and opsin (3).   In further analogy with α and β band
excitation, the resulting opsin product was able to regenerate
rhodopsin by reacting with 11-cis retinal, showing that the bleaching process is not accompanied by denaturation of the protein.
The present investigation was undertaken on the basis of these
findings, aiming to clarify the bleaching mechanism associated
with γ-band excitation.   The main open questions concerned the
exact nature of the absorbing chromophore, the identification of the
primary photoproduct and the establishment of the quantum yield
of its generation.   Such questions are relevant to the problem
of energy transfer from the protein to the retynil residue (3) and
thus to the spatial distribution of the aromatic amino acids in
visual pigments.

METHODS

Bovine rhodopsin solutions in 2% Ammonyx LO were prepared
and purified as previously described (6).   Absorbance ratios of
the order of $D(498 \text{ nm})/D(280 \text{ nm}) = 1.7-1.8$ were obtained, being
indicative of the purity of the preparations.   Continuous excitation experiments with the 253.7 line of a low pressure Hg arc
were performed using previously described actinometric procedures
(13). Pulsed laser-photolysis experiments employed the 265 nm
(20 nsec) pulses of a Q-switched, frequency quadrupled Nd laser.
The pulses were amplified to reach energies of a few mJ, thus
guaranteeing saturating excitation conditions (6).   The details
of the transient detection system were identical to those employed
with 530 nm excitation (6).

RESULTS AND DISCUSSION

Light absorption in the γ band
Opsin, the protein associated with the rhodopsin pigment,
exhibits an intense absorption band at 280 nm which is mainly
due to its aromatic amino acids, tyrosine and tryptophane (1).
Protonated Schiff bases of retinal, considered as model compounds
for visual pigments, are also characterized by a band around
280 nm (1).   It is thus feasible to assume that the UV (γ)
absorption of a molecule such as rhodopsin is due to a superposi-
tion of a band due to the aromatic amino acids in the protein
moiety and one due to the retynil-polyene residue which is responsible for both β (330 nm) and α (498 nm) bands.   For an
evaluation of the emissive and photochemical properties of

rhodopsins excited in the γ band, it is important to consider the exact relative contributions of the two chromophoric moieties at 280 nm.

The absorbance changes at 280 nm (ΔD_γ) associated with the photobleaching of rhodopsin have been investigated in several laboratories (9-14).   In most cases (9,12-14) no detectable absorbance changes around 280 nm were found to accompany the bleaching of the main α band of, e.g., bovine rhodopsin ($\Delta D_\gamma \approx 0$). In two cases (10,11) a decrease of $\sim 3\%$ was reported.   Since the net reaction associated with bleaching is represented by the process: rhodopsin $\xrightarrow{h\nu}$ opsin+(all-trans retinal), the absorbance change at 280 nm is given by $\Delta D_\gamma = D_{Op} + D_{Re} - D_{Rh}$, where D_{Op}, D_{Re} and D_{Rh} denote the absorbance of opsin, all-trans retinal and rhodopsin at 280 nm, respectively.   In contrast to the pronounced environmental effects on the fluorescence of aromatic amino acids, it appears that only little changes exist between the absorption spectra of free and various protein bound tyrosine and tryptophane molecules (15).   It is therefore plausible to assume in a first approximation that the rhodopsin absorbance may be expressed by the sum: $D_{Rh} = D_{Op} + D_{Rm}$, where D_{Rm} is the contribution of the retynil moiety.   In a characteristic bleaching experiment, associated with a concentration change of the order of $\Delta C \approx 10^{-5} M$ one may thus write:

$$\Delta D_\gamma = D_{Re} - D_{Rm} = \Delta C\{\varepsilon_{Re} - \varepsilon_{Rm}\} \approx 10^{-5}\{\varepsilon_{Re} - \varepsilon_{Rm}\}$$

Since $\varepsilon_{Re} \approx 7 \times 10^3 M^{-1} cm^{-1}$ (16), the observation that $\Delta D_\gamma \approx 0$ implies that $\varepsilon_{Rm} \approx 7 \times 10^3 M^{-1} cm^{-1}$ which is about 10% of the total extinction of rhodopsin at 280 nm ($\varepsilon_{Rh} \approx 6.9 \times 10^4 M^{-1} cm^{-1}$ at 280 nm, based on the value of 4.06×10^4 for the extinction at 498 nm (16)).   Such an estimate for ε_{Rm} is consistent with the 280 nm absorbance of a protein-free model solution of a protonated Schiff base of retinal in the presence of trichloroacetic acid, which fairly simulates rhodopsin in both extinction coefficient and wavelength of maximum α-band absorption (13).

A contribution of the same order of magnitude (i.e. of not more than 10%) of the retynil moiety to the absorbance of the pigment at 280 nm, is also obtained by summing up the absorbance values, at 280 nm, of the 8 ± 1 tryptophanes ($\varepsilon = 5690$ cm^{-1} M^{-1}) and 15 ± 2 tyrosines ($\varepsilon = 1280$ cm^{-1} M^{-1}) present per mole of rhodopsin (12,17,18).   A total extinction of $(6.5\pm0.7) \times 10^4$ M^{-1} cm^{-1} is obtained as compared with $\varepsilon_{Rh} = 6.9 \times 10^4$ M^{-1} cm^{-1}).

It may thus be concluded that most (i.e. $\sim 90\%$) of the light absorbed within the γ band of rhodopsin leads to excitation of the aromatic amino acids in the protein moiety of the pigment.

Photochemistry of rhodopsin initiated by γ·band excitation

The photobleaching of frog rhodopsin in aqueous solutions has
been previously investigated by Kropf (3), who reported the appar-
ent values of $\phi_b \simeq 0.21\pm0.07$ and $\phi_b \simeq 0.26\pm0.06$ for 254 nm and 280nm
excitation, respectively.   However, the relatively high absorb-
ance in the region of the γ band in the solutions submitted to
investigation, indicates that only 53% of the quanta at the above
wavelengths were used for exciting rhodopsin, the rest being
absorbed by opsin impurities.   (For a pure frog rhodopsin preparation $D(280\ nm)/D(502\ nm) = (8.90\times10^4/4.20\times10^4) \simeq 2$, while
the actual absorption ratio in the above solutions was ~ 4).
Thus, correcting for the inner filter effect of the impurity, a
value of $\phi_b \simeq 0.48\pm0.10$ is obtained for the photobleaching quantum
yield in the γ band (14).

Similar experiments were presently carried out by us with
aqueous,Ammonyx LO, solutions of bovine rhodopsin.   The bleaching
quantum yield measured, using 254 nm excitation, was ϕ_b = 0.4±
0.05.

In order to elucidate the bleaching mechanism initiated by
excitation in the ultraviolet, rhodopsin solutions were exposed
to the 265 nm pulse of the frequency quadrupled Nd laser.   The
main transient phenomena, shown in Fig.1, are essentially identical to those recently observed by us for the pulsed 530 nm
excitation of rhodopsin.   The first transient (Fig.1b), associated with a maximum in the difference spectrum around 570 nm,
decays with a half life of $\sim$ 100 nsec.   It may be thus identified with bathorhodopsin which has been recently shown to be the
only primary product in the 530 nm photolysis of rhodopsin at
room temperature, in complete analogy with the low temperature
sequence (1).   The difference spectrum in Fig.1 recorded after
5 msec corresponds to generation of metarhodopsin II.   The
spectrum, as well as its growing-in half life ($\tau_{1/2} \simeq$ 100 μsec),
are both essentially identical to those reported for pulsed
laser excitation in the α band (6).

A detailed analysis of the transient laser effects following 530 nm excitation (6) has indicated that all transient
phenomena could be quantitatively accounted for by the scheme:

rhodopsin $\overset{h\nu}{\underset{h\nu}{\rightleftharpoons}}$ bathorhodopsin - $\cdots$ → metarhodopsin II

$$\overset{h\nu}{\underset{h\nu}{\boxed{}}} \rightarrow \quad \text{isorhodopsin}$$

in which the photoequilibrium between rhodopsin, bathorhodopsin
and isorhodopsin is established during the (20 nsec) lifetime
of the saturating laser pulse.   The difference spectrum at

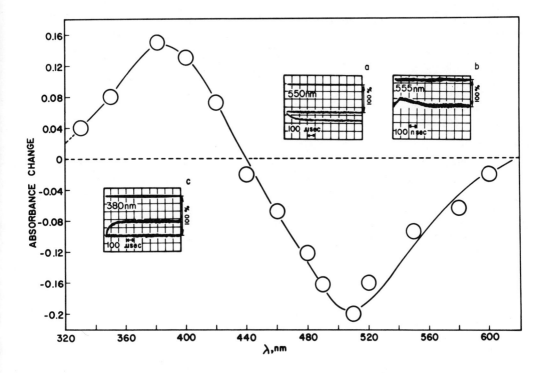

Fig.1.   Characteristic transient patterns following the
pulsed Nd laser excitation of bovine rhodopsin, D(498nm) =
0.5-0.8, in aqueous (2%) Ammonyx LO at room temperature.

Insets: Oscillograms showing the decay (a) (of metarhodopsin I) and growing-in (c) processes associated with the
generation of metarhodopsin II.   (b) represents the decay of
bathorhodopsin.   The 100% notation refers to the difference
between readings in the presence (lower horizontal trace) and
in the absence (upper trace) of the monitoring light beam.

The difference spectrum recorded 5 msec after pulsing
represents the superimposed absorbance of metarhodopsin II,
isorhodopsin and bleached rhodopsin.   (For details see ref.6).

the stage of metarhodopsin II reflects the superimposed absorption of bleached rhodopsin, isorhodopsin and metarhodopsin II (6). Its independency of the excitation wavelength (530 or 265 nm) implies that the photostationary conditions obtained during the pulse are associated with the generation of isorhodopsin, not only when using the 530 nm line, but also when exciting within the γ band at 265 nm. Further support for the generation of isorhodopsin in the present 265 nm high intensity laser excitation experiments is obtained from the absorbance changes measured at 470 nm after $\sim$ 50 nsec ($\Delta D(470$ nm$)$). The ratio $\Delta D(570$ nm$)/\Delta D(470$ nm$)$ is positive, while a negative value is expected in the absence of isorhodopsin (6).

<u>Energy transfer from the protein to the retynil chromophore</u>
The above analysis clearly shows that excitation within the γ band of bovine rhodopsin leads to a sequence of photochemical transformations, identical to that resulting from excitation within the visible α band. Such a channelling of u.v. energy to the visual photochemical path is consistent with the results of Kropf for frog rhodopsin, which showed that following γ band excitation the polyene is isomerized to the all-trans conformation, and the opsin produced regenerates the pigment by reacting with 11-cis retinal (3).

For a more quantitative analysis of the photochemical effects associated with irradiation within the γ-band, the following expression for the bleaching quantum yield (ϕ_b) should be considered:

$$\phi_b = \phi_b^D + \phi_b^{ET} = \gamma_d \zeta_b^D + \gamma_p \phi_{ET} \zeta_b^{ET}$$

ϕ_b^D and ϕ_b^{ET} are the observed bleaching quantum yields due, respectively, to direct light absorption by the retynilic moiety and to energy transfer from the protein. γ_d and γ_p ($\gamma_d + \gamma_p = 1$) are the light fractions absorbed by each of the two chromophores, ζ_b^D and ζ_b^{ET} are the intrinsic bleaching yields associated with direct excitation and energy transfer, respectively, and ϕ_{ET} is the yield of energy transfer from the protein to the retynil moiety. Setting (see above) $\gamma_d \simeq 0.1$, $\gamma_p \simeq 0.9$ and $\phi_b \simeq 0.4$ (which is only a lower limit for frog rhodopsin), with the upper values $\zeta_b^D = \zeta_b^{ET} = 1$, one obtains $\phi_{ET} = 0.33$ as a lower limit for the yield of energy transfer. Direct excitation of bovine rhodopsin in both α and β bands is associated with the wavelength (9) and temperature (19) independent bleaching quantum yield of 0.67. (The same value has also been obtained for the visible excitation of frog rhodopsin (9)). It has been argued that the above photo-chemical characteristics of rhodopsin may be accounted for by the quantitative population of a common minimum in a barrierless poten-tial energy curve along the 11-12 coordinate of the polyene moiety (19). If the same state is also reached by direct excitation in the γ band, then $\zeta_b^p = 0.67$. Since a Förster-type energy transfer from the protein tryptophanes to the polyene (see below) will in-volve maximum overlap with the β band of the chromophore as the accepting level, it is plausible that also $\zeta_b^{ET} = 0.67$. In such a case an energy transfer quantum yield of $\phi_{ET} = 0.5$ will be obtain-ed with the above values of γ_d, γ_p and ϕ_b. Even higher values ($\phi_{ET} \simeq 0.70$) are obtained for frog rhodopsin for which $\phi_b = 0.48$.

In spite of the uncertainties in the values presently avail-able for ϕ_b and γ_d/γ_p, it is evident from the above analysis that a substantial transfer of energy from the protein to the polyene takes place following γ band excitation of rhodopsin leading to the production of bathorhodopsin. It should be pointed out that subsequent energy transfer to bathorhodopsin accounts for the generation of isorhodopsin as a secondary photoproduct during the laser pulse, as in the case of direct excitation within the α band (6). This is in keeping with the observation that the ex-tinction within the β band, which is the energy accepting level of the retynilic chromophore, is essentially identical for both rhodopsin and bathorhodopsin (20).

Energy transfer from the protein to the retinal is also con-sistent with the data of Ebrey, who found a yield of $\phi_f^{Rh} = 0.02$ for the fluorescence of bovine rhodopsin excited at 280 nm (λ_{max}(fluor) = 330 nm), as compared with the value of $\phi_f^{op} = 0.1$ measured for free opsin in solution (21). A yield ratio of $\phi_f^{op}/\phi_f^{Rh} = 4$ has been recently confirmed by us (13). The wavelength independency (13,21) of the rhodopsin emission (which is characteristic of the tryptophane chromophores, is indicative of the efficiency of energy transfer from excited tyrosines to the tryptophane moieties (22). Neglecting, in a first approxi-mation, drastic environmental effects on the fluorescence yield (which may e.g., be induced by changes in the protein structure), the above numbers may be interpreted in terms of energy transfer to the retynil polyene as an important deactivation path of the tryptophane fluorescence in rhodopsin.

In principle, important information relevant to the geometri-cal distribution of tryptophanes around the polyene chromophore in rhodopsin could be derived from experimental values of ϕ_{ET}. Assuming a resonance Förster-type transfer from a donor at a distance r from the acceptor, the expressions:

$$\phi_{ET} = r^{-6}/(r^{-6} + R_o^{-6})$$

with R_o, the distance of 50% transfer efficiency, being given by:

$$R_o = (JK^2 Q_o n^{-4})^{1/6}(9.79 \times 10^3)$$

will hold (23). K^2 is the orientation factor for dipole-dipole transfer, Q_o is the emission yield of the donor in the absence of transfer, n is the refractive index of the medium and J the spectral overlap integral, given by $J = \int F(\lambda)\epsilon(\lambda)\lambda^4 d\lambda / \int F(\lambda)d\lambda$, where $F(\lambda)$ and $\epsilon(\lambda)$ are the wavelength-dependent fluorescence intensities of the donor and the extinction coefficient of the acceptor, respectively.   The main difficulties associated with a quantitative analysis of the present data, are due not only to the previously discussed uncertainty in the value of ϕ_{ET}, but mainly to the fact that with rhodopsins one deals with a number of tryptophane donors of unknown geometrical distribution, so that each may be characterized by a different r value.   More explicitly, it is at present impossible to differentiate between very efficient transfer from a limited number of tryptophanes in the neighbourhood of the retynil chromophore and a lower efficiency transfer from a large number of relatively distant aromatic donors.   On the basis of the above expression for the overlap with the pigment's β band ($J = 1.26 \times 10^{-14}$ cm^3 M^{-1}) it may however be deduced that, for the majority of the tryptophane residues, r < 30 Å. A promising tool, for a more quantitative approach to the problem, should involve an accurate analysis of protein fluorescence lifetimes in both opsin and rhodopsin.

Acknowledgement
The authors are indebted to Dr. Ch.R.Goldschmidt for his assistance in carrying out the laser experiments, and to Prof. A. Kropf for many valuable discussions.

REFERENCES

1   For a general treatise and a recent review on visual photochemistry see :
     a) Handbook of Sensory Physiology, Vol.VII/1 (Photochemistry of Vision) H.J.A. Dartnall Editor, Springer-Verlag, Berlin-Heidelberg-New York, 1972.
     b) Ebrey, T. and Honig, B. (1975) Quart. Rev. Biophys. 8, 124-184.

2   Goodeve,C.F., Lythgoe, R.J., Schneider, E.E. (1942)
    Proc.Roy.Soc. B 130, 380-395.
3   Kropf,A. (1967) Vision Res. 7, 811-818.
4   Busch, G. Applebury, M. Lamola, A. and Rentzepis, P. (1972)
    Proc.Nat.Acad.Sci. USA 69, 2802-2806.
5   Bensasson, R., Land, E.J. and Truscott, T.G. (1975)
    Nature, 258, 768-770.
6   Goldschmidt, C.R., Rosenfeld, T. and Ottolenghi, M. (1976)
    Nature, in press.
7   Rosenfeld, T., Alchalel, A. and Ottolenghi, M. (1972)
    Nature 240, 482-483.
8   Cone, R.A. (1972) Nature,New Biol. 236, 39-42.
9   Dartnall, H.J.A. (1972) Handbook of Sensory Physiology,
    Vol.VII/1 (Photochemistry of Vision), H.J.A.Dartnall Ed.,
    Springer-Verlag, Berlin-Heidelberg-New York, pp. 122-145.
10 Hubbard, R. (1969) Nature 221, 432.
11 Ebrey, T.   Private communication.
12 Cooper, A. and Hogan, M.E. (1976) Biochem.Biophys.Res.Comm.
    68, 178-182.
13 Rosenfeld, T., Ph.D.Thesis, Hebrew University, Jerusalem,
    1976.
14 Kropf, A.   Private communication.
15 Wetlaufer, D.B. (1962) Adv.Protein Chem. (Academic Press,
    New York) 17, 303-380
16 Hubbard, R., Brown, P.K. and Bownds, D. (1971)   Methods
    in Enzymology, Vitamins and co-Enzymes, Part C.
    McCromick D.B. and Wright L.D., Editors, Academic Press, N.Y.
    18, 615-653.
17 Edelhock, H. (1967) Biochemistry 6, 1949.
18 De-Grip, W.J. (1974) Thesis University of Nijmegen, The
    Netherlands.
19 Rosenfeld, T., Honig, B., Ottolenghi, M., Hurley, J. and
    Ebrey, T.G., VII IUPAC Symposium on Photochemistry, Aix-en-
    Provence, July 1976. J. Pure Appl. Chem., in press.
20 Yoshizawa, T. and Wald, G. (1963) Nature 187, 1279-1282;
    Yoshizawa, T. in ref.1, pp.146-179.
21 Ebrey, T.G., (1972) Photochem.Photobiol. 15, 585-588.
22 Longworth, J.W. and Ghiron, C.A. (1976) Proc.Symp.
    Excited States of Biological Molecules (Ed.J.B.Birks)
    John Wiley, New York, in press.
23 Förster, T. (1966) in Modern Quantum Chemistry, Istanbul
    Lectures, ed. Sinanoglu, O. (Academic Press, New York),
    sect.III-B, pp. 93-137.

DEGENERATIONS OF THE RETINA INDUCED BY LIGHT

John Marshall

Department of Visual Science, Institute of Ophthalmology

Judd Street, London WC1H 9QS

INTRODUCTION

Until relatively recent times man, like his fellow creatures, was confined to a light environment which conformed to a diurnal rhythm with photic intervals generally not exceeding sixteen hours. The use of fire and primitive oil lamps helped to extend this period of illumination, and such sources were used extensively even in the latter part of the nineteenth century. The advent of gas lighting and later of the tungsten lamp, heralded an era wherein man could control and maintain his environment at relatively high levels of illuminance. However, even these sources precluded the synthetic attainment of 'daylight' levels of illumination without undesirable levels of collateral thermal radiation. Today since the advent of the high efficiency fluorescent lamp, daylight levels of illumination are common in public buildings throughout the world and there is a trend for higher and higher illuminances [1-2]. This trend in the development of high efficiency general lighting has been paralleled in laboratories by the development of special intense light sources such as xenon arcs, or lasers [3-4]. Thus a situation now exists whereby specific artificial light sources analogue solar radiation. The retinal hazards of solar radiation have been recorded since Galen, therefore increasing numbers of laboratories have begun to investigate the effects of retinal exposure to artificial sources. The result of this work is an accumulating body of literature which attests to the deleterious effects of exposure of the retina to excessive light [5-10]. Such excessive stimuli range between short pulse duration high power density flashes, to prolonged exposure to moderate intensity commercial lighting. In the present paper the morphology of the

resultant retinal degenerations will be described and their
causative mechanisms will be discussed.

METHODS

The experimental methods employed in the present paper have
been previously described; light sources and radiometry (9-10),
histology and electronmicroscopy (11), autoradiography and biochemistry (12).

RESULTS AND DISCUSSION

In general terms there are at least three mechanisms by which
light may induce damage to the retina, and these are broadly
related to the pulse duration of the exposure. Thus in order of
increasing duration the predominant mechanisms for threshold damage
are, thermomechanical (sub nanosecond to nanosecond), thermal
(100 microseconds to seconds), and photochemical (100 seconds to
days).

The primary event in any type of radiation damage to a
biological system is the absorption of radiation by that system.
The very properties which allow the eye to be an effective
transducer of light also render it vulnerable to photic damage.
First there is an increase of power density, or irradiance, of
approximately 10^5 between the cornea and the retina, and secondly
the retina possesses two major chromatic systems whose function is
to absorb light. Both of these pigmentary systems are implicated
in light induced damage to the retina. The pigment of the pigment
epithelium is melanin and is the primary absorption system involved
in the thermoacoustic and thermal damage, while the visual pigments
in the photoreceptor cells appear to have some role in the
insidious changes associated with prolonged exposure. In general
thermal and thermoacoustic lesions are discrete highly localised
areas of degeneration whilst photochemical degenerations tend to
involve large areas or entire retinae.

A typical lesion produced by short pulse duration irradiation
(100 microseconds to seconds) will show vacuolation and nuclear
pyknosis of the pigment epithelial cells within the irradiated
area when examined under a light or electron microscope (11).
Similar damage occurs in the overlying photoreceptor cells, and
the disc membranes in their outer segments are disorientated and
disrupted. This damage arises in response to the passage of
elevated thermal transients being conducted into the surrounding
tissue and away from the melanin granules in the pigment
epithelium. The degree of involvement of inner retinal layers
remote from the pigment epithelium is a function of the size of the
area irradiated, and of the duration of the exposure. Thus for
small retinal images, and short exposure durations, e.g. most
pulsed lasers, only the pigment epithelium and the photoreceptor

cells are damaged. Large retinal images or prolonged irradiation such as exposure to continuous wave lasers, voltaic arcs or the sun, may result in extensive intra-retinal conduction of thermal energy and cause degenerative damage to all layers of the neural retina. If extremely high power densities are presented to the retina, such as Q-switched laser pulses (nanoseconds), the retinal burn resulting from the thermal degradation of absorbed light becomes secondary. In these exposures the melanin granules absorb so much energy in such a short time that the resultant phase changes in the cellular constituents are explosive. Thus retinal damage arises due to bulk physical displacement of tissue. In some instances further minor damage may result from the acoustic transients recorded in areas of retina remote from the irradiation.

Two further absorption sites facilitate thermally induced degenerations in the inner retinal layers, the haemoglobin in the retinal blood vessels (13-14), and the yellow macular pigment (15-16). The blue green emission (488nm) of the argon laser is used therapeutically to selectively seal pathologically abnormal blood vessels, however the absorption characteristics of the macular pigment (peaks, 460nm, 490nm) (17) prevent its use in close proximity to the macula.

The harmful effects of prolonged low level illumination were first demonstrated in the retinae of rats (5), and immediate discussions ensued as to whether this was an anomalous response in a predominantly nocturnal animal or a general indication that retinal function can only be maintained in photic environments with a cyclic periodicity. The primary degenerative changes in the rat were a disorientation of the photoreceptor membranes. Such changes had an identical morphology but were more extensive than those seen in short pulse damage. Experiments in our laboratory on a diurnal animal, the pigeon, showed that prolonged exposures produced retinal degeneration, and further that the cone photo-receptor cells were preferentially damaged (10). We have since tried to establish a relationship between the spectral absorption characteristics of the pigeon cones and selective cone damage by prolonged exposure to monochromatic sources. However the results of these experiments are inconclusive and difficult to interpret due to the presence of highly absorbant oil droplets in the inner segments of pigeon cones (18). Further experiments on a fish, the roach, which possesses morphologically distinct cones whose absorption characteristics have also been determined, have again proved unsuccessful. In these animals only the red responding cones degenerate, and do so regardless of the wavelength of the stimuli. However, other workers have described behavioural studies which show corresponding monochromatic colour deficiencies in animals repeatedly exposed to monochromatic sources (19).

Although the underlying mechanism of photochemical damage is unknown it appears that the dramatic response in cones is due to a lower threshold or tolerance to damage of any kind (20). This may be related to the kinetics of the cone outer segment membranes.

The outer segment membranes in rods are in a continuous state
of flux and are renewed throughout life (21). Spent discs are
ingested and phagocytosed by the pigment epithelium (22). Thus in
any traumatic situation involving the receptor outer segments,
those of rods have a good chance of recovering due to resynthesis.
The mechanism of cone membrane replacement if any is unknown, and
hence damage to this system seems to result in cone loss.
Observations on human retinae show that the membranes in cone outer
segments become fewer and more disorientated with age. Thus senile
degenerative conditions and the light history of an individual may
well be related.

REFERENCES

1. CHURCHMAN, A.T. (1971) Electronics & Power, January, 4.
2. I.E.S. Code (1968) Illuminating Engineering Society, London.
3. SCHAWLOW, A.L., & TOWNES, C.H. (1958) Phys. Rev. 112, 1940.
4. MAIMAM, T. (1960) Nature, 187, 493.
5. NOELL, W.K., WALKER, V.S., KANG, S.B., & BERMAN, S. (1966)
        Invest. Ophthal. 5, 450.
6. FRIEDMAN, E., & KUWABARA, T. (1968) Arch. Ophthal., 80, 265.
7. GORN, R.A., & KUWABARA, T. (1967) Arch. Ophthal. 77, 115.
8. HAM, W.T., WILLIAMS, R.C., MUELLER, M.A., GUERRY, D., CLARK,
        A.M., & GEERAETS, W.J. (1965) Bull. New York Acad. Sci.
        28, 517.
9. MARSHALL, J., & MELLERIO, J. (1970) Brit. Med. Bull., 26, 156.
10. MARSHALL, J., MELLERIO, J. & PALMER, D.A. (1972) Exp. Eye Res.
        14, 164.
11. MARSHALL, J. (1970) Invest. Ophthal., 9, 97.
12. KENNEDY, A.J., VOADEN, M.J., & MARSHALL, J. (1974) Nature,
        252, 50.
13. LITTLE, H.L., ZWENG, H.C., & PEABODY, R.R. (1970) Trans.
        Amer. Acad. Ophthal. Otolaryng., 74, 85.
14. MARSHALL, J. & FANKHAUSER, F. (1972) Trans. Ophthal. Soc. U.K.,
        92, 469.
15. MARSHALL, J., HAMILTON, A.M., & BIRD, A.C. (1974) Experienta,
        30, 1335.
16. MARSHALL, J., HAMILTON, A.M., & BIRD, A.C. (1975) Brit. J.
        Ophthal., 59, 610.
17. RUDDOCK, K.M. (1963) Vision Res., 3, 417.
18. MARSHALL, J., PALMER, D.A. & MELLERIO, J. (1973) In: Colour
        73 page 259. Adam Hilger, London.
19. SPERLING, H.G., & HARWERTH, R.S. (1972) Optica Acta, 19, 395.
20. TSO, M.O.M., WALLOW, I.M.L., & POWELL, J.O. (1973) Arch.
        Ophthal., 89, 228.
21. YOUNG, R.S. (1965) Anat. Rec. 151, 484.
22. MARSHALL, J. (1970) Vision Res. 10, 821.

SYMPOSIUM XV

MUTAGENIC EFFECTS OF RADIATION

MUTAGENIC REPAIR PATHWAYS IN YEAST

R.C. von Borstel and P.J. Hastings

Department of Genetics, University of Alberta

Edmonton, Alberta, Canada   T6G 2E9

ABSTRACT

In yeast, the *RAD6* pathway has been shown to be involved in the induction of transitions. Three points of evidence indicate that yeast has another mutagenic repair pathway: 1) it is needed for induction of transversions; 2) some chemicals are still mutagenic in *rad6*; 3) the mutant *rad6* is a mutator, presumably by channelling spontaneous lesions into another mutagenic repair pathway.

Since the two mutagenic repair pathways identified so far produce different predominant mutations, allele specificity in reversion studies can be used to distinguish the pathways.

INTRODUCTION

The concept has been established primarily by Witkin (1) that mutations occur when pre-mutational lesions, induced in DNA by radiation or chemicals, are repaired by enzymes with mutagenic properties. When the so-called "error-free" pathways are genetically impaired by mutants, then mutations are induced with much higher frequencies than with normal strains, because these lesions are then more frequently repaired by the mutational enzymes in a "mutagenic (error-prone) repair" pathway.

The concept of mutagenic repair has not been used previously for considering the fate of spontaneous lesions in DNA. In the decade of the 1960s the relation was established between spontaneous mutation rates and the different aspects of DNA metabolism (2),

683

namely, recombination (3), replication (4), and repair (5). A
close affinity between mutator and antimutator strains was described in replication systems by Schnaar, Muzyczka and Bessman (6)
and Hershfield and Nossal (7) who showed that defects in the editing function of the 3'-exonuclease portion of a DNA polymerase
molecule enhanced the spontaneous mutation rate, and defects in the
DNA polymerase portion itself depressed the spontaneous mutation
rate.

But the plethora of loci in a yeast cell with mutator activity (over 20) led us to seek other explanations; through an analysis
of the spontaneous mutation rates associated with radiation-
sensitive loci, it became clear that some steps in every repair
pathway have mutator activity. From this fact we have evolved the
following schema based on the concept of mutagenic repair.

REPAIR PATHWAYS AND THE CONCEPT OF CHANNELLING

By an analysis of epistatic, additive, and synergistic interactions between pairs of radiation-sensitive loci, three pathways
have been identified in *Saccharomyces cerevisiae* for repair of
lesions induced by ultraviolet radiation (8, 9). Two of these
pathways are assumed to be error-free, that is, no mutational
enzymes are encountered during the repair of UV-induced lesions.
The other pathway is assumed to be the mutagenic repair pathway
because, when this pathway is blocked genetically, the induced
mutation frequency fails to increase with dose (10, 11). The three
pathways are named for the genetic locus controlling the purported
first step of each pathway: *RAD3*, *RAD51*, and *RAD6*. The mutant
rad3 is believed to block the first step of the excision repair
pathway (12), *rad51* blocks the first step of a pathway which repairs both X-radiation-induced and UV-induced lesions (13), and
rad6, which also repairs lesions induced by both nonionizing and
ionizing radiation, is believed to block the first step in the
mutagenic repair pathway (14).

It is of interest that mutants for the first step of each
pathway exhibit enhanced spontaneous mutation rates. We believe
that mutation rate enhancement gives credence to the concept of
"channelling" of the initial induced or spontaneous lesions in the
DNA. If the lesion "substrate" cannot be accepted into one pathway because it is genetically blocked, then the substrate has an
enhanced probability of being accepted by a mutagenic repair pathway.

THE SECOND MUTAGENIC REPAIR PATHWAY

Channelling of lesions is a useful way of thinking about spontaneous and induced mutations. It leads to predictions and explains

diverse phenomena within one framework.

The channelling hypothesis provides cogent reasons for the existence of another mutagenic repair pathway besides the *RAD6* pathway: First, spontaneous mutations occur in *rad6* strains. Unless spontaneous mutations follow their own rules of occurrence, then it follows that spontaneous mutagenesis in *rad6* strains is evidence in itself for the presence of at least one more mutagenic repair pathway. Second, some chemical mutagens are not dependent on the *RAD6* pathway for mutation induction (15). On the other hand, it could be argued that the lesion induced by these chemicals is the mutation itself and mutagenic repair is not involved. Third, the *RAD6* pathway is a transition-specific mutagenic repair pathway (11), yet we know other types of mutations, i.e. transversions and frameshifts, do occur. Fourth, the *rad6* mutant is a mutator. This means that lesions now not accepted by the *RAD6* pathway are channelled into another repair pathway where mutational enzymes can act (16).

Witkin (17) has found that one of the mutants on a mutagenic repair pathway in *Escherichia coli* is a mutator strain. Therefore, it is likely that more than one mutagenic repair pathway is present in *E. coli* as well.

THE RELATION BETWEEN LIQUID HOLDING RECOVERY AND PETITE MUTAGENESIS

Once it is seen that channelling of both spontaneous and induced lesions occurs, then apparently unrelated mutagenic phenomena become explicable.

For example, liquid holding recovery (LHR) is evidence for an error-free repair pathway which operates in stationary-phase cells held in the dark (to distinguish LHR from photorecovery). LHR is seen when cells are held in buffer for protracted periods after UV irradiation. Whatever the molecular action of LHR might be, it has been observed that petite strains (strains with defective mitochondria) of yeast do not exhibit LHR (18).

On the basis of the theory of channelling, the lack of LHR in petite strains of yeast is consistent with the observation that petite strains of yeast exhibit enhanced spontaneous mutation rates (19, 20, 21, 22, 23).

It seems possible that fully functional mitochondria are necessary for LHR to operate. If the mitochondria are defective, then channelling of the lesions to other pathways takes place.

It is of interest that the extra spontaneous mutations which arise in petite strains are quite specific in nature, and this must be deduced within the framework of mutational specificity.

MUTATIONAL SPECIFICITY AS AN OUTCOME OF CHANNELLING

It would seem that specificity is conferred to only a limited degree by a mutagen, and perhaps only a little more so by the base sequences in the region of DNA being mutated (24).

There is evidence that the *RAD6* pathway in yeast is specific for transitions (11). Since transversions and frameshift mutations do occur, it seems possible that another mutagenic repair pathway might, for example, elevate the proportion of transversions to transitions. Thus, for example, a repair pathway might utilize a terminal transferase which would randomly accept any base, and thereby transversions would occur twice as frequently as transitions.

This brings us to the problem of petite mutagenesis which appears to be specific for one type of mutation. If the block of LHR channels lesions into another repair pathway, it would appear to channel them into a mutagenic repair pathway with specific attributes. This does not appear to be the case when the *RAD3* pathway is blocked at the first step, where the enhancement of spontaneous mutation rate yields an array of types of mutation.

So mutagenic repair pathways may be specific or general with respect to the types of mutations produced, and certain types of lesions seem to enter some pathways preferentially.

The theory of channelling predicts that strains with anti-mutator activity should represent genetic blocks of mutagenic repair pathways. Channelling of lesions from a nonspecific muta-genic repair pathway into a specific mutagenic repair pathway should result in a higher proportion of specific mutations, yet the net effect may be that the mutation rate would decrease because a high efficiency of channelling into error-free pathways may occur simultaneously. Likewise, it is possible that an antimutator for one type of mutation being measured may be a mutator for another type of lesion. The mutator mutant *mut1* appears to follow this rule (22, 25). The only true antimutator strain, therefore, would be one where *all* mutagenic repair pathways have been genetically blocked simultaneously. (26).

REFERENCES

1. Witkin, E.M. (1966) Brookhaven Symp. Biol., *20*, 17-55.
2. von Borstel, R.C. (1969) Japan. J. Genet., *44* (Suppl.1), 102-105.
3. Magni, G.E. and von Borstel, R.C. (1962) Genetics, *47*, 1097-1108.
4. Speyer, J.F., Karam, J.D. and Lenny, A.B. (1966) Cold Spring

    Harbor Symp. Quant. Biol., *31*, 693-697.
5. Böhme, H. (1967) Biochem. Biophys. Res. Comm., *28*, 191-196.
6. Schnaar, R.L., Muzyczka, N. and Bessman, M.J. (1973) Genetics,
    73 (Suppl.), 137-140.
7. Hershfield, M.S. and Nossal, N.G. (1973) Genetics, *73* (Suppl.),
    131-136.
8. Game, J.C. and Cox, B.S. (1973) Mutation Res., *20*, 35-44.
9. Brendel, M. and Haynes, R.H. (1973) Molec. Gen. Genet., *125*,
    197-216.
10. Lemontt, J.F. (1971) Genetics, *68*, 21-33.
11. Lawrence, C.W., Stewart, J.W., Sherman, F. and Christensen, R.
    (1974) J. Mol. Biol., *85*, 137-162.
12. Game, J.C. and Cox, B.S. (1972) Mutation Res., *16*, 353-362.
13. Cox, B.S. and Game, J.C. (1974) Mutation Res., *26*, 257-264.
14. Lawrence, C.W. and Christensen, R. (1976) Genetics, *82*, 207-232.
15. Prakash, L. (1976) Genetics, *83*, 285-301.
16. Hastings, P.J., Quah, S.-K. and von Borstel, R.C. (1976)
    (submitted for publication).
17. Witkin, E.M. (1973) Ann. Acad. Brasil. Ciên., *45* (Suppl.),
    185-191.
18. Lyman, J.T. and Haynes, R.H. (1967) Radiation Res. Supplement 7,
    222-230.
19. von Borstel, R.C., Cain, K.T. and Steinberg, C.M. (1971)
    Genetics, *69*, 17-27.
20. Flury, F. and von Borstel, R.C. (1972) Can. J. Genet. Cytol.,
    14, 727.
21. Flury, F. and von Borstel, R.C. (1973) Genetics, *74* (Suppl.),
    s81-s82.
22. von Borstel, R.C., Quah, S.-K., Steinberg, C.M., Flury, F. and
    Gottlieb, D.J.C. (1973) Genetics, *73* (Suppl.), 141-151.
23. Flury, F., von Borstel, R.C. and Williamson, D.H. (1976)
    Genetics, *83*, 645-653.
24. Sherman, F. and Stewart, J.W. (1974) Genetics, *78*, 97-113.
25. Gottlieb, D.J.C. and von Borstel, R.C. (1976) Genetics, *83*,
    655-666.
26. Research sponsored by grants from the National Research Council
    of Canada and grant 21294 from the National Institute of
    General Medical Sciences.

PATHWAYS OF UV REPAIR AND MUTAGENESIS IN
SACCHAROMYCES CEREVISIAE

B. S. COX

OXFORD UNIVERSITY

Botany School, South Parks Road, Oxford, U.K.

Starting in 1967 with work published by Nakai[1] a
number of workers have isolated in the yeast
Saccharomyces cerevisiae, mutants which were more than
usually sensitive to treatment with mutagens[2,3,4,5,6].
Generally, the phenotype sought was sensitivity to ultraviolet light of 254 nm wavelength, but many of the
mutants were found also to be sensitive to other
mutagens, such as ionising radiation[1,3,4,6,7] or
chemicals[7]. These mutants fell into a rather large
number of different loci[8,9], at least twenty-two
conferring UV-sensitivity when mutant and a further
eight loci primarily determining resistance to ionising
radiation. Subsequently, mutants isolated with other[10]
phenotypes such as resistance to UV-induced mutation
or sensitivity to methyl-methane sulphonate (Snow,
unpublished) were also found to be UV-sensitive. In
some cases, these mutations had occurred at loci
different from those previously identified.

In 1969, Game undertook the task of attempting to
assign these mutants to particular repair pathways using a genetical approach. The rationale of this
endeavour was based on the notion that repair proceeds
by a series of sequential steps, that is by a 'pathway'
of reactions and that, judging by the large number of
loci involved, there may be more than one pathway
involved in repairing UV-induced damage. If two
mutants, x and y control enzymes involved in the same
pathway, thus:-

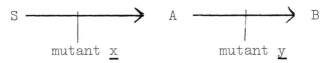

then the phenotype of the double mutant x, y will have
the phenotype of one of the single mutants. Obviously,
once a pathway has been blocked by one mutation, it can
make no difference to the repair capability of the cell
if a later step is also blocked. The first mutation
is epistatic to the second, and when UV-sensitive
mutations are combined, the interaction would be
expressed as the double-mutant having the same survival
curve as one of the single mutants.

Game described three classes of interaction in
double and multiple mutant combinations[11, 12, 13]:
epistasis, additivity and synergism. "Additivity"
describes the observation that the double-mutant
survival curve can be obtained by adding the distance
between the wild-type survival curve and one of the
mutant survival curves to the survival curve of the
other mutant. "Synergism" is the name given to the
observation that the double mutant is more sensitive
than what is expected from additivity. Synergistic
interactions are the most interesting of the three
since, whereas a variety of metabolic situations may
give rise to either the "epistatic" or "additive" type
of interaction, an observation of synergism is capable
of only a limited number of interpretations[12,13]:
namely that the two mutants affect different pathways
and that the pathways interact. One kind of interaction
might be that the two pathways concerned use a common
substrate. It was argued that when two mutants showed
a synergistic interaction in this way it was likely
that one or both of them was blocking the first
irreversible step in its pathway from the common
substrate. Using these criteria, eight mutants were
assigned to three UV-repair pathways. It was shown
that triple mutant strains in which all three putative
pathways were blocked had an LD_{37} dose of UV irradiation
sufficient to induce between one and two dimers per
haploid genome[13].

Many of the mutants have also been classified by
physiological and biochemical methods. Correspondence
between such classifications and the pathways described
by Game and Cox is quite close. In 1969, for example,
Parry and Parry [14] classified many of the mutants into

four groups depending on their response to post-UV treatment. Mutants in two of these groups (group 3 and 4) remain photoreactivable for long periods when held in non-growth conditions following UV irradiation, whereas other mutants and wild-type yeast rapidly lose photo-reactivability. The mutants which retain photoreactiva-bility, rad1, rad2, rad3, rad4 and rad16 are all epistatic to one another, and comprise a single pathway in the scheme proposed by Game.

These five mutants have also been assayed by various workers and by various methods for their abil-ity to excise dimers[15,16,17,18,19]. None of them are able to do so. It would seem that, so far, the genetic-al analysis which assigns the mutants to 'distinct' repair pathways is supported by the more detailed phenotypic criteria which can be applied by physiologic-al and biochemical techniques. Table 1 summarises the properties of a number of the mutants, and classifies them accordingly.

The phenotype which concerns this symposium, mutability, can also be used to place rad mutants into distinct groups[13,20,21,24]. The mutations seem to fall into three phenotypic classes. There are those which, like the recA and lex mutants of Escherichia coli drastic-ally reduce or abolish the induction of mutations by UV. Secondly, several rad mutations dramatically enhance the UV-mutability of strains that contain them. This is similar to the behaviour of hcr⁻ mutants of E. coli. It is interpreted as an indication that when a repair pathway such as excision is largely error-free; that is inserts no wrong bases in the course of repair; when such a repair pathway is blocked by mutation, many more of the lesions induced by UV are diverted down an alternative repair pathway which does make mistakes: is error prone[22]. Finally, there are many rad mutants which appear to have about the same number of mutations induced in them as are induced in wild-type strains by similar doses of UV. The data presented in Table 1 are compiled from the work of Lawrence and his associates[21,24]. This work is by far the most comprehensive and detailed study carried out on the effects of these and other mutations on UV mutability and makes it possible to make some sort of quantitative comparison with wild-type for well-understood revertible loci at particular doses of UV.

Three observations can be made about these data.

1) With the exception of rad16, all mutations

affecting pathway I and dimer excision show the enhanced
UV-induced mutation rates characteristic of similar
mutants in other organisms. With the exception of rad22,
no other mutant does.

2) Two mutations, rad6 and rev3 block almost
completely UV-induced reversion at both loci dealt with
here.

This holds true for nearly all other loci whose UV-
induced reversion is assayed in these mutants, and in them
it is also found that forward mutation is lowered or
prevented. The remaining rad mutations assigned to this
pathway by Lawrence and Christensen on the grounds of
their epistatic interactions with one another and their
common sensitivity to ionising radiation all block UV-
induced reversion at the locus of cyc1-9, an ochre
mutation[26]. The possible exception is rad18 whose
effect on the reversion rate is relatively small.

3) Some of the mutants of pathway III, all of which
appear to be necessary for reversions of cyc1-9, show
specificity in that they do not effect greatly the UV-
induced reversion of cyc1-131. cyc1-131 reversion
demands a change from GUG to AUG[21].

Lawrence (personal communication) has data that
show that specificity of control of UV-induced mutation
is characteristic of all the mutants of this group or
pathway, with the exception of rad6. The specificity
is not simply determined by the particular base change
required for reversion, since in rev 2, for example,
certain ochre mutations revert at a normal rate, whereas
cyc1-9 does not. He suggests that the composition of the
neighbouring bases in the DNA is likely to influence
mutability. This specificity allows one to make some
statements about UV mutagenesis, even if at present, no
detailed description can be offered. Clearly UV mutagenesis in yeast, as in bacteria, is the consequence of a
pathway of DNA repair metabolism. The evidence at present
indicates that it is the consequence of just one of the
three pathways found by Game and Cox. Lawrence and
Christensen[21] in a rad1 rad6 double mutant, and Lemontt[27]
in a rad2 rad9 double mutant - strains in which, supposed-
ly only the third, minor, pathway was open - showed that
no mutations were induced by UV among the survivors.
However, the error-prone pathway is not it would seem,
a "conventional" metabolic pathway with substrates being
converted in orderly succession until intact DNA

which may have mutations of any kind in it is formed as the final end product. The specificity of mutagenesis suggests that at certain stages in repair, specific sequences of DNA have to be recognized and perhaps modified before repair can progress. This recognition (and modification) may require the co-operation of several proteins such that at some sequences a single defective protein may prevent recognition, but that at others this defect is immaterial, and wild-type levels of repair and mutation are observed at that sequence or locus (although, of course, repair will be prevented at other damaged sequences). This predicts that as analysis of the effects of these mutants progresses, it should be possible to distinguish among "recognition" or "modifier" gene products and genes controlling more general repair functions.

Finally, it is worth considering what information this analysis of repair functions in yeast provides about spontaneous mutation. Brychy[28] has shown that certain of the rad mutations are "mutators", that is, spontaneous mutation is enhanced in the mutants. The loci involved are rad3, rad5 (rev2) and rad18. This observation suggests that both the excision-repair pathway of yeast and the error-prone pathway have functions in normal metabolism concerned with "editing" DNA. Two of these loci rad3 and rad18 are placed at the first step in one of the UV repair pathways[13]. One might suppose that in unirradiated cells, particular pre-mutational lesions are preferentially edited out by a particular one of the repair systems available. If it is blocked, then the lesions would remain and be expressed as mutations. It can be predicted that only "first step" blocks would have a mutator effect, since once repair is embarked upon it would have to be completed for the cell to survive. What is the nature of the pre-mutational lesions specific to any given repair pathway? In Table 1 have been included some comparisons compiled from data recently published by Prakash[25]. It concerns the effects of chemicals on mutagenesis in various rad strains. It is noticeable that, by and large, the mutants of the error-prone pathway III have little effect on the induction of mutations by EMS or HNO_2. (With other chemical mutagens, considerable reductions are often observed, particularly in rad6 and rad9[29]). Particularly striking is the failure of excision-defective mutants to enhance induction. In what way may chemical mutagenesis differ from UV mutagenesis? No doubt, much repair of potentially lethal damage

occurs by similar pathways and similar effects on
mutagenesis might be expected. However, whereas any
unrepaired dimer is lethal in yeast[30], not every
chemically-induced lesion is necessarily lethal:
it may remain, and in replicating DNA be the cause of
a mutation. As such it resembles a pre-mutational
lesion in an untreated cell, and may be dealt with by
the same editing processes. In Prakash's data, rad3
is conspicuous among the excision repair loci for the
enhancement of EMS-induced mutation and rad18 among the
error-prone set of loci for the enhancement of HNO_2 -
induced mutation: these are the loci described as
mutators by Brychy.

It would seem that further analysis at the level
of detail undertaken by Lawrence and Christensen and by
Prakash is likely to be very rewarding in the analysis
and understanding of both spontaneous and induced
mutagenesis.

TABLE 1

A comparison of some properties of rad mutations. The figures in the last five columns are the ratio: reversion frequency in rad/reversion frequency in RAD+ assayed in the system described at the head of the column. Dimer excision is expressed as 100 x fraction of dimers excised in rad/ fraction of dimers excised in RAD+ after a similar UV dose. (*J.C. Game, personal communication. + reversion frequency of arg4.17).

Rad Locus	Pathway	Radiobiological Group	γ-sensitivity	Dimer Excision A	Dimer Excision B	UV cycl-9 $100J.m^{-2}$ A	UV cycl-9 $100J.m^{-2}$ B	UV cycl-131 $25J.m^{-2}$	EMS cycl-131	HNO_2 cycl-131	Spont. lys1.1
Ref:-	A = 13, B = 21	14	3	17	18	21	24	21	25	25	28
1	I^A	4	+	8.5	7.3	12	44.5	5.9	0.77	2.0	1.5
2	I^A	4	+	6.4				33.3	0.31		0.96
3	I^A	4	+	6.4				67.5	5.6	3.1	5.1
4	I^A	2	+	10.6	19.3			31.7	0.48		0.85
16	I*	2	+		6.1			1.4	1.1	1.1	
15	II^A	3	SS	36.1		1.16		0.39	0.59	0.15	
50	II^A	3	SS		100			0.46			
51	II^A		SS		100						
5(rev2)	III^B	1	S		92.7	$5 \times 10^{-2+}$		0.73	2.6	1.6	6.2
6	III^A	1	SS	19.1		4.4×10^{-3}		0	0.11	0.18	0.41
8	III^B	1	S			2.3×10^{-3}					
9	III^A		S			6.6×10^{-2}		0.57	0.34	0.39	
18	III^A		SS		73.0		0.11	1.46	0.80	5.07	5.1
rev1	III^B		S		100	3.4×10^{-2}		0.56	1.8	1.6	
rev3	III^B		S		100	10^{-3}		1.5×10^{-2}	3.0	1.5	
7		1	+			1.86^+					0.96
10		1	+					0.27			
11		1	+					0.72			
12		1	+					0.96			
13			+								
14			+						3.0		
17		1	+						3.4		
19		1	+								
21		1	+								
22		1	+					8.4	2.3	0.69	

(rows 7, 10–14, 17, 19, 21, 22 — unclassified at present)

REFERENCES

1.   Nakai, S. and S. Matsumoto, 1967. Mutation Res.
          $\underline{4}$, 129.
2.   Snow, R., 1967. J. Bacteriol. $\underline{94}$, 571.
3.   Cox, B.S. and J.M. Parry, 1968. Mutation Res. $\underline{6}$,
          37.
4.   Laskowski, W., E.-R. Lochmann, S. Jannsen and E.
          Fink, 1968. Biophysik $\underline{4}$, 233.
5.   Resnick, M.A., 1969. Genetics $\underline{62}$, 519.
6.   Zakharov, I.A., T.N. Kozina and I.V. Federova,
          1970. Mutation Res. $\underline{9}$, 31.
7.   Zimmermann, F.K., 1968. Molec.gen.Genetics $\underline{102}$,
          247.
8.   Game, J.C. and B.S. Cox, 1971. Mutation Res. $\underline{12}$,
          328.
9.   Game, J.C. and R.K. Mortimer, 1974. Mutation Res.
          $\underline{24}$, 281.
10. Lemontt, J.F. 1971. Genetics $\underline{68}$, 21.
11. Game, J.C. and B.S. Cox, 1972. Mutation Res. $\underline{16}$,
          353.
12. Game, J.C. and B.S. Cox, 1973. Mutation Res. $\underline{20}$,
          35.
13. Cox, B.S. and J.C. Game, 1974. Mutation Res. $\underline{26}$,
          257.
14. Parry, J.M. and E.M. Parry, 1969. Mutation Res.
          $\underline{8}$, 545.
15. Resnick, M.A. and J.K. Setlow, 1972. J. Bacteriol.
          $\underline{109}$, 979.
16. Unrau, P., R. Wheatcroft and B.S. Cox, 1971.
          Molec.gen.Genetics. $\underline{113}$, 359.
17. Wheatcroft, R., 1973. D.Phil. thesis, Oxford Univ.
18. Ferguson, L.R., 1975. D.Phil. thesis, Oxford Univ.
19. Prakash, L. 1975. J. Molec. Biol. $\underline{98}$, 781.
20. Moustacchi, E., 1969. Mutation Res. $\underline{7}$, 171.
21. Lawrence, C.W. and R. Christensen, 1976. Genetics
          $\underline{82}$, 207.
22. Witkin, E.M., 1967. In: Recovery and Repair Mech-
          anisms in Radiobiology. Brookhaven Symp.
          Biol. $\underline{20}$.
23. Lemontt, J.F., 1972. Molec.gen.Genetics $\underline{119}$, 27.
24. Lawrence, C.W., J.W. Stewart, F, Sherman and R.
          Christensen, 1974. J.molec.Biol. $\underline{85}$, 137.
25. Prakash, L., 1976. Genetics $\underline{83}$, 285.
26. Stewart, J.W., F. Shermann, M. Jackson, F.L.X.
          Thomas and N. Shipman, 1972. J.molec.Biol.
          $\underline{68}$, 83.
27. Lemontt, J.F., 1971. Mutation Res. $\underline{13}$, 311.

28. Brychy, T., 1974. M.Sc. thesis., Univ. of
         Alberta.
29. Prakash, L., 1974. Genetics 78, 1101.
30. Wheatcroft, R., B.S. Cox and R.H. Haynes, 1975.
         Mutation Res. 30, 209.

Mitochondrial mutagenesis by 2-6-diaminopurine
in Saccharomyces cerevisiae: effect of UV light

C. Wallis and D. Wilkie

Department of Botany and Microbiology
University College London
Gower Street, London WC1E 6BT

Summary

2-6-diaminopurine (DAP), an analogue of adenine,
selectively inhibited mitochondrial biogenesis in
S.cerevisiae. This was detectable as failture of cytochromes aa_3 and b to develop leading to arrest of growth
in non-fermentable medium. The analogue induced mitochondrial mutations to antibiotic resistance, particularly
with respect to oligomycin and chloramphenicol, and to
the petite condition. The efficiency of induction was
strain dependent. DAP-treated cells when irradiated
with UV light had a greatly enhanced rate of petite
induction compared with either mutagen alone.

Introduction

Differential sensitivity of the yeast mitochondrion
to chemical mutagens such as acridine and UV light has

been known for many years with reference to the high
efficiency of induction of the deletion mutant <u>petite</u>.
Our initial finding that DAP selectively inhibited mitochondrial development in the yeast <u>S.cerevisiae</u>[1], has
been followed up by a study of the mutagenic activity of
the adenine analogue on the mitochondrial system.

Materials and Methods

Haploid strains of this laboratory were used. All
growth media buffered at pH 6.5 with Tris maleate,
contained yeast extract (0.5%) supplemented either with
2% glucose (YED) or 4% glycerol (YEG) as carbon sources.
Petite determining medium (PDM) contained 0.2% glucose
and 4% glycerol. DAP, obtained from Sigma, was added
directly as a powder to the medium before autoclaving,
up to 5 mg/ml, the maximum solubility. 2% Difco agar
was used to solidify medium. Tests of growth inhibition
using a multiple inoculation device are described elsewhere[2]. Absorption spectra were obtained in the SP1800
recording spectrophotometer. The UV source was a Philips
6 watt TUV delivering 985 ergs/cm^2/min at 9 in. (the
distance used in the irradiation), to cells suspended in
distilled water at 5×10^6/ml concentration.

Results

The minimum inhibitory concentration of DAP (MIC) to arrest growth of 11 out of 13 strains tested, was 250 µg/ml in YEG medium. MIC to inhibit growth of the other 2 strains (45B and 188) was 500 µg/ml. When fermentable YED medium was used, 11 out of the 13 strains tested, maintained growth at the maximum concentration of 5 mg/ml DAP, the remaining 2 strains (in this case, D22 and D26) were inhibited at 3 mg/ml and 2 mg/ml DAP respectively. It may be remembered that the yeast-extract medium contains adenine, estimated roughly as 100 µg/ml, so that the respective nucleic acid-synthesizing systems of nucleus and mitochondrion can discriminate between the analogue and the metabolite. Further addition of adenine relieved DAP inhibition, 100 µg/ml restoring more or less full growth in the presence of 250 µg/ml DAP in YEG medium. The results indicated that the mitochondrial system was much less able to discriminate against the analogue than the nuclear system and that the primary effect of DAP was the production of non-functional mitochondrial, hence the inability to utilize glycerol.

The antimitochondrial activity was leading to the arrest of protein synthesis in the mitochondria as seen from the absorption spectra of several strains grown in YED medium in the presence of DAP. As shown in Fig. 1, cytochromes aa_3 and b but not cytochrome c fail to develop in the presence of the analogue.

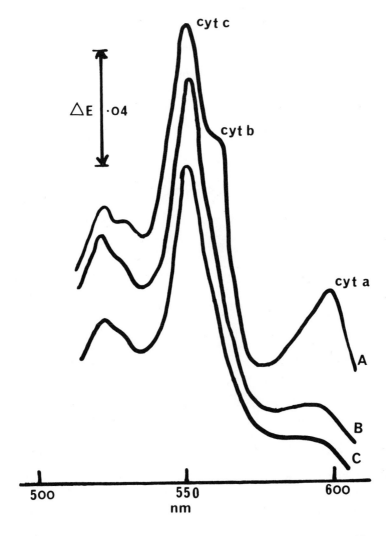

Fig.1 Absorption spectra of cells of strain B41 at
       stationary phase in glucose medium. A, control;
       B with 50μg/ml DAP; C with 100μg/ml DAP.

Mutagenic effects   (1) Antibiotic resistance. Mitochondrial mutation to antibiotic resistance is well documented in the literature[3]. In the present studies, the antibiotics erythromycin, chloramphenicol and oligomycin were used (specific inhibitors of mitochondria) to test possible induction of resistance by DAP. Table 1 gives the results for 4 strains.

Table 1. Effect of DAP on the frequencies of resistant mutants to chloramphenicol (C^R), erythromycin (E^R) and oligomycin (O^R) in 4 yeast strains[*]

Mutants per 10^7 cells

	D6		B-B		B-A		B41		
Treatment	C^R	E^R	C^R	E^R	C^R	E^R	C^R	E^R	O^R
Control	2016	20313	112	924	148	21	34	216	114
50ug/ml DAP	4212	7526	399	567	190	25	80	110	490
100ug/ml DAP	3821	5684	271	719	385	20	52	75	189

*Strains were grown for 18h in liquid YEG containing subinhibitory amounts of DAP. Cells were plated on YEG-antibiotic medium and resistant colonies scored up to 14 days after incubation.

These experiments were repeated more than once and although there was considerable variation in overall numbers of resistant mutants, the relative proportions among treated and untreated cultures were similar. The results indicated that DAP was inducing resistance to chloramphenicol and oligomycin while, at the same time, depressing the numbers of erythromycin resistant mutants. At these concentrations of DAP, there was no significant effects on nuclear mutations (reversion to prototrophy of his⁻ in strain B41 and arg⁻ in D6) nor on cell viability. It was concluded that C^R and O^R were mitochondrial mutations induced specifically by DAP. A few O^R induced mutants were tested genetically and evidence of cytoplasmid inheritance were obtained. Also, a fluctuation test[4] was carried out to establish that O^R was induced by DAP and a positive answer was obtained.

(2) Petite induction. Cells of various strains were treated with DAP as for antibiotic resistance and plated on PDM medium. Petite colony was scored on the basis of small size in the first instance and subsequently on inability to grow when transferred to YEG medium. Results (Table 2) showed a significantly higher incidence of petites among DAP-treated cells than in controls except for strain D6. These results were repeatable.

DAP-treated cells were irradiated with UV light and the effect on petite induction recorded. The results are shown in Table 3 and indicate a synergistic effect, the frequencies of petites being significantly greater than the sum of the inducing capacities of the two mutagens taken alone. There was no evidence of a synergistic

effect on cell viability.   These studies involving UV
light are still at a preliminary stage and are being
extended to cover effects on antibiotic resistance.
With respect to the latter, initial results already
indicate a synergistic effect on induction of O^R.

Table 2.   Effect of DAP treatment on petite frequency

	% petite colony*			
Treatment	B41	D6	B-B	B-A
Control	2.0	0.27	3.5	0.15
50ug/ml DAP	4.5	0.17	14.0	0.24
100ug/ml DAP	3.9	0.22	7.2	0.80

*Total colonies scored for each culture was in excess
 of 300.

Discussion

The general insensitivity of yeast cells to the mutagenic activity of DAP (i.e., insensitivity of the nucleus)
was cited by Lomax and Woods[5] who also described a strain
which had mutated to DAP sensitivity.   Our results
support these findings but demonstrate general sensitivity
of the mitochondrial system to the mutagen.   Since DAP
competes with adenine, we conclude that the effects of
the analogue result from its incorporation into nucleic

Table 3.   UV induction of petite in DAP-treated cells
           of strain B41*.

Culture	UV(min)	Viability %	Petite %
Control	O	100	3(8/278)
	1	82	7(20/290)
	3	20	23(148/638)
50ug/ml DAP	O	97	10(23/222)
	1	75	34(53/149)
	2	38	45(130/303)
	3	22	74(190/260)
100ug/ml DAP	O	98	11(34/333)
	1	80	28(83/300)
	2	49	56(162/296)
	3	27	60(141/240)

*Cultural conditions were as in Table 1.

acid.   The most striking effect is on mitochondrial
protein synthesis, leading to the conclusion that DAP
has a more deleterious action after incorporation into
RNA than into DNA in mitochondria.   For example, mRNA
containing DAP may be difficult to translate on mitochondrial ribosomes.   Another possibility is that the
enzyme system for transcription may be less able to

discriminate between analogue and base compared with that

for DNA replication, assuming that incorporation into

DNA is responsible for increased mutation rate.

The study of nucleic acid synthesis in isolated

mitochondria in the presence of DAP, both cold and

radioactive, and of DAP-resistant mutants, may resolve

these problems.

References

1. Wilkie, D. and Lee, B. (1972) Heredity, 29, 241
   (Abstract).

2. Wilkie, D. (1972) Med. Biol. Illustr., 22, 119.

3. Wilkie, D. (1975) In Methods in Cell Biology Vol.XII
   p.353 (ed. D. Prescott), Acad. Press, New York.

4. Luria, S. E. and Delbruck, M. (1943) Genetics, 28,
   491.

5. Lomax, C. A. and Woods, R. (1969) J. Bact., 100, 817.

ON THE MOLECULAR MECHANISM OF MUTAGENESIS BY CARCINOGENIC MUTAGENS

VILLANI,G., DEFAIS,M., SPADARI,S[*], CAILLET-FAUQUET,P.
BOITEUX,S. and RADMAN,M.

Département de Biologie Moléculaire, Université Libre
de Bruxelles - B 1640 Rhode-St-Genèse, Belgium
 Laboratorio di Genetica Biochimica ed Evoluzionistica
del C.N.R.,via S. Epifanio, 14 - 2700 Pavia, Italy

SUMMARY

Genetic analysis and in vivo and in vitro biochemical
studies have revealed that radiation and most chemical mutagens
induce in bacteria an error-prone DNA repair process which is
responsible for their mutagenic effect. This repair process,
called SOS repair, has been correlated with an increased cellular
capacity to replicate damaged DNA by permitting insertions of non
complementary nucleotides opposite DNA lesions. The crucial role
of the $3' \longrightarrow 5'$ exonuclease ("proof-reading") activity has been
elucidated. Unlike E.coli DNA polymerases, human α, β and γ
polymerases themselves can copy DNA post pyrimidine dimers by
inserting non complementary nucleotides.

INTRODUCTION

In relation to their effect on DNA replication, mutagens can
be divided in two groups :

(1) Mutagens which do not inhibit DNA synthesis cause subtle
    modifications of DNA bases in the template and/or in the
precursors. In this case DNA polymerases "do not know" that they
are making errors. Examples of such mutagens are ethylmethane
sulfonate, some bases analogs and deaminating agents.

(2) Mutagens which cause bulky lesions in the DNA and therefore
    inhibit DNA synthesis. Examples are ionizing and ultraviolet

(UV) radiation and a majority of chemical mutagens such as
mitomycin C, AAF, activated aflatoxin B1 and benzo(a) pyrene,etc.
Mutagens of the first class do not cause change in gene expression
such as evidenced by prophage induction in lysogenic bacteria nor
is their mutagenic effect dependent on some known bacterial gene.
Mutagens of the second class cause changes in gene expression in
bacteria (lysogenic induction, filamentous growth, etc..) and
their mutagenic effect strictly depends on a change in gene
expression i.e. induction of some cellular operons requiring
functional E.coli recA and lex genes.

The first class of mutagens are weakly or noncarcinogenic in
animals while the second classe of mutagens are typical strong
carcinogens. Since both spontaneous and induced mutations finally
occur as errors in DNA synthesis, we have analysed the extent
(both in vivo and in vitro) and the fidelity (in vitro) of DNA
synthesis on intact and UV-irradiated single stranded phage
∅X174 DNA, as well as on synthetic homopolymers. This research
can be summarised as follows :

(a) UV irradiation of E.coli causes a transient induction of a
    cellular mutagenic capacity which increases survival of the
cell and of irradiated phage infecting irradiated hosts (1, 2).
This inducible promotion of survival and mutagenesis is called
"SOS repair" (3, 4). Chloramphenicol is an inhibitor of SOS
repair (2) (Fig. 1).

(b) Inhibition of DNA replication in an E.coli dnaB mutant at
    42°C causes induction of SOS repair as evidenced by increased
mutagenesis of both irradiated and untreated phage λ and by
increased survival of the irradiated phage (Fig. 2). This
experiment (5), as well as those of E. Witkin (6) showing dnaB
(42°C)-promoted bacterial mutagenesis, demonstrate that an
abrupt blockage of DNA replication is the induction signal for
SOS repair (as well as for lysogenic induction) rather than
particular DNA lesions themselves (comp. class (2) mutagens).

(c) In vivo, none of the constitutive E.coli DNA polymerases can
    copy UV-irradiated single stranded phage ∅X174 DNA past
pyrimidine dimers. This is the reason why no mutagenesis can be
produced by irradiating ∅X174 phage only. When the host cells

---

*The essential methodology is described or cited in the legends
to figures and tables.

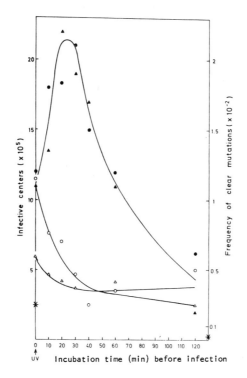

Figure 1 : Kinetics of UV induction of SOS repair and mutagenesis
          in E.coli assayed with phage λ and inhibition of the
          induction by chloramphenicol

Only results for irradiated phage λ are shown since infective
centre formation with unirradiated λ is not sensitive to the
bacterial treatment(s) described. Symbols : (1) without
chloramphenicol : infective centres (▲), frequency of clear
mutations (●) ; (2) chloramphenicol present up to the given
time-point ; infective centres (Δ), frequency of clear
mutations (O). Asterix (*) indicates respective values for
infective centres and clear mutations without irradiation of the
host cells. E.coli K12 AB1186 was grown in tryptone broth at
37°C up to 3×10⁸ bacteria per ml, half of the culture received
100 μg/ml chloramphenicol 15-20 min before harvesting and were
resuspended in 10^{-2}M MgSO$_4$ with or without 100 μg/ml chloramphenicol. The suspensions with and without chloramphenicol were
irradiated with UV doses of 9 and 7 $J.m^{-2}$ respectively to
compensate for UV shielding by the drug. The reharvested
irradiated bacteria were resuspended in tryptone broth medium,
with or without drug, incubated at 37°C from 0 to 120 min, washed
in 10^{-2} M MgSO$_4$ and infected with irradiated (50 $J.m^{-2}$) or non
irradiated phage λ at different time intervals. After adsorption
unirradiated AB 2480 uvrA6 rec13 was added as the indicator on
drug free plates. Appropriate phage dilutions were made to
determine total infective centres and the yield of λ clear
mutations.

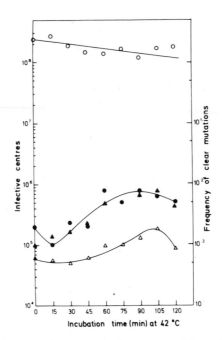

Figure 2 : Infective centres and clear mutation frequency of
           unirradiated (O, Δ) and irradiated (● , ▲) λc$^+$
           phage as a function of preincubation at 42°C of E.coli
           GY953 dnaB mutant

The E.coli dnaB ts mutant, GY953 was derived from a dnaB mutant
isolated and described by Bonhoeffer F. (Z. Vererhungslehre 98,
141–149 (1966)$_2$), was grown exponentially at 32°C in TB medium
containing 10^{-2}M MgSO$_4$ and 0.25% maltose up to 4.10^8 bacteria per
ml, harvested by centrifugation and resuspended in fresh medium.
Aliquots were withdrawn from 0 to 120 min of incubation at 42°C
and used for infection with appropriate dilution. Mutation
frequencies were obtained as a ratio of clear plaque counts on
indicator N0483 uvr$^+$rec$^+$ plated on rich medium agar to the total
plaque counts on indicator AB1157 uvr$^+$rec$^+$ plated on TA$_{12}$ medium.
AB1157 indicator has the same repair capacity as N0483. The
result of one representative experiment is presented.

have also been irradiated under optimum induction conditions·
(UV plus 30 min incubation), a substantial fraction of irradiated
ØX174 DNA molecules can be fully replicated (7). The latter is
the condition sine qua non to produce phage ØX174 mutagenesis (8).
We conclude here that SOS repair is an error-prone DNA synthesis
permitting replication of damaged DNA (Fig. 3).

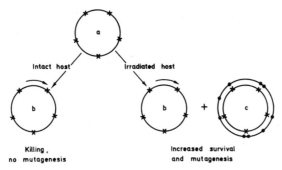

Figure 3 : A scheme illustrating the mechanism of ultraviolet-
           induced mutagenesis and of SOS repair of ⌀X174 phage
           DNA

The inducible mutagenic DNA repair (SOS repair) is interpreted
here as an error-prone DNA synthesis, which permits replication
of UV irradiated ⌀X174 DNA (i.e. survival) by inserting non
complementary nucleotides (●) opposite myrimidne dimers (X) and
also elsewhere in the DNA (mutagenesis)

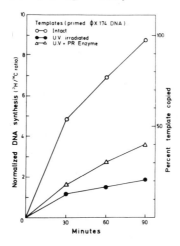

Figure 4 : Extent of DNA synthesis by total E.coli extracts on
           intact and UV-irradiated primed ⌀X174 DNA
Essentially DNA-free extracts have been prepared by a procedure
modified from Kornberg permitting in vitro synthesis of entire
⌀X174 DNA (Villani and Radman,submitted for publication). Photoreactivating enzyme purified from M.luteus was a kind gift of
Dr. Veldhuisen. A portion of UV irradiated primed ⌀X174 DNA was
treated with photoreactivating enzyme using flash photolysis.
UV irradiation fluence to template was 100 J.m^{-2}, which yields an
average of eight pyrimidine dimer per DNA molecule. Details can
be found in the full paper by Villani and Radman (submitted for
publication).

(d) <u>In</u> <u>vitro</u> DNA synthesis on primed ∅X174 DNA templates by total
    E.coli extracts stops at the first pyrimidine dimer encoun-
tered (Fig. 4).   Analysis of the hydrolysis of the newly
incorporated nucleoside monophosphates due to the 3'⟶ 5'
exonuclease ("proof-reading") activity of the polymerase I shows
that the polymerase is "idling" when it encounters DNA lesions.
Idling is due to a continuous consumption of triphosphates and
incorporation of nucleoside monophosphates, however since all
bases are mismatched to a bulky DNA lesion in the template, the
"proof-reading" activity will repeatedly excise them, thus
preventing DNA synthesis to proceed (Figs. 5 and 6).

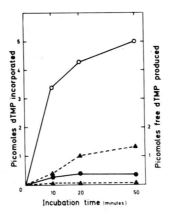

Figure 5 : Incorporation and hydrolysis of the newly incorporated
            thymidine monophosphate ("proof-reading") by purified
            DNA polymerase I (Klenow fragment) on intact and UV
            irradiated primed ∅X174 DNA

            o——o Intact template (incorporation)
            ●——● UV'd template(500 Jm⁻²)(incorporation)
            △--△ Intact template (free dNMP)
            ▲--▲ UV'd template (free dNMP)

UV irradiation fluence to the template was 500 J.m^{-2}.^{3}H-labeled
dTTP and ^{14}C-labeled ∅X174 DNA template were used. Other nucleoside triphosphates were unlabeled and present in 20 µmolar concentration. Both incorporation into the ∅X174 DNA template-primer and
the release of free dTMP were analysed on PEI-cellulose chromatography by Herschfield and Nossal's procedure (J.Biol.Chem.<u>247</u>,
3393(1972)). Production of dTMP was DNA-dependent. Only 3'   5'
exonuclease ("proof-reading") activity can account for the production of dTMP since the 5'⟶ 3' ("nick translation") activity
is missing in the Klenow enzyme (large fragment) used in our
experiments. Details can be found in the paper by Villani and
Radman (submitted for publication).

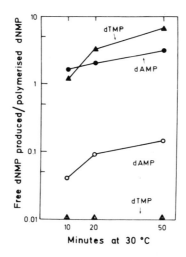

Figure 6 : DNA synthesis versus "proof-reading" exonuclease
         activity on intact and UV-irradiated ⌀X174 DNA template
         by DNA polymerase I (Klenow enzyme).
A semi-log plot of the ratio proof-reading/DNA synthesis versus
reaction time. Turnovers of both dATP and dTTP were analysed as
described in Fig. 5 and by Villani and Radman (submitted). UV
irradiation of the ⌀X174 DNA template (500 J.m^{-2}) increases the
"proof-reading" exonuclease activity relative to polymerisation
by two to three orders of magnitude.

(e) Increased misincorporation of dGMP can be detected using both
      intact and UV-irradiated polydT : oligo dA when E.coli cells
have been induced for SOS repair prior to cell breakage (ref 9
and Fig. 7). No evidence for the induction of a new error-prone
DNA polymerase (such as terminal transferase) was found, therefore we are currently searching for an inducible inhibitor of
the proof-reading activity of E.coli polymerases III and I.
The latter could account for all genetic and biochemical results.

(f) Unlike E.coli polymerases, human DNA polymerases α, β and γ
      purified from HeLa cells do not stop DNA synthesis at
pyrimidine dimers. In fact, adding more fresh enzyme and / or
permitting longer incubation times is sufficient to replicate UV-
irradiated ⌀X174 DNA. However the reaction rate is slowed-down
(Fig. 8 a b). There appears to be a major difference between
E.coli and human DNA polymerases in relation to the mechanism of
their fidelity and to their capacity to copy UV-irradiated DNA
templates. E.coli polymerases are highly accurate due to their
"proof-reading" exonuclease activity because of which they cannot
copy DNA past pyrimidine dimers. Human DNA polymerases apparently compensate the lack of exonuclease activity by a longer
checking time i.e. discrimination. Still, they are more error-
prone in that they ultimately can tolerate radiation-modified

templates (see fig. 8 a b)

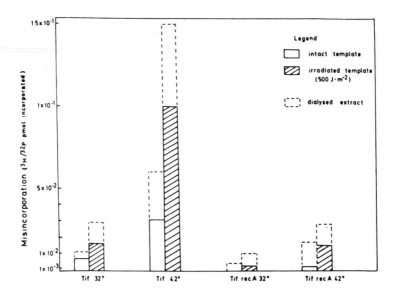

Figure 7 : Misincorporation of deoxyribonucleotides on polydT
           templates

Misincorporation of deoxyribonucleotides on intact and UV-
irradiated polydT templates is dependent on SOS induction. DNA-
free extracts were prepared by a slightly modified procedure of
Kornberg's (Kornberg, A. DNA synthesis, San Francisco, Freeman
and C°, p.217 (1974)). Tif mutants GC714 recA$^+$, UM499 recA$^-$ and
the SOS induction at 42°C are described in ref. 9. Each reaction
mixture (0.3 ml) contained 75 mM Tris hydrochloride, pH 8.6 ;
0.2 mM manganese chloride ; 9 nmol polydT$_2$: oligodA, 380 pmol
^{3}H-dDTP (specific activity 6.5 × 7 nmol ^{32}P-dATP ($\pm$50 cpm/pmol)
and 50 µl of crude extract containing approximately 50 µg of
protein. Incubations were carried out at 30°C for 45 min,
reactions were stopped by chilling and by adding 0.5 ml of 0.1 M
sodium pyrophosphate and 200 nmol of cold dGTP and dCTP. The
mixture was acid-precipitated with 2 ml of 2 N HCl in 0.1 M
sodium pyrophosphate for 15 min in ice, collected on a prewashed
GF/C Whatman glass filter, and washed with 80 ml of the same
solution. Filters were further washed with 5 ml of 98% ethanol,
dried, and the radioactivity was counted in Omnifluor in a liquid
scintillation counter. The misincorporation frequency is expressed
as the ratio of pmol of ^{3}H-dGTP and -dCTP/^{32}P-dATP incorporated
into acid-precipitable material. The background was obtained with
a parallel mixture without added template and was substracted for
each point. The base line in the figure indicates the misincorporation frequency by DNA polymerase I (Klenow fraction)
under identical conditions, and it varies from 0.6 × 10^{-3} to
1 × 10^{-3}.

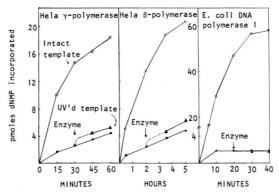

Figure 8 : Extent of DNA synthesis by human β and γ polymerases
and E.coli DNA polymerase I on intact and UV
irradiated primed ØX174 DNA

All reaction mixtures contained 50 mM Tris pH 8.5, 1 mM dithiothreitol, 10 mM MgCl$_2$, 100 μg/ml bovine serum albumin, 20 μmolar
cDTP, dATP and dGTP, 10 μmolar (^{3}H) dTTP and 0.7 μM ØX174 DNA
template. Aliquots were withdrawn at indicated times during
reaction at 36°C. Acid-precipitable radioactivity was determined
on GF/C glass fibre filters. Fresh enzyme was added as indicated
by the arrow in an equal amount to the initial enzyme. E.coli
DNA polymerase I probably stops at the first UV-induced
pyrimidine dimer (500 J.m^{-2} produce about 40 pyrimidine'dimers
per molecule) ; adding fresh enzyme does not overcome the block
of synthesis. The opposite is true for HeLa β and γ polymerases.
Furthermore they do not stop synthesis at the first pyrimidine
dimer encountered, rather their polymerizing activity is slowed
down by the presence of dimers in the template. Details will be
found in a paper by Spadari, Villani and Radman (in preparation).

(g) In connection with the preceeding paragraph, it is of
     interest that an X-ray dose optimal for in vitro
transformation of C3H 10T-1/2 mouse cells provokes the induction
of an increased activity of DNA polymerases α, β and γ. The peak
polymerase activity is attained about 10 hours after irradiation
which coincides with the optimum X-ray-induced synergism of in
vitro transformation by benzo(a) pyrene of the same cell line
(10). No evidence was found for X-ray induction of terminal
transferase or reverse transcriptase in this cell line.

(h) AMV reverse transcriptase does not stop or slow-down DNA
     synthesis on UV-irradiated ØX174 DNA or on polydC : oligo dG.
On the contrary, a range of UV doses to the template stimulates
the synthesis by reverse transcriptase (see Fig. 9). This finding
is in general agreement with studies of Loeb and coll (11) and
with the fact that r.t., unlike other polymerases, efficiently

uses methylated poly dC as the template (12) all suggesting that
the reverse transcriptase is an error-prone polymerase.

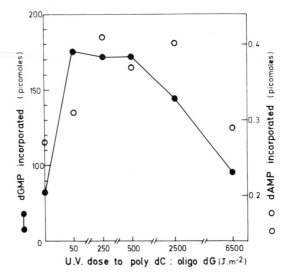

Figure 9 : Insensitivity of avian myeloblastosis virus reverse
            transcriptase to UV irradiation of the (poly dC :
            oligo dG) template : primer
Assay mixture (0.3 ml) contained 50 mM Tris pH 8.3, 6 mM
magnesium acetate, 2 mM dithiolthreitol, 60 mM NaCl, 100 pmoles
(^{14}C) ØX174 DNA template, 3 nmoles (^{3}H) dATP (5.8 x 10^5 cpm per
nmol), 3 nmoles of each dCTP, dGTP, dTTP and 37 units of AMV
reverse transcriptase, which was a kind gift of Drs. P. Baltimore
and A. Burny. Acid-precipitable radioactivity was measured
after extensive washing of excess radioactivity. Details can be
found in a paper by Villani and Radman (submitted to Nature).

(i) Beryllium is a mutagenic and carcinogenic metal. At 2 mM
    concentration, it inhibits E.coli DNA polymerase I to less
than 1% initial activity. At the same Be^{++} concentrations, human
DNA polymerases α, β and γ are only slightly inhibited but the
misincorporation of dAMP into poly dC template is greatly
increased. Irradiation of the template and the presence of Be^{++}
synergistically increase the infidelity of the three human
polymerases (Table I). The latter may be a simple model case for
the synergistic effects of different mutation and cancer
promoting treatments.

(j) We propose that the test of the fidelity of in vitro DNA
    synthesis both by bacterial and mammalian DNA replicating
enzymes can sometimes be used to reveal mutagenic potential of
radiation and chemical carcinogens.

BERYLLIUM-INDUCED INFIDELITY OF HUMAN DNA POLYMERASES α, β AND γ COPYING INTACT AND UV
IRRADIATED TEMPLATE

Primer - Template ($dG_{12}.dC_n$)	Divalent cations		DNA polymerase	Complementary nucleotide (pmol dGMP incorporated)	Non complementary nucleotide (pmol dAMP incorporated)	Error frequency
	Mg++ 10 mM	Be++ 2 mM				
Intact	+		α	81.79	0.0242	1/3379
UV-irradiated	+			58.35	0.0244	1/2391
Intact	+	+		57.88	0.1932	1/299
UV-irradiated	+	+		32.59	0.2132	1/152
Intact	+		β	270	0.0307	1/8794
UV-irradiated	+			97	0.0211	1/4597
Intact	+	+		48.51	0.0479	1/1012
UV-irradiated	+	+		19.60	0.0838	1/233
Intact	+		γ	228	0.1086	1/2103
UV-irradiated	+			188.84	0.1084	1/1742
Intact	+	+		142.8	0.1903	1/950
UV-irradiated	+	+		117.24	0.2054	1/568
Intact	+		E.coli pol I	1747	0.709	1/2464
UV-irradiated	+			1112	0.502	1/2215
Intact	+	+		5.2	–	
UV-irradiated	+	+		15		

Legend to the Table I

All reaction mixtures (total volume 100 µl) contained 50 mM Tris
pH 8.3, 1 mM dithiothreitol, 10 mM $MgCl_2$, 100 µg/ml bovine serum
albumin, 20 µmol (^{3}H) dGTP (200 cpm/pmol), 20 µmol (α ^{32}P) dATP
(10000 cpm/pmol) and 50 µmol dG_{12} : dC_n (ollaborative esearch)
hybridized in the ratio 1 : 3. Radioactive precursors were from
New England Nuclear. Per assay was added 10 µl α polymerase
(8 units per ml) 15 µl β polymerase (60 units per ml) and 10 µl
γ polymerase (31 units per ml).(one unit corresponds to incorporation of 1 nmol of dNTP's into activated DNA at 37°C for
30 min). Reactions were carried out to saturation at 36°C for
1 hour, stopped by addition of 0.2 ml 10% perchloric acid and
pelleted by centrifugation. To eliminate excess soluble radioactivity pellets were resuspended in 0.2 N NaOH, repr cipitated
twice with perchloric acid and finally spotted onto GF/C glass
fibre filters, washed according to Bollum F.J. (in "Procedures
in nucleic acid research" ed. Cantoni, G.L. and Davies, D.R. 1,
296-300 (1967)) and their radioactivity measured in an Omnifluor
Scintillation Fluid. UV-irradiation (4500 $J.m^{-2}$) of the dG_{12}.
dCn was carried out in 50 mM Tris pH 8 and 20 mM KCl, on ice, in
a thin layer using a Mineralight lamp with maximum energy output
at a wavelength of 2537 A. Fluences were measured by a Latarjet
dosimeter.

REFERENCES

1. M. DEFAIS, P. FAUQUET, M. RADMAN and M. ERRERA
   Virology 43, 495 (1971)
2. M. DEFAIS, P. CAILLET-FAUQUET, M.S. FOX and M. RADMAN
   Molec. Gen. Genet. (1976) in press
3. M. RADMAN
   in L. PRAKASH, F. SHERMAN, M.W. MILLER, C.W. LAWRENCE and
   H.W. TABER eds. Molecular and environmental aspects of
   mutagenesis, C.C. Thomas Publ., Springfield, Illinois, 1974,
   p.128
4. M. RADMAN
   in P.C. HANAWALT and R.B. SETLOW eds. Molecular mechanisms for
   repair of DNA, Plenum Press, N.Y., 1975, part A p.355
5. P. CAILLET-FAUQUET and M. DEFAIS
   Nature, submitted
6. E.M. WITKIN
   in P.C. HANAWALT and R.B. SETLOW eds. Molecular mechanisms for
   repair of DNA, Plenum Press, N.Y., 1975, part A, p. 369
7. P. CAILLET-FAUQUET, M. DEFAIS and M. RADMAN
   J. Mol. Biol. submitted
8. J.F. BLEICHRODT and W.S.D. VERHEIJ
   Molec. Gen. Genet. 135, 19 (1974)
9. M. RADMAN, P. CAILLET-FAUQUET, M. DEFAIS and G. VILLANI
   in R. MONTESANO, H. BARTSCH and L. TOMATIS eds. Screening test
   in chemical carcinogenesis, IARC Scientific Publications n°12
   Lyon, 1976, p.543
10. M. TERZAGHI and J.B. LITTLE
    in E. KARBE and J.F. PARK eds., Experimental Lung Cancer,
    Carcinogenesis and Bioassays, Springer Verlag, Berlin,
    Heidelberg, New York, 1974, p.497
11. N. BATTULA and L.A. LOEB
    J. Biol. Chem. 249, 4086 (1974)
12. G.F. GERARD, P.M. LOEWENSTEIN and M. GREEN
    Nature, 256, 140 (1975)

ACKNOWLEDGEMENTS

This work was supported by the "Fonds Cancérologique" of the
"Caisse Générale d'Epargne et de Retraite" of Belgium

CHROMOSOME EFFECTS INDUCED BY LOW LEVELS OF MUTAGENS*

Sheldon Wolff

Laboratory of Radiobiology and Department of Anatomy

University of California, San Francisco U.S.A.

INTRODUCTION

Chromosome Aberrations

In the past, the standard way to quantify the cytogenetic effects of mutagens has been to study the chromosome aberrations induced in somatic cells. The types of aberrations formed and the kinetic patterns of their induction, however, vary depending upon the mutagen and the part of the cell cycle treated,making inter-comparisons between mutagens rather difficult. For instance, with ionizing radiations chromosome aberrations that affect both chromatids of the chromosome identically are induced when G_1 cells are treated, whereas when S or G_2 cells are irradiated, chromatid aberrations are found (1). With ultraviolet light and with most chemical mutagens, on the other hand, mainly chromatid aberrations are found irrespective of the stage of the cell cycle treated. Ionizing radiations thus appear capable of breaking chromosomes at the time of treatment, whereas the other mutagens seem to induce lesions that lead to aberration formation as the cell proceeds through the S phase during which the chromosomes replicate.

With ionizing radiations it has been found that those aberrations depending upon single broken chromosomes for their formation increase approximately linearly with dose, whereas those such as translocations and dicentrics, which require the interaction of two breaks for their formation, increases approximately as the square of the dose. These radiations are relatively clean physical probes in which the problems of dosimetry are minimized. With ultraviolet radiation, on the contrary, the dosimetric problems are not quite so simple in that the radiation does not penetrate

and it can be absorbed by cytoplasmic molecules before it reaches
the chromosomes within the cell nucleus (2,3). Such self-absorp-
tion can distort the shape of dose-effect curves. Furthermore,
long mitotic delays induced by ultraviolet light can distort the
pattern of progression of cells through the mitotic cycle and lead
to the appearance at metaphase, where aberrations are scored, of
cells treated in unknown parts of the cell cycle (4). With chemical mutagens the dosimetric problems are also formidable: neither
the intracellular concentrations of the chemicals nor their biological
and chemical half-lives are known with accuracy. In whole animal
studies this is compounded even further by our lack of knowledge
regarding the diffusibility of the chemicals and their transport
to the cells in question.

The standard cytogenetic test itself suffers from being a
relatively insensitive test necessitating the tedious examination
of a large number of cells in order to detect significant effects
of low levels of mutagens (5). This latter problem is not trivial,
and consequently much effort has gone into the development of auto-
mated scoring procedures (6) for use even after cells are exposed
to ionizing radiation.

Sister Chromatid Exchanges (SCE's)

Recently, new techniques have been developed that enable us
to study one form of chromosomal effects, sister chromatid exchanges (7), with great ease. These techniques, harlequin chromosome techniques, are dependent upon the production of two chemically different sister chromatids, which can be produced by exposing the cells to a thymidine analog for one round of replication followed by a subsequent round in either the absence (7) or
the presence (8,9) of the analog. Either BrdUrd (8,9,10,11) or
IUdR (9) can be used, although the use of the former is more common. If the chromosomes are grown for only one round of replication in the presence of the analog and then for the next round in
the presence of thymidine, one of the two chromatids in each chromosome will be singly or unifilarly substituted with the analog,
whereas its sister chromatid will be unsubstituted. If the cells
are allowed to replicate for two rounds of replication in the
presence of the analog, then one chromatid in each chromosome will
be singly substituted and one will be doubly substituted. Such
chromatids have been found to stain differently when treated with
Giemsa stain (9), with the fluorescent dyes Hoechst 33258 (8) or
acridine orange (11,12,13), fluorescent dyes followed by Giemsa
stain (12,14) or a series of other dyes (15). If sister chromatid exchanges occur or are induced during either the first round
or second round of replication then they can be clearly seen in
the harlequin chromosomes that contain differentially stained sister chromatids (Fig. 1a).

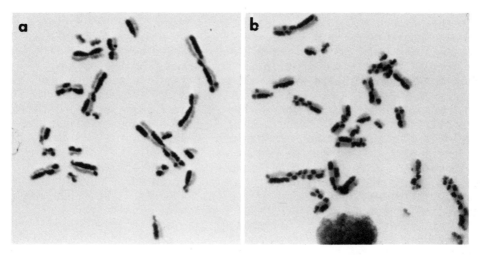

Fig. 1. Harlequin CHO chromosomes. a) Control. b) Treated
with cyclophosphamide plus Ames's S-9 mix.

SISTER CHROMATID EXCHANGE AND MUTAGEN-CARCINOGEN SCREENING

The yield of sister chromatid exchanges can be increased by a
variety of agents that damage DNA. Thus it has been found that
radiation from incorporated tritiated thymidine (16,17), X-rays
(18), UV-light (19,20,21), alkylating agents (5,22,23), or a variety of chemical-mutagen carcinogens (5,25) can increase the yield
of SCE's. In the latter case it has been found that some of the
chemical mutagens can cause a ten-fold increase in the background
level of sister chromatid exchange at concentrations of the chemical that are 100-fold less than those that can be shown to induce
standard chromosome aberrations (5), and it has been suggested
that sister chromatid exchange induction should become a standard
test for the mutagenicity of chemical compounds (5,24,25,26).

To this end when 14 proven or suspected mutagenic chemicals
(5) were screened to see if they induced SCE's in cultured Chinese
hamster cells, the only chemicals that did not cause an increase
were those that were inactive in mammals or required metabolic activation in order to become carcinogenic.

In Vivo Tests

The utility of sister chromatid exchanges to detect mutagenic
and carcinogenic compounds can be seen in studies in which we
treated rabbits with mutagens that required activation as well as
those that did not (25). In both cases, the yield of sister chromatid exchanges observed in cultured peripheral lymphocytes first

increased and then slowly returned to the control value. The
utility of such an *in vivo* test system is especially important for
those chemicals that require activation by the mixed function oxidase enzymes found in the liver, kidneys, and other tissues of the
organism. In the experiments,blood was drawn from the ears of
male New Zealand rabbits prior to their being injected with the
mutagen. Two tenths of an ml of this blood was then cultured in
5 ml of McCoy's 5A medium containing 20% fetal calf serum and 1%
Penicillin-Streptomycin along with 0.2 ml of phytohemaglutinin
(PHA-P). The cultures were maintained at 38.5°C for 18-20 hours
at the end of which time the medium was replaced with fresh medium
without PHA. BrdUrd to a final concentration of 10^{-5} M was added
at this point and the cultures incubated for 30 more hours. Col-
cemid (10^{-6}M) was present during the last 4 hours. The cells were
collected, treated with 0.075 M KCl hypotonic solution for 4 minutes and then fixed in 3:1 methanol:glacial acetic acid. The
slides were stained in Hoechst 33258, 0.5 µg/ml in M/15 Sorensen's
buffer, pH 6.8, for 20 minutes, washed in distilled water, mounted
in the same buffer with a coverslip and exposed to light from a
mercury burner for 1½ minutes. After this, the coverslips were
removed and the slides were incubated for 20 minutes in 10X SSC at
62°C before being stained in 3% Giemsa (Gurr's R66 and M/15 Sorensen's buffer, pH 6.8). This blood, which was cultured before the
rabbit was injected with the mutagen, allows each rabbit to be his
own control. There were 690 SCE's observed in 5304 chromosomes
obtained from 8 different rabbits for an average yield of 0.133
SCE per chromosome. The control range was 0.093 to 0.166 SCE per
chromosome. In blood drawn one day after intraperitoneal injection
of a rabbit with either ethyl methane sulfonate (EMS), methyl methane sulfonate (MMS), or cyclophosphamide (CP) the yield of SCE's
is increased (fig. 2). When blood is drawn from the same rabbits

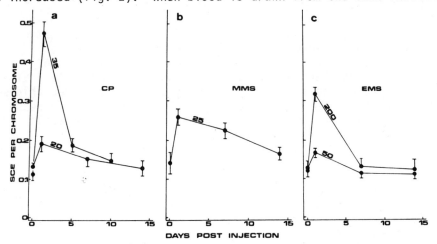

Fig. 2. SCE's in rabbit lymphocytes cultured on different days
after injection. Numbers on curves refer to mg/kg injected.

on subsequent days, the yield slowly returns to the control value.
Of these three chemicals, cyclophosphamide requires metabolic activation in order to be a mutagen-carcinogen.

Here then is a simple *in vivo* test that could be used to
screen for potential mutagen-carcinogens including those that require metabolic activation.

In Vitro Tests

An easier way to test for the activity of carcinogens and mutagens, however, is to look for SCE's induced in cultured Chinese
hamster (CHO) cells as was done by Perry and Evans (5) who had
found activity on the part of those chemicals not requiring metabolic activation. Stetka and I (26) have found, however, that the
use of the test can be expanded to the compounds requiring activation by the simple addition of rat liver microsomes as is done in
the Ames Salmonella test for mutagenicity of carcinogens (27). In
fig. 3a it may be seen that EMS dramatically increases the yield of
SCE's in CHO cells that have been treated for 20 minutes prior to
the addition of BrdUrd whereas cyclophosphamide has only a minimal
effect even at very high concentrations. Maleic hydrazide (MH),
which is not mutagenic in animal cells, has no effect. If the cells
are treated for 20 minutes with cyclophosphamide, and either a
1/60th or a 1/20th dilution of Ames's S9 Mix, which consists of liver microsomes obtained from rats induced with Arochlor, magnesium
chloride, potassium chloride, glucose-6-phosphate, NADP, and
Na_2HPO_4-NaH_2PO_4 then the compound is activated and the yield of
sister chromatid exchanges increases dramatically (fig. 3b). A
cell treated in this way is seen in Figure 1b.

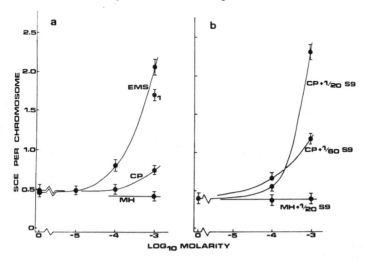

Fig. 3.   SCE's induced in CHO cells. a) no S-9. b) S-9.

SISTER CHROMATID EXCHANGE IN

REPAIR DEFICIENT XERODERMA PIGMENTOSUM CELLS

Studies with xeroderma pigmentosum cells, which are unable to carry out excision-repair of ultraviolet lesions (28), has given some clues to the nature of the lesions that can lead to ordinary chromosome aberrations on the one hand and sister chromatid exchanges on the other. Previous work investigating DNA damage and repair in xeroderma cells exposed to chemical carcinogens (29,30, 31,32) had led to these agents being classified into those in which the damage can be repaired by normal levels of excision repair and those in which it cannot, i.e., into those in which the damage is "UV-like" (e.g., caused by 4-nitroquinoline-1-oxide) and those in which it is "X-ray-like" (e.g., caused by EMS, MMS, or MNNG) (28). In XP cells, "UV-like damage" is characterized by high levels of chromosome aberrations and low levels of repair whereas "X-ray-like damage", is characterized by normal levels of chromosome aberration yields, and normal repair levels (31,33). Thus high levels of aberrations are correlated with unrepaired damage. The same correlation holds between reparability of the lesions and cell survival, as would be expected if cell death is caused by chromosome aberrations and subsequent genetic imbalance. When sister chromatid exchanges, however, are studied in either normal cells or xeroderma pigmentosum cells of complementation group A (35), it is found that the same correlation does not hold (Table 1). At very low concentrations of the chemicals, no increase in SCE's is found in normal cells after treatment with either "UV-like" or "X-ray-like" carcinogens. The xeroderma cells, however, are very sensitive to these mutagens and, contrary to the situation found when chromosome aberrations are studied, the yield of sister chromatid exchanges is increased both by 4NQO and the others (see fig. 4).

If XP cells are defective only in excision-repair then the increased numbers of SCE's would be the result of unexcised damage in DNA. The results obtained with MMS, EMS, and MNNG would, under these circumstances, indicate that there is a minor fraction of the total damage that is unexcised after treatment with these chemicals. This fraction would presumably be too small to affect the usual measurements of repair replication or unscheduled synthesis, or to give rise to higher than normal amounts of chromosome aberrations and cell killing. Increased SCE's, however, would be formed since they are very sensitive indicators of unexcised damage in DNA.

These observations and the apparent lack of correlation between SCE frequencies and either chromosome aberration frequencies or cell killing indicate that the two are different. SCE's also differ from chromosome aberrations in many other respects: they are induced at high frequencies by chemicals at concentrations that induce few aberrations (5); they saturate at low level doses of ionizing radiation (36) and tritiated thymidine (17); they react differently to postirradiation treatments with caffeine than do

CHEMICALLY-INDUCED SISTER CHROMATID EXCHANGES IN
XERODERMA PIGMENTOSUM AND NORMAL CELLS*

TREATMENT	NORMAL CELLS (GM637)		XERODERMA CELLS (XP12RO)	
	#SCE/#CHROMOSOMES	SCE PER CHROMOSOME	#SCE/#CHROMOSOMES	SCE PER CHROMOSOME
CONTROL	1369/4397	0.31	1785/4867	0.37
"UV-LIKE"				
4NQO				
4×10^{-10} M	744/2192	0.34	1764/2584	0.68
4×10^{-9} M	770/2156	0.36	2025/2271	0.89
"X-RAY-LIKE"				
MMS				
10^{-7} M	661/2141	0.31	1182/2106	0.56
10^{-6} M	710/2150	0.33	1256/2280	0.55
5×10^{-6} M	860/1981	0.43	1696/2161	0.78
EMS				
10^{-6} M	769/2299	0.33	1388/2459	0.56
10^{-5} M	797/2226	0.36	2399/2453	0.98
MNNG				
10^{-9} M	537/1726	0.31	1060/2313	0.46
10^{-8} M	875/2207	0.40	1494/2450	0.61
10^{-7} M	842/2057	0.41	4076/2519	1.62
10^{-6} M	1381/2136	0.65	TOO NUMEROUS TO BE SCORED	

*Data of Wolff, Rodin, and Cleaver.

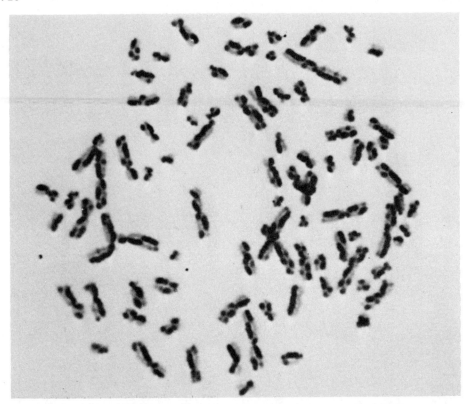

Fig. 4.   XPRO12 polyploid cell after MNNG. Note SCE's.

chromosome aberrations (29); they are not increased by low doses of
ionizing radiations (21), whereas aberrations are; and they are not
correlated in any consistent way with the increased aberrations
found in Bloom's syndrome (37), Fanconi's anemia (23), and ataxia
telangiectasia (38).

It therefore appears that SCE's are the result of fundamentally
different cellular events and lesions than are chromosome aberrations and that whereas aberrations can cause cell death, SCE's are
more representative of events, such as mutagenesis, that are com-
patible with cell survival. It should be noted that although most
of the SCE's are unlikely to be associated with mutation because
equal amounts of homologous chromatids are exchanged, in those cases
where there is a slight inequality of exchanged material, deletions,
insertions, or frameshift mutations could result as was found to occur when exchange occurs between homologous chromosomes during
meiosis (39).

SUMMARY

One of the easiest ways to quantify the effects of mutagens has

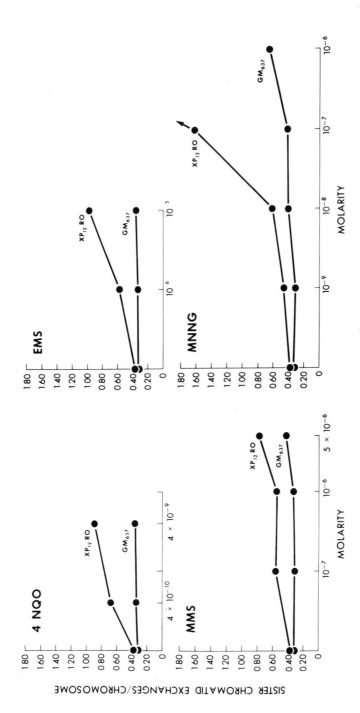

Fig. 5. SCE's induced by X-ray and UV-like mutagens in Xp and normal cells.

been to study the induction of chromosome aberrations. With ionizing radiations, which have been the most extensively studied mutagens, the dosimetric problems are far fewer than with others. Nonetheless, because of the kinetic patterns of aberration induction, there have been difficulties in assessing the effects of low level exposures even with ionizing radiations. Recently, new techniques have made it possible to study the induction of sister chromatid exchanges, which are extremely sensitive to low levels of ultraviolet light and to chemical mutagens, with great accuracy. These systems provide the most sensitive mammalian systems to screen for the effects of carcinogens and mutagens on chromosomes. Studies on the induction of sister chromatid exchange in normal as well as repair deficient xeroderma cells are now making it possible, not only to assess the effects of low level exposures, but also to gain insights into the mechanisms involved in aberration induction.

REFERENCES

1 Wolff, Sheldon: Radiation Genetics. in Mechanisms in Radiobiology (Eds. M. Errera and A. Forssberg) Academic Press, New York, 1961, pp. 419-475.
2 Wolff, Sheldon: The effect of ultraviolet radiation on genes and chromosomes. in Radiation Research, North Holland Publishing Co., Amsterdam, 1967, pp. 455-465.
3 Wolff, Sheldon: Chromosome aberrations induced by ultraviolet radiation. in Photophysiology, Academic Press, New York and London, 1972, pp. 189-204.
4 Wolff, Sheldon: Radiation effects as measured by chromosome damage. in Cellular Radiation Biology, Williams & Wilkins Co., Baltimore, 1965, pp. 167-183.
5 Perry, P. and H.J. Evans: Cytological detection of mutagen-carcinogen exposure by sister chromatid exchange, Nature 258, 1975, pp. 121-125.
6 Wald, N., R.W. Ranshaw, J.M. Herron, and J.G. Castle: Progress on an automatic system for cytogenetic analysis. in Human Population Cytogenetics (Eds. P.A. Jacobs, W. Price, and P. Law) Williams & Wilkins Co., Baltimore, 1970, pp. 263-280.
7 Taylor, J.H., P.S. Woods,and W.L. Hughes: The organization and duplication of chromosomes as revealed by autoradiographic studies using tritium-labeled thymidine, Proc. Natl. Acad. Sci. USA 43, 1957, pp. 122-138.
8 Latt, S.A.: Microfluorometric Detection of Deoxyribonucleic Acid Replication in Human Metaphase Chromosomes, Proc. Natl. Acad. Sci. USA 70, 1973, pp. 3395-3399.
9 Ikushima, T. and Wolff, S.: Sister chromatid exchanges induced by light flashes to 5-bromodeoxyuridine and 5-iododeoxyuridine substituted Chinese hamster chromosomes, Exptl. Cell Res. 87, 1974, pp. 15-19.

10 Zakharov, A.F. and Egolina, N.A.: Differential spiralisation
    along Mammalian Mitotic chromosomes: 1. BUdR - Revealed dif-
    ferentiation in Chinese hamster chromosomes, Chromosoma 38,
    1972, pp. 341-365.
11 Kato, H.: Spontaneous sister chromatid exchanges detected by a
    BUdR-labelling method, Nature 251, 1974, pp. 70-72.
12 Perry, P. and Wolff, S.: New Giemsa Method for the differential
    staining of Sister Chromatids, Nature 251, 1974, pp. 156-158.
13 Dutrillaux, B., A.M. Fosse, M. Prieur, et J. Lejeune: Analyse
    des échanges de chromatides dans le cellules somatiques hu-
    maines, Chromosoma 48, 1974, pp. 327-340.
14 Korenberg, J.R. and Freedlender, E.F.: Giemsa Technique for the
    Detection of Sister Chromatid Exchanges, Chromosoma 48, 1974,
    pp. 355-360.
15 Goto, K., T. Akematsu, H. Shimazu, and T. Sugiyama: Simple
    Differential Giemsa Staining of Sister Chromatids after Treat-
    ment with Photosensitive Dyes and Exposure to Light and the
    Mechanism of Staining, Chromosoma 53, 1975, pp. 223-230.
16 Brewen, J.G. and W.J. Peacock: The effect of tritiated thymidine
    on sister chromatid exchange in a ring chromosome, Mut. Res. 7,
    1969, pp. 433-440.
17 Gibson, D.A. and D.M. Prescott: Induction of sister chromatid ex-
    changes in chromosomes of rat kangaroo cells by tritium incor-
    porated into DNA, Exptl. Cell Res. 74, 1972, pp. 397-402.
18 Gatti, M. and G. Olivieri: The effect of x-rays on labelling
    pattern of M_1 and M_2 chromosomes in Chinese hamster cells,
    Mut. Res. 17, 1973, pp. 101-112.
19 Rommelaere, J., M. Susskind, and M. Errera: Chromosome and chro-
    matid exchanges in Chinese hamster cells, Chromosoma 41, 1973,
    pp. 243-257.
20 Kato, H.: Induction of sister chromatid exchanges by UV light
    and its inhibition by caffeine, Exptl. Cell Res. 82, 1973,
    pp. 383-390.
21 Wolff, S., J. Bodycote, and R.B. Painter: Sister chromatid ex-
    changes induced in Chinese hamster cells by UV irradiation of
    different stages of the cell cycle: the necessity for cells to
    pass through S, Mut. Res. 25, 1974, pp. 73-81.
22 Solomon, E. and Bobrow, M.: Sister chromatid exchanges - a sen-
    sitive assay of agents damaging human chromosomes, Mut. Res.
    30, 1975, pp. 273-278.
23 Latt, S.A., G. Stetten, L.A. Juergens, G.R. Buchanan, and P.S.
    Gerald: Induction by alkylating agents of sister chromatid ex-
    changes and chromatid breaks in Fanconi's anemia, Proc. Natl.
    Acad. Sci. USA 72, 1975, pp. 4066-4070.
24 Latt, S.A.: Sister chromatid exchanges, indices of human chro-
    mosome damage and repair: Detection by fluorescence and induc-
    tion by Mitomycin-C, Proc. Natl. Acad. Sci. USA 71, 1974,
    pp. 3162-3166.
25 Stetka, D.G. and S. Wolff: Sister chromatid exchange as an as-
    say for genetic damage induced by mutagen-carcinogens, Part I.

In vivo test for compounds requiring metabolic activation, Mut. Res. 1976, in press.

26 Stetka, D.G. and S. Wolff: Sister chromatid exchange as an assay for genetic damage induced by mutagen-carcinogens, Part II. *In vitro* test for compounds requiring metabolic activation, Mut. Res., 1976, in press.

27 Ames, B.N., J. McCann, and E. Yamasaki: Method for detecting carcinogens and mutagens with the *Salmonella* mammalian-microsome mutagenicity test, Mut. Res. 31, 1975, pp. 347-364.

28 Cleaver, J.E.: DNA repair with purines and pyrimidines in radiation- and carcinogen-damaged normal and xeroderma pigmentosum human cells, Cancer Res. 33, 1973, pp. 362-369.

29 Cleaver, J.E.: Repair of alkylation damage in ultraviolet sensitive (xeroderma pigmentosum) human cells, Mut. Res. 12, 1971, pp. 453-462.

30 Stich, H.F., San, R.H.C., Miller, J.A., and Miller, E.C.: Various levels of DNA repair synthesis in xeroderma pigmentosum cells exposed to the carcinogens N-Hydroxy and N-acetoxy-2-acetyl-aminofluorene, Nature New Biology 238, 1972, pp. 9-10.

31 Stich, H.F., W. Stich, and R.H.C. San: Chromosome aberrations in xeroderma pigmentosum cells exposed to the carcinogen, 4-nitroquinoline-1-oxide, and N-methyl-N'-nitro-nitrosoquanidine (37194), Proc. Soc. Exper. Biol. and Med. 142, 1973, pp. 1141-1144.

32 Regan, J.D. and Setlow, R.B.: Two forms of repair in the DNA of human cells damaged by chemical carcinogens and mutagens. Cancer Res. 34, 1974, pp. 3318-3325.

33 Sasaki, M.S.: DNA repair capacity and susceptibility to chromosome breakage in xeroderma pigmentosum cells, Mut. Res. 20, 1973, pp. 291-293.

34 Wolff, S.: Genetic effects and radiation-induced cell death. in Frontiers of Radiation Therapy and Oncology (Karger, Basel, and University Park Press, Baltimore) 6, 1972, pp. 459-469.

35 Wolff, S., B. Rodin, and J.E. Cleaver: Sister chromatid exchanges induced by carcinogens and mutagens in normal and xeroderma pigmentosum cells (submitted, 1976).

36 Marin, G. and D.M. Prescott: The frequency of sister chromatid exchanges following exposure to varying doses of ^{3}H-thymidine or X-rays, J. Cell Biol. 21, 1964, pp. 159-167.

37 Chaganti, R.S.K., Schonberg, S., and German, J.: A manifold increase in sister chromatid exchanges in Bloom's syndrome lymphocytes, Proc. Natl. Acad. Sci. USA 71, 1974, pp. 4508-4512.

38 Galloway, S.M. and Evans, H.J.: Sister chromatid exchange in human chromosomes from normal individuals and patients with ataxia telangiectasia, Cytogenet. Cell Genet. 15, 1975, pp. 17-29.

39 Magni, G.E. and R.C. von Borstel: Different rates of spontaneous mutations during mitosis and meiosis in yeast, Genetics 47, 1962, pp. 1097-1108.

The 33258 was a gift from Dr. H. Loewe, Hoechst Farbwerke.
*Work supported by U.S. Energy Research and Development Administration.

ROUND TABLE: Units, Nomenclature and Dosimetry in
                    Photobiology

DISCUSSANTS: C.S.RUPERT, R.LATARJET, G.BAUER (F.Urbach & M.Cre-
             monese, originally scheduled, were unable to attend).

     The subject introduced for discussion can be summarized as
follows:
     Photobiological effects are caused by photochemical changes
taking place in the substance of living things. The processes
producing these changes presumably obey all the well established
physical and chemical laws. In some cases (such as that of the
inactivation of viruses) the magnitude of the observed response
depends on the total amount of photochemical change produced. In
others (like vision), it depends on the rate at which photochemical change takes place. Effects may also depend on the photochemistry in more complicated ways. Photobiology tries to deal
with all these situations in a consistent, systematic way.

     Photobiological dosimetry, like all dosimetry of radiations,
is an attempt to express the relevant radiation treatment in
numerical terms. Naively, one would hope to find a numerical
measure which, either directly or through some mathematical trans
formation, is proportional to the magnitude of the effect being
studied. This is possible, for example, in ultraviolet inactivation of small, single-stranded DNA bacteriophages, where we know
of dosimetric quantities proportional to the negative logarithm
of the surviving fraction (giving exponential survival curves).
A little consideration, however, convinces one that such a relation cannot be expected in general. Biological systems are so
enormously responsive to their environments that the effects observed in them are modified by many factors besides the primary
stimulus. There is no way to make life simpler that it really is.
The dosimetric quantities commonly used in photobiology are proportional either to the amount or to the rate of photochemical
change initially induced in the living system. The relationship
between these defined quantities and the effects observed must
be determined by experiment.
     Radiation measurements always start with a physical des -
cription of the radiation itself. Different types of radiations
do not have to be described in wholly different terms. For purposes of dosimetry, all of them -- whether visible and ultra

violet light, X rays, or electron and neutron beams -- can be
described simply as a three dimensional flow of energy or particles. If enough is known about this flow, and if the appropriate
coefficients for its interaction with matter are available, then
the changes induced in the irradiated materials can be determined. Different names and definitions are used with various radiations not because of the way the world is made, but because different people working at different places and at different times
made different choices about how to proceed. The result is a
multiplicity of schemes, all individually valid, which constitute
an unncessary barrier to scientific communication, and complicate
life for the regulatory agencies responsible for radiation protection.

In 1974 a joint working group on radiation quantities was
formed by the AIP (formerly the CIP), the CIE (The International
Commission on Illumination), and the ICRU (The International Commission on Radiation Units and Measurements). This group, together with a representative of the URSI (The International Radio-
scientific Union) met at the International Bureau of Weights and
Measures in Paris the following year to formulate an approach to
a unified scheme for describing radiation fields - one which
would allow a single set of dosimetric terms to be used for all
radiation.

The Group's first effort was restricted to the description
of radiation fields, avoiding any mention of their interaction
with matter (and therefore of actual dosimetry) because agreement on this subject by the parent organizations must be reached before dosimetric matters can be considered. Reaching such
an agreement is more complicated than might at first appear.
The number of people involved in the several international organizations concerned is quite large, and some of them (say,
the illumination engineers) have little reason to converse
with people in distant sciences (like the radiologists). Those
of us who tend to work along the boundaries between sciences
are more conscious of "interscientific relations" and the need
for "interscientific law" than those who live deep in the interior. Yet the inland dwellers have their own requirements and
preferences, and their votes cannot be ignored. Many critical
consultations, like this Round Table discussion, are required
to determine the possibilities for agreement.

The essence of the Joint Working Group proposal is as follows:

1.   At the level of detail required by radiation dosimetry, any radiation can be described as either a flow of energy or of particles. (Light, for example, is treated as either a train of electromagnetic waves or a stream of photons). Since convenience alone dictates the choice, both modes of description should be provided for.

2.   The flow of energy, or of particles, in the neighborhood of a point in space is always distributed in space and direction, as well as in time and particle energy (or wave frequency). The flow representing the radiation field can be characterized by distribution functions with respect to these variables. For a given particle energy (or wave frequency -- commonly specified in the case of light by giving the vacuum wavelength) the field involves the distribution with respect to time (d/dt), with respect to volume (d/dv), with respect to the area normal to the direct propagation (d/da), with respect to the solid angle encompassed by the beam direction (d/dΩ), or with respect to two or more of these variables. Those of the resulting distribution functions which seem likely to be used most often were given names. The essential proposal, however, is not these particular names, nor use of only the distribution functions named, but the rather the concept of using the same distribution functions under the same names for describing all types of radiation fields. Different names than those listed may finally be agreed on, and still other distribution functions may be named and used.

The recommended quantities are summarized in Table I. All of them have previously been defined by international agreement, although sometimes under other names.

3.   Units used for expressing these quantities should be in the Système International (SI).

4.   The names given the quantities are those already in international usage wherever these seem suitable. However, an effort was made to eliminate names which have additional meanings in common use. "Intensity" has a widely used general meaning in addition to its use by the CIE for the quantity called "steric energy flux" in the Table. The term "density" has been used both for d/da and d/dv. Both of these words are avoided. Where a term seemed likely to be much used, a single, rather than a compound name, is convenient. Thus "flosan" (an acronym for f̲lux o̲ver

surface $\underline{area}$ $\underline{normal}$) is proposed for "fluence rate". Lack of a
simple, short name for "number of particles" presently complicat-
es nomenclature of the particle terms. Criticism of all the names
is invited.

5. Provisional symbols are shown for the quantities in Table I,
in order to allow expressing them in formulae. Some of these may
require change.

6. This scheme is suitable for a field composed of one or more
beams of incoherent radiations. However, where reflections or
scattering cause coherent radiation (from, say, a laser or micro-
wave source) to traverse a region from more than one direction,
the interference pattern peculiar to each individual situation
may impose other procedural requirements.

Table I
Proposed Radiation Quantities

Distributed with respect to → / Quantities ↓	time [s]	area* [m^2]	volume [m^3]	area* and time [$m^2 \cdot s$]	solid angle and time [$sr \cdot s$]	area* solid angle and time [$m^2 \cdot sr \cdot s$]
radiant energy $\underline{Q}$ [J]	energy flux $\underline{P} = \dfrac{dQ}{dt}$	energy fluence $\underline{\Psi} = \dfrac{dQ}{da}$	volumic energy $\underline{u} = \dfrac{dQ}{dv}$	energy flosan $\underline{\Psi} = \dfrac{d^2 Q}{da\, dt}$	steric energy flux $\underline{I} = \dfrac{d^2 Q}{dt\, d\Omega}$	energy radiance $\underline{L} = \dfrac{d^3 Q}{dt\, da\, d\Omega}$
Number of particles $\underline{N}$ [dimensionless]	particle flux $\underline{A} = \dfrac{dN}{dt}$	particle fluence $\underline{\Upsilon} = \dfrac{dN}{da}$	volumic number of particles $\underline{n} = \dfrac{dN}{dv}$	particle flosan $\underline{f} = \dfrac{d^2 N}{da\, dt}$	steric particle flux $\dfrac{d^2 N}{dt\, d\Omega}$	particle radiance $\dfrac{d^3 N}{dt\, da\, d\Omega}$

*The area $d\underline{a}$ in this table is always taken perpendicular to the direction of propagation of an elementary
radiation beam.

Note: Bracketed symbols represent abbreviations of units (in the International System, or SI) which
should be used to express the corresponding quantity or variable: [J], joules; [s], seconds;
[m], meter; [sr], steradian. The $\underline{energy\ fluence}$, for example, would be expressed in joules
per square meter [$J \cdot m^{-2}$]; the $\underline{particle\ flux}$ in reciprocal seconds [s^{-1}], etc.

The connection between these radiation field quantities and dosimetric considerations is outside the present proposal, but the meaning of the quantities may be made clearer by a few examples

The fraction f of the molecules of a photolabile substance converted to photoproduct, when the region containing them is traversed by a photon fluence Υ is

$$f = \sigma \, \Phi \, \Upsilon$$

where σ is the cross section for absorption of photons, and Φ is the quantum yield. The volumic number of photons n is related to the flosan Γ (= the fluence rate $d\Upsilon/dt$) by the equation

$$n \, c = d\Upsilon / dt,$$

c being the velocity of propagation. Consequently, we also have

$$f = \sigma \, \Phi \, c \int n \, dt.$$

Thus the photochemical effect of a radiation treatment can be related both to the fluence and to the volumic energy.

A particular distinction needs to be made between the energy fluence (or particle fluence) and the energy (or number of particles) per unit area incident on an explicitly specified surface -- a quantity already defined as the "dose" in ultra violet photobiology. These two quantities have the same dimensions, but are generally not the same. The fluence can be described as the nergy (or number of particles) entering a small area. If we set out to irradiate a complex object like human skin with ultraviolet light, the photochemical effect (say, the induction of pyrimidine dimers in DNA of a basal cell nucleus) is determined by the fluence in the locale of this nucleus. If we wish to compare effects on these basal cells with effects on another cell growing in tissue culture the same fluence would have to be applied in both cases. However, determining the fluence at a basal cell in irradiated skin would be very difficult. This fluence results from radiation which is strongly scattered from overlying corneum and cells, weakened somewhat by absorption. In such an arrangement the fluence will be approximately proportional to the energy (or number of photons) entering per unit area of skin surface; i.e., on the ultra violet dose to the skin.

This quantity is not the fluence in the locality of the basal cell, nor is it the fluence at the skin surface, unless all the

radiation happens to fall on this surface perpendicularly, but it
is the quantity which any dermatologist will use in describing
his work. If the same beam of ultraviolet radiation falls on the
surface of skin at progressively increasing angles from the perpendicular (and acts for the same length of time) the fluence at
the skin surface will remain the same. The ultra violet dose, on
the other hand, will decrease as the cosine of the angle from the
perpendicular. The fluence of scattered light at the basal cell
layer will also decrease approximately as the cosine of this angle,
as will the number of pyrimidine dimers induced in a nucleus located there.

If, instead of monochromatic radiation, a range of radiation
frequency (particle energy) exists, more care must be taken --
particularly with light, where both σ and Φ depend on wave length.

If each different wave length present acts in the same way,
but with different effectiveness, then the fraction of photolabile
molecules present which undergo reaction, becomes

$$f = \int \sigma(\lambda)\, \Phi(\lambda)\, (d\Upsilon/d\lambda)\, d\lambda\,.$$

This can be expressed as

$$f = \sigma_r\, \Phi_r \int A(\lambda)\, (d\Upsilon/d\lambda)\, d\lambda = \sigma_r\, \Phi_r\, \langle\Upsilon\rangle$$

where σ_r and Φ_r are the values of the absorption cross section and
quantum yield at some reference wave length λ_r , and A (λ) is
the relative effectiveness of various wave lengths (i.e., the action spectrum). A (λ r) = 1. We cannot calculate an effective
fluence $\langle\Upsilon\rangle$ for dosimetric purposes, analogous to the monochromatic fluence Υ, unless we also know both the action spectrum A (λ)
and the wave length distribution of the radiation embodied in
dΥ/dλ . This is a large complication over the monochromatic
case and one which is currently not well handled, but one which
we shall have to address in order to deal with many practical
problems, like sunlight irradiation.

These matters must be attended to after agreement is reached
on the starting point -- the description of a radiation field.

The discussants first reviewed particular aspects of the subject, after which the meeting was thrown open for comments from
the floor. A member of those attendingwanted reassurance that
the absence of specific dosimetric quantities from the table did
not mean that these were, in some sense, being eliminated from

use.   Considerable discussion time was spent in clarifying the
point that the Joint Working Group proposals are not changes to
present photobiological dosimetric terminology, but that they represent instead an effort to describe all radiation fields in
the same way.

Particular criticism was directed against the word "fluence",
which (because of its linguistic root suggesting a flow) tends to
be confused with flux, or some other time rate. The name originated with the ICRU, which has been the only international body, so
far, to define the quantity formally. Since the CIE is now about
to define it, this is a suitable time to reconsider its name. A
suggestion from the floor that some adjectival form of "area" be
used to represent the d/da (in the same way that "volumic" represents d/dv) did not lead to a concrete proposal during the meeting.

It was pointed out that complete acceptance of one set of
names by all organizations may be difficult, and that a few quantities may remain different for different organizations. In such
cases the other organization's names could be acknowledged parenth-
etically in the formal definitions.

At the close, the meeting expressed approval of the Joint
Working Group's proposals regarding definitions of quantities for
describing a radiation field, and the Group was encouraged to
continue in its efforts. The meeting expressed its specific
displeasure, however, regarding the name "fluence", and a wish
that some other term be found. No objections were voiced to the
other names.

ROUND TABLE: PROTECTION FROM UV AND VISIBLE RADIATIONS

Organizer: A. Castellani (Italy)
Discussion Leader: A. Hollaender (USA)
Reporter: A. Andersen (USA)

Dr. Hollaender opened the session by expressing a concern that
much of the older literature is not being used. He stated that an
effort was underway to compile such references in a computer-accessed
system similar to the Environmental Mutagen Information Center at
Oak Ridge, Tennessee, U. S. A.

Turning to questions of health hazards, Dr. Hollaender commented
that little had been presented at the Congress on effects of UV and
visible light on the eye. Dr. Lamola outlined the extensive data already available concerning effects on the eye ,including cataract,
photokeratitis and retinal lesions. With light that is not uncomfortable to view, the short-term retinal effects should be minimal. With
respect to laser uses in systems to be visualized, such as holograms,
Dr. Lamola continued with the expressed belief that such situations
would, by definition, not involve light too uncomfortable to view.
In answer to a question, he further stated that cataractogenic wavelengths include the UVB and UVC regions. A question from the audience expressed concern over the broad nature of the definitions of
UVA, UVB and UVC. Doctors Lamola and Laterjet both expressed the
belief that these terms are nonetheless useful.

From the audience, a comment was made that the United States
National Institute for Occupational Safety and Health has a set of
criteria upon which an occupational UV standard could be based.
Dr. Mathews-Roth indicated, however, that corneal burns <u>do</u> occur in
laboratories where UV is used and in hospital operating rooms where
UVC is used. Dr. Lamola indicated that <u>implementation</u> of existing
standards may be what is needed here. Dr. Faber expressed the view
that uniformity of standards is a problem in Europe. Dr. Lamola

agreed and further specified that the greatest need was for uniform
labeling of hazards that should appear on the lamps in clear terms.

Dr. Hollaender asked for comments on laser safety. Dr. Faber
indicated that this has been adequately handled through existing
standards such as the American National Standards Institute Standard
for the Safe Use of Lasers which puts lasers into classes with more
controls on use for those in higher classes. (Rapporteur's note:
On August 2, 1976, the U. S. A. Food and Drug Administration Performance Standard for Laser Products became effective; the standard provides that such products, as they are manufactured, have safety features
consistent with hazard potential.)

Dr. Faber expressed a problem in protecting public health from
sunlight in that the attitude of many people is that they should have
a tan. Dr. Lamola agreed that there is a problem of education, and
that we need to explain that sunburning is to be avoided.

Dr. Hollaender asked for comment upon the added risk of light
and drug combinations. Dr. Mathews-Roth stated that this was, of
course, a problem. There is a broad spectrum of chemicals that are
photosensitizers and that induce photoallergic reactions in human
skin. She further stated that this problem presently is handled via
the package insert that accompanies such drugs. Physician education
to heed such warnings is what may be needed. Dr. Lamola commented
that many adverse reactions are not found until after a drug is mar-
keted and that testing for phototoxicity may be desirable before drug
approval is granted by the U. S. A. Food and Drug Administration.
Dr. Lamola continued by saying that very few laboratories control
their own light environment. The exceptions are those involved in
photobiological research. He was unaware of the amount of lab-to-
lab irreproducibility of carcinogenic testing that was due to different lighting. Dr. Mathews-Roth spoke of a simple test for potential
phototoxicity developed by Dr. F. Daniels, Jr. This test involved
exposing yeast and the drug to the ambient laboratory lighting. Yeast
killing was stated by Dr. Mathews-Roth to parallel the phototoxic
effect when drug and light are tested in humans. Dr. Andersen cautioned against any assumption about the spectral distribution of
normal laboratory lighting since measurements by his colleagues show
a variability between lamps of different manufacturers, between different lots of the same lamp, and between lamps from the same lot.

Dr. Mathews-Roth asked about tests presently used for carcinogenicity. Dr. Anderson replied that a simplified version of the procedure is to give animals large amounts of drugs and to look for
toxicity, mutations and tumors. Unless there is prior knowledge of
phototoxic effect, this probably would not appear in the protocol.
At present, the U. S. A. Food and Drug Administration is developing
criteria for Good Laboratory Practices to ensure that research done

ROUND TABLE: TOPICAL PHOTOPROTECTION OF NORMAL SKIN

Discussion Leader: T. B. Fitzpatrick (USA)
Reporter: M. A. Pathak (USA)

Sun-protective topical preparations are chemical agents in suitable solutions or ointment vehicles that attenuate the deleterious
responses of normal white skin to excessive exposures of UV-B
(290-320 nm) and UV-A (320-400 nm) obtained during outdoor work or
sports activities. These deleterious effects of sunlight exposure
are cumulative and can lead to the development of degeneration of
the skin (wrinkling) or more serious health effects such as malignant melanoma and nonmelanoma skin cancer (basal and squamous cell
carcinoma). These latter effects are increasing at an alarming rate
since World War II and there is a need for a new evaluation and a
standardization of the efficacy of topical sun-protective agents,
and a need to educate the population regarding effective methods of
protection and precautions against these deleterious effects.

The Round Table discussion on "Topical Photoprotection of
Normal Skin" included over 120 participants from Europe, the United
States of America, and Japan. There was general agreement that new
technology has made available suitable ultraviolet sources and new
methods of measurement of ultraviolet flux, and that this new technology now needs to be utilized for the development and standardization
of the ultraviolet light sources to be used in the laboratory for
evaluating new preparations. It was unanimously agreed that screening of sun-protective agents can be done under laboratory conditions
using human volunteers, and that it is possible to simulate conditions
of usage in the laboratory by use of environmental chambers, tread
mills, immersion, etc. for determining the effects of sweating and
swimming. However, no preparation could be advocated for general use
without controlled field tests under conditions of high and low humidity, immersion in salt and fresh water, following vigorous exercise
with increased skin temperature and sweating. The most vigorous dis-

cussion concerned unanswered questions regarding the definition and
standardization of the erythema response (the so-called "minimal
erythema dose", or MED) and the effect of field size and the exposure
dose on the erythema response. It was felt that at this time, the
erythema response is the most reasonable estimate of ascertaining
the deleterious effects of sunlight in susceptible subjects. It
was felt also that on the basis of present knowledge, the topical
application of an effective sunscreen will afford protection against
erythemal response, skin cancer and skin degeneration following years
of sunlight exposure. The use of a "protection factor" as a means of
standardization of topical sun-protective agents was regarded as reasonable and essential. This system of labeling on the basis of protection factor has been used in Europe for several years but now
must be defined in precise terms and adopted widely.

The consensus of the Round Table discussion was that much could
be accomplished by (1) defining the important unsolved problems; and
(2) marshalling the expertise of scientists and physicians in univer-
sities and industry and using new, available technology to obtain new
data on the efficacy of present and newly-formulated, sun-protective
preparations. The definition of the problems to be solved in precise
terms could be assembled by an international task force composed of
photobiologists and photomedically-oriented physicians. This task
force could also initiate multidisciplinary collaborative studies
between laboratories.

TABLE

WORKING CLASSIFICATION OF SUN-REACTIVE SKIN TYPES
USED IN CLINICAL PRACTICE

Skin Type b	Skin Reactions[c] to First thirty-minute Exposure of the Summer
I	Always burn, never tan
II	Usually burn, tan below average (with difficulty)
III	Sometimes mild burn, tan about average
IV	Rarely burn, tan above average (with ease)

a. More precise and sensitive classification will be needed as knowledge evolves.

b. Type I and Type II persons often have light skin color, blue eyes, may have red scalp hair, and may or may not have freckling; however, some persons with dark brown hair and blue or green eyes have Type I and Type II sun reactions.

c. At age 12-40 years.

From: Halocarbons: Environmental Effects of Chlorofluoromethane Release. National Academy of Science, 1976.

APPENDIX

This series of questions was given to the participants and serves as a summary of the goals.

I.   Why? (Rationale)

   A. Evolve a rationale for sun protection in normal skin

   B. Attempt to prevent acute and chronic "sunburnism", a syndrome that comprises erythema, skin carcinoma, solar degeneration of the skin, and malignant melanoma

II. Who? (The high risk or susceptible normal light-skinned population)

   Develop a concept of an internationally acceptable classification of skin types as a practical approach to the need for photoprotection

III. How? (Methodology)

   A. Develop safe and effective sunscreen agents according to accepted protocol for determining photoprotection for normal skin against UV-B (290-320 nm) and UV-A (320-400 nm) radiation. Propose techniques for determining effectiveness:

      1. under laboratory conditions with proper light sources and dosimetry

      2. under outdoor field conditions including varying stresses (e.g. , sweating, swimming, skiing, varying humidity, effect of abrasion, etc.)

   B. Define needs:

      1. Survey Skin Types*, national and international

      2. Standardize sunscreen assay techniques for establishing effectiveness

---

* Sun-reactive Skin Types (I-IV) based on person's history (see Table).

C. Devise a simple instruction program for physician education

D. Label products for the consumer with simple, well-defined terms of "protection factor"

E. Educate the consumer concerning the hazards of sun exposure

F. Appoint an international task force from the academic institutions and sunscreen industry for evolving recommendations for standardization of topical sunscreens

REPORT ON ROUND TABLE: PHOTOCHEMOTHERAPY (PUVA) OF PSORIASIS

Organiser: G. C. Fuga (Italy)

Reporter and Discussion Leader: K. Wolff (Austria)

The purpose of this well attended round table - approximately
90 researchers participated in the discussion - was to provide
clinicians and investigators with a forum to exchange ideas in an
informal atmosphere, to weigh present experience against the initial
impressions of 1974 and 1975, and to discuss the expectations and
concern related to this treatment. Six major topics had been chosen
for the discussion; each of them was introduced by one of the members
of the panel, which consisted of G.C. Fuga (organizer), K. Wolff
(discussion-leader), T.B. Fitzpatrick, J.A. Parrish, M.A. Pathak,
F. Dall'Acqua and W. Morrison.

The topic "efficacy and short-term safety of PUVA" was
introduced by Parrish, who gave a short report on over one thousand
patients treated in the U.S. Cooperative Clinical Trial, comprising
16 university centers and 20 investigators. Parrish's report again
stressed the dramatic efficacy of photochemotherapy and the surprisingly low incidence of acute side-effects; in this large group of
patients there was no evidence of hepatotoxicity or eye changes.

In the discussion Mizuno (Japan) referred to one case of
hepatitis and retinitis out of a series of 20 patients treated with
PUVA but in view of the fact that no such side-effects had been
observed in the 300 patients treated by the Vienna group, the
similarly large series of the Boston group and the 1000 cases of the
U.S. Cooperative Trial, these findings were considered coincidental.
Fitzpatrick pointed out the fallacies of making premature conclusions
as to cause and effect with regard to such observations, and
mentioned the experience of Wolff who had observed a minute superficial spreading melanoma in a patient who was about to enter PUVA-

treatment; had this lesion not been discovered it would have been extremely difficult to disprove its relationship to photochemotherapy one year later. Frenk reported on a flare of arthropathy in a patient with psoriatic erythroderma and this stimulated the discussion of whether psoriatic arthropathy is amenable to photochemotherapy.

Thune recommended the use of Gamma-GT to monitor liver function in psoriatics treated with PUVA and the group from Leiden directed attention to the fact that edema of the lower legs and pustulation of psoriatic lesions may sometimes occur with higher doses of PUVA-therapy. Brodhagen addressed himself to the nature of the peculiar type of itching observed in approximately 5 to 10% of patients treated with PUVA - an observation shared by most participants for which, however, no satisfactory explanation was offered.

Topical treatment with 8-methoxypsoralen and UVA was extensively discussed by Schaffer who stressed that topically applied psoralens are resorbed by the skin and that no significant differences of side-effects exist between systemic and topical treatment. Fisher demonstrated results obtained with trisoralen baths (50 mg/bathtub) and irradiation given immediately thereafter using a dysprosium lamp of extremely high energy output. Fisher had obtained excellent results but Kligman, though impressed by the principle of this approach and having used a different light-source, was unable to report on similarly good results. Brodhagen and Wiskemann reported that they had stopped topical photochemotherapy; Schaffer stressed the importance of incorporating psoralens into an oily vehicle. Wolff made a plea to use the same precautionary measures in topical 8-MOP therapy as in systemic PUVA treatment because of the high percutaneous resorption rate of the drug.

A very informative introduction to the pharmacology, kinetics of psoralens, and molecular mechanisms of photochemotherapy was given by Pathak who stressed the fact that, after oral administration of 8-MOP, 99% of the drug is excreted in the urine after 24 hours. Dall'Acqua expanded on future developments in this field which should be directed towards developing psoralens with faster resorption and excretion rates, and drugs with better photosensitisation capacities. The questions of whether monofunctional adducts can cause sensitisation, or whether cross-links are required, and whether clinical photosensitisation is necessary for a clinical response, were amply discussed. Schaffer showed data demonstrating much higher tissue levels of psoralens after topical application than after systemic administration and emphasized that smaller amounts of PUVA-energy are required to treat patients sensitized topically. Pathak indicated that more cross-links are formed during topical photochemotherapy and that less biotransformation of the drug is to be

expected using this approach. The question of whether binding of
8-MOP to melanin occurs was not answered and the problem of possible
dark-reactions between psoralens and DNA was raised again; Dall'Acqua
and Pathak emphasized that, although intercalation can occur, there
is no evidence for covalent binding between DNA and psoralens without
long-wave ultra-violet light.

Wolff gave a survey of the present concerns about potential
long-term side-effects of photochemotherapy, discussing both
theoretical considerations and animal experiments performed to date;
the report dealt with potential degenerative changes in the skin,
oncogenesis, and long-term eye changes. Follow-up studies available
to date do not indicate that such changes will occur but it was
stressed that the follow-up periods available are still too short to
permit valid answers to these questions. Both Fitzpatrick and Wolff
speculated that the total cumulative energy delivered to a patient
over a long period of time may be crucial and that prospective
studies will have to determine the upper limit of the total radiation
energy load that can still be considered safe.

Other indications for PUVA therapy were discussed by Morrison
who described the results obtained by treating patients with mycosis
fungoides, atopic dermatitis, and vitiligo. There can be no doubt
as to the efficacy of PUVA in the treatment of early-stage mycosis
fungoides which was also illustrated by a poster presented at this
meeting. Similarly, the response of atopic dermatitis to PUVA was
shown to be excellent although Honigsmann cautioned that too many
maintenance treatments may be required for this disease to be kept
under control. This contrasted with the experience of the Boston
group who did, in fact, induce long-lasting remissions in these
patients. Mizuno reported about equally good results from Japan,
and Wolff mentioned that both dermatitis herpetiformis and follicular
mucinosis can also be cleared by this treatment.

The heterogeneity of diseases amenable to PUVA treatment again
raised the question pertaining to the mechanisms involved in this
therapeutic approach. Whereas there cannot be any doubt that PUVA
somehow interferes with DNA synthesis and cell turnover in psoriatic
epidermis the notion appears to be gaining ground that inflammatory
cells of dermal infiltrates, particularly lymphocytes, may also be
affected or modified by the interaction between UVA and psoralens.

In his closing remarks Fitzpatrick gave a short historical
survey of the development of PUVA therapy and the history of
psoralens in medicine, and presented a balanced account of the
problems, benefits and possible risks of this treatment, stressing
the necessity of risk-benefit ratio decisions when deciding to employ
this treatment. He contrasted the therapeutic efficacy and

potential risks of PUVA with those obtained by and expected of the
other effective therapeutic modalities available today, particularly
cytotoxic agents; he emphasized that even simple ultra-violet light
(sun-lamps), established in photomedicine for decades, are open to
criticism because of potential long-term sequelae (carcinogenesis)
and that thousands of patients today receive methotrexate for control
of psoriasis. He also called for well controlled, long-term
prospective studies, as are presently being performed both in the
U.S. Cooperative Clinical Trial and the European Cooperative Photochemotherapy Program.

ROUND TABLE: PHOTOBIOLOGY AND EDUCATION

Organizer: H. Mohr (FGR)
Discussion Leader: A. Castellani (Italy)
Reporter: W. R. Briggs (USA)

Professor Castellani served as Chairman for a workshop on
Photobiology and Education on September 2, 1976, during the 7th
International Congress of Photobiology. Participants were Drs.
W. R. Briggs, B. Kok, K. K. Rohatgi-Mukherjee, H. Mohr , R. B. Setlow,
C. S. Rupert, K. Shibata, P. S. Song, and J. D. Spikes. During
approximately 90 minutes of discussion, a number of salient points
emerged. The purpose of this brief paper is to summarize these points
and present a few recommendations for future International Congresses
of Photobiology.

It is clear that in most cases, photobiology is taught in the
context of other disciplines. For example, such phenomena as photosynthesis and phtottropism are taught in the context of plant physiology. In medical schools, photobiology topics may be taught in the
context of radiology. Several of the participants had had actual
experience with photobiology courses. In most cases, these courses
were taught by a team rather than a single individual, and it appeared
that courses taught by a single person were frequently selective in
ways reflecting the interests and/or capabilities of that person.
The traditional European or Japanese University system also makes it
difficult to teach photobiology as a single unified topic, and even
within many American University systems, the necessary cooperation
between biologists and chemists is sometimes difficult and awkward to
arrange.

The question was raised as to whether there was really a true
discipline of photobiology. In the formal sense, the answer was no.
For example, there are not departments of photobiology, and a graduate
student normally would not receive a Ph.D. in photobiology. An
additional problem raised by some participants was that there really

755

was no adequate text for an entire course (although several small volumes of recent origin were mentioned favorably, and mention was made of a major text in preparation, being edited by Dr. Kendrick Smith). The general conclusion was that in many cases, photobiologists share common tools but use them within rather different frameworks.

A two-way problem was discussed at some length: most biologists feel uncomfortable trying to teach the needed photochemistry and photophysics; and most photochemists feel uncomfortable trying to do justice to the extensive literature of photophysiology. When photobiology courses were taught by a team, there was almost always an imported specialist to teach the necessary photochemistry. The photochemistry courses normally offered by chemistry departments were usually graduate courses considered beyond the reach of the average biology undergraduate or graduate student. Furthermore, the organic photochemistry taught was not usually appropriate for the needs of the photobiologist. It was stressed that the appropriate photochemistry should involve electronic spectroscopy, and should deal in concepts of energy levels and electronic structures; e.g., the difference between ground and excited states, and how excitation alters acidity, geometry, dipole moment, etc. It should also cover energy levels, potential surfaces, and photochemical kinetics related to the lifetimes of various states. Several participants pointed out that such material could be taught at a qualitative level and still be of considerable use to those with training principally in biology. They also pointed out the danger of beginning a photobiology course with intensive photochemistry lectures. A better course would perhaps begin with phenomenology and then lead into the more physical and mathematical material as it became obvious that it was needed. Some participants stressed the importance of recruiting students into photobiology predominantly from the physical sciences and mathematics. The present situation, however, is that a great many students enter photobiology from biological backgrounds, and it is of prime importance to give them the necessary photochemistry to approach their research problems at a reasonable level of photochemical sophistication.

The participants from the United States of America mentioned the great success of the "School Sessions" held for the last four years at the annual meetings of the American Society for Photobiology. These sessions were held typically each morning before the beginning of the regular sessions, and lasted one hour. They consisted of a lecture given by a specialist in some area of photobiology; e.g., nanosecond time-resolved fluorescence spectroscopy. These lectures were given at the level of the interested nonspecialist, and have been proven to be an extremely effective device in broadening knowledge of photobiology among the widely dispersed disciplines

within which there is interest in the influence of light on living
organisms.

The recommendation was made that some such device could be a
part of the next International Congress on Photobiology, and could
also serve at meetings of other national photobiology groups. A
topic that was particularly stressed for such a "School Session"
was photochemistry for photophysiologists.

The recommendation was also made that the journal, Photochemistry
and Photobiology continue and perhaps expand its practice of review-
ing new books of interest to photobiologists, and perhaps publish
comprehensive lists of such books on occasion.

Despite concern over the various matters mentioned briefly
in this paper, there was a general feeling that photobiology faced
a vigorous future, and that the problems were not unsurmountable.
Attendance at national and international meetings on photobiology
supports this contention.

AUTHOR INDEX